APPLICATIONS OF CONTACT GEOMETRY AND TOPOLOGY IN PHYSICS

APPLICATIONS OF CONTACT GEOMETRY AND TOPOLOGY IN PHYSICS

Arkady L Kholodenko
Clemson University, USA

World Scientific

NEW JERSEY • LONDON • SINGAPORE • BEIJING • SHANGHAI • HONG KONG • TAIPEI • CHENNAI

Published by

World Scientific Publishing Co. Pte. Ltd.
5 Toh Tuck Link, Singapore 596224
USA office: 27 Warren Street, Suite 401-402, Hackensack, NJ 07601
UK office: 57 Shelton Street, Covent Garden, London WC2H 9HE

Library of Congress Cataloging-in-Publication Data
Kholodenko, Arkady Leonidovich, 1948–
Applications of contact geometry and topology in physics / by Arkady L Kholodenko (Clemson University, USA).
pages cm
Includes bibliographical references and index.
ISBN 978-981-4412-08-7 (hardcover : alk. paper)
1. Geometry. 2. Topology. 3. Mathematical physics. I. Title.
QC20.7.G44K46 2013
530.15'636--dc23

2012048838

British Library Cataloguing-in-Publication Data
A catalogue record for this book is available from the British Library.

Typeset by Stallion Press
Email: enquiries@stallionpress.com

Printed in Singapore by World Scientific Printers.

Preface

Although contact geometry and topology was briefly discussed in the book by V.I. Arnol'd "Mathematical Methods of Classical Mechanics" (Springer-Verlag, 1989, 2nd edition), it still remains the domain of research in pure mathematics, e.g. see recent monograph by H. Geiges "An Introduction to Contact Topology" (Cambridge Univ. Press, 2008). Some attempts to use contact geometry in physics were made in the well-written monograph "Contact Geometry and Nonlinear Differential Equations" (Cambridge Univ. Press, 2007). Unfortunately, even excellent style of this monograph happens to be not sufficient to attract enough attention of physics community to this type of problematics as the GOOGLE search results indicate. To our knowledge, this book is the first serious attempt to change the existing status quo. In it we demonstrate that, in fact, all branches of theoretical physics can be rewritten in the language of contact geometry and topology. From mechanics, thermodynamics and electrodynamics to optics, gauge fields and gravity. From physics of liquid crystals to quantum mechanics and quantum computers, etc. The book is written in the style of the famous Landau–Lifshitz (L-L) multivolume course in theoretical physics. This means that its readers are expected to have solid background in theoretical physics (at least at the level of the L-L course). No prior knowledge of specialized mathematics is required. All needed new mathematical information is given in the context of discussed physical problems. As in the L-L course some problems/exercises are formulated along the way. As in this course, these are always supplemented by either solutions or by hints (with exact references). Unlike the L-L course, though, some theorems (sometimes with complete proofs) and remarks are also presented. This is done with the purpose of facilitating interest of our readers in deeper study of subject matters discussed in the text.

Our readers should be aware that even in mathematics literature there is some (only formal) discrepancy in terminology which, most likely, is of historical origin. The very same discipline-contact geometry — is also known as sub-Riemannian geometry, as nil geometry, as Finslerian geometry, as theory of optimal control (used in robotics), as theory of nonholonomic mechanics, as the Cauchy–Riemann geometry, as the complex hyperbolic geometry, etc. More narrowly, it is also known as the geometry and topology of Heisenberg manifolds (used, for example, in the theory of radars, computer vision, etc.). At the phenomenological level the same discipline is also known as the theory of force-free fields.

The diversity of uses of the very same mathematical principles is in stark contrast with the fact that various groups of researches while discussing the same type of contact phenomena are not even aware of the existing interconnections. Specifically, in 1996, World Scientific published a book by Gerald E. Marsh entitled "Force-Free Magnetic Fields. Solutions, Topology and Applications". No mention of contact geometry and topology is made in this book. Our readers are encouraged to read both books for the sake of comparison and for gaining some historical background essential for better appreciation of the role of contact geometry/topology in physics. Mathematics of the force-free fields has long and rich history of several hundred years. It is quite remarkable that its connection with contact geometry and topology was recognized only very recently! Physicists familiar with the Ginzburg–Landau theory of superconductivity should be aware of the fact that this theory was actually initiated by London brothers in 1935. The essence of superconductivity (and superfluidity) is captured in just one force-free equation! Variational model of superconductivity developed by Ginzburg and Landau, now known in literature as the Abelian Higgs model, would not be possible without work by Londons' as the authors acknowledged themselves. Surely, Ginzburg–Landau model reproduces London's results. The non-Abelian extension of this model done by Higgs in early '60s of 20th century was instrumental in development of the Standard Model of particle physics. The spontaneous breaking of gauge symmetry in the Abelian (Ginzburg–Landau) version of the Higgs model is being observed as Meissner's effect. It causes the magnetic field to be expelled from the superconductor's bulk. Theoretically, this is equivalent to spontaneous acquiring of mass by the Abelian (that is Maxwellian electromagnetic) gauge field. By analogy, it is believed that the existence of Higgs field in the non-Abelian version of the Ginzburg–Landau theory is responsible for the spontaneous mass generation for all known massive particles of the Standard Model. Although Meissner's effect in superconductors is known for quite some time, catching the non-Abelian Higgs boson is still an outstanding problem.

On July 4, 2012 CERN announced the discovery of new particle (suspected to be the Higgs boson). In the mass media this particle is known as God particle http://en.wikipedia.org/wiki/Higgs_boson. It has such a name because of its unique role in the mass generation process. At the same time, as it follows from the discussion just presented, God's particle is just a reflection of God's principle — the dominant role given to the force-free condition in Nature. Equivalently, the very same dominant role is being played by the contact geometry and topology in practically all branches of theoretical physics. It can be looked upon as the universal master key for all branches of physics. The purpose of this book is to convince our readers that this is indeed the case. The diversity of applications of contact geometry and topology explains its universal value for theoretical physics. By reading this book, our readers will acquire enough expertise to be able to recognize "the same forest behind different kinds of trees".

This book is composed of nine chapters and four appendices. Some of these appendices are rather long. This makes our book similar in style not only to the L-L course but also to the book by V.I. Arnol'd "Mathematical Methods of Classical Mechanics". In it, the most important material is presented in numerous appendices. Appendices are having their own numeration of references. References in the main text are shown as numbers in square brackets, e.g. [135], while those in appendices are shown as letters and numbers, e.g. [D.15], etc. Appendices will sometimes contain references from the main text. Occasionally, the main text will contain some references designated for appendices. This should cause no confusion since we shall use such references only if we discuss some topics from respective appendices. As in the L-L course, we do not give an advise to our readers regarding how to read this book. We believe that those who will read this book, are capable of making such a decision for themselves.

This book is dedicated to author's moral teachers: Lev Davidovich Landau and Vladimir Igorevich Arnol'd both of whom the author had never met in person but who shaped his vision of, approach to and style in science. The author would also like to thank Professor Hagen Kleinert (Department of Physics, Freie Universität, Berlin) for his moral encouragement through many years, Professor Dale Rolfsen (Department of Mathematics, University of British Columbia) for exhibiting unusual calmness in the most critical moments and for his open mindedness both of which shaped up the rest of author's scientific life, Professor Lou Kauffman (Department of Mathematics, University of Illinois at Chicago) for his flamboyant temperament and relentless stride for simplicity and clarity which, in author's life experiences, is absolutely unparallel, Professor Leo Kadanoff (Department of Physics, University of Chicago) for organizing countless unforgetable seminars in all branches of theoretical physics which shaped up author's vision of this discipline. Last, but not least, the author owes thanks to his first US mentor, Professor Karl Freed (The James Franck Institute and the Department of Chemistry, University of Chicago) whose love of polymer physics was so contageous that the author had easily got infected by this kind of "sickness" for the rest of his life. Readers of this book will be able to recognize influences of the people listed above on its content. It would never be written in its current form should the author not met these people in person on numerous occasions. Surely, they are not responsible for likely inaccuracies in the book. Locally, the author would like to thank Prof. John Etnyre (Georgia Institute of technology, Atlanta, GA) for helpful consultations, Drs. Tatyana Zhebentyayeva and Brian Dominy at Clemson, for their help in preparing figures for this book. Finally, the author owes his special thanks to Dr. Zurab Silagadze (Budker Institute of Nuclear Physics, Novosibirsk, Russia) who supplied many references which otherwise would never appear in the text.

Arkady L. Kholodenko
Clemson University
Clemson, South Carolina
November 2012

Contents

APPLICATIONS OF CONTACT GEOMETRY AND TOPOLOGY IN PHYSICS

Chapter 1

Motivation and Background

1.1 General Information

It is well known that the phenomenon of superconductivity is taking place in some metals at low temperatures. Even in those cases when one is dealing with the so-called high temperature superconductors [1], one is not expecting these materials to stay superconducting way above the room temperature. At the same time, when one is talking about the controlled thermonuclear synthesis, e.g. done at tokamak installations [2], we are not expecting these machines to work at room temperatures. As is well known, these installations should be operational at temperatures of millions degrees of Kelvin. In this chapter we would like to discuss physical processes common for those in tokamaks, superconductors, colloidal suspensions, quark-gluon plasmas (hadron physics [3–5]), etc. Thus far these processes were treated separately, apparently, due to the historical peculiarity as described, for example, in the book [6] by Valery Kozlov. The peculiarity stems from the fact that Newtonian mechanics (the first edition of the Principia was published in 1687) had hard time of being accepted in continental Europe where the same problems were treated with alternatively developed concepts and methods of fluid mechanics. Among those who objected Newton were not only Leibnitz, but also Bernoulli brothers, Huygens, Pascal, Fermat and many lesser known scientists. Even in 1730, in his work which won the prize announced by the Paris Academy of Sciences, Bernoulli judged Newton's theory as flawed. It is quite remarkable that traces of these stormy relations between scientific communities of continental Europe and British Isles could still be felt even in 2009! To demonstrate this, we would like to mention the paper entitled "Fluidic electrodynamics: Approach to electromagnetic propulsion" by Martins and Pinheiro published in Physics of Fluids in 2009 [7]. In it, the authors "report on a new methodological approach to electrodynamics based on a fluidic viewpoint". Since electrodynamics is just a part of mechanics, the discovery made in 2009 can be attributed only to systematic eradication of historical facts from scientific textbooks and monographs. Conversely, there are many attempts to recast formalism of fluid mechanics into Hamiltinian mechanical formalism, e.g. see [8] and references therein. Fortunately, the book by Kozlov contains concise historical summary of major developments in respective scientific fields.

1.2 Fluid Mechanics Formulation of Hamiltonian and Jacobian Mechanics. Emergence of the Force-Free Fields

Following Kozlov [6] and using some facts from our work [9] on rheological properties of colloidal suspensions of hard spheres, we describe now the fluid mechanical formulation of Hamiltonian mechanics. For this purpose, we recall that Euler's equation for ideal fluids [10] can be written in the form

$$\frac{\partial}{\partial t}\tilde{\omega} = \nabla \times (\tilde{\omega} \times \mathbf{v}), \tag{1.2.1}$$

where the vorticity $\tilde{\omega}$ is defined by $\nabla \times \mathbf{v}$ with $\mathbf{v}$ being the velocity of fluid.[a] Using this equation it is convenient to introduce another vector $\mathbf{s} = \tilde{\omega} \times \mathbf{v}$. With its help solutions of Eq. (1.2.1) can be subdivided into two classes: (a) $\mathbf{s} \neq \mathbf{0}$ and (b) $\mathbf{s} = 0$. To facilitate finding of these solutions, it is convenient to investigate if there is a family of surfaces $f(x_1, x_2, x_3) = c, c \in \mathbf{R}$ orthogonal to the vector field $\mathbf{s}$. Mathematically, the orthogonality condition can be expressed as follows. Without loss of generality it is permissible to let $\mathbf{s} = g\nabla f$ with g being some smooth function. Then, the orthogonality condition can be written as

$$(\nabla \times \mathbf{s}) \cdot \mathbf{s} = 0. \tag{1.2.2}$$

Clearly, this equation is trivially satisfied when $\mathbf{s} = 0$. In this case there are no surfaces orthogonal to $\mathbf{s}$ and the vorticity $\tilde{\omega}$ and velocity fields are either collinear or anti-collinear, that is

$$\tilde{\omega} = \alpha \mathbf{v} \tag{1.2.3}$$

[a] Our readers are expected to be familiar with differences between the Eulerian and Lagrangian formulations of fluid mechanics [9].

with $\alpha \in \mathbf{R}$. Since the case $s \neq 0$ has been studied in the book by Kozlov, in this chapter we shall only focus on the case described by Eq. (1.2.3). The justification for such a restriction will become obvious upon reading this chapter.

Vector fields satisfying condition given by Eq. (1.2.3) are *called force-free fields.*[b] The origin of this terminology comes from electrodynamics. In the system of units in which $m = c = e = 1$ the motion of an electron in the magnetic field $\mathbf{H}$ is described by

$$\frac{d}{dt}\mathbf{v} = \mathbf{v} \times \mathbf{H}. \tag{1.2.4}$$

Evidently, if $\mathbf{v} = \alpha\mathbf{H}$ the force vanishes. Following [9] we now demonstrate that $\mathbf{H}$ in electrodynamics corresponds to $\tilde{\omega}$ in fluid mechanics. For this purpose we recall that in magnetostatics

$$\nabla \times \mathbf{H} = \mathbf{j} \text{ (Ampere's law)} \tag{1.2.5}$$

for some current $\mathbf{j}$. Let now $\mathbf{H} = \nabla \times \mathbf{A}$ with $\mathbf{A}$ being such that $\nabla \cdot \mathbf{A} = 0$. Then Eq. (1.2.5) is equivalent to

$$\nabla^2\mathbf{A} = -\mathbf{j}. \tag{1.2.6}$$

At the same time, in fluid mechanics it is possible to represent the local fluid velocity $\mathbf{v}$ as $\mathbf{v} = \nabla \times \mathbf{A}$ so that by taking into account the definition of vorticity $\tilde{\omega}$ we obtain an identity

$$\nabla^2\mathbf{A} = -\tilde{\omega}. \tag{1.2.7}$$

Comparison between Eqs. (1.2.6) and (1.2.7) leads to the following identifications: $\tilde{\omega} \rightleftarrows j$ and $\mathbf{v} \rightleftarrows \mathbf{H}$. More accurately, we can either write $\mathbf{v} = \alpha\mathbf{H}$ causing Eq. (1.2.4) to become force-free or, write equivalently

$$\tilde{\omega} = e\mathbf{v} \text{ (London's equation)}, \tag{1.2.8}$$

where e is yet another constant. This equation is the same as the force-free Eq. (1.2.3). The choice of constant depends on the problem. In particular, in [9] it is demonstrated that Eq. (1.2.8) corresponds to the famous London equation of superconductivity in which the parameter e should play a role of an electron charge (more accurately, the charge of electron pair). Upon rewriting Eq. (1.2.5) in the equivalent form

$$\nabla \times \mathbf{v} = \tilde{\omega}\ , \tag{1.2.9}$$

[b]Such a name is being used in magnetohydrodynamics, while in fluid mechanics they are known as *Beltrami* fields. These fields also arise in physics of liquid crystals.

one can think about some functional whose variation will reproduce both Eqs. (1.2.8) and (1.2.9). This is the path chosen by Landau and Ginzburg resulting in their famous Ginzburg–Landau (G-L) superconductivity functional. Detailed derivation is also provided in [9] where these results are used for description of rheology of hard sphere suspensions. Independently, a connection between the superconductivity and hydrodynamics of ideal fluids was established much earlier by Fröhlich [11]. Incidentally, Fröhlich was also the first who introduced polarons, later discussed by both Landau and Feynman.

The results just described suggest that mathematical description of various apparently very different physical phenomena — from conventional superconductivity to dynamics of colloidal suspensions, etc. — should have common origin whose key element is the free-force Eq. (1.2.3). The list of possible physical applications is much wider than those just mentioned. Before describing some of these applications, following Kozlov [6], it is helpful to demonstrate the isomorphism between the Hamiltonian mechanics and dynamics of ideal fluids. Alternative treatments can be found in [8] and references therein.

Exercise 1.2.1. In terminology of [6] the two classes (a) $\mathbf{s} \neq \mathbf{0}$ and (b) $\mathbf{s} = 0$ describe the vortex motions respectively in *strong* and *weak* sense. Since in this book we discuss only the motion of vortex fields in the weak sense, investigate various physical problems described by the condition $\mathbf{s} \neq \mathbf{0}$. Provide examples from mechanics, hydrodynamics, theory of liquid crystals, etc. In the case of fluids, investigate to what extent the results depend on incompressibility of fluids and on fluid non-ideality. Hint: look up [6].

To begin demonstration of an isomorphism, we shall assume that $\tilde{\omega}$ is time-independent. Then it is possible to introduce a (scalar) potential Φ so that Euler's Eq. (1.2.1) can be equivalently rewritten as

$$\frac{\partial}{\partial t}\mathbf{v} = \mathbf{v} \times \tilde{\omega} - \nabla\Phi. \tag{1.2.10}$$

By combining this equation with the equation of continuity for fluids, it is possible to arrive at the Lamb equation

$$\frac{\partial}{\partial t}v_i + \left[\frac{\partial v_i}{\partial x_j} - \left(\frac{\partial v_i}{\partial x_j}\right)\right] v_j = -\nabla_i \cdot \Phi. \tag{1.2.11}$$

Now let Σ_t^n be the phase space $\mathcal{P} = T^*\mathcal{M}$ of some mechanical system admitting a single-valued projection onto configurational space $\mathcal{M}$. In canonical coordinates $\mathbf{x}, \mathbf{y}$ this manifold is defined by the equation $\mathbf{y} = \mathbf{u}(\mathbf{x}, t)$. The following theorem has been proven in [6].

Theorem 1.2.1. *The manifold Σ_t^n is an invariant manifold for the canonical Hamiltonian equations with the Hamiltonian $H(\mathbf{x}, \mathbf{y}, t)$ if and only if the field $\mathbf{u}$ satisfies the Lamb-type equation*

$$\frac{\partial}{\partial t} u_i + \left[\frac{\partial u_i}{\partial x_j} - \frac{\partial u_i}{\partial x_j}\right] v_j = -\nabla_i h. \tag{1.2.12}$$

In this equation $h(\mathbf{x}, t) = H(\mathbf{x}, \mathbf{u}(\mathbf{x}{,}t), t)$ so that

$$\mathbf{v}(\mathbf{x}, t) = \left.\frac{\partial H}{\partial \mathbf{y}}\right|_{\mathbf{y}=\mathbf{u}(\mathbf{x},t)}. \tag{1.2.13a}$$

In addition,

$$\frac{d\mathbf{y}}{dt} = -\left.\frac{\partial H}{\partial \mathbf{x}}\right|_{\mathbf{y}=\mathbf{u}(\mathbf{x},t)} = \frac{\partial}{\partial t}\mathbf{u} + \frac{\partial \mathbf{u}}{\partial \mathbf{x}} \cdot \mathbf{v} = \frac{\partial}{\partial \mathbf{t}}\mathbf{u} + \frac{\partial \mathbf{u}}{\partial \mathbf{x}} \cdot \left.\frac{\partial \mathbf{H}}{\partial \mathbf{y}}\right|_{\mathbf{y}=\mathbf{u}(\mathbf{x},t)}. \tag{1.2.13b}$$

In order to use this theorem consider now an example. For the Hamiltonian given by

$$H = \frac{1}{2}\sum_{i,j=1}^{n} g_{ij}(\mathbf{x}, t) y_i y_j + V(\mathbf{x}) \tag{1.2.14}$$

Eq. (1.2.12) acquires the form

$$\frac{\partial}{\partial t} u_i + \sum_{j,k} g_{ik}\left[\frac{\partial u_i}{\partial x_j} - \frac{\partial u_j}{\partial x_i}\right] u_k = -\frac{\partial}{\partial x_i} H. \tag{1.2.15}$$

For $n = 3$ and $g_{ij} = \delta_{ij}$ we get $\mathbf{u} = \mathbf{v}$ and Eq. (1.2.15) formally coincides with the Lamb's Eq. (1.2.11), provided that we identify Φ with h. The already mentioned recent work by Martins and Pinheiro [7] provides further ramifications of the correspondence just described. Consider now some special cases.

(a) Suppose that

$$\frac{\partial u_i}{\partial x_j} - \frac{\partial u_i}{\partial x_j} = 0 \tag{1.2.16}$$

in Eq. (1.2.12). By comparing Eqs. (1.2.10) and (1.2.11) we realize that Eq. (1.2.16) is equivalent to the force-free condition, Eq. (1.2.3). Let therefore $\mathbf{u} = \nabla S$, where S is to be defined momentarily. This allows us to rewrite Eq. (1.2.11) as

$$\nabla \left(\frac{\partial}{\partial t} S + h \right) = 0. \tag{1.2.17}$$

From here we obtain,

$$\frac{\partial}{\partial t} S + h = \mathcal{F}, \tag{1.2.18}$$

where $\mathcal{F}$ is some time-dependent function. Since this function can always be removed by a gauge-type transformation

$$S \to S - \int^{t} \mathcal{F}(\tau) d\tau,$$

we end up with the Hamilton–Jacobi equation

$$\frac{\partial}{\partial t} S + H(\mathbf{x}, \nabla S, t) = 0 \tag{1.2.19}$$

in which S is the action functional, as is well known. By definition, the manifold Σ_t^n is *Lagrangian* if Eq. (1.2.16) holds, otherwise, it is called the *vortex manifold.* That is using the Lagrangian manifold is synonymous with using the force-free condition, Eq. (1.2.3). This observation is a germ of a very deep mathematics known as *contact geometry and topology* to be discussed in this book. For the time being let us consider another special case.
(b) $\frac{\partial}{\partial t} u_i = 0$. Then, instead of Eq. (1.2.12), we end up with

$$(\operatorname{rot} \mathbf{u}) \cdot \mathbf{v} = -\nabla_i h. \tag{1.2.20}$$

In this equation $\operatorname{rot} \mathbf{u} = \frac{\partial u_i}{\partial x_j} - \frac{\partial u_i}{\partial x_j}$. It can be shown [6] that by changing variables $\mathbf{x} \to \mathbf{z}$ (and by diagonalization of $\operatorname{rot} \mathbf{u}$), Eq. (1.2.20) can always be brought to the canonical Hamiltonian form

$$I \dot{\mathbf{z}} = -\frac{\partial h}{\partial \mathbf{z}} \tag{1.2.21}$$

with symplectic matrix I given in its usual form by

$$I = \begin{pmatrix} 0 & E \\ -E & 0 \end{pmatrix}.$$

For this to happen, the dimensionality n of the phase space should be even so that z_i and $z_{n/2+i}$ are canonically conjugate variables.

From these examples it follows that symplectic geometry and topology always involves even-dimensional spaces while the contact geometry and topology involves odd-dimensional spaces. If we look at the contact geometry as an extension of symplectic geometry that takes care of time, then the lowest nontrivial dimension of the contact manifold is three.

1.3 Some Basic Facts about the Force-Free Fields

To proceed, we shall now discuss some basic facts about force-free fields. For this purpose, we follow the classical work [12]. In it the authors discuss the most general form of solutions to the magnetic force-free field equation

$$\nabla \times \mathbf{H} = \alpha \mathbf{H}. \tag{1.3.1}$$

Since $\operatorname{div} \mathbf{H} = 0$, by applying operator $\nabla\times$ to both sides of the above equation we obtain

$$\nabla^2 \mathbf{H} + \alpha^2 \mathbf{H} = 0. \tag{1.3.2}$$

This result demonstrates that the curl operator plays the same role for the Laplacian as Dirac operator for the d'Alambertian. This fact is going to be used later in this chapter.

Not every solution of Eq. (1.3.2) is solution of Eq. (1.3.1) even though the converse is true. This can be seen from the following example. In the case of superconductivity one should, instead of Eq. (1.3.2), consider an equation of the type

$$\nabla^2 \mathbf{H} - \alpha^2 \mathbf{H} = 0. \tag{1.3.3}$$

It describes the Meissner effect — one of the main features of both high and low temperature superconductors. In plain English this equation tells us that the magnetic field is expelled from the superconductor [1, 9]. Also, it means that inside superconductors otherwise massless photons acquire mass. Since the electromagnetic field is just an Abelian gauge field, it is only natural to think about its non-Abelian extension. In such a form this field is used in quantum chromodynamics. The simplest solution of the scalar version of Eq. (1.3.3) is given by the Yukawa potential:

$$\psi(\mathbf{r}) = c\frac{e^{-\alpha r}}{r}. \tag{1.3.4}$$

Here c is some constant while $r = |\mathbf{r}|$. As is well known, Yukawa used/postulated this potential for description of short-range forces (meson exchange) between baryons in the nucleus. If the Abelian/electromagnetic field can acquire mass in superconductors, can the same happen in the non-Abelian case? If, indeed, the non-Abelian field can acquire mass in quark–gluon plasma, then one can recover the Yukawa potential from the first principles. This is, in part, the essence of one of the millennium prizes set up by the Clay Mathematics Institute.[c] As by-product of this recovery, one may also be able to obtain an explanation of quark confinement, that is of unproven statement that it is impossible to observe/to detect free quarks. We shall discuss some aspects of this problem later in this book.

At this point we are interested in finding connections between Eqs. (1.3.1)–(1.3.3). For this purpose let us assume that $\mathbf{v} = \pm\kappa\mathbf{A}$ where κ is some non-negative constant. In view of Eq. (1.2.3) we obtain,

$$\tilde{\boldsymbol{\omega}} = \alpha\mathbf{v} \quad \text{or} \quad \nabla\times\mathbf{v} = \alpha\mathbf{v} \quad \text{or} \quad \nabla\times\mathbf{v} = \pm\kappa\alpha\mathbf{A} \quad \text{or}$$
$$\nabla\times\nabla\times\mathbf{A} = \pm\kappa\alpha\mathbf{A}. \tag{1.3.5}$$

In view of the condition $\nabla\cdot\mathbf{A} = 0$ the last result is equivalent to

$$\nabla^2\mathbf{A} = \mp\kappa\alpha\mathbf{A}. \tag{1.3.6}$$

From here it follows that the "+" sign option should be associated with the superconducting/superfluid solution, Eq. (1.3.3), while the sign "−" option leading to Eq. (1.3.1) is typically considered in the context of plasma physics [12]. Evidently, solutions of Eqs. (1.3.1) and (1.3.3) are not the same while in both cases we had began with the same force-free Eq. (1.2.3). In our work [9] by analogy with theory of superconductivity we argued that the relation $\mathbf{v} = \pm\kappa\mathbf{A}$ can be safely replaced by $\mathbf{v} = \pm\kappa\mathbf{A}+\sigma\nabla\psi$, where σ is some constant and ψ is some harmonic function (in view of the requirement $\operatorname{div}\mathbf{v} = 0$). This follows from the fact that physically we are interested not in the vector potential $\mathbf{A}$ but in the velocity $\mathbf{v}$ (or magnetic field $\mathbf{H}$). Obtained results explain why under circumstances just described Eq. (1.2.8) can be identified with the London equation.[d]

Exercise 1.3.1. Investigate how the results just presented could be generalized to include the case of high temperature superconductivity. Investigate the mechanical analog of these phenomenological results. Hint: Use [1], [3] and [9].

---

[c] E.g. see http://www.claymath.org/millennium/Yang-Mills_Theory/

[d] E.g. read section on supercoductivity in the classical monograph [13].

To solve the vector Helmholtz Eq. (1.3.2), following [12] it is necessary to know solutions of the related scalar equation

$$(\nabla^2 + \omega^2)\varphi = 0. \tag{1.3.7}$$

Such an equation is well known from quantum mechanics (scattering theory) and quantum field theory. In the absence of boundaries it describes, say, the incoming/outgoing (standing waves). In nonrelativistic quantum mechanics. Such a solution is obtainable from the Yukawa solution, Eq. (1.3.4), by simple replacement of α by $i\alpha$. Unlike quantum mechanics, in [12] it is argued that in the theory of force-free/Beltrami fields any solution of Eq. (1.3.7) can serve as an input for obtaining solutions for the vector Eq. (1.3.2). There are three independent types of such solutions. These are $\mathbf{L} = \nabla\varphi, \mathbf{P} = \text{curl}(\varphi\mathbf{a})$ and $\mathbf{T} = \text{curlcurl}(\varphi\mathbf{a})$ where $\mathbf{a}$ is some constant vector. Solution $\mathbf{P}$ is called *poloidal* while solution $\mathbf{T}$ is called *toroidal*. These are easy to visualize. As we shall discuss below, the Beltrami/force-free fields admit the vortex-type solutions. The vortices are typically having a toroidal shape. So, the toroidal solution is appropriate for description of the circulation of the vector field inside the torus while the poloidal field describes circulation in the space outside the torus. The existence of this or that solution is determined by the physics of the problem involving the force-free fields, e.g. see [13], where such types of solutions are discussed in the context of superfluidity and superconductivity.

At this point our readers may ask: Wait a second, how one can mix/confuse quantum and classical phenomena in such an arbitrary way? Fortunately, the answer is easy and natural. When London brothers postulated [14] their (now famous) equation, they did it *exclusively at the classical level*! As a result, subsequently two volumes on superconductivity and superfluidity written by Fritz London were published [15]. These volumes were written to a large extent without uses of fine details obtainable with help of more rigorous/deeper quantum mechanical treatments. It is not too difficult after seeing works of London to re-obtain both Maxwell's (or Ampere's) and London's equations by variation of functional now known as the Ginzburg–Landau (G-L) functional. Details of such a derivation can be found in [13], in our work [9], and in Chap. 3 below. For variational reformulation of equations postulated by London brothers and for study of properties of these equations Ginzburg was awarded the Nobel prize. Subsequently, the phenomenological G-L theory of superconductivity was replaced by the detailed microscopic quantum theory reproducing the G-L results. Many fine details can be found in the classical monograph by Abrikosov, Gorkov and Dzyaloshinski [16]. Supersymmetric non-Abelian extension of the U(1) G-L model known in literature as the Seiberg–Witten (S-W) model is briefly discussed in the book by Jost [17]. Detailed mathematical treatment of the model can be found in [18]. We shall discuss some facts of S-W model in Chap. 9.

In both Eqs. (1.2.3) and (1.3.1) the parameter α is a constant, some real number. In the case if one wants to obtain Eq. (1.3.3) from Eq. (1.3.2), one can formally replace this number by purely imaginary. The difference between these two equations is coming from the difference in physics. Physics enters into play also through various symmetries and boundary conditions. Many plasma physics related solutions of these equations are discussed in the book by Marsh [20]. The same book describes some superconducting solutions anticipating results obtained later and discussed below.

Thus far we have not yet discussed the situation when α is coordinate-dependent. Evidently, for such a case the force-free condition, Eq. (1.2.3), is not violated. Clearly, it is necessary to use some additional information in order to exclude this possibility. It is helpful to discuss this issue in the context of magnetohydrodynamics where historically it was initially formulated.

Chapter 2

From Ideal Magnetohydrodynamics to String and Knot Theory

2.1 General Information

To prove that in the force-free Eq. (1.2.3) only coordinate-independent constant α is physically sensible, we follow arguments made by Taylor [21, 22] in the context of magnetohydrodynamics. This will be accomplished in the next chapter while here we provide needed background. In particular, our readers can consult [23] where some basic facts about magnetohydrodynamics are nicely summarized. The subject of magnetohydrodynamics is roughly a combination of electrodynamics and fluid mechanics. In view of results discussed in Sec. 1.2, this looks like a tautology. To show that this is indeed the case, we notice that in the present case fluid is made of charged particles so that the fluid is in plasma state. Such a state is expected to be used in installations such as tokamaks and/or spheromaks for producing energy via controlled thermonuclear fusion. It makes sense to define some kind of reference state (just like ideal gas in thermodynamics) for such plasma configurations. Traditionally, it is chosen as "superconducting" state, that is the state for which the conductivity is infinite. Not surprisingly, it is possible to discuss dynamics of such a state using the G-L theory of superconductivity [24]. To see this it is sufficient to notice that in this regime the magnetohydrodynamics can be described in terms of the Euler-type equation (e.g. see Eq. (1.2.1)) for ideal fluids [21–23]

$$\frac{\partial}{\partial t}\mathbf{H} = \nabla \times (\mathbf{v} \times \mathbf{H}). \tag{2.1.1}$$

To facilitate understanding of this equation we follow the same path as in Sec. 1.2. That is we take into account that infinitely conducting incompressible fluid should be described by the equations

$$\operatorname{div} \mathbf{v} = 0, \quad \operatorname{div} \mathbf{w} = 0, \tag{2.1.2a}$$

where $\mathbf{w} = \sqrt{\mu}\,\mathbf{H}$ and μ is constant magnetic permeability. Using Eqs. (1.2.10), (1.2.11) and assuming the steady-state for which Eq. (1.2.20) holds, we obtain

$$(\mathbf{v} \cdot \nabla)\mathbf{v} - (\mathbf{w} \cdot \nabla)\mathbf{w} = -\nabla \Pi. \tag{2.1.2b}$$

In this equation $\Pi = p + \mathbf{w}^2/2$ with p being the hydrostatic pressure. Further- more, in anticipation of the use of the force-free condition, we write as well

$$\mathcal{L}_{\mathbf{v}}\mathbf{w} = (\mathbf{v} \cdot \nabla)\mathbf{w} - (\mathbf{w} \cdot \nabla)\mathbf{v} = 0 \tag{2.1.2c}$$

for the Lie derivative.[a] Finally, let $\mathbf{v} = \sqrt{\rho}\mathbf{q}$ with ρ being constant density (incompressible fluid condition) and $\mathbf{q}$ being steady fluid velocity. Following the same reasoning as used for the steady-state approximation of Eq. (1.2.10), in accord with [26], we write: $\mathbf{q} \times \mathbf{H} = \nabla \Phi$. This implies that the fluid velocity and magnetic field are *tangential* to surfaces $\Phi = \text{const}$. Being tangential requires fulfilment of two conditions: $\mathbf{q} \cdot \nabla \Phi = 0$ and $\mathbf{H} \cdot \nabla \Phi = 0$. Such surfaces are called *Maxwellian*. Evidently, the existence of Maxwellian surfaces is linked to the existence of force-free fields. Furthermore, from the above equations it follows that the fluid velocity and magnetic field appear on equal footing. This makes the above equations invariant under substitution (rotation)

$$\begin{pmatrix} \mathbf{v} \\ \mathbf{w} \end{pmatrix} \Rightarrow \begin{pmatrix} \cosh \lambda & \sinh \lambda \\ \sinh \lambda & \cosh \lambda \end{pmatrix} \cdot \begin{pmatrix} \mathbf{v} \\ \mathbf{w} \end{pmatrix}, \tag{2.1.3}$$

provided that $\mathbf{v} \cdot \nabla \lambda = \mathbf{w} \cdot \nabla \lambda = 0$. If either $\mathbf{H}$ or $\mathbf{v}$ vanishes or if $\mathbf{v} = \alpha \mathbf{H}$ (e.g. see Eq. (1.2.3)), the above system acquires familiar Euler's hydrodynamic form:

$$\operatorname{div} \mathbf{v} = 0, \quad (\mathbf{v} \cdot \nabla)\mathbf{v} = -\nabla p. \tag{2.1.4}$$

The above invariance is not the only one present in this system. By notching that foliations basically are the flow patterns associated with solutions of ordinary differential equations, the above hydrodynamic (or magnetohydrodynamic) system introduces new element into the theory of foliations. This

---

[a] Accessible (for physicists) mathematically rigorous discussion about uses of Lie derivatives in fluid mechanics can be found in the paper by Boyland [25]. The same paper also explains the connection between the force-free (Beltrami) fields and the condition $\mathcal{L}_{\mathbf{v}}\mathbf{w} = 0$.

new element is associated with posing and solving the so-called Gillbarg problem which (in the case of hydrodynamics) is described below.

2.2 The Gillbarg Problem and the Theory of Foliations

This problem can be formulated as follows. *Under what condition the hydrodynamic flow is uniquely determined by its streamline (that is by foliation) pattern?* To answer the above question, let us consider two flows $\mathcal{F} = (\mathbf{v}, p)$ and $\mathcal{F}^* = (\mathbf{v}^*, p^*)$ having the same streamlines. For this to happen we suppose that $\mathbf{v}^* = f\mathbf{v}$ where f is some scalar function. From the incompressibility Eq. $(2.1.4)_1$, it follows that $\mathbf{v} \cdot \nabla f = 0$ and, hence, it should be a constant along the streamlines (just like Φ above). In addition, Eq. $(2.1.4)_2$ leads to $\nabla p^* = f^2 \nabla p$. Taking the curl on both sides yields

$$\nabla p \times \nabla f^2 = 0.$$

This result provides the additional condition on f. If $\nabla f = 0$, then $\mathbf{v}^* = c\mathbf{v}$ with c being some constant. If $\nabla p = 0$ we are having a flow of uniform pressure with straight streamlines. When $\nabla f \neq 0$ and $\nabla p \neq 0$, it follows that f is the function of p only. Since we already obtained that $\mathbf{v} \cdot \nabla f = 0$, the dependence of f on p leads now to the condition: $\mathbf{v} \cdot \nabla p = 0$. If now $\mathbf{v} = q\mathbf{t}$, where $\mathbf{t}$ is the unit vector tangential to the streamline and $q = |\mathbf{v}| = \text{const}$, then incompressibility Eq. $(2.1.4)_1$ is replaced by

$$\operatorname{div} \mathbf{t} = 0. \tag{2.2.1}$$

All hydrodynamic flows can now be subdivided into two classes: (a) those for which $\operatorname{div} \mathbf{t} \neq 0$ and, (b) those for which $\operatorname{div} \mathbf{t} = 0$. In the last case *the foliation pattern is not defining flow uniquely.* Instead, *there is an infinity of flows* for such a case since the system of equations Eq. (2.1.4) is invariant under substitutions $(\mathbf{v}, p) \to (\sqrt{f'(p)}\mathbf{v}, f(p))$.

The obtained results can be immediately translated into language of magnetohydrostatics [27] by making the following identifications $(\mathbf{v}, -p) \rightleftarrows (\mathbf{H}, \Pi)$ with $\Pi = p + \frac{\mathbf{H}^2}{2}$. The extension to Eqs. (2.1.2) becomes obvious now. Details are given in [28].

2.3 From String-Theoretic Lund–Regge Equation to Landau–Lifshitz Equation for the Vortex Filament

The existence of Maxwellian surfaces is consistent with Eq. (2.1.2c). This fact allows us to introduce the "adapted" coordinate system as it is done in

[26, 28]. If χ, ψ and ω are coordinates of such system then, without loss of generality, it is possible to associate $\omega =$ const. with one of such surfaces. In this coordinate system we write for the combination $\mathbf{v} \cdot \nabla = \partial_\chi$ and, accordingly, $\mathbf{w} \cdot \nabla = \partial_\psi$. Furthermore, in such coordinates uses of Eulerian coordinate vector $\mathbf{r}(\chi, \psi, \omega)$ leads to relations: $\partial_\chi \mathbf{r} = \mathbf{v}$ and $\partial_\psi \mathbf{r} = \mathbf{w}$. Such introduced vectors are tangent to the χ and ψ coordinate lines on the Maxwellian surfaces labelled by ω.

To proceed, requires us to rewrite Eqs. (2.1.2a) and (2.1.2b) in terms of coordinates of the adapted coordinate system. In particular, Eqs. (2.1.2a) acquire the following form:

$$\operatorname{div} \mathbf{v} = \operatorname{div} \mathbf{r}_\chi = \frac{|\mathbf{r}_\chi, \mathbf{r}_\psi, \mathbf{r}_\omega|_\chi}{|\mathbf{r}_\chi, \mathbf{r}_\psi, \mathbf{r}_\omega|} = 0, \quad \operatorname{div} \mathbf{r}_\psi = \frac{|\mathbf{r}_\chi, \mathbf{r}_\psi, \mathbf{r}_\omega|_\psi}{|\mathbf{r}_\chi, \mathbf{r}_\psi, \mathbf{r}_\omega|} = 0. \quad (2.3.1)$$

It is always possible to select the adapted coordinates in such a way that the volume $|\mathbf{r}_\chi, \mathbf{r}_\psi, \mathbf{r}_\omega|$ is normalized to one. Under this condition, Eq. (2.1.2b) acquires the form

$$\mathbf{r}_{\chi\chi} - \mathbf{r}_{\psi\psi} + \Pi_\chi \mathbf{r}_\psi \times \mathbf{r}_\omega + \Pi_\psi \mathbf{r}_\omega \times \mathbf{r}_\chi + \Pi_\omega \mathbf{r}_\chi \times \mathbf{r}_\psi = 0. \quad (2.3.2)$$

If the Maxwellian surfaces are the constant pressure surfaces, then $\Pi = \Pi(\omega)$. Under such an assumption Eq. (2.3.2) is transformed into

$$\mathbf{r}_{\chi\chi} - \mathbf{r}_{\psi\psi} = \Omega(\omega) \mathbf{r}_\chi \times \mathbf{r}_\psi \quad (2.3.3a)$$

so that

$$\Pi(\omega) = -\int^\omega d\tilde{\omega} \Omega(\tilde{\omega}). \quad (2.3.3b)$$

Equation (2.3.3a) is the famous Lund–Regge equation introduced by Lund and Regge in [29]. To analyze this equation further we would like to discuss physical arguments which lead Lund and Regge to the discovery of their equation. [29] can be considered as substantial refinement of earlier work by Nielsen and Olesen [30]. These authors found a remarkable relationship between the dynamics of the Nambu–Goto string model and that of the vortices in superfluid ^{4}He and in superconductors. The book by Barbashov and Nesterenko [31] provides a good follow up review of various developments in string theory initiated by Lund and Regge work. Lund and Regge assumed that the vortices do have some finite thickness. Account for this property can be made as follows. Begin with a point-like object in contact with liquid and impose the non-slip boundary condition via

$$\mathbf{v}(\mathbf{r}, t) = \frac{d\mathbf{r}}{dt} \equiv \mathbf{v}(t) \quad (2.3.4)$$

so that the velocity $\mathbf{v}(\mathbf{r}, t)$ of the fluid is the same as the velocity $\mathbf{v}(t)$ of the point-like object. For extended objects the above requirement should be modified to

$$v^j(t) = \frac{\partial r^j}{\partial t} + \frac{\partial r^j}{\partial \sigma}\frac{\partial \sigma}{\partial t}, \tag{2.3.5}$$

where $j = x, y$ or z while σ is coordinate along the vortex filament (that is ω for the Maxwellian surface). With such a picture in mind Lund and Regge introduce a vector potential

$$\mathbf{A}(\mathbf{x}) = \frac{k}{4\pi}\oint_C d\sigma \frac{1}{|\mathbf{x} - \mathbf{r}(\sigma)|}\mathbf{v}(\sigma).$$

They argue that Eq. (2.3.5) is compatible with this vector potential if and only if the condition

$$\varepsilon_{ijk}\frac{\partial r^j}{\partial \sigma}\left(\frac{\partial r^k}{\partial t} - v^k\right) = 0 \tag{2.3.6}$$

holds. Such an equation is then treated as an equation of motion by these authors. It is obtainable by minimization of the Lund–Regge action functional. Since the technical details of its minimization are involved, we refer our readers to the pedagogically written paper by Berdichevsky [32] where all technical details can be found. From this paper as well as from the original work by Lund and Regge, the final result — string-like equation of motion for the vortex filament — is obtained

$$\left(\frac{\partial^2}{\partial \tau^2} - \frac{\partial^2}{\partial \sigma^2}\right)\mathbf{r} = \mathcal{F}\left(\frac{\partial \mathbf{r}}{\partial \tau} \times \frac{\partial \mathbf{r}}{\partial \sigma}\right) \quad \text{(Lund–Regge)} \tag{2.3.7}$$

coinciding with Eq. (2.3.3). In this equation the "nonrelativistic" time t was replaced by the "relativistic" world time τ parametrizing the word-line evolution. Unlike the magnetohydrodynamics case, Eq. (2.3.7) should be supplemented by the constraints

$$\frac{\partial \mathbf{r}}{\partial \tau} \cdot \frac{\partial \mathbf{r}}{\partial \sigma} = 0 \quad \text{and} \quad \left(\frac{\partial \mathbf{r}}{\partial \tau}\right)^2 + \left(\frac{\partial \mathbf{r}}{\partial \sigma}\right)^2 = 1. \tag{2.3.8}$$

Although these constraints are not essential for magnetohydrodynamics, they can still be implemented without destroying results obtained in previous subsection as explained in [26, 28]. The price for doing so is that the new coordinate lines on the Maxwellian surfaces typically will not coincide with the streamlines and magnetic field lines. In addition, the volume form $|\mathbf{r}_\chi, \mathbf{r}_\psi, \mathbf{r}_\omega|$ is going to change its appearance. Furthermore, we have to require $\Omega(\omega)$ in Eq. (2.3.3) to be a constant since in Eq. (2.3.7) $\mathcal{F}$ is a constant. All these requirements are somewhat inconvenient since they look a bit artificial in the context of magnetohydrodynamics. To repair the situation, following Rogers and Schief [26, 28] we need to discuss the degenerate cases, e.g. $\mathbf{v} = 0$ (or $\mathbf{w} = 0$) and $\mathbf{v} = \alpha(\omega)\mathbf{w}$, first. In such cases the adapted coordinates change so that now $\mathbf{v} \cdot \nabla = \partial_\chi + \alpha\beta\partial_\psi$ and $\mathbf{w} \cdot \nabla = \beta\partial_\psi + \alpha\partial_\chi$. Accordingly, we write $\mathbf{v} = \mathbf{r}_\chi + \alpha\beta\mathbf{r}_\psi$ and $\mathbf{w} = \alpha\mathbf{r}_\chi + \beta\mathbf{r}_\psi$. In these expressions $\alpha^2 \neq 1, \beta = \text{const}$ and it is assumed that the fluid velocity $\mathbf{v}$ is nonvanishing. When $\alpha \neq 0$ and $\beta = 0$, the fluid velocity and magnetic field are aligned so that the force-free condition $\mathbf{v} = \alpha(\omega)\mathbf{w}$ is enforced. This situation differs from that for pure hydrodynamic where $\alpha = \beta = 0$. The Lund–Regge Eq. (2.3.3) is recoverable under condition $\alpha^2 \neq 1, \beta = \text{const}$. But, the force-free configuration originating from the requirement $\alpha \neq 0$ and $\beta = 0$ makes more physical sense as we would like to explain now. When the force-free condition is enforced, the volume-preserving condition $|\mathbf{r}_\chi, \mathbf{r}_\psi, \mathbf{r}_\omega| = 1$ remains unchanged while the total pressure Π is now related to $\Omega(\omega)$ as

$$\Pi = \int^{\omega} d\tilde{\omega}\Omega(\tilde{\omega})[\alpha^2(\tilde{\omega}) - 1]. \tag{2.3.9}$$

Thus, in the force-free case, the Lund–Regge Eq. (2.3.3) degenerates into

$$\mathbf{r}_{\chi\chi} = \Omega(\omega)\mathbf{r}_\chi \times \mathbf{r}_\psi. \tag{2.3.10}$$

The magnitude of the fluid velocity now is $|\mathbf{v}| = |\mathbf{r}_\chi| = v$ so that $v_\chi = 0$. In the light of Eq. (2.2.1) this result is important as we shall demonstrate shortly below. In addition, this observation allows us to introduce yet another coordinate transformation $(\chi, \psi, \omega) \to (s, b, \omega)$ via

$$s = \int \left[v d\chi + \frac{\mathbf{r}_\chi \cdot \mathbf{r}_\psi}{v} d\psi \right], \quad b = \int v d\psi \tag{2.3.11}$$

so that in these coordinates Eq. (2.3.10) is converted into

$$\Omega(\omega)\mathbf{r}_b = \mathbf{r}_{ss} \times \mathbf{r}_s, \tag{2.3.12a}$$

provided that $\mathbf{r}_s^2 = 1$ and $|\mathbf{r}_s, \mathbf{r}_b, \mathbf{r}_\omega| = 1/v^2$. Since $v_\chi = 0$, use of the results of Sec. 2.2 allows us to write $\mathbf{v} = v\mathbf{t}$ and to use the identification $\mathbf{t} = \mathbf{r}_s$ for the unit tangent vector to the streamlines. The condition $v_s = 0$ is being fulfilled now automatically (since $\chi \rightleftarrows s$) leading to the Landau–Lifshitz (L-L) equation

$$\Omega(\omega)\mathbf{t}_b = \mathbf{t} \times \mathbf{t}_{ss}. \tag{2.3.12b}$$

To obtain Eq. (2.3.12b) from Eq. (2.3.12a) it is sufficient to apply to both sides of Eq. (2.3.12a) the operator $\partial/\partial s$ and to use $\mathbf{t} = \mathbf{r}_s$ after differentiation. In addition, the L-L equation should be supplemented by the condition

$\operatorname{div} \mathbf{t} = 0$, Eq. (2.2.1), as consequence of the fact that $\mathbf{v} = \text{const}$. Fortunately, this condition is not too strong for destruction of the underlying physics. This topic is discussed in the next section.

Exercise 2.3.1. Investigate how just obtained results will change the results of the next subsection if the condition $q = |\mathbf{v}| = \text{const.}$ is violated. Investigate to what extent dissipation effects may affect the obtained results and those which follow.

2.4 Foliations of $\mathbf{R}^3$ by the Maxwellian Surfaces

In Sec. 2.2 we discussed two types of flows. For flows whose velocity magnitude is constant the L-L Eq. (2.3.12b) describes dynamics of the unit tangent vector $\mathbf{t}$ to the streamlines (leaves of foliation) located on one of the Maxwellian surfaces. Such flows are not unique. They are determined up to a scale factor that is some function f whose properties were described in Sec. 2.2. If one imposes additional (quantization) condition on the dynamics of $\mathbf{t}$, e.g. recall how such a condition is imposed on eigenfunctions of the z-th component of angular momentum in quantum mechanics, then the space $\mathbf{R}^3$ will be foliated by the Maxwellian surfaces in such a way that each surface is associated with specific quantization condition. In this (and the following) section we provide some details of how this can actually happen. In view of the constraint $\operatorname{div} \mathbf{t} = 0$, the Serret–Frenet frame composed of the vector triad $X = (\mathbf{t}, \mathbf{n}, \mathbf{b})$ in addition to familiar moving frame equation of motion

$$\frac{d}{ds} X^T = \mathbf{A} X^T \tag{2.4.1}$$

with the matrix $\mathbf{A}$ in its standard form

$$\mathbf{A} = \begin{pmatrix} 0 & \kappa & 0 \\ -\kappa & 0 & \tau \\ 0 & -\tau & 0 \end{pmatrix}, \tag{2.4.2}$$

where parameters $\kappa(s)$ and $\tau(s)$ are being respectively local curvature and torsion, should satisfy the following two equations:

$$\frac{d}{ds} X^T = \mathbf{B} X^T \tag{2.4.3}$$

and

$$\frac{d}{ds} X^T = \mathbf{C} X^T, \tag{2.4.4}$$

where

$$\mathbf{B} = \begin{pmatrix} 0 & \theta_{ns} & \Omega_b + \tau \\ -\kappa & 0 & -\operatorname{div}\mathbf{b} \\ -(\Omega_b + \tau) & \operatorname{div}\mathbf{b} & 0 \end{pmatrix} \tag{2.4.5}$$

and

$$\mathbf{C} = \begin{pmatrix} 0 & -(\Omega_n + \tau) & \theta_{bs} \\ \Omega_n + \tau & 0 & \kappa + \operatorname{div}\mathbf{n} \\ -(\Omega_b + \tau) & -(\kappa + \operatorname{div}\mathbf{n}) & 0 \end{pmatrix}. \tag{2.4.6}$$

Details of the derivation of these equations are provided in [26]. In these equations $\Omega_b = \mathbf{b}\cdot\operatorname{curl}\mathbf{b}, \Omega_n = \mathbf{n}\cdot\operatorname{curl}\mathbf{n}, \theta_{ns} = \mathbf{n}\cdot\frac{\delta\mathbf{t}}{\delta n}$, where $\frac{\delta}{\delta n} = \mathbf{n}\cdot\nabla, \theta_{bs}$ is obtained analogously by replacing $\mathbf{n} \rightleftarrows \mathbf{b}$. Here $\nabla = \mathbf{t}\frac{\delta}{\delta s} + \mathbf{n}\frac{\delta}{\delta n} + \mathbf{b}\frac{\delta}{\delta b}$ so that $\operatorname{div}\mathbf{t} = \theta_{bs} + \theta_{ns}$, $\operatorname{div}\mathbf{n} = -\kappa + \frac{\delta\mathbf{n}}{\delta b}\cdot\mathbf{b}$ and $\operatorname{div}\mathbf{b} = -\frac{\delta\mathbf{n}}{\delta n}\cdot\mathbf{b}$. We also need to define $\Omega_s = \mathbf{t}\cdot\operatorname{curl}\mathbf{t}$. It can be demonstrated that $\Omega_b + \Omega_s + \Omega_n = 2(\Omega_s - \tau)$. In these equations τ is the torsion while κ is the curvature of a curve. Suppose now that our curve lies on the surface Σ so that the vector $\mathbf{n}$ is parallel to the normal to Σ. Accordingly, the vectors $\mathbf{t}$ and $\mathbf{b}$ are tangential at each point of such a surface. The surface with the property just defined can be obtained, for instance, from the requirement

$$\Omega_n = 0 \tag{2.4.7}$$

which can be shown, e.g. see Theorem 2 of [33], to be compatible with the condition $\operatorname{div}\mathbf{t} = 0$. As a by-product, the L-L Eq. (2.3.12b) is obtained in its canonical form, that is with $\Omega(\omega) = 1$.

To demonstrate that requirement, Eq. (2.4.7), is indeed associated with the Maxwellian surfaces we recall that to obtain the Lund–Regge equation requires identification of the Maxwellian surfaces with the surfaces of constant pressure $\Pi = \Pi(\omega)$. In such a case one can use the hydrostatic–magnetohydrostatic correspondence discussed in Sec. 2.2 and, in view of this, use the results of [33]. Thus, if $\Omega_n = \mathbf{n}\cdot\operatorname{curl}\mathbf{n} = 0$, it is appropriate to represent the unit vector $\mathbf{n}$ as

$$\mathbf{n} = \psi\nabla\Phi \tag{2.4.8}$$

so that the tangent plane to $\Phi = \text{const.}$ is spanned by the vectors $\mathbf{t}$ and $\mathbf{b}$. Accordingly,

$$\mathbf{t}\cdot\nabla\Phi = 0 \quad \text{and} \quad \mathbf{b}\cdot\nabla\Phi = 0. \tag{2.4.9}$$

Hence, if $\mathbf{t}$ is tangent to the streamlines, $\mathbf{b}$ is tangent to the magnetic field lines (or vice versa). Clearly $\Phi = \text{const.}$ represents one of the Maxwellian surfaces. Such surfaces are known as *isodynamic* in view of the imposed condition $\mathbf{v} = v\mathbf{t}$ (or $\mathbf{w} = w\mathbf{t}$). To find such surfaces explicitly, consideration of symmetry are extremely helpful. Usually, one is looking for solutions with axial or spherical symmetry expecting these surfaces to be some surfaces of revolution. The most typical surfaces obtained in such a way are either nested helicoidal surfaces or nested tori [28]. The helicoidal solutions (in the context of conformational statistics of polymers) could be relevant to problems involving coiling of DNA and proteins. Other applications of such solutions, including their applications to fluid turbulence, are discussed in detail in comprehensive review by Reed [34]. In this chapter we shall only discuss the case of nested tori for the reasons explained in the next section.

2.5 The Maxwellian Tori and the Torus Knots Associated with Them

In 1984, Joel Langer and David Singer proved the following remarkable theorem, [35].

Theorem 2.5.1. *There is a countable infinity of (similarity classes of) closed non-planar elastic* curves in $\mathbf{R}^3$. *All such elasticae are embedded and lie on embedded tori of revolution. Infinitely many of these are knotted and the knot types are the* (m, n)*-torus knots satisfying* $m > 2n$. *The integers* m *and* n *determine the elasticae uniquely (up to similarity).*

Remark 2.5.1. The above theorem does not account for closed elastic curves which can self- intersect. The simplest knot with considerable symmetry, the figure eight knot, made of piece of springy wire may have self-intersection points. Langer and Singer conjectured that there should be *at least three* such points (counting multiplicity) in a knotted wire.

Remark 2.5.2. Recall that the knot K is a *torus knot* if it can be embedded into the surface of a torus [36].

Remark 2.5.3. Based on the above facts it remains to connect solutions of the L-L Eq. (2.3.12) (with $\Omega(\omega) = 1$) with the elastica solutions of Theorem 2.5.1. In order to place such solutions onto the Maxwellian tori.

The above self-intersection conjecture has been analyzed further in a number of papers. For instance, in the paper by Ivey and Singer [37]. Before discussing their findings we need to mention some results from the follow up paper by Langer and Singer published in 1996 [38]. In it the authors made a detailed comparison between the differential geometry of elastic curves and that of elastic rods. Unlike the case of curves, whose thickness is ignored, the elastic rods possess thickness which is taken into account explicitly. The limit of zero thickness should then be carefully performed to merge the results for curves with those for rods.

Remark 2.5.4. Recall that Lund and Regge also assumed that the vortices have finite thickness but they replaced infinitely thin vortex lines by surfaces. *In the context of string theory accounting for thickness is equivalent to replacing strings by membranes.*

The results of calculations naturally differ for closed and open rods. Here we shall only discuss the case of closed rods for the sake of compatibility with results of previous subsections. In both cases it is helpful to introduce the *centerline* — a unit speed curve along the axis of the rod. In the context of semiflexible polymers (the Kratky–Porod model) discussed in Sec. 8.4.2 presence of this centerline is accounted by the constraint $\int ds\delta(\mathbf{v}^2(s)-1)$ in the path integral [39] describing the conformational properties of individual semiflexible polymer chain. If such a constraint is taken into account, the equilibrium conformation of such a centerline is determined by yet another theorem proven by Langer and Singer [38].

Theorem 2.5.2. *If a curve $\gamma(s)$ has nonvanishing curvature $\kappa(s)$ and is the centerline of a (symmetric) Kirchhoff elastic rod in equilibrium, then γ is an extremum of the functional $\mathcal{F}(\gamma) = \lambda_1\mathcal{L}(\gamma) + \lambda_2\mathcal{T}(\gamma) + \lambda_3\mathcal{K}(\gamma)$, where λ_1–λ_3 are some real constants $\mathcal{L}(\gamma) = \int ds, \mathcal{T}(\gamma) = \int ds\tau(s)$ and $\mathcal{K}(\gamma) = \frac{1}{2}\int ds\kappa^2(s)$.*

Remark 2.5.5. In our work [40] on Polyakov–Kleinert rigid string [41] we discussed minimization of $\mathcal{F}(\gamma)$ in which $\lambda_2 = 0$. Such a restriction makes perfect sense for the open rods but should be used with some caution for closed rods (termed as "elasticae" by Langer and Singer [42] in yet another paper on which Theorem 2.5.1. is based). Extension of one-dimensional results for the functional $\mathcal{F}(\gamma)$ to two-dimensional surfaces leads to functional for the Polyakov–Kleinert rigid string model. Mathematically, the two-dimensional analog of $\mathcal{F}(\gamma)$ with $\lambda_1 = \lambda_2 = 0$ is known as the Willmore functional. In our work, [40], we determined conditions under which the Polyakov–Kleinert rigid string model functional coincides with the Willmore functional. Such a functional is widely used in differential geometry [43]. Minimization of such a functional produces the Willmore tori.

The result depends upon the space (or space–time) into which such tori are embedded. Initially, it was believed that the rigid string model is well-suited for description of gluonic excitations in the quark–gluon plasma of QCD. Subsequently, more detailed analysis caused replacement of the Polyakov–Kleinert model by the Faddeev Niemi model [44] producing knotted/linked gluonic configurations made of knotted tori of finite thickness. Details are discussed below.

In the paper by Ivey and Singer [37] the above Theorem 2.5.2 was used essentially. In accord with Theorem 2.5.1 obtained for smooth closed elastic rods with constant torsion, the same results for torus knots were obtained using functional $\mathcal{F}(\gamma)$. However, when the torsion is not constant, the situation is becoming more delicate. It is delicate in several ways. First, it is possible to generate *all* types of torus knots without restrictions on these knots. Second, the stability of such torus knot configurations is determined by the torsion along the centerline. In the case if the configuration is unstable, there will be just one self-intersection for the torus knots (Theorem 2.5.1). Third, in addition to torus knots, the functional $\mathcal{F}(\gamma)$ is capable of producing figure eight knots whose conformations may be stable or unstable depending upon the value of torsion along the centerline.

Remark 2.5.6. A pedagogical summary of results for elastic curves and rods can be found in the paper by Singer [45]. From this reference it follows, in particular, that the theory of dynamics of elastic rods is isomorphic to that developed for dynamics of rigid bodies by Kirchhoff in the 19th century. As a result of such an isomorphism, it has become possible to use the results for dynamics of elastic curves and rods in the theory of optimal control dealing, for instance, with dynamics of satellites [46] or with the problems of computer vision [47]. If instead of dynamics of rigid bodies we would be interested in dynamics of deformable bodies, we would eventually obtain the correspondence between the dynamics of deformable bodies and that of Yang–Mills gauge fields. Incidentally, such a correspondence was noticed first by Guichardet [48] in the context of rotational and vibrational motions of large molecules. Comprehensive review of the dynamics of deformable molecules can be found in [49]. Later in this book we shall discuss this topic further in the context of gauge fields, contact geometry and optimal control.

With all fascinating applications of elastic curves and rods just mentioned, the question still remains: How are these results related to the L-L equation describing motion of the vortex filament? Evidently, the configurations of elastic curves/rods which follow from Theorem 2.5.2 should include those described by the L-L equation. That this is indeed the case was demonstrated in the same paper by Ivey and Singer [37]. We would like to place their results into context of this book by discussing additional sources of information. Specifically, in connection with the Remark 2.5.6 we want to mention the results by Kida [50] on motion of vortex filaments without changes of their form. Comprehensive extension of Kida's results from filaments to Maxwellian surfaces is discussed in the Master's thesis by Grice [51]. Kida looked at possible solutions of the standard L-L equation, e.g Eq. (2.3.12b) with $\Omega(\omega) = 1$, describing translational, rotational and slipping motion along the filament of a filament whose shape does not change in time. Subsequently, Ricca (with collaborators) [52] studied numerically time-dependent evolution of the vortex knots with purpose of estimating the time needed for such knots to become unstable and to break. These knots are present in Kida's solutions as torus knots in accord with Theorem 2.5.1. However, Kida considered not only closed filaments forming torus knots but also open filaments. There is no need for us to discuss such configurations as results of the next chapter indicate. In the meantime, we would like to mention the following. In [53] Ricca reobtained Kida's results for torus knots by much simpler methods. Refinements of his results were made by Calini and Ivey [54–58]. A very detailed book by Kambe [59] also provides the wealth of useful related information. All these authors use essentially the Hasimoto transform. Such a transform is used for converting the L-L into the nonlinear Schrödinger equation (NSE). The NSE equation is one of the benchmark equations of the theory of solitons. In particular, the theory of finite gap solutions for this equation [60] was used along with the inverse Hasimoto transform needed for reproducing various types of torus knots, including those with self-intersections. Since initially the L-L equation was used, only the torus knots of various types were reproduced in [54–58]. Recall that in the paper by Ivey and Singer [37] the figure eight knot was obtained as well. Figure 2 of [37] provides a summary for the parameter (moduli) space of quasi-periodic elastic rods. It is depicted as a circle with x and y axes of the Cartesian coordinate system placed at the circle's center. Along the y-axis the elastic curves are located (that is curves for which $\lambda_2 = 0$). Along the x-axis the curves whose torsion is constant are located. The S-shaped curve which begins at the South pole, goes through the center of the circle and ends at the Northern pole corresponds to the self-intersecting rods. Finally, Kida's curves occupy the first and third quadrants. Clearly, curves whose parameter space lies in the second and third quadrants might represent knots other than the torus knots. Both the torus and the figure eight knots represent respectively the simplest Seifert-fibered and hyperbolic-type knots (since their complements in S^3 are either Seifert-fibered or hyperbolic spaces). All other knots can eventually be reduced (untangled) to these two [36]. From the discussion above it follows that the functional $\mathcal{F}(\gamma)$ should be capable, in principle, to reproduce these two distinct families of knots. But, in view of the Remark 2.5.1 all such knots are expected to have some self-intersections. The dissertation by Heiko von der Mosel entitled "Existence and Regularity for Nonlinear Elastic Self-Contact Problems" (Bonn University, 2001) is devoted to study of this type of problems. Preliminary results were published in 1999 [61]. Subsequently, they were refined in series of papers [62, 63]. Our readers might notice similarities between the approach taken in these papers and those used by us for knot complexity calculations [64]. At the physical level of rigor the same problem of self-intersections was studied by Pierre–Louis [65] who used functional $\mathcal{F}(\gamma)$ with $\lambda_2 = 0$ (that is for the elasticae in Langer–Singer terminology). In view of the results just presented, the results of [65] cannot be used for description of all types of knots.

Exercise 2.5.1. Investigate how Kida's results may change if we allow slow deformation of closed curves/filaments while they move in three-space. Hint: Take into account gauge fields emerging under such conditions. This problem should be qualified as variation on the theory of Berry's phase.

Chapter 3

All About and Around Woltjer's Theorem

3.1 General Information

In 1958, Woltjer wrote a paper [66] in which Eq. (1.3.1) was derived variationally. In anticipation of other physical applications, it is useful to investigate why such a derivation is better than variationally producing Eq. (1.3.2) (or (1.3.4)) and then selecting solutions satisfying Eq. (1.3.1). The situation in the present case is very similar to that encountered in quantum field theory. Indeed, as it was mentioned after Eq. (1.3.2), the difference between Eqs. (1.3.1) and (1.3.2) (or (1.3.3)) is very much the same as the difference between the Dirac equation (analog of Eq. (1.3.1)) and the Klein–Gordon equation (analog of Eq. (1.3.2) (or (1.3.3))). Field-theoretical aspects of this correspondence will be discussed in the next chapter. In this chapter we shall discuss instead the developments related to Woltjer's work.

We begin with discussion of the details of Woltjer's work first. For this purpose, we have to minimize the functional

$$\mathcal{J}(\mathbf{A}) = \int_V dV[(\operatorname{curl}\mathbf{A})^2 + \alpha(\mathbf{A}\cdot\operatorname{curl}\mathbf{A})]. \qquad (3.1.1)$$

Since

$$\delta\mathcal{F}(\mathbf{A}) = \int_V dV[2\operatorname{curl}\mathbf{A}\cdot\operatorname{curl}\delta\mathbf{A} - \alpha(\delta\mathbf{A}\cdot\operatorname{curl}\mathbf{A} + \mathbf{A}\cdot\operatorname{curl}\delta\mathbf{A})], \qquad (3.1.2)$$

integrating by parts and assuming that at the enclosing surface $\delta\mathbf{A} = 0$ produces

$$\operatorname{curl}\operatorname{curl}\mathbf{A} - \alpha\operatorname{curl}\mathbf{A} = 0. \qquad (3.1.3)$$

This equation is equivalent to Eq. (1.3.1) since $\operatorname{curl}\mathbf{A} = \mathbf{H}$. The above derivation suffers from the deficiency of assuming that the Lagrangian multiplier α is a constant. Even though Woltjer provided some plausible arguments why this assumption should be valid, this did not deter other researchers, most notably, Taylor [21], Moffatt [67] and Arnol'd [68], from further investigation of this issue.

We begin with discussion of arguments made by Moffatt. Following [67], it is convenient to introduce the helicity integral I over the volume V

$$I = \int_V \mathbf{v} \cdot \tilde{\omega}\, dV \tag{3.1.4}$$

along with two other integrals, the kinematic kinetic energy for incompressible fluid given by

$$\frac{2T}{\rho} E = \int_V \mathbf{v}^2\, dV, \tag{3.1.5}$$

where ρ is density of the fluid and T is temperature (in units in which Boltzmann's constant $k_{\mathrm{B}} = 1$), and the rotational kinetic energy Ω of fluid given by[a]

$$\Omega = \int_V \tilde{\omega}^2\, dV. \tag{3.1.6}$$

Use of the Schwarz inequality produces either

$$E\Omega \geq I^2 \quad \text{or} \quad \Omega \geq \frac{I^2}{E}. \tag{3.1.7}$$

Moffatt demonstrated that the equality is achieved only if Eq. (1.2.3) holds with α being a constant. Under such circumstances we obtain: $\alpha I = E$. We would like to explain the physical and mathematical meaning of just obtained result. For this purpose, following [9, 29] we shall assume that the vorticity $\tilde{\omega}$ in the above equations can be represented as $\tilde{\omega}(\mathbf{r}) = k \oint_C d\sigma \mathbf{v}(\sigma) \cdot \delta(\mathbf{r} - \mathbf{r}(\sigma))$, that is it vanishes everywhere, except at the closed contour C of negligibly small thickness. Then, using Eq. (1.2.7), the vector field $\mathbf{A}$ can be obtained as $\mathbf{A}(\mathbf{x}) = \frac{k}{4\pi} \oint_C d\sigma \frac{1}{|\mathbf{x} - \mathbf{r}(\sigma)|} \mathbf{v}(\sigma)$. Since

$$\begin{aligned} E &= \frac{1}{2} \int dV \mathbf{v}^2 = \frac{1}{2} \int dV (\nabla \times \mathbf{A}) \cdot \mathbf{v} \\ &= \frac{1}{2} \int dV [\mathbf{A} \cdot \tilde{\omega} + \mathrm{div}[\mathbf{A} \times \mathbf{v}]] = \frac{1}{2} \int dV \mathbf{A} \cdot \tilde{\omega}, \end{aligned} \tag{3.1.8}$$

we can rewrite this result equivalently as

$$E = \frac{k^2}{8\pi} \oint_{C_1} d\sigma \oint_{C_2} d\sigma' \frac{\mathbf{v}(\sigma) \cdot \mathbf{v}(\sigma')}{|\mathbf{r}(\sigma) - \mathbf{r}(\sigma')|}. \tag{3.1.9}$$

[a] Without loss of generality we put $\frac{2T}{\rho} = 1$ while arriving at this result.

It can be further rearranged. To do so, we notice that in view of definitions we just made, it is permissible to write

$$\oint_{C_1} d\mathbf{r}\cdot\mathbf{A} = \frac{k}{4\pi}\oint_{C_1} d\sigma \oint_{C_2} d\sigma' \frac{\mathbf{v}(\sigma)\cdot\mathbf{v}(\sigma')}{|\mathbf{r}(\sigma)-\mathbf{r}(\sigma')|} = \frac{2}{k}E. \tag{3.1.10}$$

Equation (3.1.10) can be further rewritten with help of definitions we already made. For this task we have to use the relation $\mathbf{v} = \pm\kappa\mathbf{A}$ which was postulated prior to writing of Eq. (1.3.5). Using this result in Eq. (3.1.10) we obtain, by appropriately adjusting the orientation of the contour C_1,

$$\begin{aligned}\oint_{C_1} d\mathbf{r}\cdot\mathbf{A} &= \frac{1}{\kappa}\oint_{C_1} d\mathbf{r}\cdot\mathbf{v} = \frac{1}{\kappa}\oint_{C_1} d\mathbf{r}\cdot(\nabla\times\mathbf{A})\\ &= \frac{k}{\kappa 4\pi}\oint_{C_1} d\sigma\oint_{C_2} d\sigma'[\mathbf{v}(\sigma)\times\mathbf{v}(\sigma')]\cdot\frac{(\mathbf{r}(\sigma)-\mathbf{r}(\sigma'))}{|\mathbf{r}(\sigma)-\mathbf{r}(\sigma')|^3}\\ &\equiv \frac{k}{\kappa} lk(1,2) \end{aligned} \tag{3.1.11a}$$

with $lk(1,2)$ being the linking number between contours C_1 and C_2. Using the last two equations it is possible to rewrite now the energy, Eq. (3.1.9), in terms of linking numbers. That is the hydrodynamics of ideal fluids can be reformulated as the Abelian topological field theory of the type discussed in [64]. Specifically, taking into account the non-negativity of E, we obtain:

$$E = \frac{k^2}{2\kappa}\,|lk(1,2)|. \tag{3.1.11b}$$

Equation (3.1.11a) coincides with that obtained by Moffatt and Ricca [69], where it was derived using different arguments. Furthermore, thus obtained expression for the energy coincides with that which can be found in the Landau Lifshitz textbook, [13], p. 200 (top). In the last case this result was obtained as an exercise, just for two linked contours C_1 and C_2. The result, Eq. (3.1.11b), for the energy is in accord with that found by Arnol'd, [68], p. 121. Instead of the inequality given by Eq. (3.1.7) he proved that

$$E \geq C\cdot|\mathcal{H}(\mathbf{v})|, \tag{3.1.12}$$

where $\mathcal{H}(\mathbf{v})$ is the helicity which he defined in a way equivalent to I (Eq. (3.1.4)), that is by

$$\mathcal{H}(\mathbf{v}) = \int_V \mathbf{v}\cdot\mathrm{curl}^{-1}\mathbf{v}\; dV. \tag{3.1.13}$$

To show that both results are indeed equivalent, it is sufficient to take into account that in arriving at Eq. (3.1.11b) we used both $\mathbf{v} = \text{curl}\,\mathbf{A}$ and $\mathbf{v} = \pm\kappa\mathbf{A}$, that $\mathcal{H}(\mathbf{v})$ in Eq. (3.1.12) is of the same form as the second term in the Woltjer's functional $\mathcal{F}(\mathbf{A})$, and that use of Eq. (3.1.3) back in $\mathcal{F}(\mathbf{A})$ produces result compatible with the inequality, Eq. (3.1.12). The constant C in this inequality is equal to the largest (by absolute value) eigenvalue $|\lambda_{\text{max}}|$ of the operator curl^{-1}. The inequality becomes an equality for v's satisfying equation: $\text{curl}^{-1}\mathbf{v}^* = \lambda_{\text{max}}\mathbf{v}^*$.

The relationship between the helicity and linking numbers was found by Moffatt [67]. To describe it, we need to think about the ith vortex filament as a torus of cross-sectional area S_i. Then, the flux Q_i can be defined in a usual way as

$$Q_i = \oint d\mathbf{S}_i \cdot \mathbf{v} \tag{3.1.14}$$

so that

$$\mathcal{H}(\mathbf{v}) = \sum_i lk(i,i)Q_i^2 + 2\sum_{i\neq j} lk(i,j)Q_jQ_i. \tag{3.1.15}$$

The self-linking numbers $lk(i,i)$ were discussed in [64]. In particular, for a single self-linked vortex filament for which

$$lk(i,i) = Tw(i) + Wr(i) \tag{3.1.16}$$

with twisting $Tw(i)$ and writing $Wr(i)$ as defined in [64]. Moffatt demonstrated that

$$\mathcal{H}(\mathbf{v}) = lk(i,i)Q_i^2 \equiv h_iQ_i^2. \tag{3.1.17}$$

This is just a corollary of Eq. (3.1.15).

It is instructive to provide some physical explanation of the results just presented. It comes from the following observation. In mathematics, knots are considered as just an embedding of S^1 into $\mathbf{R}^3$ (or into S^3 and sometimes, into $\mathbf{H}^3$). As such, physical thickness of the knot does not play any role. In reality, it does play a role as explained in [68] (private communication to Arnol'd by Anrei Sakharov). Following this reference, we consider the simplest Hopf link, Fig. 1. The velocity field confined to two tubes of such a link can be affected only by the volume-preserving diffeomorphisms. To minimize an energy of such confined fields is possible by shortening the length of most of trajectories. Such a reduction of orbits leads to shrinking of the velocity vectors along the streamlines. This causes fattening of the respective tori (tubes) because the diffeomorphisms are volume-preserving.

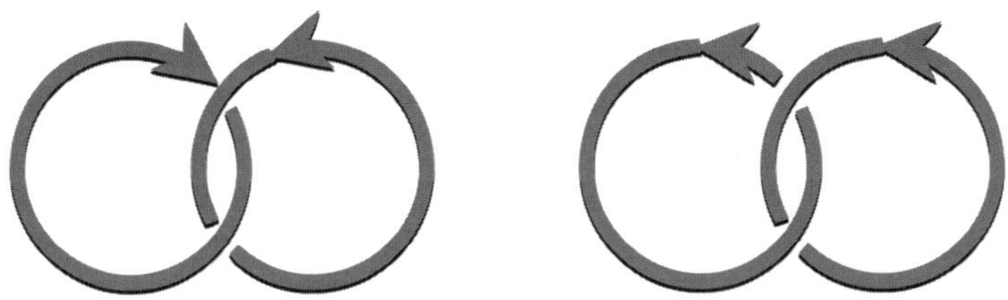

(a) (b)

Fig. 1. Two types of Hopf links: (a) the right-hand Hopf link, (b) the left-hand Hopf link.

The thickening of tubes can continue only till they reach the configuration when the cross-section of one tube will be tightly bounded by another torus tube and vice versa. This configuration bounds the energy from below. These arguments bring us to the necessity of introducing the volumes V_i of tubes for each link component. Without loss of generality, it is then possible to focus attention only at one knotted torus. Then, following Moffat and Chui [70], it is possible to replace the inequality, Eq. (3.1.12), by the lower bound for the (knot) energy

$$E_{\min} \simeq m(h_i)Q_i^2V^{-\frac{1}{3}}. \tag{3.1.18}$$

In this formula $m(h_i)$ is some non-negative function of the self-linking. It is expected that this function may possess a number of local minima

$$\{m_0(h_i), m_1(h_i), m_2(h_i), \ldots\}. \tag{3.1.19}$$

By definition, these numbers would represent the knot energy spectrum should they be known. But, they are not! Only $m_0(h_i)$ is possible to estimate with some confidence as shown in [70]. More recent work by Maggioni and Ricca [71] provides better computational scheme for determination of the "ground state" energies for many knot types. Alternative methods were developed by Friedman and He [72]. They will be discussed below in connection with the Faddeev–Skyrme (F-S) model.

Remark 3.1.1. In the same work the notion of the *asymptotic crossing number* was introduced (e.g. see our work, [9], for some physical applications of this concept). We shall provide extra details about this number in connection with our discussion of the F-S model.

Remark 3.1.2. Various physical and mathematical problems related to knot thickness had become the focus of attention recently. For example, in the work by Cantarella, Kushner and Sullivan [73], the following problem was posed and solved: If the *rope length* of a knot is defined as a quotient of its length by its thickness (the radius of the largest embedded tube around the knot), are there minimizers for this quantity for knots of any type? Alternatively, following Gonzales and Llave [74] it is possible to formulate and to solve the same problem as follows: Find the thickest curve of prescribed length amongst those of given knot type. The *thickness of such a curve is an intrinsic property of a knot*, and the curve itself provides an *ideal shape or representation of the knot type.*

3.2 Equilibria in Liquid Crystals and the Faddeev–Skyrme Model for Pure Yang–Mills Fields

Being armed with results just described, it is appropriate at this point to discuss the Faddeev–Skyrme model. Although such a model was obtained as a result of the Abelian reduction of the non-Abelian (SU(2)) pure Yang–Mills field theory [44], its mathematical description is closely related to that which we just discussed. To understand better the underlying physics of this model, surprisingly, some knowledge of the theory of liquid crystals is very helpful. In particular, it is known [75] that the distortional part f_d of the (Oseen, Frank, Eriksen) free energy density for nematics is given by

$$f_d[\mathbf{n}] = \frac{1}{2}K_1(\operatorname{div}\mathbf{n})^2 + \frac{1}{2}K_2(\mathbf{n}\cdot\operatorname{curl}\mathbf{n})^2 + \frac{1}{2}K_3(\mathbf{n}\times\operatorname{curl}\mathbf{n})^2. \quad (3.2.1)$$

Such defined free energy has obvious minimum: $\operatorname{div}\mathbf{n} = 0$, $\mathbf{n}\times\operatorname{curl}\mathbf{n}=0$ and $\mathbf{n}\cdot\operatorname{curl}\mathbf{n}=\alpha$. That is the minimum is achieved for the force-free configurations, provided that we can demonstrate that the condition $\mathbf{n}\cdot\operatorname{curl}\mathbf{n}=\alpha$ falls into the same category of the force-free fields. This can be demonstrated following [34]. According to this reference all solutions of the force-free Eq. (1.3.1) can be composed with the help of the following basic field types:

1. Solenoidal vector fields

$$\operatorname{div}\mathbf{H} = 0. \quad (3.2.2)$$

2. Complex lamellar vector fields

$$\mathbf{H}\cdot(\operatorname{curl}\mathbf{H}) = 0. \quad (3.2.3)$$

3. Beltrami, or force-free fields for which

$$\mathbf{H}\times\operatorname{curl}\mathbf{H} = 0. \quad (3.2.4)$$

Furthermore, since

$$(\mathbf{H}\cdot\text{grad})\mathbf{H} = \text{grad}\frac{\mathbf{H}^2}{2} - \mathbf{H}\times(\text{curl}\,\mathbf{H}), \tag{3.2.5}$$

the equivalent of Eq. (1.3.1) can also be written as

$$(\mathbf{H}\cdot\text{grad})\mathbf{H} = \text{grad}\,\frac{\mathbf{H}^2}{2}. \tag{3.2.6}$$

Using Eq. (1.3.1), we obtain as well

$$\mathbf{H}^2\alpha = \mathbf{H}\cdot(\text{curl}\,\mathbf{H}). \tag{3.2.7}$$

To investigate this result further we take the divergence on both sides of Eq. (1.3.1) resulting in the following equation

$$\alpha(\text{div}\,\mathbf{H}) + \mathbf{H}\cdot(\text{grad}\,\alpha) = 0. \tag{3.2.8}$$

Using Eq. (3.2.2) we end up with

$$\mathbf{H}\cdot(\text{grad}\,\alpha) = 0 \tag{3.2.9}$$

causing us to require α = const. Even though we encountered analogous situation already in Sec. 2.2 of Chap. 2, we can still arrive at the same conclusion differently. For this purpose taking curl on both sides of Eq. (1.3.1) we obtain:

$$\text{curl}\,\text{curl}\,\mathbf{H} - \alpha\ \text{curl}\,\mathbf{H} = \mathbf{H}\times(\text{grad}\,\alpha) \tag{3.2.10}$$

causing us again to require α = const. to be in compliance with Eq. (3.1.3). Thus, both $\mathbf{H}$ and curl $\mathbf{H}$ are Beltrami fields with *the same* constant α. Such type of vector field is called *Trkalian field*, after Trkal, a Russian physicist who first discovered such a field. Evidently, all successive curls of Trkalian fields are Trkalian fields with the same coefficient α.

Even though a very detailed analysis of these fields was made in [76] (and references therein), fortunately, this topic is still open for discussion. It will be performed later in this book when we shall introduce some basics facts from contact geometry. In the meantime, going back to Eq. (3.2.7) and requiring $\alpha = 0$ we arrive at the condition for the complex lamellar vector field, Eq. (3.2.3). In the context of liquid crystals the condition $\alpha \neq 0$ corresponds to *cholesterics*, [75], p. 287, while the condition $\alpha = 0$ corresponds to *smectics A*, [75], p. 339. In 1972 De Gennes discovered the remarkable analogy between smectics A and conventional superconductors [77]. In 1988 Lubensky and Renn [78] extended his results to describe the so-called twist grain boundary (TGB) phases in liquid crystals. We shall re-obtain de Gennes superconductivity result for smectics in the context of magnetohydrodynamics in the next subsection. Results by Lubensky and Renn will be discussed in Chap. 7, in the context of contact geometry.

Consider now the free energy $\mathcal{F}_d[\mathbf{n}]$ for the case when $K_1 = K_2 = K_3 = K$. In this case the distortional free energy acquires the form known for the classical version of three-dimensional Heisenberg ferromagnet, [75], pp. 104–105,

$$\mathcal{F}_d[\mathbf{n}] = \int f_d[\mathbf{n}]d\mathbf{r} = \frac{K}{2}\int d\mathbf{r}[(\operatorname{div}\mathbf{n})^2 + (\operatorname{curl}\mathbf{n})^2]. \tag{3.2.11}$$

This result is useful to compare with Woltjer's functional, Eq. (3.1.1). Such a comparison can be facilitated as follows. In Woltjer's functional $\mathcal{F}(\mathbf{A})$ the minimization is made for fields for which $\operatorname{div}\mathbf{A} = 0$. This condition can be included into $\mathcal{F}(\mathbf{A})$ by the method of Lagrangian multipliers being used for dealing with systems with constraints. At the same time, if we are looking for the extrema of $\mathcal{F}_d[\mathbf{n}]$ only for fields for which $\operatorname{div}\mathbf{n} = 0$, then $\mathcal{F}_d[\mathbf{n}] = E$, e.g. see Eq. (3.1.8). As a result, we can apply the Arnol'd inequality, Eq. (3.1.12), and the rest of results of the previous section. These conclusions are immediately transferable to the F-S model. Indeed, following work by Vakulenko and Kapitanskii [79] we define the F-S functional for the static variant of this model given by

$$\begin{aligned}\mathcal{F}_{\text{F-S}}[\mathbf{n}] &\equiv \mathcal{F}_1[\mathbf{n}] + \mathcal{F}_2[\mathbf{n}]\\ &= \int_{\mathbf{R}^3} d\mathbf{r}\left[\frac{1}{2\lambda^2}\partial_i n^a \partial_i n^a + \frac{\epsilon^2}{2}\sum_{i,k=1}^{3}(\varepsilon_{abc}\partial_i n^a \partial_k n^b n^c)^2\right].\end{aligned} \tag{3.2.12}$$

As usual, the summation over the repeated indices is assumed. As in the case of liquid crystals, the functional, Eq. (3.2.12), should be minimized under the constraint: $n^a n^a = 1$. Then, for $\epsilon = 0$ the F-S model coincides with that used for the liquid crystals, that is with Eq. (3.2.11), e.g. see p. 104 of [75]. The second term $\epsilon \neq 0$ is added in order to be in compliance with Derrick's theorem [80]. It can be formulated in the present context as follows. Rescale the coordinates: $\mathbf{r} \to \rho\mathbf{r}$ so that the F-S functional is rewritten as

$$\mathcal{F}_{\text{F-S}}[\mathbf{n}, \rho] \equiv \rho\mathcal{F}_1[\mathbf{n}] + \rho^{-1}\mathcal{F}_2[\mathbf{n}]. \tag{3.2.13}$$

Then, the necessary condition for existence of stable (particle/solitonic-like) solutions is expressible as the condition

$$\frac{\partial}{\partial\rho}\mathcal{F}_{\text{F-S}}[\mathbf{n}, \rho] \mid_{\rho=1} = 0. \tag{3.2.14}$$

In the present case such a condition leads to the requirement:

$$\mathcal{F}_1[\mathbf{n}] = \mathcal{F}_2[\mathbf{n}]. \tag{3.2.15}$$

Instead of looking at Derrick's sufficient condition

$$\frac{\partial^2}{\partial \rho^2}\mathcal{F}_{\text{F-S}}[\mathbf{n},\rho]\mid_{\rho=1}\geq 0, \tag{3.2.16}$$

Vakulenko and Kapitanskii were looking for the lower bound for the $\mathcal{F}_{\text{F-S}}[\mathbf{n}]$. They began with the following crucial observations. If we can construct the vector function $\mathbf{H}$ such that $H_l = \varepsilon_{ikl}\varepsilon_{abc}\partial_i n^a \partial_k n^b n^c$, then it can be shown that $\operatorname{div}\mathbf{H} = 0$. And, if this is so, then $\mathbf{H} = \operatorname{curl}\mathbf{A}$. Now, take into account, that for $\epsilon = 0$ the F-S model is described by the functional $\mathcal{F}_d[\mathbf{n}]$, Eq. (3.2.11). Since both $\mathbf{n}$ and $\mathbf{A}$ are vectors, we can make them proportional to each other. This makes sense in view of condition given by Eq. (3.2.15). Thus, we end up with the familiar hydrodynamic-type functional:

$$\mathcal{F}_{\text{F-S}}[\mathbf{n}] = \frac{1}{2}\int_{\mathbf{R}^3} d\mathbf{r}\left[\frac{1}{\lambda^2}(\operatorname{div}\mathbf{n})^2 + \epsilon^2\mathbf{H}^2\right] \tag{3.2.17}$$

in which the incompressibility condition is accounted using as before the method of Lagrangian multipliers. Vakulenko and Kapitanskii then follow the same path as was used by Arnol'd who arrived at the inequality, Eq. (3.1.12). Because in the present case the situation is complicated by the presence of the nonlinear sigma model-type constraint $n^a n^a = 1$, there is a need in some refinements of Arnold's result which are of interest to discuss. We shall discuss these refinements in context of the benchmark work by Freedman and He [72]. Before discussing this topic in some detail, it is worth mentioning that the helicity integral, Eq. (3.1.4), can be written in the form suggested by Moffatt, [67], p. 119,

$$I = \int_V \mathbf{v}\cdot\tilde{\omega}\, dV = \frac{1}{4\pi}\int_{V(1)} dV(1)\int_{V(2)} dV(2)\frac{\mathbf{r}_{12}}{\|\mathbf{r}_{12}\|^3}\cdot[\dot{\mathbf{r}}(1)\times\dot{\mathbf{r}}(2)]. \tag{3.2.18}$$

As it was proven by Arnol'd [68], the R.H.S. defines the so-called *asymptotic linking number.* It is a topological invariant but is not too restrictive/sensitive. Freedman and He argue that it is zero, for instance, for the nontrivial two-component Whitehead link. They suggest to use instead the *asymptotic crossing number* $Cr(1,2)$, which, in view of Arnol'd's inequality, Eq. (2.6.12), is given by

$$Cr(1,2) = \frac{1}{4\pi}\left|\int_{V(1)} dV(1)\int_{V(2)} dV(2)\frac{\mathbf{r}_{12}}{\|\mathbf{r}_{12}\|^3}\cdot[\dot{\mathbf{r}}(1)\times\dot{\mathbf{r}}(2)]\right|. \tag{3.2.19}$$

As before, it is possible then to introduce the asymptotic self-linking and self-crossing numbers. Next, these authors prove that

$$E_2[\mathbf{v}] = \int dV\,|\mathbf{v}|^2 \geq \frac{1}{V^{\frac{1}{3}}}\left(\int dV\,|\mathbf{v}|^{\frac{3}{2}}\right)^{\frac{4}{3}} = \frac{(E_{\frac{3}{2}}[\mathbf{v}])^{\frac{4}{3}}}{V^{\frac{1}{3}}}, \tag{3.2.20}$$

where

$$E_{\frac{3}{2}}[\mathbf{v}] \geq \left(\frac{16}{\pi}\right)^{\frac{1}{4}} Cr(1,1)^{\frac{3}{4}}. \tag{3.2.21}$$

This result is now useful to compare with that obtained by Vakulenko and Kapitanskii. These authors obtained

$$\mathcal{F}_{\text{F-S}}[\mathbf{n}] \geq \kappa\frac{\epsilon}{\lambda}Cr(1,1)^{\frac{3}{4}}, \tag{3.2.22}$$

where κ is some positive constant whose exact value is obtained in [81, 82]. Taking into account Eq. (3.2.15), the above result can be effectively rewritten as

$$\mathcal{F}_2[|\nabla \times \mathbf{A}|] \geq \tilde{\kappa}Cr(1,1)^{\frac{3}{4}} \tag{3.2.23}$$

which would coincide with Arnol'd estimate given by Eq. (2.6.12) should the exponent for $Cr(1,1)$ be equal to one but it is 3/4! At the same time by combining inequalities given by Eqs. (3.2.20) and (3.2.21) we obtain back Arnol'd-type inequality, Eq. (3.1.12),

$$E_2[\mathbf{v}] \geq \frac{(E_{\frac{3}{2}}[\mathbf{v}])^{\frac{4}{3}}}{V^{\frac{1}{3}}} \geq \left(\frac{16}{\pi V}\right)^{\frac{1}{3}} Cr(1,1). \tag{3.2.24}$$

Hence, the inequality Eq. (3.2.21) implies both Arnold's inequality and the lower bound for the F-S functional.

Exercise 3.2.1. Since in both cases the knotted structures are involved, how are these structures related to each other? Hint: Take a look at [83].

In obtaining the estimate, Eq. (3.2.22), the nonlinear sigma model constraint $n^a n^a = 1$ has not been taken into account explicitly in [79]. Thus, it appears that this constraint happens to be nonessential. Such a conclusion is superficial as we would like to explain now.

We begin with the observation that for $\epsilon = 0$ the F-S model is just the well-known nonlinear O(3) σ-model. Mathematically, it describes the harmonic map from $\mathbf{R}^3$ to S^2 [17]. Already in [44] the suggestion was made that one can safely replace $\mathbf{R}^3$ by S^3. For actual calculations this happen to be essential because in such a case one obtains the well-known Hopf map discussed, for example, in the book by Arnol'd and Khesin [68], p. 128.

This circumstance was fully exploited by Battye and Sutcliffe [84] in their benchmark work on minimization of the F-S model in search of "hopfions" that is of knotted and linked soliton-like solutions of the F-S model. In the meantime, to explain the inequality, Eq. (3.2.22), obtained by Vakulenko and Kapitanskii, we need to discuss the logic behind Freedman and He inequality, Eq. (3.2.21), first. As it was mentioned already, the Arnol'd inequality, Eq. (3.1.12), and Freedman–He inequality, Eq. (3.2.21), are compatible with each other, e.g. see Eq. (3.2.24). However, Arnol'd inequality, Eq. (3.1.12), suffers from one essential drawback noticed in [72]. It can be easily understood based on simple dimensional arguments. Indeed, consider the velocity entering $E_2[\mathbf{v}]$. Its dimensionality is L/T so that altogether the dimensionality of $E_2[\mathbf{v}]$ is L^5/T^2. At the same time, the dimensionality of $\mathcal{H}(\mathbf{v})$ is the same as $Cr(1,2)$ defined in Eq. (3.2.19) that is L^6/T^2. Such a mismatch of dimensionalities can be avoided by introducing 3/2 energy, that is $E_{\frac{3}{2}}[\mathbf{v}]$. Then, in the left- and right-hand sides of Freedman–He inequality, Eq. (3.2.21), the dimensionalities are in agreement. Let us now look at the estimates made by Kapitanskii and Vakulenko. Since the constraint $n^a n^a = 1$ makes the vector $\mathbf{n}$ dimensionless, to make $\mathcal{F}_1[\mathbf{n}]$ and $\mathcal{F}_2[\mathbf{n}]$ dimensionless requires constants λ and ϵ to have both the same dimensionality $L^{\frac{1}{2}}$ so that their ratio in Eq. (3.2.22) is dimensionless. Now notice that the dimensionality of the "magnetic field" $\mathbf{H}$ in their paper is L^{-2} so that the dimensionality of the field $\mathbf{A}$ is L^{-1}. Using these results we conclude that the helicity $\mathcal{H}$ used in their work is dimensionless. Because of this, the rationale for the exponent to be 3/4 in inequality Eq. (3.2.22) remains obscure. To clarify this situation, it is appropriate to mention two alternatively used models: one by Nicole [85], and the other one by Aratyn, Ferreira and Zimmerman (AFZ) [86]. The first model uses the functional $\mathcal{F}_{\text{Nic}} = \int_{\mathbf{S}^3} d\mathbf{r}[\partial_i n^a \partial_i n^a]^{\frac{3}{2}}$ without the dimensional prefactor while the second uses the functional $\mathcal{F}_{\text{AFZ}} = \int_{\mathbf{S}^3} d\mathbf{r} \sum_{i,k=1}^{3} [(\varepsilon_{abc} \partial_i n^a \partial_k n^b n^c)^2]^{\frac{3}{4}}$ again, without the prefactor. Since $\mathbf{n}$ is dimensionless, both models are manifestly dimensionless, and hence, both are conformally invariant. Both are used for minimization instead of the F-S model functional. With this information it makes sense to conjecture that

$$\mathcal{F}_{\text{Nic}} \geq \mathcal{F}_{\text{AFZ}} \tag{3.2.25}$$

in accord with the Freedman–He inequality, Eq. (3.2.21). In view of Eqs. (3.1.8), (3.2.15), it is also plausible to assume that (up to some dimensionless constants)

$$\mathcal{F}_{\text{F-S}} \geq \mathcal{F}_{\text{Nic}} \geq \mathcal{F}_{\text{AFZ}} \tag{3.2.26}$$

in accord with the Vakulenko–Kapitanskii main result, inequality, Eq. (3.2.22).

Exercise 3.2.2. Prove the inequalities Eqs. (3.2.25) and (3.2.26). Hint: Proving these inequalities is equivalent to explanation of the exponent 3/4.

To conclude this subsection, it is appropriate to mention that at the moment of writing of this book, in accord with the inequalities Eq. (3.2.26), all analytically known hopfions obtainable by minimization of these three models are in agreement with each other [87].

3.3 Refinements of Woltjer's Theorem. Implications for Magnetohydrodynamics, Superconductivity and Liquid Crystals

The paper by Taylor [21] (see also his more extended version in [22]) discusses general problem of equilibration dynamics of plasma in toroidal (tokamak) and spherical (spheromak) geometries which can be formulated as follows.

Suppose that the initial state of plasma is arbitrary except that both the magnetic field and current are tangential to the conducting wall. The system is not in stable equilibrium and, when released, will move and dissipate energy before coming to rest. Only when its energy reaches the minimum it is incapable of further movement. Hence, the problem lies in finding the final state minimizing energy under some imposed constraints on plasma's motion.

In [21, 22] Taylor argues that the final state of plasma should be force-free state in which the constant α is coordinate-independent. Only in such a state particles of plasma cannot dissipate any energy through action of the Lorentz force but still can lower the total energy further by radiation. Since Taylor's analysis accounts for finite conductivity of plasma, it supplies more value to results presented above (valid for infinite conductivity) since even if the conductivity is finite, it is still possible to use the infinite conductivity paradigm.

There are many refinements of Woltjer's and Taylor's results. For instance, in [88] much more careful analysis than in Woltjer's paper of the boundary conditions was made. In the Ph.D. dissertation by Cantarella [89] and, subsequently, in the number of papers, most notably in [90], the textbook quality analysis of the role of topology in minimization problems of the Woltjer-type was made while in [91] an up-to-date supplement to Taylor's work can be found. Mathematically rigorous results supporting Taylor's relaxation hypothesis had been recently obtained by Nishiyama [92]. They provide the strongest to date support of recent results by di Vita [93] who is using the infinite conductivity assumption in magnetohydrodynamics of plasma for development of Ginzburg–Landau-type description of plasmas based on classical work by Fröhlich's [11].

It is of interest to discuss now several ramifications of di Vita's results. This makes sense for several reasons. Firstly, we would like to explain the "superconducting rationale" behind solution characterized by the condition $\Omega_n = 0$, e.g. see Eq. (2.4.7). It was left unexplained in papers by Rogers and Schief discussed earlier. Secondly, we would like to extend the superconducting plasma results of [94]. Third, along the way we shall recover de Gennes superconducting result for smectics A (previous subsection) by using arguments entirely different from the original.

We begin by raising the following question: What can happen with the type of fields discussed in Sec. 2.2 of Chap. 2, if they are substituted into Eq. (3.2.7)? In such a case our field is given by $\mathbf{H} = H\mathbf{t}$, where $H = \text{const.}$ so that its substitution into Eq. (3.2.7) produces

$$\mathbf{t} \cdot \operatorname{curl} \mathbf{t} = \alpha. \tag{3.3.1}$$

In Sec. 2.4, we introduced $\Omega_n = \mathbf{n} \cdot \operatorname{curl} \mathbf{n}$ and $\Omega_s = \mathbf{t} \cdot \operatorname{curl} \mathbf{t}$. Now we have to require

$$\Omega_n = 0 \tag{3.3.2}$$

to be in compliance with Eq. (2.4.7). To reproduce de Gennes superconducting result for smectics A and to be in accord with [94] for plasma, we have to discuss as well the full implications of the alternative constraint

$$\Omega_s = 0 \tag{3.3.3}$$

either for plasmas or for smectics.

For this purpose we need to reproduce some results from the Ginzburg–Landau (G-L) theory of superconductivity following [13]. In the simplest form, we are confronted with the task of reproducing one of Maxwell's equations (Ampere's law)

$$\operatorname{curl} \mathbf{H} = \frac{4\pi}{c}\mathbf{j} \tag{3.3.4}$$

along with the London's equation

$$\operatorname{curl} \mathbf{j}_s = -\frac{e^2 n_s}{mc}\mathbf{H} \tag{3.3.5a}$$

or, equivalently,

$$\mathbf{j}_s = \frac{e\hbar n_s}{2m}\left(\nabla\Phi - \frac{2e}{\hbar c}\mathbf{A}\right) \tag{3.3.5b}$$

as a result of variation of some functional, now known as the G-L functional. The exact meaning of constants in these equations is not too important for us since it differs from problem to problem. Nevertheless, for the sake of arguments we define these constants in the superconducting language. So, we have: n_s is the density of superconducting condensate (Cooper pairs), c is the speed of light, $2m$ is the mass of the Cooper pair, e is electron's charge, $\hbar$ is Planck's constant. What is important for us is the explicit form of this functional. It is given by (e.g. see [13], Eq. (45.10))

$$\mathcal{F}_{\text{G-L}} = \int dV \left\{ \frac{\mathbf{H}^2}{8\pi} + \frac{\hbar^2}{4m} \left| \left(\nabla - \frac{2ie}{\hbar c} \mathbf{A} \right) \psi \right|^2 + a|\psi|^2 + \frac{1}{2} b[|\psi|^2]^2 \right\}. \tag{3.3.6}$$

Such an expression is somewhat inconvenient for rigorous mathematical analysis as explained, for example, in [17]. But it is looking very much the same (leaving some technicalities aside, e.g. the non-Abelian character of the field **H**, the spin connection and so on) as the Seiberg–Witten (S-W) functional (defined, in general, on some 3 or 4 dimensional manifold) [17]. Because of this apparent similarity, it is evident that all results associated with the G-L functional, in principle, can be recovered from the S-W functional. In this book we tend to avoid complications associated with such recovery. Nevertheless, we shall mention some consequences of this similarity later in the book. Thus, the corrected G-L functional is given by [13, 17]:

$$\mathcal{F}_{\text{G-L}}[\psi, \mathbf{A}] = \int dV \left\{ \frac{(\nabla \times \mathbf{A})^2}{8\pi} + \frac{\hbar^2}{4m} \left| \left(\nabla - \frac{2ie}{\hbar c} \mathbf{A} \right) \psi \right|^2 + \frac{g}{4} (\sigma - |\psi|^2)^2 \right\}. \tag{3.3.7}$$

It needs to be varied over ψ and **A**. Variation over **A** reproduces Eq. (3.3.4) in which the current is given by

$$\mathbf{j} = -\frac{ie\hbar}{2m} (\psi^* \nabla \psi - \psi \nabla \psi^*) - \frac{2e^2}{mc} |\psi|^2 \mathbf{A}. \tag{3.3.8}$$

If we make σ in Eq. (3.3.7) to be equal to $n_s/2$ and write $\psi = \sqrt{n_s/2}\exp(i\Phi)$, then use of such an ansatz in Eq. (3.3.8) brings it back to Eq. (3.3.5b) while use of the operator curl to Eq. (3.3.5b) brings it back the London Eq. (3.3.5a). Since variation over ψ in G-L functional is of no immediate interest, we are not reproducing it here. Evidently, use of the above ansatz for ψ causes the last term in Eq. (3.3.7) to disappear. The second term in Eq. (3.3.7) vanishes as well if

$$\mathbf{A} = \frac{\hbar c}{2e}\nabla\Phi. \tag{3.3.9}$$

Using this result in the first term of Eq. (3.3.7) makes it zero as well. Hence, under conditions just described, the G-L functional indeed reaches its absolute minimum.

Now we are ready to look at these results from the point of view of force-free fields. Since for the superconducting current $\mathbf{j}_s = en_s\mathbf{v}_s$ and since $\mathbf{v}_s = \frac{\hbar}{2m}\nabla\Phi$, Eq. (3.3.9) acquires the form

$$\mathbf{P} = \mathbf{p}_s - \frac{e}{c}\mathbf{A} = \mathbf{0}. \tag{3.3.10}$$

That is the generalized momentum of the Cooper pair in the ground state is zero. But we can now also define the generalized velocity as $\mathbf{V} = \mathbf{v}_s - \frac{1}{c}\mathbf{A}$. Since we have to require it to be also zero, we obtain, $\mathbf{v}_s = \frac{1}{c}\mathbf{A}$, to be compared with the superconducting option in the relation $\mathbf{v} = \pm\kappa\mathbf{A}$ which was *postulated* prior to Eq. (1.3.5). Consider now the generalized vorticity $\boldsymbol{\Omega} = \operatorname{curl}\mathbf{V}$. The force-free condition in the present case becomes a statement: $\mathbf{0} = \alpha\mathbf{0}$, where α has the same meaning as in Eq. (3.2.7). Evidently, α can also be put equal to zero as it is done in [94]. If in Eq. (3.2.7) we select this option, then we end up with results of the Theorem 1 of [33] dealing with the complex lamellar vector fields for which $\Omega_s = 0$, e.g. see Eq. (3.2.3). Recall that the same condition is required for smectics A, [75], p. 339. According to [33], instead of the L-L equation, we shall now obtain the L-L-type equation

$$\mathbf{t}_b = \frac{\partial}{\partial s}\left(\mu\mathbf{t} \times \frac{\partial\mathbf{t}}{\partial s}\right) \tag{3.3.11}$$

with $\mu = \mu(s)$ in the most general case. When $\mu(s)$ is linear in s, we end up with exactly integrable equation again [33]. Clearly, it is possible to restrict μ's to the constant values based on physical considerations. Whether or not we can use $\mu(s)$ with linear dependence on s is determined by the consequences of Theorem 2.5.2.

Exercise 3.3.1. Investigate if such an equation is obtainable from the theory of elastic rods.

Hint: Look at [16–18] mentioned in [33].

Thus, Theorem 1 of Rogers and Schiff, [33], indeed describes the superconducting-type solution for plasma as well as for smectics A in accord with de Gennes [77] whose derivation of this result is entirely different. The question still remains: Can the condition $\Omega_n = 0$ also describe the superconducting state? We provide some arguments in favor of positive answer to this question in the next subsection.

3.4 Proca's Massive Electrodynamics and Stueckelberg's Trick

Consider once again the G-L functional $\mathcal{F}_{\text{G-L}}[\psi, \mathbf{A}]$ defined by Eq. (3.3.6). In the London limit, that is for $\psi = \sqrt{n_s/2}\exp(i\Phi)$, this functional acquires the form

$$\mathcal{F}_{\text{G-L}}[\psi, \mathbf{A}] = \int dV \left\{ \frac{(\nabla \times \mathbf{A})^2}{8\pi} + \frac{\hbar^2}{4m}(n_s/2)\left|\left(\nabla\Phi - \frac{2e}{\hbar c}\mathbf{A}\right)\psi\right|^2 \right\}. \tag{3.4.1}$$

Consider next a gauge transformation $A_\mu \to A_\mu + \frac{\hbar c}{2e}\partial_\mu \Phi$. It transforms the above functional into

$$\mathcal{F}_{\text{G-L}}[\psi, \mathbf{A}] = \int dV \left\{ \frac{(\nabla \times \mathbf{A})^2}{8\pi} + \frac{e^2 n_s}{2mc^2}\mathbf{A}^2 \right\}. \tag{3.4.2}$$

Let $\frac{4\pi e^2 n_s}{mc^2} = \delta^{-2}$, then $\mathcal{F}_{\text{G-L}}[\psi, \mathbf{A}]$ can be rewritten in the form

$$\mathcal{F}_{\text{G-L}}[\psi, \mathbf{A}] = \frac{1}{8\pi}\int dV\{(\nabla \times \mathbf{A})^2 + \delta^{-2}\mathbf{A}^2\} \tag{3.4.3}$$

coinciding with Eq. (48.13) of Landau and Lifshitz book, [13]. This equation is useful to compare with Woltjer's Eq. (3.1.1). Variation of Eq. (3.1.1) leads to the force-free Eq. (3.1.3). At the same time, variation of Eq. (3.4.3) leads to Eq. (1.3.5) with $\pm\kappa\alpha$ replaced by $-\delta^{-2}$. That is we re-obtain Eq. (1.3.3) in which $-\alpha^2$ is replaced by $-\delta^{-2}$. The functional, Eq. (3.4.3), can be looked upon as time-independent version of the so-called Proca functional describing the massive Abelian gauge vector field [95].

To introduce the Proca equation and Proca functional, recall that in four-dimensional notations Maxwell's equations can be written in compact and suggestive form as follows:

$$\partial_\mu F^{\mu\nu} = j^\nu \quad \text{and} \quad \partial_\mu \tilde{F}^{\mu\nu} = 0, \tag{3.4.4}$$

where the tensor $F^{\mu\nu} = \partial^\mu A^\nu - \partial^\nu A^\mu$, and its dual $\tilde{F}^{\mu\nu}$ is $*F^{\mu\nu} = \frac{1}{2}\varepsilon^{\mu\nu\rho\sigma} F_{\rho\sigma}$. j^ν is the four-dimensional current $j = (\rho, \mathbf{j})$. In the Lorentz gauge, that is when $\partial_\nu A^\nu = 0$, the first of Maxwell's equations acquires the form

$$\Box A^\nu = j^\nu, \tag{3.4.5}$$

where $\Box$ is the d'Alambertian, that is $\Box = \partial_t^2 - \nabla^2$. Proca's equations generalize Maxwell's equations in an obvious way

$$\partial_\mu F^{\mu\nu} + m^2 A^\nu = 0. \tag{3.4.6}$$

With the Lorentz gauge imposed, they acquire the form:

$$(\Box + m^2) A^\nu = 0. \tag{3.4.7}$$

For the time-independent fields Eq. (3.4.7) coincides with the already known Eqs. (1.3.2) and (1.3.3) for the force-free fields for suitably chosen mass parameter. Although we shall discuss this topic further in some detail in Chap. 4, the obtained result is helpful for comparison between the Proca and the G-L functional. Specifically, following [96], the Lagrangian density $\mathcal{L}$ for the Proca functional in the Lorentzian four-space of signature $(+---)$ is given by[b]

$$\mathcal{L}_{\text{Proca}} = -\frac{1}{2} F^\dagger_{\mu\nu} F^{\mu\nu} + m^2 V^\dagger_\mu V^\mu \tag{3.4.8}$$

so that, indeed, the static version of this Lagrangian coincides (with proper identification) with expression under the volume integral in Eq. (3.4.3). Furthermore, for the Proca fields Maxwell's equations acquire the following form[c] [97]

$$\nabla \cdot \mathbf{H} = 0, \quad \nabla \times \mathbf{E} = -\frac{\partial \mathbf{H}}{\partial t} \tag{3.4.9a}$$

and

$$\nabla \cdot \mathbf{E} = \rho - m^2 V, \quad \nabla \times \mathbf{H} = \mathbf{j} + \frac{\partial \mathbf{E}}{\partial t} - m^2 \mathbf{A}, \tag{3.4.9b}$$

where $\mathbf{E} = -\nabla V - \frac{\partial \mathbf{A}}{\partial t}$ as usual. For static fields and in the absence of currents and charges the last two equations produce the Yukawa-type solutions, Eq. (1.3.4). The same result is obtainable by varying the G-L functional, Eq. (3.3.7). Thus, Proca electrodynamics is electrodynamics of superconductivity. Surprisingly, this conclusion was officially reached only in 2008 [98]!

Exercise 3.4.1. Check the calculations of [98].

With all successes of Proca approach to superconductivity, there remains two major issues: gauge invariance and renormalizability. The

[b] Here the symbol † means complex conjugation if the fields are not quantized and Hermitian conjugation if the fields are quantized.

[c] All constants, except m, are suppressed.

functional Eq. (3.4.3) is not gauge invariant, that is it is not invariant with respect to substitution $\mathbf{A} \to \mathbf{A} + \nabla *$. There are two ways to restore the gauge invariance. One is to consider the so-called Higgs mechanism of mass generation which in the present case boils down to using the full G-L functional, Eq. (3.3.7), which is manifestly gauge invariant. Another is to consider the so-called Stuekelberg trick [96]. Instead of Proca's Lagrangian, Eq. (3.4.8), Stueckelberg suggested to use

$$\mathcal{L}_{\text{Stueck}} = -\frac{1}{2}F^{\dagger}_{\mu\nu}F^{\mu\nu} + m^2\left(V^{\dagger}_{\mu} - \frac{1}{m}\partial_{\mu}B^{\dagger}\right)\left(V^{\mu} - \frac{1}{m}\partial_{\mu}B\right) \\ - (\partial_{\mu}V^{\dagger\mu} + mB^{\dagger})(\partial_{\mu}V^{\mu} + mB). \tag{3.4.10}$$

If for a moment we would ignore the last term in this expression, in the static case we would re-obtain back the London approximation to the G-L functional, that is Eq. (3.4.1). Clearly, this result is not going to change if we put $A_{\mu} = V^{\mu} - \frac{1}{m}\partial_{\mu}B$ in Eq. (3.4.10). Thus, we obtain

$$\mathcal{L}_{\text{Stueck}} = \mathcal{L}_{\text{Proca}} - (\partial_{\mu}A^{\dagger\mu} + mB^{\dagger})(\partial_{\mu}A^{\mu} + mB). \tag{3.4.11}$$

The above Lagrangian is (gauge) invariant under transformations

$$A_{\mu} \to A'_{\mu}(x) = A_{\mu}(x) + \partial_{\mu}\Lambda, \quad B_{\mu} \to B'_{\mu}(x) = B_{\mu}(x) + m\partial_{\mu}\Lambda,$$

provided that scalar fields Λ and B obey the same type of differential equations

$$(\partial^2 + m^2)\Lambda = 0, \quad (\partial^2 + m^2)B = 0. \tag{3.4.12}$$

The last term is helping to preserve the gauge invariance. It can (and should) be considered as the gauge-fixing term [96] if the model based on the Stuekelberg Lagrangian is to be quantized. Quantization procedures used for both the Proca and Stuekelberg Lagrangians thus far were unsuccessful [96]. Therefore, we shall discuss another way of doing things in Chap. 4. In the meantime, we need to make a connection between the results of this section with those discussed previously. In particular, in the previous subsection all superconducting results were obtained as a consequence of Eq. (3.3.9). This equation is a bit contradictory since the vector field $\mathbf{A}$ is a pure gauge meaning that the magnetic field $\mathbf{H} = \operatorname{curl}\mathbf{A}$ apparently should be zero. Nevertheless on p. 177 of [13] we find the following result

$$\oint_C \mathbf{A}\cdot d\mathbf{l} = \iint_S (\operatorname{curl}\mathbf{A})\cdot d\mathbf{S} = \iint_S \mathbf{H}\cdot d\mathbf{S} = \frac{\hbar c}{2e}\oint d\mathbf{l}\cdot\nabla\Phi = n\phi_0 \tag{3.4.13}$$

with $\phi_0 = \frac{hc}{2|e|}$. This result is being interpreted as quantization of the magnetic flux penetrating superconducting ring. To correct the above paradox on the same page we find the following paragraph.

"The quantization of magnetic flux has another aspect: it causes the values of the total current J that can flow along the superconducting ring (in the absence of an external magnetic field) to be discrete. This current J creates a magnetic flux trough the ring equal to LJ/c where L is the self-inductance. Equating this to $n\phi_0$, we find the possible values of the current $J = cn\phi_0/L$. In contrast to the magnetic flux quantum, the "quantum of the total current", like the self-conductance L, depends on the shape and size of the ring."

Remark 3.4.1. The paradox with curl $\mathbf{A}$ can be resolved as follows. The curl $\mathbf{A}$ should be zero only in the simply connected domain. In the present case, the topology of the ring domain is that of the punctured disc. This observation causes us to use the same reasoning as was used in Eq. (3.1.11) and for the same reasons. Thus, quantization of flux in Eq. (3.4.13) is caused by the topological reasons. Quantum mechanics is not essential. Incidentally, under such circumstances study of the minima for the G-L functional, Eq. (3.3.7), becomes more complicated. In physical terms these complications could be attributed to the monopole contributions to the G-L free energy functional to be considered further below.

Remark 3.4.2. Since the superconducting current can change, reasonings of [94] based on the validity of Eq. (3.3.10) should be modified. This brings us closer to the affirmative answer to the question posed in the previous subsection: Can the condition $\Omega_n = 0$ also describe the superconducting state? Affirmative arguments resolving this issue are provided in the next remark.

Remark 3.4.3. Explanation of the magnetic flux quantization in [13], even though correct, is much less illuminating as that given by Feynman in his famous "Feynman's Lectures in Physics", [99], e.g. see comments to his Fig. 21.4. In particular, the current quantization in the ring-shaped superconductor begins with the following experimental setup. The ring is placed in the magnetic field above the temperature of the superconducting transition. The magnetic field is then capable of penetrating the body of the ring. Next, the temperature is gradually lowered. Below the transition temperature the *external* magnetic field is removed while the magnetic flux through the hole gets trapped because the induced superconducting current flowing at the thickness of about δ from the surface of the ring creates its own magnetic field through the hole. Since the current is quantized, the flux is quantized as well and vice versa. This experimental setup was suggested by Fritz London, [15], in 1950. It was experimentally confirmed in 1961.[d] In the same section 21-8 on superconductivity, Feynman notices that the description of superconductivity is the same as the description of hydrodynamics of charged fluids (that is of magnetohydrodynamics, of dynamics of colloidal suspensions, etc.). To our knowledge, Feynman had made no attempts to develop this idea.[e] He expressed it only in connection with his brief description of the dynamics of superconductivity from which it follows that the force-free condition holds even in non-equilibrium situations. This observation allows us to conclude that the condition $\Omega_n = 0$ is also capable of describing the superconducting state in which dynamics of vortices is described by the L-L Eq. (2.3.12b) (with $\Omega(\omega) = 1$). Based on information already provided, this can also be seen from reading of pp. 865 and 866 of [33].

Exercise 3.4.2. Results of [94] are based on the assumption of the validity of Eq. (3.3.10). This assumption is too restrictive as we just explained. Try to develop an extension of the results of [94]. Hint: Use Feynman-style arguments, e.g. read pp. 303–306 of [100].

3.5 New Interpretation of the Dirac Monopole and its Use in the Problem of Quark Confinement

Let us now talk about the monopoles. The Gauss law for magnetism

$$\nabla \cdot \mathbf{H} = 0 \tag{3.5.1}$$

indicates that there are no free magnetic charges (monopoles). Following Dirac, by analogy with electrostatics, we assume, nevertheless, that

$$\nabla \cdot \mathbf{H} = 4\pi g \delta^3(\mathbf{r}), \tag{3.5.2}$$

where g is the magnetic charge. Then, by analogy with electrostatics, where the Gauss law reads as

$$\nabla \cdot \mathbf{E} = 4\pi e \delta^3(\mathbf{r}), \tag{3.5.3}$$

[d] E.g. see references in [99].
[e] See, however, the exercises in Chap. 4.

we obtain for the field strength **H** of a magnetic monopole the anticipated result:

$$\mathbf{H} = g\frac{\mathbf{r}}{r^3} = -\nabla\frac{g}{r} \tag{3.5.4}$$

implying that the familiar relation $\mathbf{H} = \operatorname{curl}\mathbf{A}$ is no longer valid. This is so because

$$\oint d\mathbf{S}\cdot\operatorname{curl}\mathbf{A} = \oint dvol\nabla\cdot\operatorname{curl}\mathbf{A} = 0. \tag{3.5.5}$$

This is surely not the case now since Eq. (3.5.2) implies

$$\oint d\mathbf{S}\cdot\mathbf{H} = \oint dvol\nabla\cdot\mathbf{H} = 4\pi g. \tag{3.5.6}$$

The question arises: What kind of the vector potential **A** will reproduce Eq. (3.5.6)? This is a topological-type of question [90]. To avoid singularity (at $\mathbf{r} = 0$) of the Dirac monopole, e.g. see Eq. (3.5.4), Wu and Yang [101] constructed a covering space, a fiber bundle, on which **A** is nonsingular.

Instead of following the traditional path of constructing such a fiber bundle (which incidentally is going to be constructed anyway later below), we would like to make a little detour in order to look at the above results from the alternative point of view. Let us first visualize the topology of the underlying problem by developing ideas of Remark 3.4.1. In view of this remark, we are dealing with the Hopf-type links, that is with two linked rings (e.g. see Fig. 1). We place one of the rings on some plane while the other perpendicular to this plane. In such a case we are dealing with the winding problem that requires us to introduce the 1-form Ω

$$\Omega = \frac{1}{2\pi}\frac{xdy - ydx}{x^2 + y^2}. \tag{3.5.7}$$

It is convenient to use it on the circle S^1 so that it can be temporarily identified (up to a constant) with the combination $\mathbf{A}\cdot d\mathbf{l}$ in Eq. (3.4.13). In polar coordinates this form simplifies to $\Omega = \frac{d\Phi}{2\pi}$. Clearly, $\oint d\mathbf{l}\cdot\nabla\Phi = \oint d\Phi$ so that the rest of the results of Eq. (3.4.13) follow. The form Ω is closed but is not exact if the contour surrounds a hole. In the book by Fulton [102] it is shown how using this 1-form it is possible to develop all concepts of homotopy, homology and cohomology generalizable to higher dimensions. In particular, the theory of Čech homology and cohomology uses exactly the same patching and with the same results as obtained in the paper by Wu and Yang where the language of fiber bundles was used [101].

Exercise 3.5.1. 1. Read [102], Chap. 15, and investigate if there is any difference between the fiber bundle construction of Wu and Yang and that required by the Čech cohomology methods.[f] 2. Think about why these methods are not used for electric charges.

Since these methods are not used for electric charges, there is an apparent asymmetry between the static electricity and magnetism with or without monopoles. This can be seen by studying the Aharonov–Bohm effect for charged particle moving in space where the constant magnetic field is confined to some space domain.

Consider the form Ω from the point of view of theory of foliations. For this purpose we need to introduce a smooth vector field $\boldsymbol{\xi} = \{\xi^\alpha(\mathbf{x})\}$, $\alpha = 1, \ldots, n$ assigned, say, on some domain U of the Euclidean space $\mathbf{R}^n$ (notice: addition of just one point at spatial infinity converts $\mathbf{R}^n$ to S^n) so that $\mathbf{x} = \{x^1, \ldots, x^n\}$. Using such defined vector field we introduce a unit vector $\mathbf{n}(\mathbf{x}) = \frac{\boldsymbol{\xi}(\mathbf{x})}{|\boldsymbol{\xi}(\mathbf{x})|}$. This vector is defining the system of coordinates with origin at $\mathbf{x}$. It is also defining the Gaussian map $f \colon U \to S^{n-1}$. If Q is any closed hypersurface, $Q \subset U$ then, the degree of mapping $\deg f\mid_Q$ can be defined. In particular, if Q is a sphere $S^{n-1} \subset \mathbf{R}^n$, in the domain $\mathbf{R}^n \backslash 0$ there is a closed differential form Ω of rank $n-1$ (a volume form) given by [102, 103]

$$\Omega = \frac{1}{\gamma_n} \frac{\sum_{i=1}^{n}(-1)^i x^i \wedge dx^1 \wedge \cdots \wedge d\hat{x}^i \wedge \cdots \wedge dx^n}{((x^1)^2 + \cdots + (x^n)^2)^{\frac{n}{2}}}, \tag{3.5.8}$$

where the normalizing factor γ_n is defined by the requirement

$$\int_{S^{n-1}} \Omega = 1. \tag{3.5.9}$$

Evidently, the one-form Ω, Eq. (3.5.8), is the Ω form defined by Eq. (3.5.7) when $n = 2$. Unfortunately, the form, Eq. (3.5.7), cannot be used for our case because it is only valid in the plane with a hole. For the monopole such a form is not good since on any plane above or below the hole it becomes exact. In the plane though it allows us to define the winding number $W(C, p)$ as follows:

$$W(C, p) = \int_C \Omega \tag{3.5.10}$$

---

[f] See also the textbook by Shabat [103] where the same ideas are presented in a somewhat different context but very nicely.

for any contour C enclosing point p in the punctured plane $\mathbf{R}^2 \backslash p$. In two dimensions we can add a point at infinity to make S^2 from $\mathbf{R}^2$. Then, the winding number $W(C, p)$ is just the degree of mapping f: $S^2 \rightarrow S^1$. In general, we can write

$$\int_U f^* \Omega = (\deg f \mid_Q) \int_Q \Omega \tag{3.5.11}$$

where f^* is the pull-back map. Since normally we take Q as S^{n-1}, Eq. (3.5.9) makes the integral on the R.H.S equal to one. This then provides us with the definition of the degree of mapping.

Such a scheme of calculating degree can be slightly modified. Specifically, for $\mathbf{R}^3 \backslash 0$ we can think about constructing the Hopf map f: $S^3 \rightarrow S^2$ with the degree of mapping being the same as in the Faddeev–Skyrme model discussed in Sec. 3.2. Hence, the degree of mapping must coincide with the expression for the helicity given in Sec. 3.1. The Hopf map $S^3 \rightarrow S^2$ was used by Ryder, [104] for description of the Dirac monopole.

Exercise 3.5.2. By combining results of Ryder's paper and those in the book by Arnol'd and Khesin, [68], demonstrate that expression for helicity can be obtained with help of the integral $\int_{S^2} \Omega = 1$. Hint: Look at Example 1.19 on p. 127 of [68].

Given all these facts, we are ready to demonstrate that Eq. (3.4.13) can be interpreted in terms of the Dirac monopole formalism. This becomes apparent in view of the discussion on p. 127 of the book by Arnol'd and Khesin about construction of the Hopf map and its role in hydrodynamics of incompressible ideal fluids. The construction of such a map begins with a sphere S^2 surrounding 0 in $\mathbf{R}^3 \backslash 0$ space. On S^2 the 2-form Ω (up to a normalization constant) is given by $\Omega = \sin\theta d\theta \wedge d\phi$. It is closed but not exact. Nevertheless $\int_{S^2} \Omega = 4\pi$. Next, $\mathbf{R}^3$ is compactified to S^3 by adding one point at spatial infinity. The pre-images of any two generic points on S^2 correspond to the Hopf link in S^3 (Fig. 1). That is to say: each point on S^2 corresponds to a ring in S^3 so that two generic points on S^2 correspond to one copy of linked rings in S^3. Such made Hopf link can be brought back to $\mathbf{R}^3$ via stereographic projection: from S^3 to $\mathbf{R}^3$. Thus, we just constructed a fiber bundle. In fact, S^3 is the Hopf fiber bundle. It is made of S^2 as the base space and S^1 (one of the rings belonging to the Hopf link) as the fiber. That is locally one can represent S^3 as the product $S^2 \times S^1$ (surely not globally!). In view of Eq. (3.5.6) we can formally write $\mathbf{H} = g\Omega$. But, since Ω is not an exact form we cannot write $\mathbf{H} \simeq dA$ on S^2. However, this becomes a possibility on S^3 [104] where the vector potential 1-form A is given by $A = -(d\psi + \cos\theta d\varphi)$. Here φ and θ are the polar angles on the base space S^2 while ψ is the angle on the S^1 fiber.

Since, according to [68], the helicity $\mathcal{H}(\xi)$ of the *null-homologous* vector field ξ on some 3-manifold M is given by

$$\mathcal{H}(\xi) = \int_M \alpha \wedge d\alpha \tag{3.5.12}$$

and, since the divergence-free vector field ξ is *null-homologous* if the 2-form ω_ξ associated with it is obtainable from the *globally defined* 1-form α, that is $\omega_\xi = d\alpha$, we obtain: $M = S^3, \alpha = -(d\psi + \cos\theta d\varphi)$, $H = gd\alpha = g\sin\theta d\theta \wedge d\varphi$. At the same time, this 2-form on S^3 is obtainable via pull back of the 2-form "living" on S^2, that is $H = g\ f^*\Omega = gd\alpha$, where $\Omega = \sin\theta d\theta \wedge d\varphi$. Hence, on S^3, as it was just demonstrated, the form $f^*\Omega$ is exact. Thus, we have either

$$\oint_{S^2} d\mathbf{S} \cdot \mathbf{H} = 4\pi g \tag{3.5.13a}$$

on S^2 or, by temporary suppressing constants,

$$\oint_{S^3} \alpha \wedge f^*\Omega = \oint_{S^3} \alpha \wedge d\alpha = \mathcal{H}(\xi) \tag{3.5.13b}$$

on S^3. This result is a generalization of Eq. (3.5.11) describing the degree of mapping. Since the degree of mapping is an integer, Arnol'd and Khesin, [68], pp. 141–152, proved that the helicity $\mathcal{H}(\xi)$ is also an integer. In Chap. 5 we shall provide alternative, more physically motivated, proof of this result. In fact, in view of the arguments just presented, it should be clear without any mathematical proofs that it can be written in terms of the linking (or self-linking[g]) numbers defined by Eq. (3.1.11) so that it is an integer.

Thus, the results just presented strongly suggest that physical properties of the Dirac monopole can be interpreted in terms of the helicity for the Hopf-type links made of superconducting currents (that is by the linked currents subject to the force-free condition). That this is the right condition can be seen from Feynman's discussion of superconductivity in [99]. From this reference we find that London's Eq. (3.3.5a) can be written as $\nabla \times \mathbf{v} = -\frac{e}{m}\mathbf{H}$ (e.g. see our Eq. (1.2.8)) so that the Lorentz-force $\mathbf{v} \times \mathbf{H}$ is indeed zero. To finalize our arguments interpreting interlocked rings as monopole

[g]Linking is for two vector fields ξ and η while self-linking is for just one vector field.

we follow the same chain of equalities as exhibited already in Eq. (3.4.13) for interlocked rings.

Thus, we begin with $\oint_{S^2} d\mathbf{S}\cdot\mathbf{H} = 4\pi g$ for the monopole. But then in the monopole calculations [104] the sphere S^2 is dissected into two discs $D_i, i = N, S$, so that instead of having $\iint_S(\operatorname{curl}\mathbf{A})\cdot d\mathbf{S}$ in the superconducting ring case we end up with

$$\oint_{\mathbf{S^2}} \mathbf{dS}\cdot\mathbf{H} = \iint_{D_N}(\operatorname{curl}\mathbf{A}^N)\cdot d\mathbf{S} - \iint_{D_S}(\operatorname{curl}\mathbf{A})\cdot d\mathbf{S}$$
$$= \oint_C \mathbf{A}^N\cdot d\mathbf{l} - \oint_C \mathbf{A}^S\cdot d\mathbf{l} \equiv \oint_C \mathbf{A}^{\text{tot}}\cdot d\mathbf{l} = 2g\oint_C d\mathbf{l}\cdot\nabla\Phi = 4\pi g,$$

where C is the contour along the equator of S^2. Comparison with Eq. (3.4.13) indicates a complete agreement. Furthermore, because of this agreement, it is possible now to obtain Schwinger's condition [104] for quantization of the magnetic charge. Indeed, by looking at Eq. (3.4.13) we obtain the following identification: $\frac{\hbar c}{2e}2\pi n = 4\pi g$.

To make next step we notice that in the superconducting case the charge of the Cooper pair is $2e$. Hence, it is appropriate to replace $2e$ by the renormalized charge e. In such a case we obtain: $\frac{2ge}{\hbar c} = n$. Schwinger's result $ge = \hbar c$ is obtained for $n = 2$ in accord with [104]. If we keep n as an arbitrary integer we obtain instead the Dirac quantization condition for the monopole

$$ge = \frac{n}{2}\hbar c$$

as promised. Now, we can do much more. For instance, consider the linking number $lk(1,2)$, defined by Eq. (3.1.11a). We have

$$lk(1,2) = \frac{1}{4\pi}\oint_{C_2} d\sigma'\mathbf{v}(\sigma')\cdot\oint_{C_1} d\sigma\left[\frac{(\mathbf{r}(\sigma)-\mathbf{r}(\sigma'))}{|\mathbf{r}(\sigma)-\mathbf{r}(\sigma')|^3}\times\mathbf{v}(\sigma)\right]$$
$$\equiv \oint_{C_2} d\sigma'\mathbf{v}(\sigma')\cdot\mathbf{A}[\mathbf{r}(\sigma')]$$

to be compared with Eq. (3.4.13). Such a comparison indicates that the combination $\mathbf{A}/h\phi_0$ in Eq. (3.4.13) should be identified with $\mathbf{A}[\mathbf{r}(\sigma')]$ since $lk(1,2) = n$. Incidentally, the vector potential $\mathbf{A}[\mathbf{r}(\sigma')]$ happens indeed to coincide with that produced by the Dirac monopole, [105], p. 15.

Remark 3.5.1. Since the result, Eq. (3.5.12), for the helicity $\mathcal{H}(\xi) = lk(1,2)$ was initially obtained for description of dynamics of incompressible ideal fluids as exhibited by Eqs. (3.1.11) and (3.1.12), all Euler (ideal-type) fluids, e.g. magnetohydrodynamics (in a certain limit), colloidal suspensions, liquid crystals, quantum chromodynamics (that is the F-S model), etc. can be described in terms of models involving statistical mechanics of monopole gases. We shall return to this topic in Chap. 6 in order to re-obtain these (monopole) results using different mathematical tools which are of independent interest.

With these accomplishments, the issue of major importance still remains to be resolved. For electric charges the electroneutrality is guaranteed by the requirement of global gauge invariance as is well known from any textbook on field theory. The fiber bundle formalism of Wu and Yang and that of Ryder in its original form does not contain the analog of electroneutrality for monopoles. In the paper by Marsh, [106], such an analog was formally found.

It is instructive to augment formal derivation by Marsh by explaining the "electroneutrality"[h] for monopoles using some topological arguments based on physics. We begin with the observation that there are actually two types of Hopf links: the left-hand and the right-hand Hopf links. These are depicted in Fig. 1.

Clearly, one of them can be associated with the monopole while the other with the anti-monopole. Once such an identification is made, it is necessary to think about modification of Wu–Yang result to account for differences between Hopf links and, by doing so, for the charge neutrality for monopoles.

The book by Fulton, [102], is perfect point of departure for this task. From it, we find that

$$\omega = \frac{dz}{z} = \frac{dx + idy}{x + idy} = \frac{xdx + ydy}{x^2 + y^2} + i\frac{xdy - ydx}{x^2 + y^2} = d(\ln r) + 2\pi i \Omega. \tag{3.5.14}$$

Accordingly, the integral $\int_C \omega$, where $C(t) = p + re^{it}, 0 \leq t \leq 2\pi$, equals to $2\pi i$. Now let U be $n+1$ connected region of $\mathbf{R}^2$ (that is $\mathbf{R}^2$ with n deleted holes) and ω be any closed form on U. Define a *period* $\mathfrak{p}_i(\omega)$ around p_i (a location of ith hole) as

$$\mathfrak{p}_i(\omega) = \int_{C_i} \omega \tag{3.5.15}$$

[h]We write "electroneutrality" since in the present case we are dealing with magnetic charges.

(with C_i being a circle around ith hole). For any closed 1-chain C and any closed 1-form ω on U we have

$$\int_C \omega = \sum_{i=1}^{n} W(C_i, p_i)\mathfrak{p}_i(\omega). \tag{3.5.16a}$$

Since we can always write $\omega = f(z)dz$, the above result can be rewritten in alternative, more familiar, form as

$$\frac{1}{2\pi i}\int_C f(z)dz = \sum_{i=1}^{n} W(C_i, p_i)\mathrm{Res}_{p_i}(f). \tag{3.5.16b}$$

Furthermore, if we compactify $\mathbf{R}^2$ by adding a point at ∞ so that it becomes S^2, we obtain as well

$$\sum_{i=1}^{n} W(C_i, p_i)\mathrm{Res}_{p_i}(f) + W(C_\infty, p_\infty)\mathrm{Res}_{p_\infty}(f) = 0. \tag{3.5.16c}$$

This is well-known result from the complex analysis. Fortunately, these results are transferable to higher dimensions. This transfer requires us to identify $\deg f|_Q$ (which is $W(C_i, p_i)$) with the index of the vector field [102] having isolated singularities, say, in $\mathbf{R}^n$. In such a case the following theorem can be used [107].

Theorem 3.5.1. *Let $\mathcal{F}$: $P^{n+1} \to M^n$ be a smooth map between oriented smooth manifolds with M^n being compact and connected. If $X \subseteq P$ be a compact domain with smooth boundary $N^n = \partial N$, and suppose N is made of a disjoint union of submanifolds $N_1^n, \ldots, N_k^n$. If $f_i = \mathcal{F}\mid_{N_i}$, then*

$$\sum_{i=1}^{k} \deg(f_i) = 0. \tag{3.5.17}$$

Proof. Let $f = \mathcal{F}\mid_N$ so that

$$\deg(f) = \sum_{i=1}^{k} \deg(f_i). \tag{3.5.18}$$

Let also $\omega \in \Omega^n(M)$ be an n-form such that $\int_M \omega = 1$, then

$$\deg(f) = \int_N f^*(\omega) = \int_X d\mathcal{F}^*(\omega) = \int_X \mathcal{F}^*(d\omega) = 0, \tag{3.5.19}$$

where the first equation is just restatement of Eq. (3.5.11), second is just the Stokes' theorem, while the third comes from the fact that $d\omega = 0$ on X (since the dimensionality of X is one degree higher). □

Remark 3.5.2. Evidently, in our case $n = 2$, $P^{n+1} = S^3$, $M^2 = S^2, N_i^2 = S_i^2$. Thus, since we are allowed to identify the degree with the index, Eq. (3.5.19) is just the statement of electroneutrality for monopoles.

Remark 3.5.3. It is essential that the components of ∂N are located *inside* of S^3. Since the ith sphere S_i^2 is associated with the ith monopole, the monopole gas is naturally confined. *If monopoles can be associated with quarks, then the condition under which the electroneutrality holds is also assuring the quark confinement.*

Such an identification of monopoles with quarks indeed was made by Mandelstam in 1975 [108]. His treatment follows that by Nambu [109] whose work is similar to that by Lund and Regge [29]. In [109] only the Abelian superconducting model is considered in which the monopoles were introduced by hand following Dirac's prescription [110]. They are located at the ends of the vortex line. This line is interpreted as Nambu–Dirac string. Since in [108] the non-Abelian version of the Nambu–Dirac model was considered, the Nambu–Dirac-type strings were transformed from the Abelian case to the non-Abelian case. Since such strings are artefacts of the original Dirac treatment of monopoles, as it was explained by Wu and Yang, it is essential to provide an alternative explanation of these string-like objects.

Emergence of string-like objects having monopoles at their ends can be given natural mathematical explanation. It is given in the book by Schwarz [111]. Since the idea of his explanation is exactly the same as known from the theory of functions of one complex variable, he uses essentially the same topology as in Theorem 3.5.1 above. That is we have to consider again $n = 2$, $P^{n+1} = S^3$, take $M^2 = S^2 \equiv S$ and $N_i^2 = S_i^2$, where now $i = 1, 2$. This is depicted in Fig. 2(a). As before, now let $\mathcal{F}$: $P^{n+1} \to M^n, n = 2$. Take into account that M^2 is simply connected, that $f_i = \mathcal{F}\mid_{N_i}$ and $f_0 = \mathcal{F}\mid_M$.

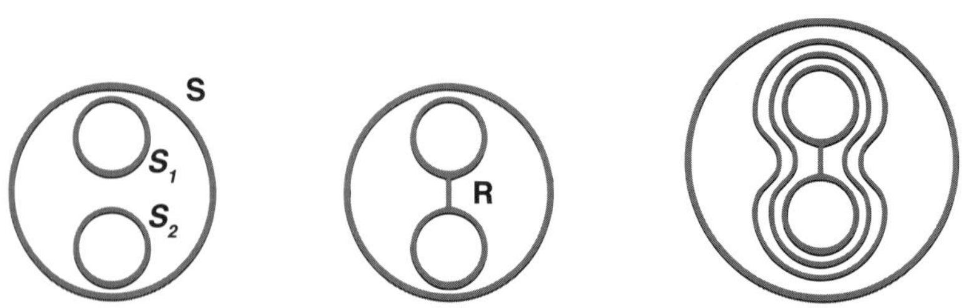

(a) (b) (c)

Fig. 2. Homotopy operations associated with electroneutrality. Refer to text for details.

Next, choose some point x_1 on $S_1^2 \equiv S_1$ and x_2 on $S_2^2 \equiv S_2$. Connect x_1 with x_2 by the "string" R as depicted in Fig. 2(b). Denote the space made of S_1 and S_2 and the string R by W. Evidently, the space W is simply connected. This gives us an opportunity to homotopically deform the outer sphere M^2 into W as depicted in Fig. 2(c).

This means that the map $M^2 \to W$ is an identity and its degree is zero (as before). At the same time, if the sphere M^2 is represented analytically as $x_0^2 + x_1^2 + x_2^2 = 1$, we can subdivide it, almost as in the case of a single monopole [104], into three parts: M_1, M_2 and M_3, as follows. Let the part M_1 be associated with the domain $x_0 > 1/2$ of M^2 and, respectively, M_3 with the domain $x_0 < -1/2$. Thus, M_2 is associated with the domain $-1/2 \leq x_0 \leq 1/2$. After such subdivision, we can map M_1 into S_1 with point x_1 deleted and, respectively, M_3 into S_2 with point x_2 deleted. After that, M_2 can be mapped into a string R joining S_1 with S_2. Since the degree of mapping we can identify with the index, we recover back the result given by Eq. (3.5.11) adopted for the case $k = 2$. Because of this, the arguments just presented can be extended to $k > 2$ which brings us back to configurations discussed in Mandelstam's work, [108].

Remark 3.5.4. Figure 2 makes us think about hadrons (nucleons or baryons) "living" in a bag in which monopoles (made of quarks) are connected by strings to each other. Such bag model was indeed proposed in 1974 in [112, 113] and is known in literature as the MIT bag model.

Exercise 3.5.3. Theorem 3.5.1 and its implications are helpful only for treatment of the Abelian monopoles. Already in their seminal paper, [101], Wu and Yang extended their fiber bundle construction: from the Abelian to non-Abelian case. Since Polyakov and 't Hooft obtained the non-Abelian monopoles using entirely different arguments [114], questions arise: (a) How the non-Abelian monopole of Wu and Yang is related to that of 't Hooft and Polyakov? (b) In addition to monopoles by 't Hooft and Polyakov there are monopoles by Bogomolny, Prasad and Sommerfield (BPS monopoles). How then is Wu and Yang non-Abelian monopole related to the BPS monopole? (c) How construction of a single monopole of each type can be extended to multimonopole configurations? How the fiber bundle construction of Wu and Yang can be extended to the multimonopole configurations? Hint: To answer the last question, Chan and Tsou wrote the book [115].

Remark 3.5.5. The original Dirac theory of Abelian monopoles was built under the assumption that monopoles can carry only magnetic charge.

That is they cannot carry both the electric and the magnetic charges [116]. This then allows us to develop theory in a completely symmetric fashion with respect to interchanges between the electric and magnetic charges. Such a symmetry is somewhat formal however since the estimated Dirac monopole mass is much larger than that for, say, proton [116]. The question then emerges: Can such formal symmetry (duality) be extended to the non-Abelian case? It turns out that the non-Abelian situation is much more complicated as explained recently in the paper by Kapustin and Witten [117] whose work, in turn, was inspired by work of Goddard, Nuyts and Olive, summarized in [116], and that by Montonen and Olive [118]. Since the paper by Kapustin and Witten requires uses of supersymmetry[i] while the results obtained by Montonen and Olive still remain as a plausible conjecture, we shall make no attempts to discuss the results of [117] any further in this book. Nevertheless, the non-Abelian extension of the Dirac monopole model brings into play yet another feature-dyons. These hypothetical particle-like formations suggested by Schwinger can carry both the electric and magnetic charges simultaneously [116]. This feature makes dyons very different from other monopole types.

Indeed, it is known that already at small distances away from the well-localized 't Hooft–Polyakov monoploles these objects act as if they are much more familiar Dirac monopoles [114]. That is at such distances the non-Abelian Yang–Mills field is acting as if it is Abelian [114, 116]. Thus, if we would begin with the non-Abelian version of the G-L model, the sort of Abelian reduction just described would produce us an assembly of the Dirac-like monopoles. Such a conclusion makes perfect sense in the light of results of this section. We shall discuss further these and other aspects of monopoles in Sec. 5.6.

Exercise 3.5.4. Investigate if dyons could be abelianized. Hint: Use results of this subsection. For recent official status of this problem see [105, 119].

[i]The possibility of supersymmetry in Nature very recently (as of the end of 2012) was put into serious doubt by experimental data coming from the LHC (Large Hadron Collider).

Chapter 4

Topologically Massive Gauge Theories and the Force-Free Fields

In Sec. 3.4 we discussed Proca's massive electrodynamics. We found that the Lagrangian for the massive Proca fields is not gauge-invariant. This defect of Proca's model can be repaired either by Stueckelberg's trick or by use of the Higgs mechanism of mass generation which is essentially the same thing as our treatment of the G-L functional (discussed in Sec. 3.4) leading to the Meissner's effect. Such an effect is associated with the mass generated by spontaneous (rotational) symmetry breaking. Stueckelberg's trick is working well at the classical level but is having problems at the quantum level. At the quantum level models using this trick are non-renormalizable as it was mentioned in Sec. 3.4. Use of the Higgs mechanism for the mass generation is also problematic. It is problematic because the Higgs field can be made out of the pure Yang–Mills gauge fields as explained in Sec. 5.6.2 below. And, if this is the case, it cannot be considered as fundamental. Fortunately, there is yet another way of mass generation. In fact, we have already considered it at the classical level! Gauge-invariant Woltjer's functional, Eq. (3.1.1), is an excellent example. It can be looked upon as functional for the static Maxwellian field in which the force-free condition is taken into account via method of Lagrangian multipliers. Its minimization produces the force-free Eq. (3.1.3) which can be interpreted as the Maxwellian analog of Dirac's equation for relativistic electron. This is explained in Sec. 1.3, e.g. see Eqs. (1.3.1) and (1.3.2) and comments related to these equations.

Totally independent from the force-free problematic the Minkowski-space analog of the Woltjer's functional was discussed in physics literature along with its non-Abelian version [120, 121].[a] The initial motivation in physics comes from the belief that such three-dimensional topological gauge field theories can be viewed as high temperature limits of the corresponding four-dimensional gauge theories. Not only these three-dimensional field theories are much easier to analyze but, in addition, they provide new feature: *topological mass generation*, absent in four dimensions. Significance of the topological mass generation was immediately recognized in [120–122] where it was used not only in the non-Abelian Yang–Mills field theories but also in the gauge reformulated theory of gravity. Here we shall discuss only the Abelian version of the model discussed in [120, 121]. It is also known as the Maxwell–Chern–Simons (MCS) model. In literature sometimes the G-L model with an extra Chern–Simons term is also called MCS model [123]. Since the London limit of such extended MCS models was studied in [124] and the results are not much different from those obtained in [120, 121], we shall comment only on the original formulations [120, 121].

Exercise 4.1. Read [123, 124] and decide for yourself if such a restriction is sufficient. Hint: Read also discussions below, in this chapter.

Thus, we adopt Woltjer's Lagrangian, e.g. see Sec. 3.1, to (2 + 1)-dimensional Minkowski space of signature $(+,-,-)$. By choosing such a signature we follow [123, 124]. In such a case the Lagrangian of Deser *et al.* [121, 125] coincides (up to notation of symbols) with that by Woltjer. Explicitly,

$$\mathcal{L} = -\frac{1}{4}F_{\mu\nu}F^{\mu\nu} + \frac{1}{4}m\varepsilon_{\mu\nu\rho}F^{\mu\nu}A^{\rho}, \tag{4.1}$$

where $F_{\mu\nu} = \partial_\mu A_\nu - \partial_\nu A_\mu$ as usual. Variation of this functional leads to the following equations of "motion"

$$\partial_\mu F^{\mu\nu} + \frac{m}{2}\varepsilon^{\nu\alpha\beta}F_{\alpha\beta} = 0. \tag{4.2}$$

These are manifestly gauge invariant with respect to gauge transformations of the type: $A_\mu \to A_\mu + \partial_\mu f$. At the same time, the Lagrangian, Eq. (4.1), changes by the total derivative under the same gauge transformation. Such a derivative is normally discarded if the appropriate boundary conditions on the fields are imposed. This procedure may be problematic in the multiply

[a] For the latest efforts in this direction, please consult [122].

connected spaces. Since the same is true in Woltjer's case for which results of [89, 90] apply,[b] there should be a way to deal with such type of problems in the Minkowski space, e.g. see [127].

Relabel the combination $\frac{1}{2}\,\varepsilon^{\nu\alpha\beta}F_{\alpha\beta}$ as $*F^{\mu}$, where $*F^{\mu}$ is the dual of $F_{\alpha\beta}$, accordingly, $F^{\mu\nu} = \varepsilon^{\mu\nu\alpha} * F_{\alpha}$. Then, following [128], we introduce the three-dimensional analog of the four-dimensional self-duality equation. It is given formally by the equation

$$A_{\mu} = \frac{1}{2m}\varepsilon_{\mu\nu\rho}F^{\nu\rho} \tag{4.3}$$

which is instantly recognizable as the force-free condition in space–time of Minkowski signature. Notice that in four dimensions in the Riemannian space (that is in space with Euclidean signature) the self-duality condition is given by $F_{\alpha\beta} = \pm * F_{\alpha\beta}$ [114]. Using the force-free Eq. (4.3) in Eq. (3.4.6) we re-obtain Eq. (4.2). Thus, Eq. (4.2) is equivalent to Proca's Eq. (3.4.6)! The same Eq. (4.3) is compatible with the Lorentzian gauge condition $\partial_{\nu}A^{\nu} = 0$ — an equivalent of the standard incompressibility condition in three dimensions — as required. Notice that, unlike the even-dimensional self-duality condition which is manifestly gauge-invariant, its odd-dimensional analog, Eq. (4.3), is not! Only when the field A is Trkalian, e.g. see Eq. (3.2.10), Eq. (4.3) becomes gauge-invariant. We shall from now on assume that this is the case. This then allows us to introduce the mass topologically without destroying gauge invariance and without need for the Higgs boson.[c]

Hence, following the same logic as was used by Woltjer (but not being aware of Woltjer's work!), the authors of [128] posed and solved the variational problem of finding the action functional whose variation produces the "self-duality" Eq. (4.3). The Lagrangian for such functional is given by

$$\mathcal{L} = \frac{1}{2}A_{\mu}A^{\mu} - \frac{1}{2m}\varepsilon^{\mu\nu\rho}A_{\mu}\partial_{\nu}A_{\rho}. \tag{4.4a}$$

It differs from that by Deser *et al.*, e.g. see Eq. (4.1)[d] and, according to the authors of [128], is not gauge-invariant. This non-invariance is corrected by using the Trkalian-type fields as was explained already. That this difference

[b] See also [126].

[c] This fact is not in contradiction with the latest results coming from the LHC announced on 4 July 2012. We explain this fact in Chap. 9, Sec. 9.7.4.

[d] Here we use signature of the paper [124] making the corresponding Eq. (4) of [128] to look a bit different. But, surely, with account of signature differences Lagrangians in both cases are the same.

from Eq. (4.1) is only apparent was also demonstrated by Deser and Jackiw in [125]. Evidently, such a difference will resurface in the Euclidean versions of these functionals as well. Therefore, it is useful to re-obtain Deser's and Jackiw's results using the Euclidean version of the Deser–Jackiw functional. It is essentially the same as discussed by Woltjer, e.g. Eq. (3.1.1). Using Eq. (3.1.3) we obtain: $\operatorname{curl}\mathbf{A} = \alpha\mathbf{A}$. Hence, Woltjer's (and, therefore, Deser–Jackiw) Lagrangian with account of this Trkalian condition acquires the form:

$$\mathcal{L} = \mathbf{A}^2 + \frac{1}{\alpha}(\mathbf{A}\cdot\operatorname{curl}\mathbf{A}) \tag{4.4b}$$

coinciding with the Euclidean version of Eq. (4.4a) obtained in [128]. Arguments of Deser and Jackiw are essentially the same. They also use equations of motion for rewriting the Lagrangian in alternative fashion.

In both [121] and [128] the task was to obtain the equation of motion for massive particle whose mass is created topologically. Thus, in both cases the problem was reduced to study of solutions of the force-free Eq. (4.3). Since we discussed it in the Euclidean context in Chap. 3, it is easy to generalize the obtained results to the Minkowski space–time. By keeping in mind our convention for the signature, the Minkowski analog of Eq. (1.3.2) can be written now as

$$(\Box + m^2)\mathbf{A} = 0, \tag{4.5a}$$

where d'Alambertian $\Box$ is defined by

$$\Box = \frac{\partial^2}{\partial t^2} - \nabla^2 \tag{4.5b}$$

with ∇^2 being the two-dimensional Laplacian. To convert this result to Euclidean space requires us to make a switch from zero to finite temperature field theory. In the present case, it is sufficient just to make a replacement $t \to it$ in the d'Alambertian so that Eq. (4.5a) is converted into Eq. (1.3.3) describing the Meissner effect in the Abelian G-L model. Correctness of such a transition into Euclidean space is supported by the following quotation from [121], e.g. see p. 433 of this reference:

"Since a [Euclidean] three-dimensional theory describes a four-dimensional one at high temperature, perhaps the topological mass is a phenomenological description of magnetic screening due to topological excitations in four dimensions. Note also that massive particles in two and three dimensions emerge for topological reasons. One may therefore speculate that a topological mass generating mechanism — as yet undiscovered — can also operate in four dimensions. This would provide a welcome alternative to the conventional Higgs procedure."

Exercise 4.2. In Sec. 3.3, we noticed that the relation between the force-free condition, Eq. (1.3.1), and the Helmholtz Eq. (1.3.2) is very much the same as between the Dirac and the Klein–Gordon equations. It happens, that one can go further along these lines. Thus, the task lies in describing both the force-free (and, thus, the Hemholtz) and the Dirac (and, thus, the Klein–Gordon) equations as special cases of the same master equation. Use supersymmetric formalism to achieve this goal. Hint: Look at [129] as an example.

Exercise 4.3. At the end of Sec. 3.4, in Remark 3.4.3, we mentioned that the description of superconductivity can be made the same as the description of charged fluids (Feynman). In [9] we demonstrated that this is indeed the case. It turns out that it is possible to move things much further using this analogy by applying it to the description of the quantum Hall effect. Think about this possibility based on what was discussed already in this chapter. Hint: look at [130].

Remark 4.1. Lately quantum Hall effect had been used in some designs of quantum computers (employing principles of the topological qauantum computation). More about this is discussed in Chap. 8 and Appendices C and D.

Chapter 5

Contact Geometry and Physics

5.1 General Information

The logical path described thus far is nontraditional. Indeed, we had begun with classical mechanics. Then we reformulated it in the form used for description of ideal incompressible fluids. From incompressible fluids we went on to magnetohydrodynamics and to standard superconductivity, then from superconductivity to Dirac monopoles, and finally, we ended up in the domain of science associated with the hadron physics in which the non-Abelian Yang–Mills (NAYM) fields play the key role. Such a path from what is traditionally considered as classical to what is known as quantum is striking. Thus, our journey into classical-quantum correspondence had already began. In this and the following chapters by exploiting the results already discussed, we shall be in the position to blur the boundaries of classical-quantum correspondence further. This is done with several purposes in mind. First, the formalism we are about to describe is perfectly suited for explanation of results by Lubensky [78] on the TGB phases in liquid crystals. Second, in [83] it is mentioned that the Faddeev–Skyrme hopfions can be modelled using theory of elastic rods. From this identification with rods the following problem emerges: If physically interesting results coming from the NAYM field theory are obtainable by the method of Abelian reduction, can such a reduction keep some memory about its non-Abelian origin? Stated differently, from results of our review paper, [64], it follows that the Abelian version of Chern–Simons (ACS) topological field theory is seemingly capable of reproducing only the Hopf-type simple links. Nevertheless, recently it was demonstrated [131] that the ACS is capable of describing the nontrivial knots and links in the sense that with its help it is possible to reconstruct the Kauffman and the Jones knot/link polynomial invariants used for recognition of nontrivial knots and links. Beginning from this chapter, we shall discuss another route of reproducing nontrivial knots and links, perhaps, sometimes with self-intersections. To obtain these results requires some basic knowledge of contact and symplectic geometry and topology. Our journey into contact geometry begins in the next section.

5.2 Some Basic Facts about Contact Geometry and Topology

Previously, we had briefly mentioned about some elements of the theory of foliations. Many results of this theory can be traced back to the Frobenius theorem which we would like to reproduce here. For this purpose, let us consider a 1-form α defined, say on $\mathbf{R}^n$, which does not vanish at 0. Suppose that we are given two functions f and g. Then, we can ask: Under what conditions $\alpha = fdg$? That is we are looking for an integrating factor for the Pfaffian equation $\alpha = 0$. If, indeed, $\alpha = fdg$, then $d\alpha = df \wedge dg = df \wedge f^{-1}\alpha = f^{-1}df \wedge \alpha \equiv \theta \wedge \alpha$. Using this result we obtain:

$$\alpha \wedge d\alpha = 0. \tag{5.2.1}$$

This result is known in literature as the *Frobenius theorem*. The meaning of this theorem can be discussed from the point of view of theory of dynamical systems. It can also be discussed in the context of thermodynamics. Let us discuss this in some detail.

Specifically, consider a 1-form "living" in $\mathbf{R}^3$

$$\alpha = A(x,y,z)dx + B(x,y,z)dy + C(x,y,z)dz. \tag{5.2.2}$$

Using it we obtain

$$\begin{aligned}\alpha \wedge d\alpha = \bigg[A\left(\frac{\partial C}{\partial z} - \frac{\partial B}{\partial y}\right) + B\left(\frac{\partial A}{\partial z} - \frac{\partial C}{\partial x}\right) \\ + C\left(\frac{\partial B}{\partial x} - \frac{\partial A}{\partial y}\right)\bigg] dx \wedge dy \wedge dz.\end{aligned} \tag{5.2.3}$$

If the Frobenius theorem holds true, the above 3-form should be equal to zero. This can be restated as the already familiar lamellarity condition $\mathbf{H} \cdot \operatorname{curl} \mathbf{H} = 0$, e.g. recall Eq. (3.2.3), for some vector $\mathbf{H} = \{A, B, C\}$. Such a condition was discussed in Sec. 3.3 in the context of superconductivity, magnetohydrodynamics and liquid crystals.

Consider now the 1-form $\alpha = dz + xdy$ also defined, say, in $\mathbf{R}^3$. Since $d\alpha = dx \wedge dy$, we obtain

$$\alpha \wedge d\alpha = dx \wedge dy \wedge dz \neq 0 \tag{5.2.4}$$

since the volume form $dx \wedge dy \wedge dz$ is nonzero. Thus, if the condition Eq. (5.2.1) is the condition of complete integrability (assuring existence of foliations), the condition Eq. (5.2.4) is the condition of complete non-integrability. It guarantees the existence of contact structures on a 3-manifold e.g. see Fig. 4 and compare it with Fig. 9 of Lubensky's paper, [78].

Remark 5.2.1. Notice that when we were talking about the Dirac monopole we used the helicity $\mathcal{H}[\alpha] = \int_M \alpha \wedge d\alpha$, Eq. (3.5.12). Evidently, the combination $\alpha \wedge d\alpha$ under the integral is expected never to be zero. This explains relevance of the (contact) result, Eq. (5.2.4), to previous discussions.

Following [132, 133] we would now like to explain why, indeed, the result given by Eq. (5.2.4) can be called "*contact*". In doing so, we shall assume that our readers are familiar with the very basic facts about linear and nonlinear partial differential equations of the first order, e.g. like those used in the context of the Hamilton–Jacobi equations.

Thus, we begin our discussion by considering the simplest representative and familiar to every physicist example: the one-dimensional harmonic oscillator. In the appropriate system of units it is described by the Hamiltonian $H(p, q) = \frac{1}{2}(p^2 + q^2)$. Using it, the Hamilton–Jacobi Eq. (1.2.19) is given now by

$$\frac{\partial S}{\partial t} + \frac{1}{2}\left(\left(\frac{\partial S}{\partial q}\right)^2 + q^2\right) = 0. \tag{5.2.5}$$

Because the Hamiltonian is time-independent, the above equation can be simplified to

$$\left(\frac{\partial S}{\partial q}\right)^2 + q^2 = C, \tag{5.2.6}$$

where C is some constant. In this equation $\frac{\partial S}{\partial q} = p$ as required. For future use, it is convenient to rewrite this result as the 1-form α, that is as

$$\alpha = dS - pdq. \tag{5.2.7}$$

Evidently, this 1-form α must vanish on real (physically realizable) trajectories. For such trajectories α is known in mathematics literature as the *Cartan distribution.* It is of the same type as the already introduced $\alpha = dz + xdy$. For the sake of argument, by employing the appropriate rescaling in Eq. (5.2.6), the constant C can be made equal to one. Then, instead of Eq. (5.2.6), we obtain

$$p = S' = \frac{\partial S}{\partial q} = \pm\sqrt{1-q^2}. \tag{5.2.8}$$

Consider this result from general standpoint of the implicit function theorem. In particular, consider the Euclidean space $\mathbf{R}^3$ made out of triples q, S and p and let $F(q, S, p) = 0$ describe some surface $\mathcal{E}$ in this space. To describe $\mathcal{E}$ in some detail we notice that Eq. (5.2.6) can also be written as $F(q, S, S') = 0$, where

$$S = S(q) \quad \text{and} \quad p = S'(q) \equiv \frac{\partial S}{\partial q}. \tag{5.2.9}$$

From here, it follows that the coordinate q can serve as ("time") parameter in the equation $F(q, S, S') = 0$ as long as the relationship $S = S(q)$ is invertible. This may not always be possible. For example, in Eq. (5.2.8), the derivative $\frac{\partial^2 S}{\partial q^2}$ becomes infinite for $q = 1$. By excluding such points we observe that functions S and p are *not* arbitrary: the latter is the derivative of the former.

Remark 5.2.2. In mathematics, this observation leads to introduction of spaces of 1-jets defined as follows. Let M be an n-dimensional manifold. Then, the set of 1-jets $J^1(M, \mathbf{R})$ is determined by the point $\mathbf{q} = \{q_1, \ldots, q_n\} \in M$, by the value of $S(\mathbf{q}) \in \mathbf{R}$ and, by the set $\mathbf{p} = \{p_1, \ldots, p_n\}$ for which $p_{k\cdot} = \partial S/\partial q_k, k = 1-n$. Thus defined $(2n+1)$-dimensional manifold of 1-jets is known as *contact manifold.* It should not be confused with more familiar $2n$-dimensional *symplectic manifold.* Incidentally, n-dimensional manifolds defined by $p_{k\cdot} = \partial S(\mathbf{q})/\partial q_k, k = 1 - n$, are known as *Lagrangian* manifolds [134].

Going back to our example of harmonic oscillator we have to distinguish between the equation $F(q, S, S') = 0$ and the surface $\mathcal{E}$: $F(q, S, p) = 0$, placed in $\mathbf{R}^3$. Every solution of $F(q, S, S') = 0$ can be visualized as some curve on $\mathcal{E}$. In our case, the surface $\mathcal{E}$ is defined by the equation $p^2(q) + q^2 = 1$. This is an equation for a cylinder whose (cylindrical) coordinates x, y and z are determined respectively by q, p and S. The "solution curve" on such

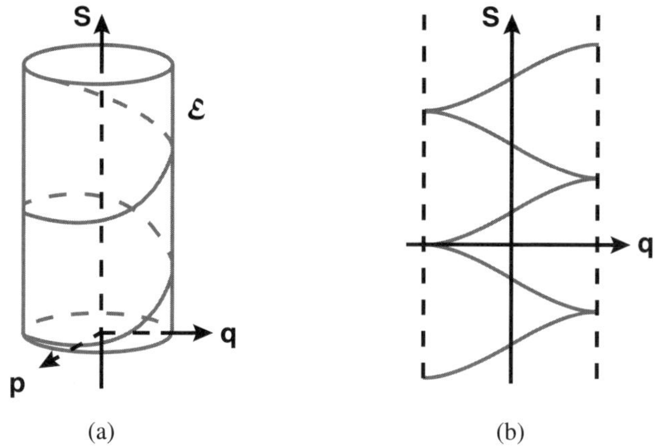

Fig. 3. (a) Representative integral curve on $\mathcal{E}$-surface. (b) Projection of this curve into (q, S)-plane.

a cylinder is determined in parametric form as follows:

$$z = S, y = p(S), x = q(S). \tag{5.2.10}$$

At this point it is possible to define the *integral curves* of the equation $F(q, S, S') = 0$ as subset of curves on $\mathcal{E}$ parametrized as in Eq. (5.2.10) and such that their projection to q-axis is diffeomorphism. Figure 3(a) depicts surface $\mathcal{E}$ along with the representative integral curve while Fig. 3(b) depicts the projection of such integral curve into (q, S)-plane. Such projections will be discussed in Chap. 9 in connection with Legendrian knots/links.

Singular points at which the diffeomorphism is lost are cusps either in (q, S) (or in (p, S)) planes. They can be determined as follows. Suppose $w = \{q_0, S_0\}$ is one of such points in (q, S)-plane. Then, for this point $F(q_0, S_0, p_0) = 0$. Therefore, in the vicinity of such a point

$$F_q dq + F_S dS + F_p dp = (F_q + p_0 F_S + p_0' F_p) dq = 0, \tag{5.2.11}$$

where $F_q = \partial F / \partial q$, etc. But since $dp = [\frac{\partial^2 S}{\partial q^2}|_{q=q_0}] dq \equiv p_0' dq$, for singular points, this equation should be decomposed as

$$\frac{\partial F}{\partial q} + p \frac{\partial F}{\partial S} = 0 \quad \text{and} \quad \frac{\partial F}{\partial p} = 0. \tag{5.2.12}$$

These general results we can now be applied to the harmonic oscillator where the equation $F(q, S, S') = 0$ is given by $(\frac{\partial S}{\partial q})^2 + q^2 - C = 0$ and the surface $\mathcal{E}$ is given by $p^2 + q^2 - C = 0$. This equation for $\mathcal{E}$ is satisfied identically if we write $p = \sqrt{C}\sin(\varphi + c)$ and $q = -\sqrt{C}\cos(\varphi + c)$ with c being yet another constant and φ being an angular parameter. We shall suppress c by redefining φ in what follows. In view of Eq. (5.2.9) physically acceptable solutions are those for which $\frac{\partial S}{\partial q} = \frac{\partial S}{\partial \varphi}\frac{\partial \varphi}{\partial q} = p$. From here, we obtain: $p = \frac{1}{\sqrt{C}\,\sin\varphi}\frac{\partial S}{\partial \varphi}$. Since we know already that $p = \sqrt{C}\,\sin\varphi$, we obtain as well: $\frac{\partial S}{\partial \varphi} = C\,\sin^2\varphi$. That is, $S = \frac{C}{2}(\varphi - \frac{1}{2}\sin 2\varphi)$. The condition $\alpha\mid_{\mathcal{E}} = 0$ is in accord with just obtained results as can be directly checked.

Next, we look for singular points using Eq. (5.2.12). Since the surface $\mathcal{E}$ does not contain S explicitly, Eqs. (5.2.12) acquire the following form:

$$\frac{\partial F}{\partial q} = 0, \quad \frac{\partial F}{\partial p} = 0. \tag{5.2.13}$$

Equation (5.2.8) along with the already introduced paramerization for p and q helps us to rewrite the above equations in the form

$$\frac{dq}{dp} = \frac{\sin\varphi^*}{\cos\varphi^*} = 0 \tag{5.2.14}$$

implying that singular values for φ^* are: $0, \pm\pi, \pm 2\pi, \ldots$. This is in accord with the results depicted in Fig. 3(b). In Chaps. 6–9 we shall extend these simple results to less trivial physically interesting situations.

Exercise 5.2.1. Derive the Hamiltonian equations for harmonic oscillator based on the results just described. Hint: Follow the logic of [135], pp. 81, 82, or [136], Appendix $4K$.

5.3 Contact Geometry of Thermodynamics

Obtained results can now be broadly generalized. First of all, we would like to illustrate them using another example familiar to physicists — thermodynamics. Let U, S, T, P and V be respectively the internal energy, the entropy, the temperature, the pressure and the volume. The 1-form α is defined now as

$$\alpha = dU - TdS + PdV. \tag{5.3.1}$$

It vanishes on the Legendrian surface (submanifold), an analog of $\mathcal{E}$, which in the present case represents the equation of state. Such a surface is obtained by using the analogs of $p_{k\cdot} = \partial S/\partial q_k$ relations in thermodynamics. In the present case these are given by the relations $T = (\frac{\partial U}{\partial S})_V$ and

$P = -(\frac{\partial U}{\partial V})_S$. These two equations allow, in principle, to eliminate S thus forming the equation of state $\mathcal{F}(P, V, T) = 0$.

Exercise 5.3.1. The above analogy between thermodynamics and mechanics can be extended further. In fact, using just obtained results it is possible to derive the analog of the Hamilton–Jacobi equation for thermodynamics. Hint: Read [136], paragraph 47.

Summarizing. Thus far we obtained the following chain of correspondences: classical mechanics$\rightleftarrows$thermodynamics$\rightleftarrows$electrodynamics$\rightleftarrows$ geometrical optics [136]$\rightleftarrows$hydrodynamics$\rightleftarrows$magnetohydrodynamics$\rightleftarrows$superconductivity$\rightleftarrows$non-Abelian Y-M fields (via Abelianization) — all derivable from contact geometry/topology. In the rest of this book we shall enlarge this list of correspondences by including in it quantum mechanics and general relativity. The correspondences just mentioned provide sufficient motivation for study of contact geometry/topology in detail. The purpose of this book is just to point our readers into this direction leaving details for more specialized books and research papers. With this in mind, we proceed further.

5.4 Contact and Symplectic Geometry and Liquid Crystals

Thus, let us consider once again the 1-form defined in Eq. (5.2.7) and, by analogy with symplectic geometry, let us consider the transformation: $(q, S, p) \to (q_1, S_1, p_1)$ of $\mathbf{R}^3$ such that

$$dS_1 - p_1 dq_1 = \rho(q, S, p)(dS - pdq), \tag{5.4.1}$$

provided that the function ρ: $\mathbf{R}^3 \to \mathbf{R}$ is nowhere zero. If such transformation exists, it is called *contact transformation and the respective 1-forms are called contactomorphic.* It carries the *integral curves* of the equation $F(q, S, S') = 0$ to the integral curves of the transformed equation $F(q_1, S_1, S_1') = 0$. In view of this result, it is becoming obvious that the 1-form α is defined up to a common multiplier ρ and the result given by Eq. (5.2.4) is not going to change in the sense that now

$$\rho\alpha \wedge d(\rho\alpha) = \rho^2 \alpha \wedge d\alpha \neq 0. \tag{5.4.2}$$

To demonstrate that this is the case, we need only to calculate $d(\rho\alpha)$, that is

$$d(\rho\alpha) = d\rho \wedge \alpha + \rho d\alpha.$$

But the product $d\rho \wedge \alpha$ is zero since ρ is function "living" on $\mathbf{R}^3$. Generalization of these results to $(2n + 1)$-dimensional spaces, $n > 1$, with coordinates $\mathbf{q} = \{q_1, \ldots, q_n\} \in M$, $\mathbf{p} = \{p_1, \ldots, p_n\}$ and S is obvious now. In particular, the multidimensional analog of 1-form defined in Eq. (5.2.7) is given by

$$\alpha = dS - \mathbf{p} \cdot d\mathbf{q}. \tag{5.4.3}$$

From here we obtain more familiar symplectic 2-form

$$d\alpha = d\mathbf{q} \wedge d\mathbf{p} \equiv \sum_{i=1}^{n} dq_i \wedge dp_i. \tag{5.4.4}$$

Conversely, it is possible to obtain the contact 1-form from the symplectic 1-form $\beta = \mathbf{p} \cdot d\mathbf{q}$. Indeed consider a hyperplane defined by the equation

$$\sum_{i=1}^{n} p_i dq_i = 0. \tag{5.4.5}$$

This is so because in $\mathbf{R}^3$ in the x, y, z the system of coordinates a plane Π going through some point $\mathbf{r} = \{x_0, y_0, z_0\} \in \mathbf{R}^3$ is described with help of the vector $\mathbf{N} = \{A(x,y,z), B(x,y,z), C(x,y,z)\}$ such that $\mathbf{N} \cdot (\mathbf{r} - \mathbf{r}_0) = 0$.

Let, for instance, $p_1 \neq 0$ in Eq. (5.4.5), then we can rewrite it as follows

$$dq_1 + \frac{p_2}{p_1} dq_2 + \cdots + \frac{p_n}{p_n} dq_n = 0. \tag{5.4.6}$$

Notice that this result remains unchanged if, instead of p_i we would have $\lambda p_i \forall\, i$ for some non-negative λ. Clearly, Eq. (5.4.6) is not uniquely defined since instead of $p_1 \neq 0$ we could choose any other nonzero p_i and do the same. Such a circumstance makes the contact hyperplane a fiber bundle with n-dimensional base space made of tuples $\{q_1, \ldots, q_n\}$ and the fiber — the $(n-1)$-dimensional projective space $\{p_1 : \ldots : p_n\} \in \mathbf{RP}^{n-1}$. Such a projective space can be constructed as it is usually done by patching $\mathbf{RP}^{n-1}$ by affine charts and by defining the transition functions between these charts. Equation (5.4.6) is written in one of these charts in which we can introduce the new momenta $\tilde{p}_{i-1} = p_i/p_1, i = 2, \ldots, n$ representing a point $\{\tilde{p}_1, \ldots, \tilde{p}_{n-1}\}$ in such a space.

Using the fact that Eq. (5.4.5) represents the hyperplane in *q-space*, we realize that $\alpha \mid_{\mathcal{E}} = 0$ is also a hyperplane. It is easier to visualize this hyperplane by rewriting it in a more familiar notations, e.g. $\alpha = dz + xdy = 0$. Since in $\mathbf{R}^3$ all hyperplanes are described by the equation

$$Adx + Bdy + Cdz = 0, \tag{5.4.7}$$

we deduce that for our case either $C = 1, B = x$ and $A = 0$ or $dx = 0$ and A arbitrary. Physically interesting is the first option. In such a case

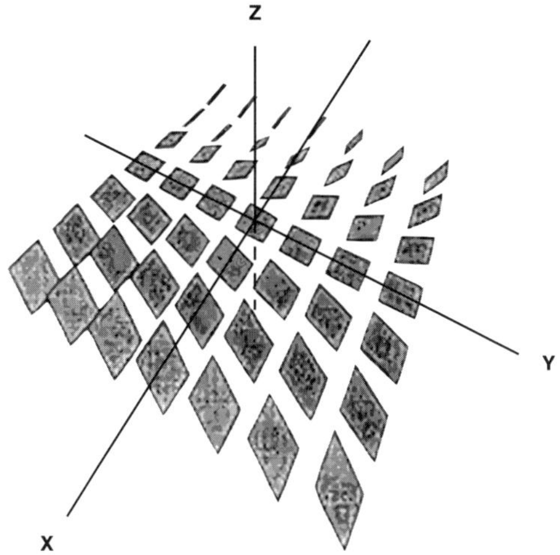

Fig. 4. The standard contact structure on $\mathbf{R}^3$.

the vector **N** is parametrized by the slope x of the line in $(y - z)$-plane. Since by construction this line is always lying in this plane and should be perpendicular to α-plane, this means that α-plane is located parallel to the x-axis. The plane is rotating around the x-axis. The "speed" of rotation is determined by x. All this is depicted in Fig. 4.

Such a behavior is observed in liquid crystals, e.g. in the already mentioned TGB phases described in the paper by Lubensky, e.g. see Fig. 9 of [78]. Since in his paper the physical properties of such phases are described in terms of the G-L theory of superconductivity, the task remains to demonstrate the same by employing the formalism of contact geometry. Although in full, it is described in Chap. 7, it is also helpful to do so through the already noticed correspondence between hydrodynamics of incompressible ideal fluids and the G-L theory of superconductivity. Thus, we need to discuss now the dynamics of ideal fluids in the context of contact geometry and topology.

5.5 Force-Free (Beltrami) Fields and Contact Geometry and Topology of Hydrodynamics and Electromagnetism

Beginning with Woltjer's theorem discussed in Chap. 3 we learned that the presence of force-free (Beltrami) fields is always associated with some kind of twisting. The quantitative measure of this twisting is helicity. Naturally, the question arises about the connection between these fields and the contact geometry. This question was resolved affirmatively in [137]. In order to discuss this paper we have to discuss a few related problems. In particular, our 1-form $\alpha = dS - pdq$ was determined in the Euclidean space $\mathbf{R}^3$ made out of triples q, S and p. The Euclidean space $\mathbf{R}^3$ can be compactified by adding one point at infinity. Such compactified space becomes S^3. In view of its relevance to monopoles, knots, etc., the question arises: Is there a difference between the 1-form defined on $\mathbf{R}^3$ and on S^3? In addition, by recalling that in statistical mechanics one is placing a statistical system in the box with periodic boundary conditions it is also of interest to look at 1-forms "living" on T^3 and see if they differ from those on $\mathbf{R}^3$ or S^3.

Exercise 5.5.1. Think about embedding of T^3 into $\mathbf{R}^5$ and look for contact structures in $\mathbf{R}^5$. Hint: Read [138]. In view of Theorem 5.5.1 below it is possible also to relate contact structures in $\mathbf{R}^5$ to those in S^5. Alternatively, the Beltrami fields "living" on T^3 were discussed in some detail in [68], Chap. II. They are surely not the same as those 'living" on S^3. Think about the likely implications of this fact to statistical mechanics.

Knowledge of solution to the above exercise leaves us only with the discussion about the differences between 1-forms "living" respectively on $\mathbf{R}^3$ and S^3. Following [139] it is known that the contact 1-form on $\mathbf{R}^3$ can be written either as $\alpha = dz + xdy$ or as $\alpha = dz + xdy - ydx = dz + r^2 d\varphi$, where in the last equality we used the polar coordinates in the (x, y)-plane. Consider as well the 1-form in $\mathbf{R}^5$. Evidently, it is given by

$$\alpha = dz + \sum_{i=1}^{2}(x_i dy_i - y_i dx_i). \tag{5.5.1}$$

Next, we consider S^3 embedded into $\mathbf{R}^4$. According to [139] we obtain the following result for the 1-form[a]

$$\alpha = \sum_{i=1}^{2}(x_i dy_i - y_i dx_i) \tag{5.5.2}$$

naturally coinciding with that in [104] where it was used for construction of the Hopf fibration for the Dirac monopole as discussed in Sec. 3.5.

[a]It should be noted that in [139] the standard 1-form on S^3 is given with the prefactor 1/2 in front, e.g. see p. 24, while later on, for S^{2n+1} , it is given as in Eq. (5.5.2), e.g. see p. 53 of this reference.

The following theorem is of profound physical importance for what will follow and for the monopole theory in particular.

Theorem 5.5.1. *For any point $p \in S^3$, the two contact manifolds $S^3\backslash\{p\}$ and $\boldsymbol{R}^3$ are contactomorphic.*

Proof. ([139][b]) Let (x_1, y_1, x_2, y_2) be the coordinates in $\mathbf{R}^4$ while (u, v, w) be the coordinates in $\mathbf{R}^3$. Write $\mathcal{F} : S^3\backslash\{p\} \to \mathbf{R}^3$ for the stereographic projection with p chosen as the North pole. Then,

$$u = \frac{x_1}{1 - y_2}, \quad v = \frac{y_1}{1 - y_2}, \quad w = \frac{x_2}{1 - y_2}$$

so that the inverse map is given by: $x_1 = \lambda u, y_1 = \lambda v, x_2 = \lambda w, y_2 = 1 - \lambda$, where $\lambda = 2/(1 + u^2 + v^2 + w^2)$. Thus, if the canonical 1-form on S^3 is given by Eq. (5.5.2), the canonical 1-form on $\mathbf{R}^3$ is given by $\alpha|_{\mathbf{R}^3} = dw + udv - vdu$. Next, using the stereographic projection it is possible to project α defined in Eq. (5.5.2) into $\mathbf{R}^3$ with the result: $\check{\alpha} = \lambda^2\tilde{\alpha}$, where

$$\tilde{\alpha} = \left(r^2 d\varphi + wrdr + \frac{1}{2}(1 + w^2 - r^2)dw\right).$$

To complete the proof, it is necessary to find such a diffeomorphism F of $\mathbf{R}^3$ that $F(\alpha|_{\mathbf{R}^3}) = \tilde{\alpha}$. It happens that such a diffeomorphism is possible to construct explicitly as explained in [139], pp. 56 and 57. Thus, the theorem is proved. □

The proof of this theorem allows us now to develop the theory of incompressible fluids on S^3 and then to reformulate everything in the superconducting G-L-type formalism as it was done in Sec. 3.3. The seminal work by Fröhlich, [11], is very helpful for such a task. With help of this reformulation it should become possible to describe the TGB liquid crystalline phases in the way found originally in the work by Lubensky [78].

Thus, we begin with reformulation of dynamics of incompressible fluids on S^3 following [137]. In doing so we shall also use some results from [140]. The first task lies in reformulating the force-free (Beltrami) condition $\nabla \times X = fX$ in the language of differential forms.[c] This requires us to find an equivalent of curl operator for differential forms on a 3-manifold (e.g. on S^3).

[b]It should be noted that in [139] the proof is given for any S^{2n+1} and $R^{2n+1}, n = 1, 2, \ldots$. This allows one to use the results of this theorem for solution of the preceding exercise.

[c]To avoid confusion, here f has the same meaning as α in Eq. (1.3.1).

For some Riemannian manifold M^n equipped with the metric 2-form $g = g_{ij}dx^i \otimes dx^j$ one can obtain using such a metric a 1-form by contracting this 2-form with the vector field $X = X^i \frac{\partial}{\partial x^i}$. This leads to the result: $X^\flat = \imath_X g = g(X, \cdot) = g_{ij}X^i dx^j$. This $\flat$-mapping transforms vector fields into 1-forms. It also has an inverse. Specifically, if $X^\flat$ is the 1-form, then $X^\sharp = g^{ij}X_i \frac{\partial}{\partial x^j} = X$. Recall that the Hodge star operator $*$ on a manifold M^n is the linear operator mapping p-forms Ω^p into $n-p$ forms Ω^{n-p} "living" on M^n. Explicitly, for the basis elements of $\Omega^p(M)$ it is acting as follows:

$$*(dx^{i_1} \wedge \cdots \wedge dx^{i_p}) = \frac{\sqrt{|g|}}{(n-p)!} g^{i_1 l_1} \cdots g^{i_p l_p} \varepsilon_{l_1 \cdots l_p l_{p+1} \cdots l_n} dx^{l_{p+1}} \wedge \cdots \wedge dx^{l_n} \tag{5.5.3}$$

with $|g| = \det g_{ij}$ as usual. If id is the identity operator, then $** = (-1)^{p(n-p)}$ id. For $n = 3$, this gives $** = \mathrm{id} \forall p = 0, \ldots, 3$. With such definition in our hands, the following result obtained in [141], Proposition (6.2.12),

$$\begin{aligned} \alpha \wedge *\beta &= g(\alpha, \beta)dV, \alpha, \beta \in \Omega^p(M), \quad p = 1, 2, \ldots; \\ dV &= \sqrt{|g|}dx^1 \wedge \cdots \wedge dx^n, \\ g(\alpha, \beta) &= \frac{1}{p!}\alpha_{l_1 \cdots l_p}\beta_{k_1 \cdots k_p} g^{l_1 k_1} \cdots g^{l_p k_p} \end{aligned} \tag{5.5.4}$$

is helpful. To connect the contact and the Riemannian structures, say on 3-manifold M^3, Chern and Hamilton proved the following remarkable theorem, [142]:

Theorem 5.5.2. *Every contact form on a 3-manifold has the adapted Riemannian metric g.*

In view of Eq. (5.5.4), this means that for any contact 1-form α it is possible to find the *adapted* metric g such that:

$$(a)\ \alpha \wedge d\alpha = \alpha \wedge *\alpha = dV \equiv \mu \text{ implying } (b)\ g(\alpha^\sharp, \alpha^\sharp) = 1. \tag{5.5.5}$$

In contact geometry these results can be reformulated with help of the Reeb vector field(s) X associated with 1-form α. By definition,

$$\imath_X \alpha = 1 \quad \text{and} \quad \imath_X d\alpha = 0 \quad \text{(Reeb field).} \tag{5.5.6a}$$

This convention serves as normalization for α-forms. Such a normalization is purely conventional though. In fact, it is sufficient to write instead

$$\imath_X \alpha > 0 \quad \text{and} \quad \imath_X d\alpha = 0 \quad \text{(Reeb field).} \tag{5.5.6b}$$

Since, as defined above, $X^\flat = \imath_X g = g(X, \cdot) = g_{ij}X^i dx^j$, it makes sense to look at what happens if X belongs to the Reeb field X, that is if we let

$\alpha = \imath_X g$ with X being the Reeb vector field. The first Reeb condition then reads: $\imath_X\ \imath_X g = 1$, that is $g_{ij}X^i X^j = 1$ in accord with Eq. (5.5.6a). The second Reeb condition requires us to evaluate $d\alpha = d(\imath_X g)$. The easiest way to do so is by considering a specific example exibiting all needed features.

Thus, we adapt the above general results to S^3 for which the 1-form is given by Eq. (5.5.2). This time, however, we shall conveniently divide the R.H.S. of Eq. (5.5.2) by the factor of 2. This is needed in view of the fact that the symplectic 2-form ω on $\mathbf{R}^4$ given by

$$\omega = dx_1 \wedge dy_1 + dx_2 \wedge dy_2 \tag{5.5.7}$$

can be adapted to the unit 3-sphere $x_1^2 + y_1^2 + x_2^2 + y_2^2 = 1$ by considering a contraction $\imath_Y\omega$, where the radial vector field $Y = \frac{1}{2}r\partial_r = \frac{1}{2}(x_1\partial_{x_1} + y_1\partial_{y_1} + x_2\partial_{x_2} + y_2\partial_{y_2})$ can "live" both on $\mathbf{R}^4$ for which $r^2 = x_1^2 + y_1^2 + x_2^2 + y_2^2$, and on S^3 for which $r^2 = 1$. A straightforward computation of $\imath_Y\omega$ produces then the 1-form α, Eq. (5.5.2), divided by the factor of 2. The Reeb vector field R_α associated with such 1-form is given by

$$R_\alpha = 2(x_1\partial_{y_1} - y_1\partial_{x_1} + x_2\partial_{y_2} - y_2\partial_{x_2}). \tag{5.5.8}$$

This is so because we have now: $\frac{1}{2}\alpha(R_\alpha) = x_1^2 + y_1^2 + x_2^2 + y_2^2 = 1$, in accord with the first condition given by Eq. (5.5.6). The second condition now reads: $\imath_{R_\alpha} \circ d[\frac{1}{2}\alpha] = \imath_{R_\alpha}\omega = -2(x_1dx_1 + y_1dy_1 + x_2dx_2 + y_2dy_2) = -2rdr$. Since $r^2 = 1$, $2rdr = 0$ on S^3 thus proving us with the second requirement given by Eq. (5.5.6).

The vector field Y is such that $\mathcal{L}_Y\omega = \omega$ as can be seen by direct computation. In this expression the Lie derivative $\mathcal{L}_Y$ is defined as usual by the Cartan formula [136]:

$$\mathcal{L}_Y = d \circ \imath_Y + \imath_Y \circ d. \tag{5.5.9}$$

The vector field Y for which $\mathcal{L}_Y\omega = \omega$ where ω is a *symplectic* 2-form is called "the *Liouville vector field*" while the 1-form $\alpha = \imath_Y\omega$ is called "*the contact form*", e.g. see Eq. (5.5.2). The Reeb field R is living on the tangent bundle to the hypersurface $\mathcal{W}$. This bundle is defined at each point of $\mathcal{W}$ by the (hyperplane) condition $\ker \imath_Y\omega$. The hypersurface $\mathcal{W}$ is "living" in the symplectic manifold $\mathcal{M}$ (whose 2-form is ω). The hypersurface $\mathcal{W}$ is transverse (that is orthogonal) to Y.

In the present case, the Y-field is orthogonal to the vector fields living in the plane $\ker \imath_Y\omega$. These results provide a useful link between the symplectic and contact geometries. The symplectic interpretation of 2-form ω requires $d\omega = 0$. From here it follows that: $\mathcal{L}_Y\omega = d\circ\imath_Y\omega + \imath_Y\circ d\omega = \omega$ or $d\circ\imath_Y\omega = \omega$.

Indeed, since $\imath_Y\omega = \alpha$ application of operator d to 1-form, Eq. (5.5.2) (divided by the factor of 2), reproduces ω, Eq. (5.5.7), as required.

If the symplectic manifold $\mathcal{M}$ is of dimension $2n$ and is having the volume form ω^n, then the contact volume form is $\alpha \wedge (d\alpha)^{n-1}$. It is of dimension $2n-1$ equal to that for $\mathcal{W}$. In the present case we have $\mathbf{R}^4$ as symplectic manifold and S^3 as contact hypersurface ω in $\mathbf{R}^4$. These two volume forms are related to each other via an elegantly looking formula: $\alpha \wedge (d\alpha)^{n-1} = \imath_Y\omega \wedge (d \circ \imath_Y\omega)^{n-1} = \imath_Y\omega \wedge \omega^{n-1} = \frac{1}{n}\imath_Y(\omega^n)$. Since $\mathcal{L}_Y\omega^n = \omega^n$, and ω^n is symplectic volume, this result provides justification for the vector field Y to be called Liouville.

Remark 5.5.1. From the above derivation it follows that the requirement $\imath_X\alpha = 1$ is not essential and, indeed, it can be replaced by a requirement $\imath_X\alpha > 0$ since we can work with the 3-sphere of any radius.

Remark 5.5.2. The above derivation of the Reeb condition provides us with yet another way of constructing contact manifolds from symplectic. Some basic information about this type of construction can be found in the classical book by Arnol'd [136].

Remark 5.5.3. The Reeb condition $\imath_X d\alpha = 0$ is equivalent to the incompressibility condition for fluids. Indeed, in [137], the condition $\mathcal{L}_{R_\alpha}\mu = 0$ replaces the conventional definition of incompressibility. This means that, given the contact form α, and the contact volume form $\mu = \alpha \wedge d\alpha$, we have to demonstrate that $\mathcal{L}_{R_\alpha}\mu = 0$. Using Eq. (5.5.9) we obtain,

$$\mathcal{L}_{R_\alpha}\mu = (d \circ \imath_{R_\alpha})\mu + \imath_{R_\alpha} \circ d\mu = (d \circ \imath_{R_\alpha})\mu = (d \circ \imath_{R_\alpha})\alpha \wedge d\alpha. \quad (5.5.10)$$

In this chain on equalities we took into account that $d\mu = 0$ since μ is the volume 3-form "living" on 3-manifold. Next, taking into account that $\mathcal{L}_{R_\alpha}(\alpha \wedge d\alpha) = (\mathcal{L}_{R_\alpha}\alpha) \wedge d\alpha + \alpha \wedge (\mathcal{L}_{R_\alpha}d\alpha)$, the above chain can be extended further as follows:

$$\mathcal{L}_{R_\alpha}\mu = (d \circ \imath_{R_\alpha}\alpha) \wedge d\alpha + \alpha \wedge (d \circ \imath_{R_\alpha}d\alpha) = 0. \quad (5.5.11)$$

This result is in compliance with both Reeb conditions which we have just checked.

Remark 5.5.4. The Liouville vector field Y is the vector field satisfying the equation $\mathcal{L}_Y\omega = \omega$. Since ω can be any form, we can take it locally as some function $\mathfrak{f}$. Then, this equation becomes just the Euler-type equation for homogenous functions, e.g. we have

$$\sum_i x_i \frac{\partial \mathfrak{f}}{\partial x_i} = \mathfrak{f}. \quad (5.5.12)$$

This explains why the vector field $Y = \frac{1}{2}r\partial_r$ is of Liouville-type.

Remark 5.5.5. The integral curves associated with the Reeb vector field R_α are circles. They represent the fibers of the Hopf fibration (see Sec. 3.5). Furthermore, the integral curves associated with the Reeb vector field are always transverse to the Cartan distributions that is to planes defined by the condition: $\alpha = 0$. Notice that this condition is not in contradiction with the Reeb normalization condition $\imath_X \alpha = 1$ as we just demonstrated using the specific example.

With all these results in our hands, we are now very close to introducing the curl operator acting on forms. In fact, such an operator was already introduced in [141, 143]. For the sake of uninterrupted reading, we would like to illustrate its action on forms following [143].

We begin with the vector $\mathbf{F} = \sum_{i=1}^{3} F_i \mathbf{e}_i$. The $\flat$-operator now converts it into the 1-form $(\mathbf{F})^\flat = \sum_{i=1}^{3} F_i dx_i$. From it, we obtain as well:

$$d(\mathbf{F})^\flat = \left(\frac{\partial F_2}{\partial x_1} - \frac{\partial F_1}{\partial x_2}\right) dx_1 \wedge dx_2 + \left(\frac{\partial F_1}{\partial x3} - \frac{\partial F_3}{\partial x_1}\right) dx_3 \wedge dx_1 + \left(\frac{\partial F_3}{\partial x_2} - \frac{\partial F_2}{\partial x_3}\right) dx_2 \wedge dx_3.$$

Next, we obtain

$$*d(\mathbf{F})^\flat = \left(\frac{\partial F_2}{\partial x_1} - \frac{\partial F_1}{\partial x_2}\right) dx_3 + \left(\frac{\partial F_1}{\partial x3} - \frac{\partial F_3}{\partial x_1}\right) dx_2 + \left(\frac{\partial F_3}{\partial x_2} - \frac{\partial F_2}{\partial x_3}\right) dx_1,$$

and, finally, we obtain:

$$(*d(\mathbf{F})^\flat)^\sharp = \left(\frac{\partial F_2}{\partial x_1} - \frac{\partial F_1}{\partial x_2}\right) e_3 + \left(\frac{\partial F_1}{\partial x3} - \frac{\partial F_3}{\partial x_1}\right) \mathbf{e}_2 + \left(\frac{\partial F_3}{\partial x_2} - \frac{\partial F_2}{\partial x_3}\right) \mathbf{e}_1 = \operatorname{curl} \mathbf{F}.$$

Using this result and following [140], we write

$$(\nabla \times X)^\flat = *dX^\flat. \tag{5.5.13}$$

In view of this definition it is possible to present the *Beltrami* 1-form $X^\flat$ as

$$*dX^\flat = fX^\flat. \tag{5.5.14a}$$

From here we conclude that the 1-form $\alpha = X^\flat$ on M^3 is *Beltrami* if

$$*d\alpha = f\alpha. \tag{5.5.14b}$$

Remark 5.5.6. This equation for differential forms is equivalent to the force-free Eq. (1.3.1). Since solutions of the steady-state Euler's Eq. (1.2.1) coincide with solutions of the force-free Eq. (1.3.1), Eq. (5.5.14b) contains, in principle, all information about the steady-state of Eulerian-type fluids. Naturally, the solution of this equation can be obtained only if the 3-manifold is specified.

Exclusion of the lammellar-type fields (e.g. of type given by Eq. (3.2.3)) in the present case means that the function f is nowhere zero. In such a case the 1-form α is called *rotational Beltrami* 1-form. Since for $n = 3$ we have $** = \text{id}$ this allows us to rewrite Eq. (5.5.14b) equivalently as

$$d\alpha = f * \alpha. \tag{5.5.14c}$$

For contact 1-form $\alpha = X^\flat$ we can choose the Riemann metric g which is *adapted* in the sense of Eq. (5.5.5). Since $X^\flat = \imath_X g = g_{ij}X^i dx^j$, as we already know, Eq. (5.5.14) acquires the form

$$*d\alpha = f\alpha \rightarrow d\alpha = f * \alpha \rightarrow d\alpha = f\imath_X \mu \tag{5.5.15a}$$

in which the technicalities of calculation of $\imath_X \mu$ are explained in detail in [144], pp. 90 and 91. The legitimacy of such an identification of $*\alpha$ with $\imath_X \mu$ can be seen from Eq. (5.5.5). Indeed, in the present case (e.g. in view of Eq. (5.5.15a)) the volume 3-form μ can be constructed as follows.

$$\mu = \alpha \wedge d\alpha = f\alpha \wedge *\alpha = f dV. \tag{5.5.16}$$

The factor f in front of dV can be eliminated by taking into account Theorem 5.5.2, Since $dV = \sqrt{|g|}dx^1 \wedge \cdots \wedge dx^n$, the combination $fdV = d\tilde{V}$ is obtainable from dV by conformal transformation, i.e. $|g| \rightarrow |\tilde{g}|$ so that $f^2|g| = |\tilde{g}|$. Because $*dV = 1$, this means that upon such a transformation we have to redefine the Hodge star operation as well, e.g. $* \rightarrow \tilde{*} = \frac{*}{f}$. Then, $\tilde{*}d\tilde{V} = *dV = 1$. Thus, if we were initially working using the metric $\tilde{g}$, then written in the metric g, Eq. (5.5.15a) acquires the form

$$d\alpha = *\alpha \rightarrow d\alpha = \imath_X \mu. \tag{5.5.15b}$$

Just obtained results are in accord with those summarized in the review paper by Ghrist, [145], where they are given without derivation. The derivation can be found in [137] where it is proven that.

Theorem 5.5.3. *Any rotational Beltrami field on a Riemannian 3-manifold is a Reeb-like and vice versa.*

Since we came to the same conclusions by analyzing in detail the contact structures on S^3, it is of interest to demonstrate that the same results hold for other 3-manifolds by simplifying arguments used in [137].

For this purpose, let X be a Reeb-like field, then $X^\flat = \imath_X g = \alpha$. According to Eq. (5.5.6) we should have either $\imath_X \alpha = 1$ or $\imath_X \alpha > 0$. But since $\imath_X \imath_X g = 1$ that is $g_{ij} X^i X^j = 1$ (as already shown above), the first Reeb condition is satisfied. In view of Eq. (5.5.15b), the second Reeb condition reads: $\imath_X(\imath_X \mu) = 0$. To justify this result we notice that the operator $\imath$ is the inverse of the operator d and, while acting on differential forms, $d^2 = 0$, the same should be true for the $\imath^2 = 0$. Details can be found in [146], p. 4. From the same reference it follows that such an operator identity works only on differential forms and not on quadratic forms defining the Riemannian metric. Indeed, the metric defining quadratic form is given by $g_{ij} dx^i dx^j = dx^i dx_i$ with $g_{ij} = g_{ji}$, while the analogous 2-form is given by $\alpha_{ij} dx^i \wedge dx^j$ with $\alpha_{ij} = -\alpha_{ji}$.

From Sec. 3.1, it is obvious that the steady-state ideal hydrodynamics and the time-independent electrodynamics admit the same mathematical description. We would now like to strengthen this observation by reformulating Eqs. (1.3.1) and (1.3.2) in the language of differential forms. In fact, we already found an analog of Eq. (1.3.1). It is given by Eq. (5.5.14). The analog of Eq. (1.3.2) is obtainable along the same lines, e.g. we have $(\nabla \times (\nabla \times X))^\flat = *d(\nabla \times X)^\flat = *d * dX^\flat \equiv \delta d\alpha = k^2 \alpha$. By comparing this result with Eq. (1.3.2), the (Hodge-type) Laplacian can be identified now as $\nabla^2 = -\delta d$ in accord with the definition given in [147]. In Sec. 3.3, we introduced the Trkalian-type fields. It is of interest now to reformulate these results in terms of differential forms.

In particular, if Eq. (5.5.14b) represents an equation for some Trkalian 1-form β, that is $*d\beta = f\beta, f = \text{const.}$, then β is also a solution of the Helmholtz equation $*d*d\beta = f^2\beta$ since from $*d\beta = f\beta$ we obtain: $*d*d\beta = *df\beta = f^2\beta$. Conversely, if $*d*d\beta = f^2\beta$, then there is 1-form $\alpha = *d\beta + f\beta$ such that $*d\alpha = *d*d\beta + *df\beta = f(*d\beta + f\beta) = f\alpha$.

Consider now the source-free Maxwell's equations written in the language of differential forms. We have

$$dE = -i\omega B \quad \text{and} \quad dH = i\omega D. \tag{5.5.16a}$$

They must be supplemented with the usual-type relations

$$B = \mu * H \quad \text{and} \quad D = \varepsilon E \tag{5.5.16b}$$

between B and H and D and E in isotropic medium. Let $k^2 = \omega^2 \sqrt{\mu\varepsilon}$, then using the above equations we obtain:

$$*d * dE = k^2 E \quad \text{and} \quad * d * dH = k^2 H. \tag{5.5.17}$$

From here with help of Eq. (5.5.16) we obtain the analog of α 1-forms

$$\alpha_E = *dE + kE = k(E - i\eta H);$$
$$\alpha_H = *dH + kH = k(H + (i/\eta)H)$$

in which $\eta = \sqrt{\mu/\varepsilon}$. Both of them are Trkalian by construction since $*d\alpha_E = k\alpha_E$ and $*d\alpha_H = k\alpha_H$ as can be easily checked.

It is possible now to add some finer details to the above grand picture. Consider, say, a plane wave 1-form:

$$E = (Adx + Bdy)e^{ikz} \tag{5.5.18a}$$

with $A = A_1 + iA_2$ and $B = B_1 + iB_2$ being some complex constants determining polarization of the wave. Accordingly,

$$H = \frac{i}{\omega\mu} * dE = \frac{1}{\eta}(Bdx - Ady)e^{ikx}. \tag{5.5.18b}$$

With such defined E and H 1-forms the contact form α_E can be constructed as follows:

$$\begin{aligned}\alpha_E &= \mathrm{Re}\{E - i\eta H\}\\ &= \mathrm{Re}\{[(A_1 + B_2)dx + (B_1 - A_2)dy)\\ &\quad + i((A_2 - B_1)dx + (A_1 + B_2)dy]e^{ikz}\}.\end{aligned}$$

Consider now a special case: $A = 1, B = 0$, then, the above general expression is simplified to the so-called *overtwisted* contact form given by

$$\alpha_E = \cos kzdx - \sin kzdy \tag{5.5.19a}$$

which for $kz \sim 0$ is reduced to more familiar form

$$\alpha_E = dx - kzdy \tag{5.5.19b}$$

e.g. see Eq. (5.2.7). Such a reduction to a 1-form resembling that given in Eq. (5.2.7) is done exclusively for illustrative purposes. In fact it is mathematically incorrect since such reduced form is no longer Trkalian.

Exercise 5.5.2. Check this statement.

At the same time, the form α_E given by Eq. (5.5.19a) is the correct contact form. Indeed, $d\alpha_E = -k\,\sin kzdz \wedge dx - k\,\cos kzdz \wedge dy$ so that

$$\begin{aligned}\alpha_E \wedge d\alpha_E &= (\cos kzdx - \sin kzdy) \wedge (-k\sin kzdz \wedge dx - k\cos kzdz \wedge dy)\\ &= -k\cos^2 kzdx \wedge dz \wedge dy + k\sin^2 kzdy \wedge dz \wedge dx\\ &= kdx \wedge dy \wedge dz\end{aligned}$$

as required.

The above discussion was made with the following purpose in mind. In mathematics literature a great deal of attention is being paid to differences between the *overtwisted* and *tight* contact structures. On three-dimensional manifolds it is proven that only these two classes of contact structures can exist [139]. In particular, Bennequin demonstrated that *the standard contact structure*, Eq. (5.5.2) *is tight*. As a corollary, in view of Theorem 5.5.1, we conclude that the standard contact structure (that is either $\alpha = dz + xdy$ or $\alpha = dz + xdy - ydx = dz + r^2 d\varphi$) on $\mathbf{R}^3$ is also tight. These facts happen to have physical significance to be discussed below.

5.6 Many Facets of the Abelian Chern–Simons Functional and Their Relation to Monopoles, Dyons and the Faddeev–Skyrme Model

5.6.1 *General Information*

In Sec. 5.1, we noticed that in recent [131] it was demonstrated that the ACS functional is capable of describing nontrivial knots and links. Specifically, with its help it is possible to construct the Kauffman and Jones knot/link polynomial invariants capable of recognizing nontrivial knots and links. The purpose of this and the following sections is to discuss another route (recently developed by mathematicians) for reaching the same conclusions. As a by-product, we obtain some new interpretations of monopoles. These are being discussed in the context of high-energy and condensed matter physics.

We begin with Eqs. (3.1.12) and (3.1.13) which are reproduced here for reader's convenience. Thus, we have

$$E[\mathbf{v}] \geq C \cdot |\mathcal{H}(\mathbf{v})|, \tag{3.1.12}$$

where the helicity $\mathcal{H}(\mathbf{v})$ is given by

$$\mathcal{H}(\mathbf{v}) = \int_V \mathbf{v} \cdot \mathrm{curl}^{-1}\mathbf{v}\, dV \tag{3.1.13}$$

or, alternatively, by Eq. (3.5.12), that is by the ACS functional.

It is useful to rewrite these results in notations of the previous section, that is in terms of differential forms. So, for the helicity we already have Eq. (3.5.12), while for the energy we obtain,

$$E[\alpha] = \int_M \alpha \wedge *\alpha. \tag{5.6.1a}$$

This functional can be looked upon as 3-dimensional version of the functional for pure Yang–Mills (Y-M) gauge fields. In view of the inequality, Eq. (3.1.12), such an identification seems to be appropriate since in four dimensions the Y-M action functional is bounded from below by the non-Abelian helicity, that is by the non-Abelian version of the Chern–Simons functional. For the sake of arguments which follow, we reproduce the key results leading to such a bound. According to [148] the Euclideanized Y-M action functional $S[\mathbf{F}]$ defined in $\mathbf{R}^4$ (leading to the instanton solutions) is given by

$$S[\mathbf{F}] = -\mathrm{tr}\int_{\mathbf{R}^4} \mathbf{F} \wedge *\mathbf{F}. \tag{5.6.1b}$$

This functional can be equivalently rewritten as

$$S[\mathbf{F}] = -\frac{1}{2}\mathrm{tr}\int_{\mathbf{R}^4} (\mathbf{F} + *\mathbf{F}) \wedge (\mathbf{F} + *\mathbf{F}) + \mathrm{tr}\int_{\mathbf{R}^4} \mathbf{F} \wedge \mathbf{F} \geq \mathrm{tr}\int_{\mathbf{R}^4} \mathbf{F} \wedge \mathbf{F} \tag{5.6.2}$$

with equality achieved for the (anti)-self-dual solutions: $\mathbf{F} = -*\mathbf{F}$. It can be shown [149] that for calculation of $\mathrm{tr}\int_{\mathbf{R}^4} \mathbf{F} \wedge \mathbf{F}$ it is sufficient to decompose $\mathbf{R}^4$ as $\mathbf{R}^3 \times [0,1]$ or, after one-point compactification, as $S^3 \times [0,1]$. In such a case, we obtain [149]:

$$\mathrm{tr}\int_{S^3\times[0,1]} \mathbf{F} \wedge \mathbf{F} = \mathrm{tr}\int_{S^3} \left(\mathbf{A} \wedge d\mathbf{A} + \frac{2}{3}\mathbf{A} \wedge \mathbf{A} \wedge \mathbf{A}\right) \doteqdot CS(\mathbf{A}), \tag{5.6.3}$$

where the symbol $\doteqdot$ means "up to a constant". The decomposition $S^3 \times [0,1]$ reflects the fact that the Chern–Simons functional $CS(\mathbf{A})$ is defined up to a number, mod $\mathbf{Z}$, that is gauge transformations make this functional to be defined only up to mod $\mathbf{Z}$. Next, since $S[\mathbf{F}] \geq CS(\mathbf{A})$ the minima of $S[\mathbf{F}]$ are determined by the minima of $CS(\mathbf{A})$. These are obtained as solutions of zero curvature equation

$$\mathbf{F} = d\mathbf{A} + \mathbf{A} \wedge \mathbf{A} = 0. \tag{5.6.4}$$

For generic SU(2) case, using Eq. (5.6.4) we obtain: $d\mathbf{A} = -\mathbf{A} \wedge \mathbf{A}$. Substitution of this result back into $CS(\mathbf{A})$ leads to the well-known result:

$$\frac{1}{8\pi^2}\int_{S^3} \mathrm{tr}\left(\mathbf{A} \wedge d\mathbf{A} + \frac{2}{3}\mathbf{A} \wedge \mathbf{A} \wedge \mathbf{A}\right) = -\frac{1}{24\pi^2}\int_{S^3} \mathrm{tr}(\mathbf{A} \wedge \mathbf{A} \wedge \mathbf{A}). \tag{5.6.5}$$

It can be shown [111], that the integral $\frac{1}{24\pi^2}\int_{S^3} \mathrm{tr}(\mathbf{A} \wedge \mathbf{A} \wedge \mathbf{A})$ is an integer — a winding number for the map $f\colon S^3 \to \mathrm{SU}(2)$. Evidently, the equation $d\mathbf{A} = -\mathbf{A} \wedge \mathbf{A}$ can be used differently, e.g. we can write as well

$$\frac{1}{8\pi^2}\mathrm{tr}\int_{S^3}\left(\mathbf{A}\wedge d\mathbf{A}+\frac{2}{3}\mathbf{A}\wedge\mathbf{A}\wedge\mathbf{A}\right)=\frac{1}{24\pi^2}\mathrm{tr}\int_{S^3}\mathbf{A}\wedge d\mathbf{A}. \quad (5.6.6)$$

Then, instead of Eq. (5.6.2), we obtain the result analogous to the inequality Eq. (3.1.12), e.g.

$$S[\mathbf{F}] \geq C \cdot |\mathcal{H}(\mathbf{v})|. \quad (5.6.7)$$

5.6.2 *From Instantons to Monopoles*

Such a comparison is plausible but is not fully legitimate. Indeed, in the first case we are dealing with the steady-state solutions for ideal fluid. This means that time is not present explicitly in the expressions we are considering. In the second case use of Euclidean time, instead of "true" (or Minkowski) time, is still not eliminating time dependencies of the fields. Much more meaningful is the comparison of the London limit of G-L functional, Eq. (3.4.1), against its non-Abelian counterpart known in literature as Bogomolny–Prasad–Sommerfeld (BPS) functional.

Remark 5.6.2.1. This comparison makes sense for several reasons. *First*, because of noticed correspondence between descriptions of superconductivity, steady-state dynamics of incompressible ideal fluids and magnetohydrodynamics, the G-L functional, Eq. (3.4.1), can be analyzed instead of energy $E[\alpha]$ for ideal fluids. *Second*, this then allows us to make a comparison between the G-L (Abelian) functional with its non-Abelian version that is with the BPS functional. *Third*, although in literature the BPS functional is capable of producing the Abelian Dirac monopoles, no such feature is officially known for the Abelian G-L functional (the discussion presented in Sec. 3.5 is aimed at removing such a discrepancy). *Fourth*, this means that even though the non-Abelian gauge theories could be abelianized, such an Abelian limit thus far could not be matched with the results originating from the Abelian G-L functional unless Dirac monopoles are put by hand into this functional as it was done, for example, by Nambu, [109]. *Fifth*, because such a deficiency was corrected in Sec. 3.5, it is important to establish new links between treatments of Abelian and non-Abelian G-L-type functionals.

Thus, because of its usefulness for description of non-Abelian monopoles [114], we shall compare the results originating from the BPS functional against those for Abelian (Dirac) monopoles (discussed in Sec. 3.5). In doing so, we follow mainly the original works, [150, 151], as well as the monographs [105, 114, 152].

As in the Abelian case, where the static 3-dimensional G-L functional, Eq. (3.4.1), is used, in the non-Abelian case the gauge fields also depend only on spatial coordinates x_1, x_2 and x_3. Following [150, 151] this fact allows us *to get rid of the Higgs field entirely* by writing $\phi^a = A_4^a$ implying that $F_{i4}^a = \partial_i A_4^a - \varepsilon^{abc} A_i^b A_4^c \equiv D_i \phi^a$. If in four dimensions the (anti)self -duality equations $F = \pm * F$ relate the 2-forms to 2-forms, in three dimensions the 2-forms are related to 1-forms, e.g. see Eq. (4.3). Evidently, $D_i\phi^a$ is equivalent to 1-form so that the (anti)self-duality condition acquires the following form:

$$F_{ij}^a = \mp \varepsilon_{ijk} D_k \phi^a. \tag{5.6.2.1a}$$

Alternatively, by taking into account that $\varepsilon_{ijk}\varepsilon_{ijn} = 2\delta_{kn}$ (here we used Einstein's summation convention), we obtain

$$B_k^a = \mp D_k \phi^a \quad \text{or} \quad * \mathbf{F} = \mp D\phi, \tag{5.6.2.1b}$$

where the magnetic field $\mathbf{B}$ (or $\mathbf{H}$) is given by the relation $B_i^a = \frac{1}{2}\varepsilon_{ijk} F_{jk}^a$ as usual. Equation (5.6.2.1) happen to be the famous Bogomoleny equations for the non-Abelian monopoles. *By analogy with Eq.* (4.3) *they can be considered as the non-Abelian version of the force-free condition.* Following [114] it is possible using these equations to re-obtain the non-Abelian analogs of Maxwell's Eqs. (3.3.4) and (3.3.8) as well as the "incompressibility" result $\operatorname{div} \mathbf{H} = 0$.

For this purpose, let us recall [146] that four Maxwell's equations can be conveniently rewritten as two equations for differential forms:

$$d\mathbf{F} = 0 \text{ (Bianchi identity) and } * d * \mathbf{F} = \mathbf{j}. \tag{5.6.2.2}$$

As in general relativity, the transition to the non-Abelian version of these equations causes us to replace the exterior derivative d by the exterior covariant derivative D which, incidentally, we just have defined above. Thus, the analog of $\operatorname{div} \mathbf{H} = 0$ in the present case is

$$D_k D^k \phi^a = 0 \quad \text{or} \quad D\mathbf{B} = 0. \tag{5.6.2.3}$$

The analog of $*d * \mathbf{F} = \mathbf{j}$, that is of Eqs. (3.3.4) and (3.3.8), in the present case is given by

$$D_i F^{aij} = \varepsilon^{abc} \phi^c (D^j \phi^b) \quad \text{or} \quad * D * \mathbf{F} = [\phi, D\phi]. \tag{5.6.2.4}$$

This result is also obtainable using Bogomolny's Eq. (5.6.2.1) as demonstrated in [114]. Once these equations are obtained, we can think of the non-Abelian analog of the G-L functional which upon variation reproduces these equations. Following [152] such a functional can be written as follows:

$$\mathcal{A}(\mathbf{A},\phi)=\frac{1}{2}\mathrm{tr}\int_{\mathbf{R}^3}\mathbf{F}\wedge*\mathbf{F}+\frac{1}{2}\mathrm{tr}\int_{\mathbf{R}^3}D\phi\wedge*D\phi. \tag{5.6.2.5}$$

In the case of SU(2) Lie group the non-Abelian gauge fields ϕ and $\mathbf{A}_\mu$ should actually be read as $\phi=\phi^a t^a$ and $\mathbf{A}_\mu=A^a_\mu t^a$. The matrices $t^a=i\sigma^a$ with σ^a being Pauli's matrices obey the commutation relations $[t^a,t^b]=-2\varepsilon^{abc}t^c$ and, in addition, they satisfy the orthogonality requirement: $\mathrm{tr}(t^a t^b)=-2\delta^{ab}$. These results cause changes of sign in front of terms in Eq. (5.6.2.5). Since arguments which will follow are not limited to SU(2) Lie algebra and not affecting the final inequality, Eq. (5.6.2.8), we shall ignore the likely sign changes in front of terms entering the functional $\mathcal{A}(\mathbf{A},\phi)$. Therefore, Eq. (5.6.2.5) can be rewritten as

$$\mathcal{A}(\mathbf{A},\phi)=\frac{1}{2}\mathrm{tr}\int_{\mathbf{R}^3}(*\mathbf{F}\mp D\phi)\wedge*(*\mathbf{F}\mp D\phi)\pm 4\pi N, \tag{5.6.2.6}$$

where the winding number N is defined by

$$N=\frac{1}{4\pi}\mathrm{tr}\int_{\mathbf{R}^3}\mathbf{F}\wedge D\phi. \tag{5.6.2.7}$$

Evidently,

$$\mathcal{A}(\mathbf{A},\phi)\geq 4\pi|N|. \tag{5.6.2.8}$$

Equivalence of Eqs. (5.6.2.5) and (5.6.2.6) is assured by the fact that for scalar-type products $(\omega,\eta)=\int\omega\wedge\eta$, the Hodge star operation is an isometry. That is $\int\omega\wedge\eta=\int*\omega\wedge*\eta$. In view of Eq. (5.6.2.1b) we can rewrite Eq. (5.6.2.7) as

$$N=\frac{1}{4\pi}\mathrm{tr}\int_{\mathbf{R}^3}\mathbf{F}\wedge*\mathbf{F}. \tag{5.6.2.9}$$

With one-point compactification of $\mathbf{R}^3$, this result acquires the form

$$N=\frac{1}{4\pi}tr\int_{S^3}\mathbf{F}\wedge*\mathbf{F} \tag{5.6.2.10}$$

which can then be compared with Eqs. (3.1.12) and (5.6.1a). Since in the monopole physics no such comparison was made to our knowledge, it is instructive to describe the existing alternative paths of logic in some detail.

5.6.3 *Topology and the Non-Abelian Monopoles*

We begin with the reminder: The goal of alternative path is to recover the Dirac-like monopole from the non-Abelian functional, e.g. from $\mathcal{A}(\mathbf{A},\phi)$. Somewhat surprisingly, by developing this path the results we are about to obtain come out in accord with earlier stated Theorem 3.5.1. For the sake of space, we shall restrict our discussion only to one monopole relying mainly on mathematically rigorous treatment of this topic given in [152]. The same reference also contains results for multimonopoles consistent with Theorem 3.5.1. At the physical level of rigor the multimonople solutions are described, for example, in [105, 114]. Just by looking at the functional $\mathcal{A}(\mathbf{A},\phi)$ it is clear that its finiteness is assured if for $r = |\mathbf{x}| \to \infty$ both $F_{\mu\nu} \to 0$ and $D\phi \to 0$. In particular, if $D\phi = 0$ for $r \to \infty$ the relationship $\partial_i\phi^a - \varepsilon^{abc}A_i^b\phi_4^c \equiv D_i\phi^a$ causes gauge fields $\mathbf{A}_i$ and (pseudo) Higgs fields ϕ^a to be coupled. The requirement $F_{\mu\nu} \to 0$ causes these gauge fields to be in the form $\mathbf{A}_i = -\partial_i g(\mathbf{x})g(\mathbf{x})^{-1}$ for some $g(\mathbf{x}) \in \mathrm{SU}(2)$. That is the gauge field must be a pure gauge. By additional gauge transformation it is possible to make $\mathbf{A}_i$ vanish. As a result, ϕ^a becomes a constant. It can be chosen as, say, $t^a \to t^3$ where, as before, $t^a = i\sigma^a$. Under such circumstances the initial SU(2) symmetry becomes spontaneously broken to U(1) so that the factor group SU(2)/U(1) is now made only of gauge transformations $g(\mathbf{x})$ satisfying $g(\mathbf{x})t^3g(\mathbf{x})^{-1} = t^3$. Since both the (pseudo) Higgs field ϕ and the gauge transformations $g(\mathbf{x})$ depend upon $\mathbf{x}$, when $r \to \infty$ their behavior is expected to be nontrivial. This means the following.

The functional $\mathcal{A}(\mathbf{A},\phi)$, Eq. (5.6.2.5), is the limiting case ($\lambda = 0$) of more general G-L-type functional

$$\mathcal{A}(\mathbf{A},\phi,\lambda) = \int_{\mathbf{R}^3}\left[\frac{1}{2}\mathrm{tr}(\mathbf{F}\wedge *\mathbf{F}) + \frac{1}{2}\mathrm{tr}(D\phi\wedge *D\phi) + \frac{\lambda}{4} * (1-|\phi|^2)\right]. \tag{5.6.3.1}$$

The nontriviality is caused by the fact that, just like in condensed matter physics, the system "remembers its past". E.g. if we use the magnetic language, the magnetic system at temperatures T lower than some critical T_c remembers the direction of the auxiliary magnetic field which was gradually switched off. In the present case, it is expected that, for instance, $x^3 \to \infty, \phi(0,0,x^3) \to t^3$ and, accordingly, in the same limit $g(\mathbf{x}) \to g(0,0,x^3)\exp\{\alpha t^3\} \to \exp\{\alpha t^3\}$. Since topologically $\mathrm{SU}(2) \sim S^3$ and $\mathrm{SU}(2)/\mathrm{U}(1) \sim S^2$, we are dealing here with the already familiar Hopf map $f\colon S^3 \to S^2$. Nevertheless, the existing treatments of monopoles, e.g. [152], proceed in a fashion different from that employed for the Dirac monopole (discussed in Sec. 3.5). Specifically, on p. 35 of [152] we find the following statement:

Let $|\phi|$ be continuous SU(2)-valued function satisfying

$$\lim_{r\to\infty}\sup_{|\mathbf{x}|=r}|1-|\phi||=0. \tag{5.6.3.2}$$

Since by definition, $|\phi|^2=-\frac{1}{2}\mathrm{tr}\phi^2$ and, since $\phi=\phi^a t^a$ and $\mathrm{tr}(t^a t^b)=-2\delta^{ab}$, we indeed obtain the needed limit. Next, we construct the ratio

$$\mathbf{e}(\mathbf{x})=\frac{\phi(\mathbf{x})}{|\phi(\mathbf{x})|} \tag{5.6.3.3}$$

restricted to $|\mathbf{x}|=R$. Then, for $r>R$ the vector $\mathbf{e}(\mathbf{x})$ is continuous and defines the homotopy class $[\phi]$ in $\pi_{d-1}(S^{n-1})$.

Remark 5.6.3.1. According to [152], p. 34, the condition given by Eq. (5.6.3.2) is *sufficient* for determining of the homotopy class. The *necessary* condition is *still unknown.*

In the present case we are dealing with $n=3$ and with $\mathrm{SU}(2)/\mathrm{U}(1)\sim S^2$ while the dimensionality of space $d=3$ so that $\pi_2(S^2)=\mathbf{Z}$ and the homotopy class $[\phi]$ defining the integer N, Eq. (5.6.2.9) (or, may be, (5.6.2.10) below), called in [152] "the *monopole number.*" In fact, such a number in physics literature is identified with the *monopole charge* [148].

Just obtained results for one monopole are extendable to multi-monopoles [152] and are based on the assumption that the monopole charge N is a winding number for the homotopy group $\pi_2(S^2)=\mathbf{Z}$. This fact is compatible with earlier stated Theorem 3.5.1. Nevertheless, the dilemma of choosing between Eqs. (5.6.2.9) and (5.6.2.10) is far from over as we shall explain further below.

Before doing so, we need to discuss several related issues. For instance, we need to discuss calculation of the monopole number, Eq. (5.6.2.9). This requires us to use the fact that asymptotically $\mathrm{tr}(\mathbf{F}\wedge D\phi)=\mathrm{tr}d(\phi\cdot\mathbf{F})$ [152]. Indeed, for $r>R$ we have $D\phi=d\phi$ and, also $D\mathbf{F}=d\mathbf{F}=0$ (Bianchi identity). Therefore, $\mathrm{tr}d(\phi\cdot\mathbf{F})=\mathrm{tr}(d\phi\cdot\mathbf{F})+\mathrm{tr}(\phi\cdot d\mathbf{F})=\mathrm{tr}(\mathbf{F}\wedge D\phi)$. With this result proven, use of Stoke's theorem produces:

$$N=\frac{1}{4\pi}\mathrm{tr}\int_{\mathbf{R}^3}\mathbf{F}\wedge D\phi=\frac{1}{4\pi}\mathrm{tr}\int_{S^2_\infty}\mathbf{F}\cdot\phi. \tag{5.6.3.4}$$

To deal with the combination $\mathrm{tr}(\mathbf{F}\cdot\phi)$ requires some additional work. First, this result should be considered as asymptotic since it "lives" at the "sphere at infinity" S^2_∞. Second, we have to use the Bogomolny Eq. (5.6.2.1b) also in this limit. This requires some ingenuity. Indeed, if in this limit we would use $\phi(0,0,x^3) \to t^3$, this would cause Eq. (5.6.2.1b) to become trivial. Then Remark 5.6.3.1 leaves us some room for another choices. For instance, to comply with Eq. (5.6.3.2) it is sufficient to write

$$\phi^a = \frac{x^a}{r} h(r) \tag{5.6.3.5}$$

so that $h(r)$ is subject to the boundary condition $h(r) \to 1$ for $r \to \infty$. Evidently, the explicit form of the function $h(r)$ is determined by the Bogomolny equations. Since we are only interested in the $r \to \infty$ limit, we shall not discuss the explicit form of $h(r)$ in non-asymptotic regime. By making such a choice for ϕ^a we can still determine the homotopy class. However, *we cannot talk about spontaneous symmetry breaking any more.* Nevertheless, Eqs. (5.6.3.2) and (5.6.3.3) can be used for determination of the homotopy class. In this case it is required to identify the isotopic indices $a = 1$–3 with the spatial indices labelling x, y, z dimensions. Under such circumstances the vector $e^a(\mathbf{x}) = \frac{x^a}{r}$ with a being the spatial index. Thus, again we obtained the mapping SU(2)/U(1) $\sim S^2$ which should be understood as follows. Condition $\lim_{r\to\infty} \sup_{|\mathbf{x}|=r} |1 - |\phi|| = 0$ defines a mapping of spatial sphere $|\mathbf{x}| = R$ at infinity S^2_∞ into a sphere S^2 defined by the condition $[\phi^1]^2 + [\phi^2]^2 + [\phi^3]^2 = 1$ in isotopic space. Since SU(2) $\sim S^3$, the removal of just one point converts S^3 to $\mathbf{R}^3$. Thus, we just arrived at the situation totally analogous to that discussed for the Faddeev–Skyrme model in Sec. 3.2. In it we had the classical spin condition/constraint: $n^a n^a = 1$. Since $n^a = n^a(\mathbf{x})$, such a condition/constraint caused us to consider a map from $\mathbf{R}^3$ to S^2. One-point compactification was then used to replace $\mathbf{R}^3$ by S^3 so that the winding number for the F-S model is just the helicity $\mathcal{H}$, Eq. (3.1.13) (or (3.5.12)), normally used for description of Hopf maps from S^3 to S^2. Nevertheless, in literature on monopoles Eq. (5.6.2.9) is traditionally used instead of Eq. (5.6.2.10) causing uses of the homotopy group $\pi_2(S^2) = \mathbf{Z}$ instead of $\pi_3(S^2) = \mathbf{Z}$ used routinely for Hopf maps [105]. Thus, *we just found the exact location of the difference in opinions — about uses and needs of one-point compactification of* $\mathbf{R}^3$ — between the monopole calculations described in literature and those for the F-S model. Exactly the same situation exists in hydrodynamics of ideal fluids [68]. In view of far reaching significance of Theorem 5.5.1 for fluids, the above dilemma for fluids is positively resolved in favor of S^3. Further below we shall provide convincing arguments in favor of the same resolution for the monopole case. In the meantime, we have to finish our calculations of the monopole number, Eq. (5.6.2.9). Thus, using Eq. (5.6.2.1) we have (by ignoring the sign): $*\mathbf{F}^a = D\phi^a \rightarrow \mathbf{F}^a = *D\phi^a \rightarrow t^a d\phi^b \wedge d\phi^c$. The combination $\mathrm{tr}(\mathbf{F}\cdot\phi)$ can now be written as (again, by ignoring the overall sign) $\varepsilon_{abc}\phi^a d\phi^b \wedge d\phi^c$.[d] A moment of thought allows us to change this result into $\varepsilon_{abc}e^a(\mathbf{x})de^b(\mathbf{x})\wedge de^c(\mathbf{x})$ in view of the asymptotic boundary condition $|\phi(\mathbf{x})| = 1$. This allows us to consider the 2-form $\Omega = \frac{1}{4\pi}\varepsilon_{abc}e^a(\mathbf{x})de^b(\mathbf{x})\wedge de^c(\mathbf{x})$ identical with that introduced previously in Eq. (3.5.8). Hence, N is indeed the winding number of $\pi_2(S^2) = \mathbf{Z}$ homotopy group, as required, since the normalization factor $\frac{1}{4\pi}$ makes N integer by construction.

From algebraic topology, e.g. see the book by Bott and Tu, [153], it is known that $\pi_k(S^k) = \mathbf{Z}\ \forall k \geq 1$. Furthermore, $\pi_k(S^3) = \pi_k(S^2)\ \forall k > 2$. In particular, $\pi_3(S^3) = \pi_3(S^2) = \mathbf{Z}$. $\pi_3(S^2)$ describes the Hopf fibration as discussed already. From these results it follows that, in principle, the correspondence: $\pi_2(S^2) \rightleftarrows \pi_3(S^3) \rightleftarrows \pi_3(S^2)$ makes sense. Below we shall discuss examples from physics literature where such correspondence was actually used.

Before doing so, we need to demonstrate that Eq. (5.6.3.4) can be identified with earlier obtained Eq. (3.5.13a) for the Dirac monopole. The result for Ω we just obtained happens to be sufficient for this task. This can be immediately seen from pedagogically written review, [154], pp. 232 and 293. The monopole 2-form $\mathbf{F}$ is given by

$$\begin{aligned}\mathbf{F} &= \frac{1}{2r^3}(xdy \wedge dz - ydx \wedge dz + zdx \wedge dy)\\ &\equiv \frac{1}{2}(e^x de^y \wedge de^z - e^y de^x \wedge de^z + e^z de^x \wedge de^y),\end{aligned} \tag{5.6.3.6}$$

where $e^x = e^x(\mathbf{x}) = \frac{x}{r}$ and so on so that, for example, $de^y = \frac{dy}{r}$, etc. In [155] it is demonstrated that this is the correct representation for the 2-form on S^2.[e] Thus, up to a constant factor, the 2-form $\mathbf{F}$ coincides with Ω. The rationale for the factor 1/2 in $\mathbf{F}$ is the following. The dual of $\mathbf{F}$ (that is $*\mathbf{F}$) is the magnetic field $\mathbf{H}$. In the present case it is given by *the same* Eq. (3.5.4) as for the Abelian Dirac monopole! This means that, at least formally, one can use the same Wu–Yang-type fiber-bundle construction, e.g. read [104] once again. For the Dirac monopole such fiber bundle is

[d] Obtained result is in agreement with Eq. (1.10), p. 103 of [153] where it was obtained with help of different arguments.

[e] In fact, in this reference general result is proven for S^{n-1}, $n \geq 1$. It coincides with Eq. (3.5.13a) as required.

characterized by the first Chern class according to [154], p. 308, that is by

$$-\mathrm{tr}\int_{S^2}\frac{\mathbf{F}}{2\pi}=n. \tag{5.6.3.7}$$

This observation explains the use of the factor of 1/2 in Eq. (5.6.3.6) thus converting the combination $\frac{\mathbf{F}}{2\pi}$ into Ω. The "−" sign in front of Eq. (5.6.3.7) is compensated by taking the trace operation, as before. The integer n comes from the fact that for the Abelian Dirac monopole instead of homotopy group $\pi_2(S^2)$ it is sufficient to use the group $\pi_1(S^1)$, e.g. see both Eq. (3.4.13) and [154], p. 309. Thus, we just demonstrated that, at least for the SU(2) Lie algebra, the non-Abelian monopole calculations can be eventually reduced to those for the Abelian Dirac monopole leading to equivalence: $\pi_1(S^1)\rightleftarrows\pi_2(S^2)\rightleftarrows\pi_3(S^2)$. If this is so, it is possible to think about the equivalence $\pi_3(S^2)\rightleftarrows\pi_3(S^3)$ in terms of mappings of 3-sphere to 3-sphere with subsequent Hopf-type reduction to the base space S^2. Such a procedure was indeed recently discussed in physics literature [156]. Mathematically rigorous proof of this fact can be found in [111].

Exercise 5.6.3.1. Use spherical coordinates in order to demonstrate that: (a) on the surface of a sphere S^2 of arbitrary radius r the 2-form $\mathbf{F}$ can be written as $\mathbf{F}=\frac{1}{2}\sin\theta d\theta\wedge d\varphi$; (b) since $\mathbf{F}=dA$ demonstrate that the vector potential $A_\pm$ for the Dirac string given by ([154], p. 232)

$$A_\pm=\frac{1}{2r}\frac{1}{z\pm r}(xdy-ydx) \tag{5.6.3.8}$$

on S^2 can be rewritten as

$$A_\pm=\frac{1}{2}(\pm1-\cos\theta)d\varphi \tag{5.6.3.9}$$

so that, indeed, $\mathbf{F}=dA_\pm$.

Remark 5.6.3.2. As it is demonstrated in [104] the 2-form $\mathbf{F}=\sin\theta d\theta\wedge d\varphi$ is obtainable as well directly from the 1-form, Eq. (5.5.2), defined on S^3. In [104] this 1-form was obtained in spherical coordinates on S^3 with the result:

$$A=-(d\psi+\cos\theta d\varphi). \tag{5.6.3.10}$$

The local section $\psi=\pm\varphi$ converts this 1-form into that in Eq. (5.6.3.9). Since the 2-form $\mathbf{F}$ is not sensitive to such a procedure it can be made both ways: that is by using results living either on S^3 or on S^2. This observation brings us back to the helicity $\mathcal{H}$. We would now like to argue that such outcome is not coincidental.

5.6.4 *Hydrodynamics and the Faddeev–Skyrme model*

This is so because the previously discussed inequalities for ideal fluids, Eq. (3.2.21), and for the Faddeev–Skyrme model, Eq. (3.2.22), are exactly the same! From Sec. 3.2, it follows that the F-S functional is nothing but reformulated in terms of cleverly chosen variables pure SU(2) Yang–Mills static action functional. These functionals must be minimized and the results of minimization should be interpreted either in terms of Dirac-like monopoles (e.g. see Sec. 3.5) which we used for fluids/superconductors or in terms of knots/links for the F-S models (e.g. see Sec. 3.2). The question naturally arises: How are monopoles related to knots and links? To answer this question, we need to compare inequalities (3.1.12) and (5.6.2.8). In the case if $|N|$ in Eq. (5.6.28) can be replaced by $|\mathcal{H}|$, such a comparison is assured. In physics literature, while using arguments very different from those in Sec. 3.5, authors of [157] came to the same conclusion. That is: *In the F-S model the linked magnetic tubes play the role of monopoles.*

As we already noticed in Sec. 3.2, such tubes in addition to being linked can also be knotted, e.g. read [83–85]. While this issue is discussed from another point of view in Chap. 9, our immediate task is to finish our discussion of the Abelian Chern–Simons functional.

Remark 5.6.4.1. [157] also discusses various applications of ideas and methods of monopole physics to various problems in condensed matter physics and, reciprocally, the role of ideas and methods of condensed matter physics in physics of monopoles. Other condensed matter applications of monopoles are discussed, for example, in [158].

Exercise 5.6.4.1. Demonstrate that: (a) the above monopole discussion *is not* limited to su(2) Lie algebra. Hint: Read [105, 119], (b) the F-S model can also be extended to su(N) with subsequent variational treatment analogous to su(2). Hint: Read [159, 160].

Remark 5.6.4.2. Usefulness of the F-S model comes from its direct link with the Skyrme model — the most reliable and well-studied model of hadron and nuclear physics [161–163].

5.6.5 *Helicity and Monopoles*

Being armed with the already obtained results, we are now in a position of investigating the helicity $\mathcal{H}$ further. In particular, we notice that the inequality, Eq. (3.1.12), can be obtained without use of the Woltjer's theorem, it is surely in agreement with it. Because of this, it is possible to re-obtain the same results. That is, instead of proving the inequality, Eq. (3.1.12), we would like to consider a variation of the functional $\mathcal{F}[\alpha]$ given by

$$\mathcal{F}[\alpha] = E[\alpha] + \nu\mathcal{H}[\alpha], \tag{5.6.5.1}$$

where $\mathcal{H}[\alpha] = \int_M \alpha \wedge d\alpha$. Such a variation yields the Beltrami Eq. (5.5.15) as required. Evidently, by rewriting already known results using different formalism we are not adding much to our understanding of these results. To obtain new results, following [68] we shall discuss other ways of interpreting the inequality, Eq. (3.1.12). In particular, let us consider the eigenvalue problem of the type

$$\mathrm{curl}^{-1}\mathbf{v} = \lambda\mathbf{v}. \tag{5.6.5.2}$$

Clearly, this problem has a solution if the force-free Eq. (1.3.1) has solution. The conditions assuring this possibility were investigated in full generality in [164] while in [89, 90], based on results of [164], the eigenvalue problem for Eq. (5.6.5.2) was studied in detail for domains of various topology. When solution of the eigenvalue problem described by Eq. (5.6.5.2) does exist, the eigenvalues can be arranged in accordance with their magnitudes. In particular, it is expected that for 3-manifolds there always exists an eigenvalue $\check{\lambda}$ of largest magnitude $|\check{\lambda}|$. If this is the case, an inequality, Eq. (3.1.12), can be equivalently rewritten as

$$E[\mathbf{v}] \geq \frac{1}{|\check{\lambda}|}\,|\mathcal{H}(\mathbf{v})| \tag{5.6.5.3}$$

so that the minimum for $E[\mathbf{v}]$ is achieved for $\check{\mathbf{v}}$ associated with $\check{\lambda}$. Having in mind physical applications, we would now like to illustrate this general result by using a specific example. For this purpose we choose the Reeb vector field defined by Eq. (5.5.8). From this equation we obtain the first order partial differential equation

$$[x_1\partial_{y_1} - y_1\partial_{x_1} + x_2\partial_{y_2} - y_2\partial_{x_2}]\mathcal{F} = 0 \tag{5.6.5.4a}$$

which is equivalent to the following system of ordinary differential equations

$$\dot{y}_1 = x_1, \dot{x}_1 = -y_1; \quad \dot{y}_2 = x_2, \dot{x}_2 = -y_2. \tag{5.6.5.4b}$$

These are just equations for two uncoupled harmonic oscillators whose solutions describe (linked) circles associated with Hopf fibration [143]. Since the components of the velocity vector $\mathbf{v} = \{x_1, -y_1, x_2, -y_2\}$ are subject to the constraint: $x_1^2+y_1^2+x_2^2+y_2^2=1$, we obtain for the helicity $\mathcal{H}$ the following result:

$$\mathcal{H}(\mathbf{v}) = \int_{S^3} \mathbf{v}\cdot \mathrm{curl}^{-1}\mathbf{v}\, dV = \frac{1}{2}\int_{S^3} \mathbf{v}\cdot\mathbf{v} dV = \frac{1}{2}\int_{S^3} dV = \frac{2\pi^2}{2} = \pi^2. \tag{5.6.5.5}$$

In arriving at this result we took into account that on S^3 the largest eigenvalue of $\mathrm{curl}^{-1}\mathbf{v}$ is $1/2$ and that the volume of the 3-sphere S^3 is $2\pi^2$ [68].

Exercise 5.6.5.1. Demonstrate that, indeed, this is the case. Hint: Use [164–166] as well as [68], Chap. 2, as the starting point.

Remark 5.6.5.1. The obtained result, Eq. (5.6.5.5), requires some explanation. Proposition 1.20 of [68], p. 127, states that $\mathcal{H}(\mathbf{v})$ is an integer. Unfortunately, instead of proving this result the authors only provide a hint. Below, we shall demonstrate how this can actually be demonstrated. This proof also provides an opportunity to use the already obtained results in a new setting. The proof is inspired by the book of Bott and Tu, [153].

Let us begin with the Dirac monopole 2-form $\mathbf{F}$, Eq. (5.6.3.6), defined on the 2-sphere S^2 of unit radius and replace the normalization factor $1/2$ by $1/4\pi$. On S^2, because of the constraint $x^2+y^2+z^2=1$, it is possible to rewrite this 2-form as follows:

$$\mathbf{F} = \frac{1}{4\pi}\frac{dx\wedge dy}{z}. \tag{5.6.5.6}$$

Use of complex variables allows one to rewrite this form as

$$\mathbf{F} = \frac{i}{2\pi}\frac{dz\wedge d\bar{z}}{1+|z|^2}. \tag{5.6.5.7}$$

Since the complex variable z can be represented as z_0/z_1 (the Hopf map) so that complex variables z_0 and z_1 are satisfying the equation

$$|z_0|^2+|z_1|^2=1, \tag{5.6.5.8}$$

describing S^3, it becomes possible to lift $\mathbf{F}$ from S^2 to S^3 so that on S^3 it acquires the following form:

$$\mathbf{F} = \frac{i}{2\pi}(z_1 dz_0 - z_0 dz_1)\wedge(\bar{z}_1 d\bar{z}_0 - \bar{z}_0 d\bar{z}_1). \tag{5.6.5.9}$$

If $z_0 = x_1+iy_1$ and $z_1 = x_2+iy_2$, then, when written in these new variables, the 2-form $\mathbf{F}$ acquires the following look:

$$\mathbf{F} = \frac{1}{\pi}(dx_1\wedge dy_1 + dx_2\wedge dy_2). \tag{5.6.5.10}$$

But we already encountered such a 2-form! Up to a constant $\mathbf{F} \simeq \omega$, e.g. see Eq. (5.5.7). At the same place we also defined the radial vector field $Y : Y = \frac{1}{2}(x_1\partial_{x_1} + y_1\partial_{y_1} + x_2\partial_{x_2} + y_2\partial_{y_2})$ so that $\imath_Y\omega = \frac{1}{2}\alpha$, where the 1-form α "living on S^3" is given by Eq. (5.5.2). Bott and Tu use instead of α the following 1-form

$$\tilde{\alpha} = \frac{1}{\pi}(x_1dy_1 + x_2dy_2). \tag{5.6.5.11}$$

Evidently, both $\frac{1}{2\pi}\alpha$ and $\tilde{\alpha}$ produce $\mathbf{F}$ upon action of the operator d on them. The reason for this is given by Theorem 5.5.1. Indeed, the form $\mathbf{F}$ is working in both $\mathbf{R}^4$ and S^3 [139]. Furthermore, for $\mathbf{R}^4$ [153], it is given by

$$\mathbf{F} = \frac{1}{\pi}\frac{dx_1 \wedge dy_1 + dx_2 \wedge dy_2}{(x_1^2 + y_1^2 + x_2^2 + y_2^2)^2}.$$

coinciding with Eq. (5.6.5.10) when restricted to S^3. Thus, if we use $\tilde{\alpha}$ given by Eq. (5.6.5.11) and take into account that $d\tilde{\alpha} = \mathbf{F}$ defined by Eq. (5.6.5.9) and use both results in the expression for the helicity, Eq. (3.5.12), we obtain,

$$\begin{aligned}\mathcal{H} &= \int_{S^3} \tilde{\alpha} \wedge d\tilde{\alpha} = \int_{S^3} \tilde{\alpha} \wedge \mathbf{F} \\ &= \frac{1}{\pi^2}\int_{S^3} [x_1dy_1 \wedge dx_2 \wedge dy_2 + x_2dx_1 \wedge dy_1 \wedge dy_2]. \end{aligned} \tag{5.6.5.12}$$

Comparison of this result with Eq. (5.6.5.5) leads to the cancellation of π^2 factor thus making $\mathcal{H}$ an integer as required.

Next, from calculations we just made and those done in the preceding subsection we realize that (up to a constant factor) the combination $\tilde{\alpha} \wedge d\tilde{\alpha} \equiv \mu$ is the volume 3-form. Using Eq. (5.5.15b) and following [68] it becomes possible now to rewrite the helicity as follows:

$$\begin{aligned}\mathcal{H} &= \int_{S^3} \alpha \wedge d\alpha = \int_{S^3} \alpha \wedge f^*\Omega = \int_{S^3} \alpha \wedge \imath_X\mu \\ &= \int_{S^3} \alpha(X) \wedge \mu = \int_{S^3} \mathbf{v} \cdot \mathrm{curl}^{-1}\mathbf{v}\, dV \end{aligned} \tag{5.6.5.13}$$

with X being the Reeb vector field. From here, we obtain the remarkable corollaries:

$$\text{(a) } \alpha(X) = \mathbf{v} \cdot \mathrm{curl}^{-1}\mathbf{v} \quad \text{and} \quad \text{(b) } f^*\Omega = \imath_X\mu. \tag{5.6.5.14}$$

The first one is connecting quantum mechanics-like expressions on the R.H.S. with the purely mechanical L.H.S. "scalar product". Broad generalization of such type of ideas can be found, for example, in [167]. The second one is connecting the 2-form (obtained as a pullback f^* — from S^2 to S^3 — with the contact volume form μ "living" on S^3. Furthermore, since $d\alpha = \imath_X \mu$ knowledge of the 1-form α on S^3, Eq. (5.5.2), determines $\mathcal{H}$ since $\alpha \wedge d\alpha \neq 0$ determines the contact structure on 3-manifolds, e.g. see Eq. (5.2.4) and comments which follow this equation.

5.6.6 *Some Comments on Dyons and Their Classical Analogs*

Before finishing this chapter we still need to discuss several topics. For instance, we would like to make some comments on dyons, e.g. see Remark 3.5.5. These are hypothetical particles suggested by Schwinger which are capable of carrying both the electric and magnetic charges simultaneously [105]. We would like to argue that should these particle exist, this would require us to redo the whole theory of foliations. This is so for the following reasons.

In Eq. (5.6.3.3) we introduced the unit vector $\mathbf{e}(\mathbf{x})$ allowing us to define the homotopy class(es).[f] In fact, Eq. (5.6.3.3) defines the Gauss map from $\mathbf{R}^3$ (where $\mathbf{x}$ "lives") to S^2. Surely, we can either compactify $\mathbf{R}^3$ to S^3 or choose the sphere at infinity S^2_∞ as it was just discussed. This time, however, we shall fix our attention on the fact that the vector field $\mathbf{e}(\mathbf{x})$ is well-defined on the sphere as long as $\phi(\mathbf{x}) \neq 0$. Points at which $\phi(\mathbf{x}) = 0$ are called *singular.* The Poincaré–Hopf index theorem connects topology of the manifold with indices of vector field "living" on it. E.g. Euler's characteristic of two-dimensional manifolds is equal to the sum of indices of vector field "living" on such manifold. Notice that this theorem is valid for manifolds of *any dimensionality* and is independent of the choice of vector field on it [168]. The index of the vector field is the same as the degree of mapping (defined in Sec. 3.5) or as the winding number, e.g. Eq. (5.6.3.4) [111, 169]. The indices can and should be identified with charges, both electric and magnetic. If there are dyons in Nature, then on the same manifold we shall have two types of vector fields — each having their own indices in such a way that the Poincaré–Hopf index theorem is obeyed in both cases. Furthermore, in both cases the location of singular points associated with these indices must coincide. Accordingly, one and the same singular point is going to be characterized in general by two indices. Clearly, such a situation does not have the Abelian analogue. It does not fit the original Dirac's idea about the electric–magnetic charge duality and it is incorrect mathematically.

[f] See also discussion related to Eq. (3.5.8).

Indeed, following [134, 170] it is known that:

(a) Any compact smooth manifold $\mathcal{M}$ can be smoothly embedded into Euclidean space $\mathbf{R}^k$ for sufficiently large k.
(b) If in k-dimensional Euclidean space such a manifold is realized as, say, $(k-1)$-dimensional hypersurface, e.g. $\mathrm{F}(x^1, \ldots, x^k) = 0$ or, *locally* as $x^i = x^i(y^1, \ldots, y^{k-1}), i = 1 - k$, then the $(k-1)$-dimensional form for the Gaussian curvature K at the point $y^1, \ldots, y^{k-1} \in \mathcal{M}$ is defined by

$$K d\sigma \equiv K\sqrt{g} dy^1 \wedge \cdots \wedge dy^{k-1}.$$

(c) If we construct the Gauss map $\psi : \mathcal{M} \to S^{k-1}$ via $n^i(\mathbf{y}) = \frac{x^i(\mathbf{y})}{|\mathbf{x}(\mathbf{y})|}$, then it is possible to prove the following:

Theorem 5.6.6.1. *For any* $k \geq 2$

$$K d\sigma = \psi^*(\Omega),$$

where Ω *is defined by Eq.* (3.5.8) (*in which one should put* $n = k$).

Proof. Read pp. 253–254 of [134]. □

Corollary.

$$\int_{\mathcal{M}} K d\sigma = \gamma_n \deg \psi \mid_{\mathcal{M}}. \tag{5.6.6.1}$$

Proof. See Eqs. (3.5.8)–(3.5.11). □

Remark 5.6.6.1. In the monopole example discussed above we can choose $k = 4$ so that the hypersurface is S^3 or we can choose $k = 3$ so that the hypersurface is S^2. Since K is scalar for a given point of hypersurface, it is defined uniquely. Thus, it is impossible to have dyons whose magnetic charge differs from the electric charge and, even if they could be of the the same magnitude, there will be no Abelian analog of dyons as we shall demonstrate now.

Although the Dirac's idea about the electric–magnetic duality seems very reasonable, there is some asymmetry in treatments of magnetic and electric charges in physics literature [105]. For instance, Dirac monopoles should be introduced by hand into Abelian electrodynamics, they are attached to strings which sometimes are used in physical arguments and sometimes are not used. Nothing of this kind seemingly exist for electric charges. They do exist in Nature but, apparently, are not associated with the Dirac-like strings so that the Wu–Yang-type fiber bundle construction [101] for electric charges seems unnecessary. This is mathematically illogical.

Following [111, 144], we now demonstrate that this traditional thinking is actually incorrect and that in the Abelian electrodynamics there is a complete symmetry/duality between the electric and magnetic charges in the spirit of Dirac but technically slightly different.

We begin with the 2-form $*\mathcal{E}$ for the electrostatic field

$$*\mathcal{E} = \imath_{\mathbf{E}} vol^3 = \sqrt{g}(E^1 dx^2 \wedge dx^3 + E^2 dx^3 \wedge dx^1 + E^3 dx^1 \wedge dx^2]. \tag{5.6.6.2}$$

Using it, the Gauss law can be written as

$$d * \mathcal{E} = 4\pi\sigma^3, \tag{5.6.6.3}$$

where the charge σ^3 is given by $\sigma^3 = \rho vol^3$. Analogously, with help of 1-form $*\mathcal{B}$ given by

$$*\mathcal{B} = B_1 dx^1 + B_2 dx^2 + B_3 dx^3 \tag{5.6.6.4}$$

the Ampere law can be written as

$$d * \mathcal{B} = 4\pi j^2, \tag{5.6.6.5}$$

where the current 2-form j^2 is given by

$$j^2 = \rho[v^1 dx^2 \wedge dx^3 + v^2 dx^3 \wedge dx^1 + v^3 dx^1 \wedge dx^2]. \tag{5.6.6.6}$$

Using these results the continuity equation is written as

$$\frac{\partial \sigma^3}{\partial t} + dj^2 = 0. \tag{5.6.6.7}$$

In spherical coordinates defined in $\mathbf{R}^3 \backslash \mathbf{0}$ the vector $\mathbf{E}$ is given by $\mathbf{E} = \frac{e}{r^2}\frac{\partial}{\partial r}$, while the vol^3 by

$$vol^3 = r^2 \sin\theta dr \wedge d\theta \wedge d\varphi.$$

Therefore,

$$*\mathcal{E} = \imath_{\mathbf{E}} vol^3 = \imath\left(\frac{e}{r^2}\frac{\partial}{\partial r}\right) r^2 \sin\theta dr \wedge d\theta \wedge d\varphi = e \sin\theta d\theta \wedge d\varphi. \tag{5.6.6.8}$$

Thus if, for example, we take $\Psi = -e\cos\theta d\varphi$ as potential, we obtain $*\mathcal{E} = d\Psi$. Since the spherical coordinates misbehave not only at the origin but also at $\theta = 0$ and π, the potential Ψ is defined everywhere, except z-axis. Since the potential is always defined up to a constant it is possible to choose instead $\Psi = e(1 - \cos\theta)d\varphi$. Since $1 - \cos\theta = 0$ for $\theta = 0$ this newly defined potential is well-behaved, except at the negative z-axis. Here we are having the Dirac-like string for the Coulombic charge! Using Eq. (5.6.6.2) we obtain an analog of the monopole Eq. (3.5.13a), that is

$$\oint_{S^2} *\mathcal{E} = 4\pi e. \tag{5.6.6.9}$$

Furthermore, in exact analogy with Eqs. (3.5.2)–(3.5.6) using Eq. (5.6.6.2) we can write as well

$$\oint d\mathbf{S} \cdot \mathbf{E} = 4\pi e$$

as anticipated. However, we can also rewrite the same result using Eqs. (3.5.8), (3.5.9). This means that we are back with the monopole form $\Omega = \frac{1}{4\pi}\varepsilon_{abc}\ e^a(\mathbf{x})de^b(\mathbf{x}) \wedge de^c(\mathbf{x})$, in accord with [111], so that the Coulombic charges can and should be treated in exactly the same way as Dirac monopoles! Incidentally, for magnetic monopoles the above steps can be repeated word-for-word. Details are given in [144]. Finally, using Eq. (5.6.6.5) the Ampere law can be rewritten in its more traditional form

$$\oint_C \mathbf{B} \cdot d\mathbf{l} = 4\pi \int_S d\mathbf{S} \cdot \mathbf{j}. \tag{5.6.6.10}$$

Here C is some closed contour which, without loss of generality, can be made planar. Then, S is the area of a circle enclosed by C and $\mathbf{j}$ is the current density passing through S. Let the current $\mathbf{j}$ be flowing through some wire which itself makes a contour Γ. This is possible when the current is superconducting [23]. Then the situation becomes completely symmetric with respect to contours C and Γ. In particular, the magnetic field $\mathbf{B}$ (or $\mathbf{H}$) defined with help of the potential $\mathbf{A}$ defined before Eq. (3.1.8) leads to the well-known and discussed Biot–Savart law [23, 90, 111]:

$$\mathbf{B}(\mathbf{r}) = \oint_\Gamma d\sigma' \left[\mathbf{v}(\sigma') \times \frac{(\mathbf{r}-\mathbf{r}')}{|\mathbf{r}-\mathbf{r}'|^3}\right] \tag{5.6.6.11a}$$

that is to the relation:

$$\mathbf{B} = 4\pi\ \mathrm{curl}^{-1}\mathbf{j}. \tag{5.6.6.11b}$$

Using this result in Eq. (5.6.6.10) leads to the expression for linking number, Eq. (3.1.11a), and, hence, to the Hopf (or Abelian) Chern–Simons invariant $\mathcal{H}[\alpha] = \int_M \alpha \wedge d\alpha$. From here, based on arguments of this subsection, the connection with Dirac monopoles is assured with interpretation given in Sec. 3.5.

Remark 5.6.6.2. In both Eqs. (5.6.6.3) and (5.6.6.5) we used Maxwell's equations with external sources when we had discussed monopoles coming from the non-Abelian Y-M fields there were no external sources. The situation in this case strongly resembles that which was discussed in Chap. 4 where we discussed the topologically massive gauge theories. In both cases the masses were generated with help of the force-free conditions on fields, e.g. see the duality Eq. (4.3) in the Abelian case and Bogomolny Eq. (5.6.2.1) in the non-Abelian case. These results do not remove entirely the following issue. If Eqs. (5.6.6.3) and (5.6.6.5) of Abelian time-independent electrodynamics involve external sources and if the non-Abelian Yang–Mills fields are just extensions of classical electrodynamics (electrostatics, magnetostatics), there should be a way to write the non-Abelian version of Eqs. (5.6.6.3) and (5.6.6.5) and to see if it is possible to obtain the non-Abelian monopoles, both electric and magnetic, by purely classical means. Evidently, if we shall follow the protocol of Yang and Wu, [101], there should be no problem for doing this. The problem will be in extending such results to multimonopoles and relating them to results obtained by other means, e.g. those discussed in Secs. 5.6.2 and 5.6.3. At the classical level, the non-Abelian Yang–Mills static "electrodynamics" was discussed in the seminal paper by Jackiw, Jacobs and Rebbi, [171]. This paper prompted publication of a large number of papers whose content is impossible to summarize in this remark. The problems with treatment of the extended bodies in Yang–Mills theory are as difficult as analogous treatments of extended bodies in general relativity initiated much earlier by Papapetrou, [172]. We bring this information to the spotlight with hope that our readers may be able to find better, more elegant, solutions/(resolutions of) to the existing difficulties with description of extended bodies in both theories since they are interrelated [146]. Some examples of resolutions of these difficulties can be found in [181], [443] and [444].

Chapter 6

Sub-Riemannian Geometry, Heisenberg Manifolds and Quantum Mechanics of Landau Levels

6.1 Motivation

In the previous chapter we provided a detailed study of various uses of helicity $\mathcal{H}[\alpha] = \int_M \alpha \wedge d\alpha$. These studies were facilitated by the fact that the contact volume form $\alpha \wedge d\alpha$ is nowhere zero, e.g. see Eq. (5.2.4). This property defines the contact structure on 3-manifold M. Even though the standard contact 1-form α on $\mathbf{R}^3$ is conventionally chosen as $\alpha = dz + xdy$ it is not the most convenient form for physical applications. We can use the contactomorphic transformations, see Eq. (5.4.1), to bring α to the form more convenient for study. This study, in part, is also needed for finishing, in Chap. 7 the proof (alternative to that given in [78]) relating the TGB phases in liquid crystals to Abrikosov vortex lattices.

To begin, we need to bring standard 1-form $\alpha = dz + xdy$ to the 1-form $\alpha = -\frac{1}{4}dt + \frac{1}{2}(ydx - xdy)$ used in [167]. This can be achieved by means of the transformation: $(x, y, z) \rightarrow (x, y, \frac{1}{2}xy - z)$ with subsequent replacement of z by $t/4$. Such 1-form fixes the form of two tangent vectors X_i. They are given by

$$X_1 = \partial_x + 2y\partial_t \quad \text{and} \quad X_2 = \partial_y - 2x\partial_t \tag{6.1.1}$$

since the equations $\alpha(X_i) = 0$, $i = 1, 2$ are defining the 2-planes. The Reeb vector R is given now by

$$[X_1, X_2] = -4\partial_t \equiv R \tag{6.1.2a}$$

so that the Reeb conditions $\alpha(R) = 1$ and $\imath_R d\alpha = \imath_R(dy \wedge dx) = 0$ are satisfied. The commutator relationship, Eq. (6.1.2a), looks similar to the basic Heisenberg's quantization condition

$$[q, p] = i\hbar\mathbf{I}. \tag{6.1.2b}$$

In fact, it can be brought to exactly the same form if we work in system of units for which $\hbar = 1$ in Eq. (6.1.2a) and use the space of wave functions of the type $\psi = \varphi(x, y)\exp(-\frac{i}{4}t)$. That is to say, in both the traditional/conventional and the contact-type quantum mechanics time is not quantized. In view of results just described and the fact that $\alpha \wedge d\alpha(X_1,X_2,R) \neq 0$, e.g. read [167], pp. 182 and 191, the contact manifolds with such properties are equivalently called as *Heisenberg manifolds. That is quantum mechanics can be looked upon as applied branch of contact geometry and topology.*

Such a conclusion is not totally surprising in the light of already noticed connections between, say, descriptions of (magneto)hydrodynamics of ideal fluids and that of superconductivity or, as in Sec. 1.2, between mechanics and hydrodynamics, etc. Based on results just presented, it should be clear that *the essence of conventional quantum mechanics is captured by three equations:*

$$\alpha(X_i) = 0, \quad i = 1, 2$$

and

$$\alpha \wedge d\alpha(X_1, X_2, R) \neq 0 \textit{ all originating in contact geometry!}$$

In addition to quantum mechanical, these equations also have mechanical and geometrical meaning. In quantum mechanics the commutator, Eq. (6.1.2b), *is postulated without any reference to specific dynamical system.* The same is true in the present case, except that the major commutator relationship is no longer postulated. It is derivable from contact geometry as we have just explained. Furthermore, as we explained in [173], the commutator relation, Eq. (6.1.2b), is equivalent to the Bohr–Sommerfeld (B-S) quantization condition $\oint \mathbf{p} \cdot d\mathbf{x} = nh$. It was derived from this condition by Heisenberg in his seminal paper on quantum mechanics written in 1925. Below, we shall demonstrate that this interrelationship also exists in the context of sub-Riemannian geometry.

6.2 The Benchmark Example

In order to use either the commutator or the B-S condition we have to think about some dynamical system where such relations can be used.

The simplest system relevant to our discussion is given by the following Lagrangian

$$\mathcal{L} = \frac{1}{2}(\dot{x}^2 + \dot{y}^2) + \frac{\lambda}{2}(\dot{t} - A_x\dot{x} - A_y\dot{y}) \tag{6.2.1}$$

describing a particle of unit mass moving in the plane $\mathbf{R}^2$ perpendicular to the constant magnetic field $\mathbf{H}$ whose vector potential $\mathbf{A} = \{A_x, A_y, 0\} = \{2y, -2x, 0\}$. The Euler–Lagrange equation for t yields $\lambda =$ const. while two other equations are given by [174]

$$\begin{aligned} \ddot{x} &= \frac{\lambda}{2}\left(\frac{\partial A_y}{\partial x} - \frac{\partial A_x}{\partial y}\right)\dot{y}, \\ \ddot{y} &= \frac{\lambda}{2}\left(\frac{\partial A_x}{\partial y} - \frac{\partial A_y}{\partial x}\right)\dot{x}. \end{aligned} \tag{6.2.2a}$$

Since $\mathbf{H} = \nabla \times \mathbf{A}$, the above two equations reduce to

$$\begin{aligned} \ddot{x} &= \frac{\lambda}{2}\mathrm{H}_z\dot{y}, \\ \ddot{y} &= -\frac{\lambda}{2}\mathrm{H}_z\dot{x} \end{aligned} \tag{6.2.2b}$$

or, written in standard form, they are given by

$$\frac{d\mathbf{v}}{dt} = \frac{\lambda}{2}\mathbf{H} \times \mathbf{v}. \tag{6.2.2c}$$

The motion is planar because of the presence of nonholonomic constraint $\dot{t} = 2(\dot{x}y - x\dot{y})$ in the Lagrangian, Eq. (6.2.1). The definition of such a constraint compatible with the results already explained is the following.

Let α be a non-integrable 1-form $\alpha = dt - 2(ydx - xdy)$ and let x, y and t depend on parameter s. Then, the *nonholonomic constraint* condition is given by $\dot{t} - 2(\dot{x}y - x\dot{y})$ or, more generally,

$$\alpha(\dot{\mathbf{r}}) = \sum_{i=1}^{3} \alpha_i dx_i(\dot{x}_i) = \sum_{i=1}^{3} \alpha_i \dot{x}_i = 0. \tag{6.2.3}$$

Should the respective 1-form be integrable (e.g. see Eq. (5.2.1)), then the constraints on velocities would become holonomic.

The transition from the Lagrangian to Hamiltonian formalism in the case of nonholonomic constraints is typically more involved than that described in traditionally written textbooks for physicists. But in the present case, when $\mathcal{L}$ is given by Eq. (6.2.1), it is elementary so that it can be skipped. Details can be found in [174], p. 355. The easiest way to arrive at the correct result for the Hamiltonian is through use of quantum mechanics. Indeed, using Eqs. (6.1.1) we obtain at once

$$\mathcal{H} = \frac{1}{2}(p_x + 2yp_t)^2 + \frac{1}{2}(p_y - 2xp_t)^2. \tag{6.2.4}$$

Here p_t is a constant which can be identified with λ, i.e. with the electric charge, in accord with Eqs. (6.2.2). Thus, using Eqs. (6.6.1) we can proceed with the standard quantum mechanical treatment of the Schrödinger-type problem with Hamiltonian $\mathcal{H}$. Since this is a well-studied problem, there is no need to reproduce it here. Instead, here we are interested to find out if quantum mechanical results can be obtained using classical mechanics defined in the sub-Riemannian space. For this purpose we shall temporarily keep p_t instead of λ in $\mathcal{H}$. Then, following [167, 175], we rewrite the Hamiltonian $\mathcal{H}$ in equivalent form as follows:

$$\mathcal{H} = \frac{1}{2}\sum_{i,j=1}^{n} h^{ij}(\mathbf{x})p_i p_j, \tag{6.2.5}$$

where the matrix $h^{ij}(\mathbf{x})$ can be read off from Eq. (6.2.4) and is given by

$$h^{ij}(\mathbf{x}) = \begin{pmatrix} 1 & 0 & 2y \\ 0 & 1 & -2x \\ 2y & -2x & 4(x^2+y^2) \end{pmatrix}. \tag{6.2.6}$$

It is essential that $\det[h^{ij}(\mathbf{x})] = 0$ for any x and y. This means that such a matrix does not have an inverse. If h^{ij} is playing a role of the metric tensor, then the fact that $\det[h^{ij}(\mathbf{x})] = 0$ makes it impossible to think about uses of Riemannian geometry for this dynamical problem. This fact is in contradistinction with the interpretation of mechanics given by Maupertuis' principle nicely explained in the book by Arnol'd [136]. For the sake of uninterrupted reading we reproduce this principle here. It states that instead of using, say, the action functional based on full Lagrangian, for the energy-conserving system it is sufficient to use the truncated action functional

$$S[\gamma] = \int_\gamma \mathbf{p}\cdot d\mathbf{q}, \tag{6.2.7}$$

where $\mathbf{q}(\tau) = \gamma(\tau)$ is the curve connecting two pre-assigned points $\mathbf{q}_0$ and $\mathbf{q}_1$. Solution(s) of the variational problem for $S[\gamma]$ yields extremal curve(s) $\gamma(\tau)$ between these points. In fact, these curves are geodesics on some Riemannian manifold. To see this, we notice that, since $\mathcal{H} = K + U$ and $\mathcal{L} = K - U$, for the fixed total energy h the kinetic energy K is given by: $2K = \frac{\partial \mathcal{L}}{\partial \dot{\mathbf{q}}} \cdot \dot{\mathbf{q}} = \sum_{i,j} h^{ij}(\mathbf{q})\dot{q}_i\dot{q}_j = 2(h-U) \equiv (\frac{ds}{d\tau})^2$. Next, by redefining time as $d\tau = \frac{ds}{\sqrt{2(h-U)}}$ we obtain:

$$\int_\gamma \mathbf{p} \cdot d\mathbf{q} = \int_\gamma \frac{\partial \mathcal{L}}{\partial \dot{\mathbf{q}}} \cdot \dot{\mathbf{q}} d\tau = \int_\gamma \sqrt{2(h-U)} ds \equiv \sqrt{2} \int_\gamma d\rho. \tag{6.2.8}$$

Thus, the original Riemannian metric $h^{ij}(\mathbf{q})$ in this case is replaced by $\sqrt{2(h-U)}h^{ij}(\mathbf{q})$ which is still a Riemannian metric since the determinant of such new metric is nonzero. This means that it is possible to use it to raise and lower the indices (e.g. $p_i = \sum_j h_{ij}\dot{q}^j$, etc.) — an impossible task for the matrix given by Eq. (6.2.6). The last relation is used essentially for demonstration of the fact that $\gamma(\tau)$ is indeed a geodesic.

Remark 6.2.1. When mechanics is reformulated with the help of Maupertuis' principle (developed, incidentally, between 1741 and 1746[a]) equations of classical mechanics become geodesics on some Riemannian manifold whose properties are determined by the system in question. If we replace the Euclidean signature by the Lorentzian, such geodesics become those known in general relativity (developed only in 1916!). Noticed in Sec. 1.2 correspondence between the hydrodynamics of ideal fluids and classical mechanics becomes more transparent in the Maupertuis' formalism since in both cases one is dealing with the motion along the geodesics [136]. In general relativity and in fluid dynamics it is necessary to deal with observables invariant under diffeomorphisms, e.g. read [68], pp. 255 and 256. Apparently, this feature makes the description of both relativity and fluid dynamics more complicated than many-body classical mechanics. This is not quite the case though. In many instances, e.g. when the estimates like those given by Eq. (3.1.12) are available, the situation can be brought under control, that is reduced to that known in classical mechanics, with help of results from knot/link theory. The same remains correct for gravity if it is formulated as loop quantum gravity (LQG) [176].

Remark 6.2.2. In the case if geodesics are closed, Eq. (6.2.8) becomes eligible for the B-S quantization rule. In fact, such a quantization takes place almost automatically. This can be seen from the fact that the reduced phase

[a]E.g. see the entry "Pierre Louis Maupertuis" in Wikipedia.

space of exactly integrable systems is typically made of tori or collection of tori as explained in [136]. This is correct, for example, for the configuration space of rigid body rotating without external forces around fixed point (located at its center of mass). For such a system it is a torus. The geodesics on such a torus may or may not be closed. But when they are closed, they form torus knots. Thus, quantization in such a case is caused by the existence of torus knots. Evidently, different energy levels may correspond to different torus-type knots. This will be illustrated below.

Exercise 6.2.2.1. Uses of knots and links in atomic/molecular physics was initiated by Tait, Kelvin and Maxwell. However, their approach to problems of atomic/molecular physics was not taken seriously enough so that it was abandoned when quantum mechanics was developed (starting from Bohr's atomic model). Nevertheless, we recommend to investigate if the treatment of Hydrogen atom with help of the Maupertuis principle leads to correct quantum energy spectrum (this time explained in terms of knotted geodesics). Try to do the same for celestial mechanics, e.g. consider the motion of major planets around our Sun ignoring interplanetary interactions. Hint: Read our work, [177], and read the rest of this book, especially, Appendix D, Sec. D.2.18.

Study of dynamics of rotating bodies is very useful since description of dynamics of almost all known exactly integrable systems of classical mechanics can be traced back to dynamics of rigid bodies [178]. Because of already discussed mechanics–hydrodynamics correspondence, these results can also be applied to the description of dynamics of ideal fluids [68, 136]. These studies of exactly integrable systems are facilitated by uses of the Maupertuis principle [178]. Not surprisingly, study of geodesic flows on Riemann manifolds has its uses, besides mechanics and hydrodynamics, in general relativity [179–181].

Remark 6.2.3. Although we shall not use the Finsler geometry in this book, Chern in [182] demonstrated that such geometry is essentially equivalent to contact geometry.

6.3 Basics of Sub-Riemannian Geometry

In the light of results just presented we are now in the position of providing some basic facts about the sub-Riemannian geometry. To do so, we notice that in Maupertuis' formalism it is sufficient only to deal with the kinetic energy, that is with the Hamiltonian of the type given by Eq. (6.2.5). Thus, from now on we shall think about all possible Hamiltonians as those which are given by Eq. (6.2.5). Since the tangent vectors X_j can be written in the form

$$X_j = \sum_{l=1}^{n} a_j^l(\mathbf{x})\partial_{x_l} \equiv a_j^l(\mathbf{x})\partial_{x_l}, \quad j = 1, \ldots, k, \tag{6.3.1}$$

it can be proven [167] that

$$h^{lm} = \sum_{j=1}^{k} a_j^l a_j^m. \tag{6.3.2}$$

By definition, the *normal geodesic* $\mathbf{x}(s)$ is a solution of Hamiltonian equations

$$\dot{x}^i(s) = \frac{\partial \mathcal{H}}{\partial p_i} \quad \text{and} \quad \dot{p}_i(s) = -\frac{\partial \mathcal{H}}{\partial x^i}, \quad i = 1, \ldots, n. \tag{6.3.3}$$

In Riemannian geometry it is always possible to write $p_j(s) = h_{ij}\dot{x}^i(s)$ so that the Hamiltonian can be rewritten as

$$\mathcal{H} = \frac{1}{2} h_{kl}\dot{x}^k \dot{x}^l \equiv \frac{1}{2} |\dot{\mathbf{x}}(s)|^2. \tag{6.3.4}$$

In mechanical language this is just the kinetic energy of a particle moving along the geodesic

$$\ddot{x}^i(s) + \Gamma^i_{jk}\dot{x}^j(s)\dot{x}^k(s) = 0 \tag{6.3.5}$$

obtained by using Eqs. (6.3.3). In the present case it is easy to demonstrate that instead of geodesic given by Eq. (6.3.5) we have to deal with the *sub-Riemannian equations for normal geodesics*

$$\ddot{x}^i(s) = \Gamma^{iab} p_a p_b. \tag{6.3.6}$$

Here the analog of Christoffel's symbol is given by: $\Gamma^{iab} = \frac{1}{2}(\frac{\partial h^{ia}}{\partial x^r} h^{rb} + \frac{\partial h^{ib}}{\partial x^r} h^{ra} - \frac{\partial h^{ab}}{\partial x^j} h^{ij})$. If $\mathbf{x}(s)$ is the normal geodesic then, using Eq. (6.2.5) in Eq. (6.3.3), the first of Hamiltonian's equations is given by

$$\dot{x}^i(s) = h^{lm} p_m \quad \text{or} \quad \dot{x}^i(s) = a_l^i a_l^m p_m. \tag{6.3.7}$$

Differentiating this result once again over s and using the second of Hamiltonian's equations we obtain after some calculations the equation for normal geodesics, e.g. see Eqs. (6.3.3). Next, by analogy with Eq. (6.3.1) we rewrite the horizontal 1-forms in a similar fashion

$$\omega_\alpha = \theta^i_\alpha dx_i \quad \text{so that} \quad \omega_\alpha(X_j) = \theta^i_\alpha a^i_j = 0. \tag{6.3.8}$$

Using Eq. (6.3.8), we obtain : $\omega_\alpha(\dot{\mathbf{x}}(s)) = (\theta^i_\alpha dx_i)(\dot{x}^i(s)\partial_{x_i}) = \theta^i_\alpha \dot{x}^i(s) = \theta^i_\alpha a^i_l a^m_l p_m = 0$. Thus, *normal geodesics are horizontal curves, i.e.* such curves for which $\omega_\alpha(\dot{\mathbf{x}}(s)) = 0$.

These general results we shall illustrate by discussing some specific example. For instance, let $\alpha = -\frac{1}{4}dt + \frac{1}{2}(ydx - xdy)$ and $X_i, i = 1, 2$, be as in Eq. (6.1.1). Then, we obtain:

$$\begin{aligned}\dot{\mathbf{x}}(s) &= \dot{x}\partial_x + \dot{y}\partial_y + \dot{t}\partial_t \\ &= \dot{x}(\partial_x + 2y\partial_t) - 2\dot{x}y\partial_t \\ &\quad + \dot{y}(\partial_y - 2x\partial_t) + 2x\dot{y}\partial_t + \dot{t}\partial_t \\ &= \dot{x}X_1 + \dot{y}X_2 - \tfrac{1}{4}(\dot{t} + 2x\dot{y} - 2y\dot{x})R.\end{aligned} \tag{6.3.9}$$

Here in the last line we used Eq. (6.1.2a) in order to replace ∂_t by R. Evidently, *the curve* $\mathbf{x}(s)$ *is horizontal* since $\dot{t} = 2y\dot{x} - 2x\dot{y}$ so that, in fact,

$$\dot{\mathbf{x}}(s) = \dot{x}X_1 + \dot{y}X_2. \tag{6.3.10a}$$

Furthermore, since $\frac{d}{ds}[\dot{t} = 2y\dot{x} - 2x\dot{y}]$ produces $\ddot{t} = 2y\ddot{x} - 2x\ddot{y}$, the same type of calculations as in Eq. (6.3.9) produces as well

$$\ddot{\mathbf{x}}(s) = \ddot{x}X_1 + \ddot{y}X_2. \tag{6.3.10b}$$

Being armed with these results we are ready now to take another look at the sub-Riemannian analogs of Eqs. (6.3.4) and (6.3.5). This is accomplished in two steps. First, using Eqs. (6.2.5)–(6.2.6) the Hamiltonian equations are obtained:

$$\begin{aligned}\dot{x} &= \frac{\partial \mathcal{H}}{\partial p_x} = p_x + 2yp_t, \\ \dot{y} &= \frac{\partial \mathcal{H}}{\partial p_y} = p_y - 2xp_t, \\ \dot{t} &= \frac{\partial \mathcal{H}}{\partial p_t} = 2y(p_x + 2yp_t) - 2x(p_y - 2xp_t) \\ &= 2(y\dot{x} - x\dot{y}).\end{aligned} \tag{6.3.11}$$

Second, using the matrix $h^{ij}(\mathbf{x})$ given by Eq. (6.2.6) we have to calculate the Christoffel-like symbols Γ^{iab}. Fortunately, this was done in [175], p. 38.

Using these results for Γ^{iab}, Eqs. (6.3.6) for normal geodescs acquire the following form

$$\begin{aligned}\ddot{x} &= 4p_t(p_y - 2xp_t),\\ \ddot{y} &= -4p_t(p_x + 2yp_t),\\ \ddot{t} &= 8p_t(xp_x + yp_y).\end{aligned} \tag{6.3.12}$$

These can be simplified with the help of just derived Hamiltonian equations. Thus, we obtain:

$$\begin{aligned}\ddot{x} &= 4p_t\dot{y},\\ \ddot{y} &= -4p_t\dot{x}.\end{aligned} \tag{6.3.13}$$

Obtained equations coincide with Eq. (6.2.2) if we take into account that $p_t = \text{const.}$ (since $\dot{p}_t = -\frac{\partial H}{\partial t} = 0$). This allows us to identify $4p_t$ with $\frac{\lambda}{2}H_z$. Furthermore, the third of Eqs. (6.3.12) is the same as the already obtained $\ddot{t} = 2y\ddot{x} - 2x\ddot{y}$. This can be seen by replacing $\ddot{x}$ and $\ddot{y}$ in just stated equation by the first and second of Eqs. (6.3.12) and simplifying results using the first two of Eqs. (6.3.11). Using Eqs. (6.3.11) and (6.3.12) we can still write

$$\begin{aligned}p_x &= \dot{x} - 2yp_t,\\ p_y &= \dot{y} + 2xp_t,\\ p_t &= \text{const.} = c.\end{aligned} \tag{6.3.14}$$

From here, using Eqs. (6.2.4), (6.3.3), (6.3.11) and (6.3.12) we obtain as well

$$\begin{aligned}\dot{p}_x &= -\frac{\partial \mathcal{H}}{\partial x_x} = 2p_t(p_y - 2xp_t) = 2p_t\dot{y},\\ \dot{p}_y &= -\frac{\partial \mathcal{H}}{\partial p_y} = 2p_t(p_x + 2yp_t) = 2p_t\dot{x},\\ \dot{p}_t &= 0.\end{aligned} \tag{6.3.15}$$

Using the first two of Eqs. (6.3.11) allows us to rewrite the Hamiltonian, Eq. (6.2.4), as

$$\mathcal{H} = \frac{1}{2}(\dot{x}^2 + \dot{y}^2) = \frac{1}{2}g_{ij}\dot{x}_i\dot{x}_j, \tag{6.3.16}$$

where $\dot{x}_i = \{\dot{x}, \dot{y}\}$ and the metric g_{ij} is given by

$$g_{ij} = \begin{pmatrix} 1 & 0 & 0\\ 0 & 1 & 0\\ 0 & 0 & 0\end{pmatrix}. \tag{6.3.17}$$

This is the metric of sub-Riemannian space defined by the horizontality condition $\omega_\alpha(\dot{\mathbf{x}}(s)) = 0$. In view of Eq. (6.3.10a) this result can be equivalently rewritten in terms of the metric tensor

$$g(X_i, X_j) = \delta_{ij} \quad \forall\ X_i \text{ for which } \alpha(X_i) = 0, \ \ i = 1, 2. \tag{6.3.18}$$

Such a rewriting in terms of velocities makes use of the word "Hamiltonian" questionable. It would be better to call it "Lagrangian". Since we follow [167], p. 113, we shall still call it Hamiltonian. What these authors call a Lagrangian can be written as follows.

$$\mathcal{L}[\mathbf{x}(\tau)] = \frac{1}{2} g_{ij}\dot{x}_i\dot{x}_j - \lambda_j\omega_j(\dot{\mathbf{x}}). \tag{6.3.19}$$

Clearly, this Lagrangian is of the same type as that given by Eq. (6.2.1). The equation

$$\frac{\partial \mathcal{L}}{\partial \lambda_j} = 0, \quad j = 1, \ldots, k \tag{6.3.20}$$

provides the horizontality conditions $\omega_j(\dot{\mathbf{x}}) = 0$. Without loss of generality we shall assume that the tensor g_{ij} is x-dependent, that is it is given by $g_{ij}(\mathbf{x})$, then variation of the action functional $S[\mathbf{x}] = \int_0^T \mathcal{L}[\mathbf{x}(\tau)]d\tau$ yields equations for the usual/regular (or Riemannian) geodesics, Eq. (6.3.5). At the same time, using the Euler–Lagrange equations in order to recover the Hamiltonian and Eqs. (6.3.6), it is not difficult to demonstrate that the same Riemannian geodesics are also normal geodesics $\mathbf{x}(s)$ for the already known Hamiltonian, Eq. (6.2.5). Thus, for dynamical problem given by the Lagrangian, Eq. (6.2.1), equations determining normal geodesics are given by Eqs. (6.3.11) and (6.3.14). Define now the kinetic energy $\mathcal{E}$ in terms of normal geodesics using Eq. (6.3.16)

$$\mathcal{E}(\gamma(\tau)) = \frac{1}{2}\int_0^T d\tau(\dot{x}^2 + \dot{y}^2). \tag{6.3.21}$$

Also, define the length $l(\gamma(\tau))$ associated with such an energy

$$l(\gamma(\tau)) = \int_0^T d\tau\sqrt{(\dot{x}^2 + \dot{y}^2)}. \tag{6.3.22}$$

Because of the Cauchy–Schwarz inequality

$$\int fg \leq \sqrt{\int f^2}\sqrt{\int g^2}$$

it is possible to write

$$l(\gamma(\tau)) \leq \sqrt{T}\sqrt{2\mathcal{E}(\gamma(\tau))}.$$

Then, such defined normal geodesic $\mathbf{x}(\tau) = \{x(\tau), y(\tau)\}$ in the plane can be linked to the integral curve/geodesic $\gamma(\tau) = \{x(\tau), y(\tau), t(\tau)\}$ on 3-manifold M, e.g. like that depicted in Fig. 3(a)). The t-coordinate of such a manifold (in the case of Fig. 3(a) it is S-coordinate) is entirely determined by the planar motion via relationship

$$t(T) - t(0) = 2\int_0^T d\tau(x\dot{y} - y\dot{x}). \tag{6.3.23}$$

Because of this relationship, the geodesics in the $x - y$ plane are automatically also the geodesics in M. In view of this, it is possible to define the Carnot–Caratheodory (C-C) distance $l(p, q)$ in such a sub-Riemannian space as:

$$l(p, q) := \inf\{l(\gamma(\tau)) \mid \gamma \textit{ horizontal curve joining } p \textit{ and } q\}. \tag{6.3.24}$$

Remark 6.3.1. Notice that both $p = \{x, y, t\}$ and $q = \{x', y', t'\}$ live in 3-manifold M known as *nilmanifold* according to Thurston's classification of 3-manifolds [183]. It is contact manifold which is equivalently called the Heisenberg manifold $\mathbf{H}_1$ since motions/isometries of such a manifold are determined by the composition law (left translation):

$$(x', y', t') \circ (x, y, t) \to (x + x', y + y', t + t' + 2(x'y - xy')). \tag{6.3.25}$$

These are not the only isometries possible. Other, e.g. rotations, will be introduced below. They are playing an important role in view of the relationship Eq. (6.3.23). This relationship is nontrivial. While the dimensionality of $x(\tau)$ and $y(\tau)$ is length $[L]$, the dimensionality of $t(\tau)$ is $[L^2]$ as can be seen from Eq. (6.3.23). In view of this, the Heisenberg dilations are defined as follows:

$$\delta_\lambda(x, y, t) = (\lambda x, \lambda y, \lambda^2 t), \quad \lambda > 0, \tag{6.3.26}$$

causing introduction of yet another distance, equivalent to C-C distance. To introduce this distance it is convenient to improve the already introduced notations. In particular, the translations defined by Eq. (6.3.25) can be written as

$$\tau_{\mathbf{h}}(z, t) = (\zeta, \tau) \circ (z, t), \tag{6.3.27}$$

where $\mathbf{h} = (\zeta, \tau) \in \mathbf{H}_1$, $z = x + iy, \zeta = x' + iy', \tau = t'$. The new distance $\tilde{d}(p, q) \equiv \tilde{d}((z, t), (\zeta, \tau))$ must be invariant under translations, i.e.

$$\tilde{d}((z, t), (\zeta, \tau)) = \tilde{d}(\tau_{\mathbf{h}}(z, t), \tau_{\mathbf{h}}(\zeta, \tau)) \quad \forall\, p \text{ and } q \in \mathbf{H}_1. \qquad (6.3.28a)$$

In addition, it should behave as

$$\tilde{d}(\delta_\lambda(z, t), \delta_\lambda(\zeta, \tau)) = \lambda \tilde{d}((z, t), (\zeta, \tau)) \qquad (6.3.28b)$$

under dilations. Evidently, that dilations are *not* isometries and that $\tilde{d}$ must also be rotationally invariant, that is invariant under transformations of the type: $\rho_\theta(z, t) = (e^{i\theta} z, t)$. The obtained results are helpful for determining $\tilde{d}(p, q)$ explicitly. To finish this task we notice that if in the composition law Eq. (6.3.25) we put both x' and y' to zero the composition law will affect only the t-direction. Thus, if we denote such transformation as $\tau(t)_{\mathbf{h}}(z, t)$, then it is clear that it is not affecting transformations in the z-plane. Therefore, instead of the group $\mathbf{H}_1$ it is sufficient to consider its quotient $\mathbf{H}_1/\Gamma_t$ where Γ_t are transformations of the type $\tau(t)_{\mathbf{h}}$. Such a quotient leaves the quadratic form $Q = xx' + yy' \equiv \langle \mathbf{x} \cdot \mathbf{x}' \rangle$ defined in the plane $\alpha(X_i) = 0$, $i = 1, 2$, invariant. Thus, the Heisenberg manifold can be locally represented as a product $\mathbf{R}^2 \times \mathbf{R}$ or, upon compactification, by $S^2 \times S^1 \simeq S^3$.

Consider first the product $\mathbf{R}^2 \times \mathbf{R}$ and determine the distance for such a product. Let $\tilde{d}((\zeta, \tau), (z, t)) = \tilde{d}((0, 0), (z, t))$ then, with accuracy up to a constant, the distance $\tilde{d}(p, q)$ is determined by

$$\tilde{d}(z, t) = (|z|^4 + t^2)^{\frac{1}{4}}. \qquad (6.3.29)$$

In view of Eq. (6.3.28b) the constant can always be put to one. In such a form the distance was discovered by Folland and Stein, [184]. By design $\tilde{d}(z, t)$ satisfies the scaling property given by Eq. (6.3.28b). Clearly, this is not the Euclidean distance. At the same time, the distance is Euclidean when $t = 0$ and for $z = 0$ the squared distance $\tilde{d}$ is Euclidean. Generalization for $\tilde{d}(p, q)$ is now straightforward. If we write $p = (x, y, t), q = (x', y', t')$, so that the composition is $p \circ q$ while the inverse element of p is $p^{-1} = (-x, -y, -t)$, then $\tilde{d}(p, q) = \tilde{d}(p^{-1} \circ q)$. It can be shown [185] that

$$Cl(p, q) \leq \tilde{d}(z, t) \leq C^{-1} l(p, q), \qquad (6.3.30)$$

where C is some constant.

6.4 Glimpses of Quantum Mechanics

With these results introduced, we are ready to inject some physics into our discussion. For this purpose, following [174], p. 354, we can reformulate dynamical problem described by the Langrangian, Eq. (6.2.1), using terminology of the theory of optimal control. In this language one is searching for two time-dependent control functions u and v minimizing the (energy) functional

$$\mathcal{F}[u,v] = \int_0^T d\tau(u^2 + v^2), \tag{6.4.1}$$

provided that these functions control the motion described by the following equations:

$$\dot{x} = u, \quad \dot{y} = v, \quad \dot{t} = A_x u - A_y v. \tag{6.4.2}$$

Clearly, such a reformulation appears to be helpful if one is interested in studying problems of optimal control in macroscopic dynamics. This assumption happens to be not entirely true since classical theory of optimal control has been very recently used, practically unchanged, in the theory of quantum control. We shall further discuss this topic in Chap. 8. In the meantime, much more relevant for us now is to observe that Eq. (3.4.13) can be written in the style of sub-Riemannian geometry. Indeed, using these equations we obtain:

$$\oint \mathbf{A} \cdot d\mathbf{l} = \frac{\hbar c}{2e} \oint d\mathbf{l} \cdot \nabla\Phi = \frac{\hbar c}{2e}[\Phi(l) - \Phi(0)]. \tag{3.4.13}$$

Given that this equation was derived in connection with the problem about quantization of constant magnetic field flux penetrating superconducting ring, we notice that Eq. (3.4.13) is equivalent to Eq. (6.3.23). Normally, in the case of superconductivity one chooses $\Phi(0) = 0$ and $\Phi(l) = 2\pi n$ so that $\oint \mathbf{A} \cdot d\mathbf{l} = hn\phi_0$ as already stated in Eq. (3.4.13). This result can be equivalently rewritten as $\iint \mathbf{H} \cdot \mathbf{dS} = hn\phi_0$ or, in more convenient form, as $H\pi r^2 = nh$ (in the system of units where $\phi_0 = 1$). Since $\dot{\mathbf{x}} = \omega r$ and since $\omega \simeq H$ (in the same system of units), the kinetic energy $\mathcal{E} = \frac{1}{2}\dot{\mathbf{x}}^2 (m = 1)$ can be equivalently rewritten as $\mathcal{E} = \frac{1}{2}(\omega r)^2 \simeq nh\omega$. This is familiar quantum mechanical result for harmonic oscillator. It was obtained in 1930 by Landau who considered quantum mechanical problem for an electron moving in the constant magnetic field, [186].

The above back of the envelope calculations suggest that they admit "classical" interpretation if, instead of the usual Euclidean geometry, we would use the sub-Riemannian one. We want now to demonstrate that this is indeed possible. For this purpose, let us consider first the energy $\mathcal{E}(\gamma(\tau))$, Eq. (6.3.21). Using Eqs. (6.3.13) we obtain

$$\begin{aligned} \ddot{x} &= 4p_t\dot{y} \to \dot{x} = 4p_t y + c_1, \\ \ddot{y} &= -4p_t\dot{x} \to \dot{y} = -4p_t x + c_2, \end{aligned} \tag{6.4.3}$$

where c_1 and c_2 are some constants. Substitution of this result into $\mathcal{E}(\gamma(\tau))$ and taking into account that the orbits are closed, we obtain

$$\mathcal{E}(\gamma(\tau)) = p_t t(T) \tag{6.4.4}$$

in view of Eq. (6.3.23) and the fact that $t(0) = 0$ by design. Since $p_t \sim H \sim \omega$, it remains to demonstrate that $t(T) \sim H\pi r^2 \sim \omega T r^2 \sim \omega\pi r^2$. This can be done with account of the fact that the combination ωT is dimensionless so that $[t] = [L^2]$ as required. Then, since the R.H.S. of Eq. (6.3.23) is reparametrization-invariant we can always select $T = \pi$ as the proper time. Without loss of generality we can put both c_1 and c_2 in Eqs. (6.4.3) to zero. In such a case we obtain for the circular orbit

$$t(T) = -8p_t \int_0^T d\tau(x^2 + y^2) \sim -\omega\pi r^2. \tag{6.4.5}$$

The sign "$-$" is unimportant since it can be changed by reversing the orientation of the integration contour/magnetic field. Thus, we just demonstrated that, indeed, $t(T) \sim \omega\pi r^2$. Substitution of this result into $\mathcal{E}(\gamma(\tau))$ produces $\mathcal{E} \sim (\omega r)^2$ as before. Notice that the obtained result for $t(T)$ is in accord with the dilation property, Eq. (6.3.26), for the Heisenberg group. This means that we can always find λ which will eliminate the factor $\omega\pi$ in $t(T)$. This also means that in order to obtain the correct quantum mechanical results we have to match thus obtained result with its quantum counterpart, Eq. (3.4.13), causing r^2 to become respectively $0, 1, 2, 3, \ldots$. In view of the requirement $\lambda > 0$, we shall not treat the trivial case: $r^2 = 0$. Furthermore, it is of great interest to re-obtain this result for r^2 without a hint coming from Eq. (3.4.13). To do so requires us to make a number of steps. To begin, taking into account that

$$\tilde{d}(\lambda z, \lambda^2 t) = \lambda(|z|^4 + t^2)^{\frac{1}{4}} \tag{6.4.6}$$

and using this result for normal geodesics parametrized by the magnetic field, it is clear that for the pre-assigned set of parameters, the scaling factor $\lambda \geq 1$. The value $\lambda = 1$ defines the *fundamental* C-C distance so that if we agree that any other distance in the Heisenberg manifold is measured in units of this distance, then the quantization of $t(T)$ comes out automatically. Next, this result can be obtained as well using the fiber bundle formalism. In fact, the already mentioned horizontality condition is not intrinsic for the sub-Riemannian geometry only. It is routinely used in the theory of fiber bundles.

Remark 6.4.1. Since in physics literature the fiber bundle theory is used typically for description of the Yang–Mills gauge fields, it should not be totally surprising that such fields naturally emerge in the formalism of non-holonomic mechanics [174] and/or sub-Riemannian geometry. Thus, quantum mechanics, gauge fields and gravity can and should be considered as branches of applied contact geometry and topology.

6.5 Fiber Bundle Reformulation of Sub-Riemannian Geometry and Classical-Quantum Correspondence. Connection with Dirac Monopoles

The above quantization problem, even though done classically, requires some nontrivial assumptions about topology associated with particle dynamics. Naturally, such assumptions should not look artificial. They must emerge based on reasonable physical considerations. For this purpose we shall follow pedagogically written Ph.D. thesis by Rupert Way [187] in which the details of the fiber bundle construction needed for completion of our quantization program are presented in the most accessible way. This will allow us to be brief. Our readers should be able to fill in the gaps in our presentation by consulting the original manuscript. In our presentation we shall assume that our readers are familiar with the most basic facts about the theory of fiber bundles, e.g. see [111, 146].

Thus, let (P, M, G, π) be the *principal fiber* bundle with the total space P and the base manifold M with the Lie group G acting to the right on P. Next, if $q \in P$, then $x = \pi(q)$, $x \in M$ and $p \in \pi^{-1}(x)$ as usual. Let Φ denote the automorphism $\Phi : G \times P \to P$ which we shall write as $\Phi_g(p) = p \cdot g$, $g \in G$. The group G pushes points in the bundle around bundle along the fibers. At any point q the tangent bundle T_qP can be decomposed into horizontal and vertical parts as follows:

$$T_qP = H_qP \oplus V_qP. \tag{6.5.1}$$

Thus, $(\Phi_g)_*(H_qP) \equiv H_{\Phi_g(p)}P$, that is $(\Phi_g)_*$ pushes $H_qP \subset T_qP$ along the fiber. Clearly, the result should be the same if we would first push q along the fiber using Φ_g to the point $\Phi_g(p)$ and then form the subspace $H_{\Phi_g(p)}P$ at that point. In our case a decomposition into horizontal and vertical parts is given by Eq. (6.3.9). If we are talking about the principal bundles, the horizontal space must be invariant with respect to the action of G. In our case this means that the base space M must be in the form $\mathbf{H}_1/\Gamma_t$ so that the group G should be chosen as Γ_t. This result is not entirely correct, however. It is not correct for a number of reasons. For one thing, if we focus our attention on the horizontal subspace only, then the already defined quadratic form Q living in the horizontal space (plane) is invariant with respect to translations (defined by Eq. (6.3.25)) and rotations $\rho_\theta(z,t) = (e^{i\theta}z, t)$ since these are symmetries and isometries of the Euclidean plane.[b] Transformations along t-axis will also leave this form unchanged. Thus, if we want to construct the principal fiber bundle, we should choose between these three groups of motions. To do so we have to apply some physical considerations. Specifically, in view of Eqs. (6.2.2) we know that the horizontal projection of normal geodesic is just a circle. Let such planar curve be denoted as $\gamma(\tau)$ and its lift to the third dimension as $\tilde{\gamma}(\tau)$. This lift is subject to the obvious constraint: $\pi(\tilde{\gamma}(\tau)) = \gamma(\tau)$. Clearly, the normal geodesic will obey this requirement. However, *other* curves will also be projected to the plane with the same outcome. Among such curves we shall select only those which are generated by rotations in horizontal plane. Evidently, the normal geodesic will be one of them. For the sake of argument we can choose a point $(r, 0)$ in the $x - y$ plane and to make a full circle by applying rotation $\rho_\theta(z,t)$. While doing so, the t-coordinate will be determined by Eq. (6.3.23). Thus, applying rotations to horizontal plane we actually generate the normal geodesics. But such an operation is not limited to normal geodesics since we can continue our rotation and, since we are interested only in circular projections, the curve $\tilde{\gamma}(\tau)$ obtained by rotating, say, by 720° still obeys the projection requirement: $\pi(\tilde{\gamma}(\tau)) = \gamma(\tau)$. However, if we use Eq. (6.3.23) the obtained area is going to be twice that for the normal geodesic. Clearly, the same result for the area can be obtained by using the dilation operation. In such a case, in view of Eq. (6.3.28b), the original normal geodesic is transformed into dilated normal geodesic. In this way one can construct a discrete spectrum of normal geodesics in $\mathbf{H}_1$. But since the dilation is *not* an isometry of $\mathbf{H}_1$ we can think instead about the set of different Heisenberg spaces parametrized by λ factor. The discreteness of λ requires some physical justification, e.g. like that which was already supplied. Such a justification only superficially explains Eq. (3.4.13) since the connection with the Dirac monopole discussed in conjunction with Eq. (3.4.13) is not at all obvious at this stage.

[b]This result is a special case considered in the theory of space-forms. Invariance of quadratic forms with respect to some symmetry operations plays central role in this theory. All 3-manifold geometries [183, 188] are obtained by studying such an invariance.

Use of the fiber bundle methods allows us to make such a connection much more convincing as we shall explain below.

To do so we have to resolve the major issue: how the fiber bundle P is related to the Heisenberg space $\mathbf{H}_1$? The answer to this question is not a simple one [189]. To avoid technicalities such as tangent cones (e.g. read Chap. 8 of [189]), etc. we shall be using methods more familiar/traditional employed in physics literature.

Exercise 6.5.1. Those of our readers who are interested in the forefront of mathematical research on this subject, are encouraged to read Chap. 8 of [189] and, subsequently, the original work by M. Gromov on which the results of Chap. 8 are based. After this, those of our readers who are capable of successfully passing this intellectual barrier (just like Landau's qualifying exam) will be ready for Chap. 9 of [189] where the discreteness of λ is explained based on much more delicate results from the theory of infinite discrete groups.

Remark 6.5.1. By rephrasing in physical language the Definition 1.3, p. 5, of the book by Montgomery, [189], the Heisenberg group $\mathbf{H}_1$ can be defined as follows.

Let $\mathbf{R}^3$ be endowed with the relation (the distribution)

$$\oint \mathbf{A} \cdot d\mathbf{l} == \frac{\hbar c}{2e}[\Phi(l) - \Phi(0)]$$

in which the vector field $\mathbf{A}$ is given by $\mathbf{A} = \{A_x, A_y, 0\} = \{2y, -2x, 0\}$ so that the scalar product in the plane defined by the above distribution is $Q = xx' + yy' \equiv \langle \mathbf{x} \cdot \mathbf{x}' \rangle$, then the above two properties define the Heisenberg group.

Exercise 6.5.2. Check using [189] and, may be, other sources if, indeed, the Definition 1.3 by Montgomery is equivalent to more traditional definition of Heisenberg group summarized in Eqs. (6.3.25–6.3.29). Hint: After this checking, read Appendix A of this book.

The above remark implies that from now on instead of traditionally postulated commutator, Eq. (6.1.2b) (or, equivalently, the postulated B-S quantization condition), we can use the above *experimentally obtained*, Eq. (3.4.13), as the basis of quantum mechanics. But this equation is describing *macroscopically observed phenomenon*! Thus, quantum mechanical behavior is not limited/confined to a very small space–time (atomic and smaller) scales as it is commonly believed! We hope that after reading this book (till this point) this fact should not come as a big surprise.

Remark 6.5.2. Uses of the Heisenberg group are not limited to quantum mechanics. Applications to many other fields ranging from genetics — to radar and signal processing and imaging, from exactly integrable systems — to number theory and thermodynamics, etc. place Heisenberg group in a unique position among all other Lie groups. It is not our purpose to list or to discuss all these applications in our book. Many of the applications just mentioned are described in [190–194]. We urge our readers not to ignore these references.

Going back to our fiber bundle discussion we notice that to some extent $\mathbf{H}_1$ resembles $\mathbf{R}^3$. But $\mathbf{R}^3$ can be compactified to S^3 by adding just one point at infinity! The same is true about the t-direction in $\mathbf{H}_1$. By adding one point at infinity we can replace the t-direction in $\mathbf{H}_1$ by S^1. Next, we can think about the Hopf fiber bundle by compactifying the horizontal $\mathbf{R}^2$ plane so that it becomes S^2. In which case there is a chance for conversion of the principal circle bundle for $\mathbf{H}_1$ into the Hopf bundle. This possibility is facilitated by the already stated Theorem 5.5.1. Now we would like to add few details.

To begin, we consider again a circular curve $\gamma(\tau)$ located in the (horizonal) compactified plane, that is in S^2. Such a curve is a sub-Riemannian geodesic [195]. When lifted to S^3, the image of such geodesic is normal geodesic $\tilde{\gamma}(\tau)$ that is horizontal curve. Such a curve may or may not be closed. In fact, if we act by analogy with $\mathbf{H}_1$, then the geodesic should be open while projecting it into plane, the circle is recovered. Nevertheless, for the sake of completness of our presentation, we shall consider both closed and open geodesics in S^3 and closed and open curves which are not geodesics.

We begin with discussion of closed geodesics in S^3. Clearly, they satisfy the projection requirement: $\pi(\tilde{\gamma}(\tau)) = \gamma(\tau)$. However, more work is needed in order to obtain additional information about these geodesics. For instance, consider a family of circles on a sphere S^2 of unit radius. Let us begin with the case when the representative circle *is not* a diameter of S^2.

Interestingly enough, in full generality such a problem is still a subject of intensive research [196]. In this area of research many famous mathematicians, e.g. Poincaré, Birkhoff, Lusternik and Schnirelman, Arnol'd, Novikov, etc. had made significant contributions [197]. Since to describe this field of research would require a separate monograph, we shall only discuss topics of immediate relevance.

In particular, we need to discuss the notion of *geodesic curvature* κ_g of the curve $\gamma(s)$ constrained to S^2. This concept should not be confused with that for the curvature of the curve. It is, in fact, similar to the concept of order parameter in Landau's theory of phase transitions. For instance, if the curve on the surface is geodesic, its geodesic curvature is zero. Then the non-geodesic surface curves will have nonzero geodesic curvatures. It can be shown that κ_g is given by

$$\kappa_g = \frac{\gamma''(s)(\mathbf{n} \times \gamma'(s))}{\|\gamma'(s)\|^3}, \tag{6.5.2}$$

where $\mathbf{n}$ is the normal (in our case) to S^2. To calculate κ_g using Eq. (6.5.2) we have to take into account that in spherical coordinates $\mathbf{n} = \{\cos\varphi \sin\theta, \sin\varphi \sin\theta, \cos\theta\}$. The circle on S^2 is generated by intersecting the sphere $x^2+y^2+z^2 = 1$ with the plane $z = h, h \in (-1, 1)$. As a result, we obtain the following coordinates for $\gamma(s) := \{\sqrt{1-h^2}\cos s, \sqrt{1-h^2}\sin s, h\}$.[c] With help of these results the geodesic curvature κ_g is found to be

$$\kappa_g = \pm\frac{h}{\sqrt{1-h^2}}. \tag{6.5.3}$$

From here we obtain as well

$$h = \frac{\kappa_g}{\sqrt{1+\kappa_g^2}}. \tag{6.5.4}$$

Evidently, when we switched from plane $\mathbf{R}^2$ to S^2 by adding just one point at infinity we expected $|h| \ll 1$. At the same time, $\kappa_g = 0$ when the "motion" of $\gamma(s)$ takes place on great circles of S^2 since great circles are geodesics for S^2. The lift of $\gamma(s)$ to S^3 generates the great circles on S^3 which also of zero geodesic curvature since they are geodesics in S^3. For $h \neq 0$ the geodesics may or may not be closed. In [195] *it is demonstrated that only when h is rational number the lift of $\gamma(s)$ produces closed geodesic in S^3 which arc torus knots*, just like those, discussed in Chaps. 2 and 3[d]! Incidentally, as results of [167] indicate, the fibers of Hopf fibration are the integral curves of the Reeb vector field, Eq. (5.5.8), see also Eqs. (5.6.5.4 a,b). How such curves are related to t-direction in $\mathbf{H}_1$? How these curves are related to geodesics in S^3? Surprisingly, to answer each of these questions requires writing of full research paper, e.g. see [198]. To avoid details, we

[c] These are known as the Darboux coordinates. They are being used also in Appendix D.

[d] The torus knots are discussed also in Appendix D in the context of quantum entanglement problem.

follow [195]. In this reference the authors take a note that the Clifford torus inside the sphere S^3 of unit radius is described in terms of pair of complex numbers (z_1, z_2) such that $|z_1|^2 = \rho^2$ and $|z_2|^2 = 1 - \rho^2, \rho \in (0, 1)$.

Remark 6.5.3. From the construction of Clifford torus it is clear that one-point $\rho = 1/2$ plays a special role. The Clifford torus with $\rho = 1/2$ is called *flat torus*. Due to obvious symmetry between z_1 and z_2 in such a case one can talk about the Heegaard splitting (or *Heegaard decomposition* [199]) of S^3 into two solid tori. Since the 3-sphere is the simplest 3-manifold, it can be demonstrated that every 3-manifold can be created by appropriately gluing two handlebodies whose surfaces are Riemannian surfaces of genus g (depending on the 3-manifold in question).

Next, consider the geodesic circle $\gamma(s)$ on S^2 and lift it to S^3. Then, the lift $\tilde{\gamma}(\tau)$ is winding around the Clifford torus with the outcome depending upon whether h is rational or not. If it is rational, as it is demonstrated in [195], a countable infinity of sub-Riemannian geodesics parametrized by rational h is obtained. These results are in qualitative accord with more recent [198]. By looking at Eqs. (5.6.5.4) we notice that Eqs. (5.6.5.4b) actually describe integral curves on the Clifford torus, e.g. similar to those depicted Fig. 3(a) so that the Reeb vector field is tangent to these integral curves. In hydrodynamics such a field provides minimum for the energy functional, e.g. see Eq. (5.6.1a) and [145]. From here we obtain the connection between the sub-Riemannian torus knots and hydrodynamics.

In the meantime, from what was just presented it follows that the connection with $\mathbf{H}_1$ is easily established if, instead of two oscillators in Eq. (5.6.5.4b), we shall consider only one (that is we select only trivial solution for another). Such physical conclusion is in accord with that our readers will find in [167], p. 247. To make a connection with the Dirac monopole, such asymmetry is undesirable. It is undesirable because of the following.

[167] contains analogs of Hamiltonian Eqs. (6.3.14), (6.3.15) in S^3. Calculations presented in detail in this reference indicate that the most physically important Eqs. (6.3.13) and (6.3.21) stay exactly the same. The question still remains: What is the analog of the last of Eqs. (6.3.11) in the case of S^3? Fortunately, we have obtained the answer already, e.g. see Eq. (5.5.2) and compare it with Eq. (5.5.1) valid for $\mathbf{H}_1$. By obtaining the answer, we now have to find some kind of an analog for Eq. (6.3.23). In doing

so, we shall follow prescriptions by Montgomery, [189], refining those which we have already made. On p. 20 of [189] we find the following paragraph:

"Start with the Heisenberg group. Replace the base space, which is the Euclidean plane, by the round two-sphere. Replace t-variable by a circular variable θ that evolves according to spherical area swept out. Scale the sphere so that the total area is 2π. This scaling is consistent with θ taking values in the unit circle. The resulting global sub-Riemannian geometry is that induced from the Hopf fibration: $S^1 \to S^3 \to S^2$". Incidentally, very nice (picturesque) description of the Hopf fibration is given in [200].

The above mathematical suggestions by Montgomery we shall now supplement by some physics. We begin by noticing that the Larmor radius ρ on the sphere S^2 of unit radius is given by ($\lambda/2 = 1, m = 1, c = 1$):

$$\rho^2 = \frac{2\mathcal{E}/H_z^2}{1 + 2\mathcal{E}/H_z^2} \tag{6.5.5a}$$

to be compared with the flat $\mathbf{R}^2$ result

$$\rho_\infty^2 = 2\mathcal{E}/H_z^2, \tag{6.5.5b}$$

where $\mathcal{E}$ is the nonrelativistic kinetic energy as before. These results were obtained in [201].

Remark 6.5.4. The result, Eq. (6.5.5a) was obtained in [201] with the help of the following equation of motion

$$\ddot{\mathbf{r}} = H_z(\dot{\mathbf{r}} \times \mathbf{r}) + \lambda \mathbf{r} \tag{6.5.6}$$

where the Lagrangian multiplier $\lambda = -\langle \dot{\mathbf{r}} \cdot \dot{\mathbf{r}} \rangle$ is enforcing the round sphere constraint.

Remark 6.5.5. For $H_z = 0$ the same equation was used in polymer physics for description of conformations of semiflexible polymers, e.g. read [39] and Chap. 8 below. Presence of the magnetic field can also be interpreted in polymer language. Indeed, on one hand, Eq. (6.5.6) coincides with Eq. (5) of the benchmark paper, [202], by E. Witten where it was interpreted as an equation of motion for the charged particle in the presence of the Dirac monopole. On the other hand, in [203], Witten's result found its uses in polymer physics. The monopole-like problem was discussed in the context of the so-called hairpin formations in conformations of semiflexible polymers.[e]

Witten was able to broadly generalize the monopole problem. This eventually resulted in what is now known as the Wess–Zumino–Novikov–Witten

[e]This topic will be further discussed in Chap. 8.

(WZNW) model — basis for all two-dimensional conformal field theory models [204]. Naturally, generalization proceeds by extending results valid for curves to those valid for surfaces as discussed in Chap. 2.

Thus, Eq. (6.5.5a) implicitly contains information related to the Dirac monopole. More work is still needed to recover this information explicitly. For this purpose we need to re-obtain the parameter h defined before Eq. (6.5.3). Since we are working with the sphere of unit radius, we have: $\rho^2 + h^2 = 1$. From here we get:

$$h = \sqrt{\frac{1}{1+\rho_\infty^2}}. \tag{6.5.7}$$

Since h is a rational number, the same is true for $h^{-1} \equiv m$. In addition, since we had begun with the flat R^2 surface to which we added one point at infinity to get S^2 we expect that $\rho_\infty^2 \ll 1$ in Eq. (6.5.5a). By keeping this in mind when dealing with Eq. (6.5.7), we obtain asymptotically

$$m \simeq 1 + \frac{1}{2}\rho_\infty^2$$

or, which is the same,

$$(m-1)\, H_z^2 = \mathcal{E}. \tag{6.5.8}$$

Using once again qualitative arguments related to Eqs. (6.4.4) and (6.4.5) we rewrite the just obtained result as follows:

$$(m-1)\,\omega^2 = \mathcal{E} \simeq (\omega r)^2 \quad \text{or} \quad m - 1 = r^2. \tag{6.5.9}$$

Since m is the rational number which, according to Eq. (6.5.7) is non-negative, it follows that r^2 is also the non-negative rational number. Previously, using physical arguments we came to the conclusion that r^2 should take respectively values $0, 1, 2, 3, \ldots$. Clearly, if the appropriate system of units is used, this result shall be recovered from Eq. (6.5.9) as well. We obtained this asymptotic result without explicit uses of the monopole formalism. It is of interest to re-obtain it using the monopole formalism which involves utilization of the Hopf fiber bundle. In such a case, it will become evident that just obtained asymptotic result is, in fact, exact. We shall use works by Chern [205], Pincall [206], Singer [207] along with the already cited [187] to complete our fiber bundle analysis.

We begin with observation that the Hopf fibration is a circle bundle over S^2 [189, 205]. Thus, we have to remind our readers some results about circle bundles first. These are nicely summarized in lecture notes by Chern [205] and Singer [207]. In particular, following Singer we realize that for two-dimensional surfaces it is always possible to construct a circle bundle. Indeed, if M is such two-dimensional manifold and $T(M)$ is its tangent bundle, then tangent vectors $\mathbf{v} \in T(M)$ are normalized so that $\langle \mathbf{v} \cdot \mathbf{v} \rangle = 1$. Introduction of such defined tangent vectors allows us to consider the three-dimensional manifold $S(M)$ via the following identification:

$$S(M) = [(m, \mathbf{v}), m \in M, \mathbf{v} \in T(M); \langle \mathbf{v} \cdot \mathbf{v} \rangle = 1]. \tag{6.5.10}$$

The third dimension is actually made by function $f(m, \mathbf{v}) = \langle \mathbf{v} \cdot \mathbf{v} \rangle$ such that $df \neq 0$ whenever $f = 1$. The circle $S^1 = [z \in C, |z| = 1]$ is the Lie group U(1) under complex multiplication. Since $e^{i\theta_1} \cdot e^{i\theta_2} = e^{i(\theta_1 + \theta_2)}$, it is just the group of rotations of the oriented (complex) plane. This group acts on $S(M)$ via $S^1 \times S(M) \to S(M)$. Specifically, if $g = e^{i\theta}$ and $\mathbf{v} = \{\mathbf{v}_1, \mathbf{v}_2\}$ is any oriented orthonormal basis for $T(M)$, then using an isomorphism between groups U(1) and SO(2) we obtain

$$g\mathbf{v} = (c_1 \cos\theta - c_2 \sin\theta)\mathbf{v}_1 + (c_1 \sin\theta + c_2 \cos\theta)\mathbf{v}_2, \tag{6.5.11}$$

where $\mathbf{v} = c_1\mathbf{v}_1 + c_2\mathbf{v}_2$. If π: $S(M) \to M$, is the projection, then $\pi^{-1}(m)$ is just the unit circle in $T(M)$. It is clear that if $(m, \mathbf{v}_1)$ and $(m, \mathbf{v}_2)$ are any two elements of $\pi^{-1}(m)$, then there is some $g \in S^1$ such that $(m, \mathbf{v}_2) = (m, g\mathbf{v}_1)$. Thus, locally $S(M)$ is a product space. This is not true globally though. To demonstrate this requires a number of steps. For instance, this requires an introduction of the notion of parallel transport. A parallel translation of the unit vector along the curve $\gamma(t), t \in [a, b]$ (so that the curve $\gamma(t)$ is the map $[a, b] \to M$) determines the unit tangent vector $\mathbf{v} \in T(M)$ for each $t \in [a, b]$. Evidently, this vector can be looked upon as tangent vector $\tilde{\mathbf{v}}$ to the curve $\tilde{\gamma}(t) : [a, b] \to S(M)$. Consider now two tangent vectors $\tilde{\mathbf{v}}$ and $\tilde{\mathbf{v}}_1$ to $\tilde{\gamma}(t)$. Let $\tilde{\mathbf{v}} = \mathbf{v}(a)$ and $\tilde{\mathbf{v}}_1 = \mathbf{v}_1(b)$ so that the curve $\tilde{\gamma}(t)$ will be determined by parallel translating vector $\tilde{\mathbf{v}}$. Evidently, this process can be associated with some group element g such that $\tilde{\mathbf{v}}_1 = g\tilde{\mathbf{v}}$. To illustrate these general statements, we adopt them for $\mathbf{H}_1$ first. The key equations for this case are Eqs. (6.3.10)–(6.3.15). In addition, since the energy, Eq. (6.3.21), is conserved, it is always possible to enforce normalization $\langle \mathbf{v} \cdot \mathbf{v} \rangle = 1$. By analyzing these results, the following conclusions can be drawn.

By identifying $\gamma(t)$ with $\mathbf{x}(s)$ it should be obvious that $\dot{\mathbf{x}}(s) = \{\dot{x}_x(s), \dot{x}_y(s)\} = \{\mathbf{v}_1, \mathbf{v}_2\} = \mathbf{v}$. The fiber bundle $S(M)$ defined by Eq. (6.5.10) in physical language is made of two-dimensional configuration space serving as the base M and the fibers are rotations leaving the quadratic form $\langle \mathbf{v} \cdot \mathbf{v} \rangle$ — the energy-unchanged. At this point we are not yet in the position to find out if such a fiber bundle is trivial or not. Notice though that the relationship $\tilde{\mathbf{v}}_1 = g\tilde{\mathbf{v}}$ can be extended to curves in $S(M)$, that is we can write as well

$$\tilde{\beta}(t) = g(t)\tilde{\gamma}(t). \tag{6.5.12}$$

Both curves are such that $\pi\left(\tilde{\beta}(t)\right) = \gamma(t) = \pi(\tilde{\gamma}(t))$. However we shall assume that $\tilde{\gamma}(t)$ is the normal geodesic, that is $\alpha(\frac{d}{dt}\tilde{\gamma}(t)) = 0$ (where the 1-form α was defined before Eq. (6.2.3)) while $\tilde{\beta}(t)$ may or may not be the normal geodesic. In such a case, we obtain:

$$\frac{d}{dt}\tilde{\beta}(t) = \dot{g}(t)\tilde{\gamma}(t) + g(t)\frac{d}{dt}\tilde{\gamma}(t).$$

In view of the fact that $\alpha(\frac{d}{dt}\tilde{\gamma}(t)) = 0$ this result can be rewritten as follows:

$$\frac{d}{dt}\ln\tilde{\beta}(t) = \dot{g}(t)\frac{1}{g(t)} = i\frac{d\theta}{dt} \tag{6.5.13a}$$

or, equivalently,

$$\tilde{\beta}(b) = \exp\left(i\int_a^b d\theta\right)\tilde{\beta}(a), \tag{6.5.13b}$$

to be compared with Eq. (6.5.12). Next, we shall assume that the curve $\tilde{\beta}(t)$ is closed. More accurately, that its projection to the base space M is closed. Then Eq. (6.5.13b) coincides with Eq. (14) obtained by Chern in [205] as the basic patching condition used in construction of fiber bundles. It can also be interpreted as the equation for parallel transport across the bundle. For completeness, we shall briefly review both of these issues now. We begin with the patching conditions.

Let the base manifold M has a finite open covering $\mathcal{U} = \{U, V, W, \ldots\}$ such that $\pi^{-1}(U)$ is a product $U \times S^1$, etc. This means that in local coordinates $S(M)$ can be written as $(m, z_U), m \in U, z_U \in S^1$. If $U \cap V \neq \emptyset$, the points of $\pi^{-1}(U \cap V)$ thus have two sets of local coordinates: (m, z_U) and (m, z_V) such that $m \in U \cap V$ while $z_U, z_V \in S^1$ and such that

$$z_U = g_{UV}(m)z_V, \tag{6.5.14}$$

where $g_{UV}(m)$ is the complex number whose absolute value is 1. In order for such a relation to make sense, the following conditions should be imposed on transition functions $g_{UV}(m)$

$$\begin{aligned} g_{UU}(m) &= 1, m \in U, \\ g_{UV}(m)g_{VU}(m) &= 1, m \in U \cap V, \\ g_{UV}(m)g_{VW}(m)g_{WU}(m) &= 1, m \in U \cap V \cap W. \end{aligned} \tag{6.5.15}$$

The last condition is known as the cocycle condition. It is having an origin in the theory of (Čech) cohomology [102, 103].[f] Using Eq. (6.5.14) we obtain

$$d\ln z_U = d\ln g_{UV} + d\ln z_V. \tag{6.5.16}$$

In view of Eq. (6.5.13b), it is convenient to re-define $d\ln g_{UV}$ as follows:

$$\theta_U = \theta_V - id\ln g_{UV}. \tag{6.5.17}$$

By combining Eqs. (6.5.16) and (6.5.17) we obtain the 1-form

$$\omega = id\ln z_U + \theta_U = id\ln z_V + \theta_V. \tag{6.5.18}$$

In view of [139], we already used the fact that

$$\alpha = dz + xdy \quad \text{or} \quad \alpha = dz + xdy - ydx = dz + r^2 d\varphi,$$

e.g. see discussion prior to Eq. (5.5.1). Thus, indeed, the above form α is the standard horizontal contact form. It is of interest to derive the same result using the notion of parallel transport. By definition, a vector $\mathbf{v} \in T(M)$ is parallel transported along $\tilde{\gamma}(t)$ if it always remains horizontal as it is transported along $\tilde{\gamma}(t)$. Explicitly, if $\gamma(s) = \mathbf{x}(s)$, then $\tilde{\gamma}(t) = (\mathbf{x}(s), t(s)) \to (\mathbf{x}(s), z(s))$ so that, as in Eq. (6.3.9), we obtain

$$\begin{aligned} \frac{d}{ds} &= \dot{x}_i(s)\frac{\partial}{\partial x_i} + \dot{z}(s)\frac{\partial}{\partial z} \\ &:= \beta^i\left(\frac{\partial}{\partial x_i} - iA_i z\frac{\partial}{\partial z}\right), \end{aligned} \tag{6.5.19}$$

where the second line should be considered just as the definition. Using it in the condition of horizontality causes us to require $\dot{x}^i(s) = \beta^i, i = 1, 2$ so

[f] E.g. see Exercise 3.5.1.

that Eq. (6.5.19) can be rewritten as

$$\dot{z}(s) + \dot{x}^i(s)iA_i z(s) = 0. \tag{6.5.20a}$$

Explicitly,

$$\dot{z}(s) + i(2x\dot{y} - 2y\dot{x})z(s) = 0. \tag{6.5.20b}$$

From here, it follows that $z(s) = \exp(it(s))$ so that $t(s) \rightleftarrows \theta(s)$. In the case if we are dealing with circular orbits the combination $2x\dot{y} - 2y\dot{x} = 2r^2\omega\dot{\varphi}$ so that Eq. (6.5.20) is equivalent to

$$\int dt = 2r^2\omega \int d\varphi \tag{6.5.20c}$$

which is exactly the same as Eq. (3.4.13)! Now we would like to prove that the combination $2r^2\omega$ should be an integer (because for circular orbits Eq. (6.5.20c) is $4\pi r^2\omega$). For this purpose we notice that the cocycle condition $g_{UV}(m)g_{VW}(m)g_{WU}(m) = 1$ can be equivalently rewritten as

$$\ln g_{UV} + \ln g_{VW} + \ln g_{WU} = 2\pi i k_{UVW}, \tag{6.5.21}$$

where k_{UVW} is some integer. In view of Eq. (6.5.13b) it makes sense to write as well

$$\ln \tilde{g}_{UV} = \ln g_{UV} - 2\pi i k_{UV}, \tag{6.5.22}$$

where k_{UV} is yet another integer. Such a definition leads to

$$\ln \tilde{g}_{UV} + \ln \tilde{g}_{VW} + \ln \tilde{g}_{WU} = 2\pi i k_{UVW} - 2\pi i k_{UV} - 2\pi i k_{VW} - 2\pi i k_{WU}. \tag{6.5.23}$$

If the R.H.S is zero, then $\ln \tilde{g}_{UV}$ must be represented by

$$\ln \tilde{g}_{UV} = g_U - g_V \tag{6.5.24}$$

which is equivalent to Eq. (6.5.17).[g] Thus, indeed, the combination $2r^2\omega$ is an integer as we wanted to demonstrate. It remains only to demonstrate that if we replace $\mathbf{H}_1$ by S^3 the obtained results remain unchanged.

We begin with observation that the integral $2\int_0^T ds(x\dot{y} - y\dot{x}) = 2\int_C(xdy - ydx) = 2\iint dS = 2\pi R^2 \equiv 2\pi\omega r^2$ present in calculations for $\mathbf{H}_1$ can be re-obtained by finding analogous integral in S^3. In fact, we have obtained such integral already, e.g. see Eqs. (5.6.3.6)–(5.6.3.10). The two-form $\mathbf{F}$, Eq. (5.6.3.6) produces exactly half of the surface area of S^2

[g]The cohomological implications of this result are explained in [103].

(e.g. see Exercise 5.6.3.1) in accord with the already cited quotation from Montgomery [189] on p. 119. Even though his remarks are helpful, we still have to find the analog of the patching condition, Eq. (6.5.13b), for the Hopf fiber bundle. For this purpose we notice that to establish the invariance of scalar product $\langle \mathbf{v} \cdot \mathbf{v} \rangle$ with respect to rotations we have to take into account known isomorphism between the U(1) and SO(2) groups. This relationship is perfectly fine for $\mathbf{H}_1$ but inconvenient for S^3. In the last case uses of complex numbers and quaternions are much more advantageous. Thus, as the first step, we replace the Euclidean scalar product $\langle \mathbf{v} \cdot \mathbf{v} \rangle = v_1^2 + v_2^2$ by the Hermitian, i.e. $\langle \mathbf{v} \cdot \mathbf{v} \rangle = \bar{z}z$, where $z = v_1 + iv_2$. The multidimensional generalization of this result is immediate as well as its invariance with respect to the action of U(1). In particular, in such a way we can construct the defining equation for S^3, that is $|z_1|^2 + |z_2|^2 = 1$. For $(z_1, z_2) \in S^3$ the action of U(1) group takes place diagonally, that is via correspondence $(z_1, z_2) \to (z_1 e^{i\theta}, z_2 e^{i\theta})$. S^3 can be looked upon as a group manifold for SU(2) Lie group. This can be seen as follows. The pair (z_1, z_2) can be represented either by the vector $\mathbf{z}$ or by the matrix g, e.g. either by

$$\mathbf{z} = \begin{bmatrix} z_1 \\ z_2 \end{bmatrix} \quad \text{or by } g = \begin{bmatrix} z_1 & -\bar{z}_2 \\ z_2 & \bar{z}_1 \end{bmatrix}, \tag{6.5.25}$$

provided that $\deg g = 1$ implying $|z_1|^2 + |z_2|^2 = 1$. The diagonal action of U(1) on SU(2) can be looked upon as follows. Let $g_0 \in \mathrm{U}(1)$ be represented as a matrix

$$g_0 = \begin{bmatrix} e^{i\theta} & 0 \\ 0 & e^{-i\theta} \end{bmatrix},$$

then the diagonal action of U(1) on SU(2) can be represented either as

$$\mathbf{z}g_0 = \begin{bmatrix} z_1 e^{i\theta} \\ z_2 e^{i\theta} \end{bmatrix} \quad \text{or as} \quad gg_0 = \begin{bmatrix} z_1 e^{i\theta} & -z_2 e^{i\theta} \\ z_2 e^{i\theta} & \bar{z}_1 e^{-i\theta} \end{bmatrix}.$$

If $\boldsymbol{\sigma} = \{\sigma_1, \sigma_2, \sigma_3\}$ are the Pauli matrices (that is quaternions), then the Hopf map $S^3 \to S^2$ is given by

$$\mathbf{x} := \mathbf{z}^{\dagger}\boldsymbol{\sigma}\mathbf{z} \equiv \{x_1, x_2, x_3\} = \{\bar{z}_2 z_1 + \bar{z}_1 z_2, i(\bar{z}_2 z_1 - z_2 \bar{z}_1), |z_1|^2 - |z_2|^2\} \tag{6.5.26}$$

so that $\mathbf{x}^2 = 1$ by design. Just defined results allow us to construct needed patching for the Hopf circle bundle [208]. The SU(2) group can be parametrized in terms of Euler's angles (ψ, θ, φ) so that any matrix g can be represented by

$$g = \exp\left\{\frac{i}{2}\sigma_3\varphi\right\}\exp\left\{\frac{-i}{2}\sigma_2\theta\right\}\exp\left\{\frac{i}{2}\sigma_3\psi\right\}$$
$$= \begin{bmatrix} \cos\left(\frac{\theta}{2}\right)e^{i(\psi+\varphi)/2} & -\sin\left(\frac{\theta}{2}\right)e^{-i(\psi-\varphi)/2} \\ \sin\left(\frac{\theta}{2}\right)e^{i(\psi-\varphi)/2} & \cos\left(\frac{\theta}{2}\right)e^{-i(\psi+\varphi)/2} \end{bmatrix}. \quad (6.5.27)$$

By comparing this result against Eq. (6.5.25), we find

$$z_1 = \cos\left(\frac{\theta}{2}\right)e^{i(\psi+\varphi)/2}, \quad z_2 = \sin\left(\frac{\theta}{2}\right)e^{i(\psi-\varphi)/2} \quad (6.5.28)$$

or, since $z_1 = y_1 + iy_2, z_2 = y_3 + iy_4$, by using Eq. (6.5.28) we can obtain $y_1, \ldots, y_4$ explicitly as well. Furthermore, using Eq. (6.5.28) we can also obtain the already known result for $\mathbf{x}$. Explicitly, $\mathbf{x} = \{\sin\theta\cos\varphi, \sin\theta\sin\varphi, \cos\theta\}$. Even though the expression for g, Eq. (6.5.27), seemingly represents the decomposition of SU(2) as $S^2 \times \mathrm{U}(1)$, this is not true since, as we had discussed, e.g. read Sec. 5.6.6, the angle φ becomes undetermined in Eq. (6.5.27) for $\theta = 0$ and $\theta = \pi$. To fix this problem we can define two sections (patches) via equalities $\psi = \pm\varphi$ as it was done for the Dirac monopole, e.g. read Remark 5.6.3.2. With such a convention use of Eqs. (6.5.27) produces

$$\mathbf{z}_+ = \begin{bmatrix} \cos\left(\frac{\theta}{2}\right) \\ \sin\left(\frac{\theta}{2}\right)e^{-i\varphi} \end{bmatrix} \quad \text{and} \quad \mathbf{z}_- = \begin{bmatrix} \cos\left(\frac{\theta}{2}\right)e^{i\varphi} \\ \sin\left(\frac{\theta}{2}\right) \end{bmatrix} \quad (6.5.29)$$

so that the analog of Eqs. (6.5.13) and (6.5.14) is given now by

$$\mathbf{z}_- = e^{i\varphi}\mathbf{z}_+. \quad (6.5.30)$$

Thus, the Hopf fiber bundle is indeed the circle fiber bundle in accord with general theory of circle bundles [205–207].

Now, it remains to re-obtain the same result using methods and results of sub-Riemannian geometry. We notice that Theorem 5.5.1 provides us with the assurance that if we were able to obtain the needed result for $\mathbf{H}_1$, the same should be true for S^3. This observation does not give us the constructive way of proving Eq. (6.5.30) though. Fortunately, most of the needed auxiliary results are already available. Indeed, using [167] we begin with comparisons. In Sec. 5.5 we introduced two vector fields: the Reeb field R_α, Eq. (5.5.8), and the radial vector field Y defined before Eq. (5.5.8). Apart from factor 2 (or 1/2), etc. the same vector fields in [167], p. 232, are written as follows:

$$Y \rightleftarrows X_1 = x_1\partial_{x_1} + x_2\partial_{x_2} + x_3\partial_{x_3} + x_4\partial_{x_4}, \tag{6.5.31a}$$

$$R_\alpha \rightleftarrows X_2 = x_1\partial_{x_2} - x_2\partial_{x_1} + x_4\partial_{x_3} - x_3\partial_{x_4}. \tag{6.5.31b}$$

In addition, though, in [167] another two vector fields are obtained. These are

$$X_3 = -x_3\partial_{x_1} - x_4\partial_{x_2} + x_1\partial_{x_3} + x_2\partial_{x_4} \tag{6.5.31c}$$

and

$$X_4 = -x_4\partial_{x_1} + x_3\partial_{x_2} - x_2\partial_{x_3} + x_1\partial_{x_4}. \tag{6.5.31d}$$

It should be noted that when comparing different reference sources the labeling of coordinates of S^3 may differ. In the present case, for the sake of comparison, we follow the notations of [167]. In particular, let $X_1 = N$, $X = -X_2$, $Z = -X_4, T = -X_3$. Next, we relabel x_3 by $y_1$and x_4 by y_2. Then, the 1-form α, Eq. (5.5.2), can be used unchanged. In such a case we obtain the results of fundamental importance, e.g. see [167], p. 235,

$$\alpha(T) = -1, \quad \alpha(X) = 0, \quad \alpha(Z) = 0, \quad \alpha(N) = 0 \tag{6.5.32}$$

indicating that, instead of two horizontality conditions, e.g. see discussion following Eq. (6.1.1), we now have three: $\alpha(X) = 0, \alpha(Z) = 0, \alpha(N) = 0$. In [167], it is being argued that the radial vector Y (or N) should be excluded from the consideration, e.g. read p. 233 of [167]. If this is so, the distinction between the sub-Riemannian calculations for $\mathbf{H}_1$ and S^3 becomes almost non-existent. Below, we shall explain this in some detail.

We start with the observation that in the present case instead of commutator, Eq. (6.1.2a), there is an analogous commutator

$$[X, Z] = -2T \tag{6.5.33}$$

so that the vector fields X and Z are horizontal while the vector field T is vertical. Because of the noticed ambiguity between various ways of labeling of coordinates of S^3, the authors of [167] also consider the commutators

$$[X, T] = -2Z \quad \text{and} \quad [T, Z] = -2X \tag{6.5.34 a,b}$$

which they say can be used equivalently. Since from the sub-Riemannian point of view they are apparently equivalent, only the commutator given by Eq. (6.5.33) shall be used. This has to be done with caution because of the following. The Reeb field defined by Eq. (5.5.8) is such that $\frac{1}{2}\alpha(R_\alpha) = 1$. At the same time, according to Eq. (6.5.32), $\alpha(X) = \alpha(X_2) = \alpha(R_\alpha) = 0$. Instead, the role of the Reeb field is played now by the vector $-2T$. Clearly, should the authors of [167] stick to commonly accepted conventions, then the commutator $[T, Z] = -2X$ is the only one which is correct. But then, the results given in Eq. (6.5.32) should be redone. Incidentally, in the case of $\mathbf{H}_1$ our readers will be unable to spot such an ambiguity.

Exercise 6.5.3. Check that the only one of three commutators given above leads to the correct quantum mechanics. The choice between commutators depends on the correct choice of the Reeb vector field forcing the rest of the fields to be horizontal. Hint: Use the Heisenberg case (Sec. 6.1) as guidance.

Next, without loss of generality we need to look at the analog of Eq. (6.3.9). According to [167] it is by given

$$\dot{\mathbf{x}}(s) = aX + bZ + cT. \tag{6.5.35}$$

Components of the velocity vector $\dot{\mathbf{x}}(s)$ are given respectively by

$$a = \dot{x}_1 x_2 - \dot{x}_2 x_1 - \dot{y}_1 y_2 + \dot{y}_2 y_1, \tag{6.5.36a}$$

$$b = \dot{x}_1 y_2 - \dot{x}_2 y_1 - \dot{y}_1 x_2 + \dot{y}_2 x_1, \tag{6.5.36b}$$

and

$$\begin{aligned} c &= \dot{x} y_1 - \dot{y}_1 x_1 + \dot{x}_2 y_2 - \dot{y}_2 x_2 \\ &\equiv \langle \dot{x}, y\rangle - \langle \dot{y}, x\rangle. \end{aligned} \tag{6.5.36c}$$

The equivalent of horizontality condition $\dot{t} = 2y\dot{x} - 2x\dot{y}$ for $\mathbf{H}_1$ now reads

$$\langle \dot{x}, y\rangle - \langle \dot{y}, x\rangle = 0. \tag{6.5.37}$$

The equivalent of the energy functional, Eq. (6.3.21), is now given by

$$\mathcal{E}(\gamma(\tau)) = \frac{1}{2}\int_0^T d\tau (a(\tau)^2 + b(\tau)^2). \tag{6.5.38}$$

The associated equivalent of the C-C length is given by

$$l(\gamma(\tau)) = \int_0^T d\tau \sqrt{(a(\tau)^2 + b(\tau)^2)}. \tag{6.5.39}$$

The crucial Eqs. (6.3.13), or, better, Eqs. (6.2.2b) also look the same as in $\mathbf{H}_1$ case:

$$\dot{a} = -2\lambda b, \quad \dot{b} = 2\lambda a. \tag{6.5.40}$$

From these equations by multiplying the first by a and the second by b, we obtain

$$a\dot{a} + b\dot{b} = 0$$

implying

$$a^2 + b^2 = \omega r^2 = \text{const.} \tag{6.5.41}$$

Using this result in Eq. (6.5.38) produces already known result: $\mathcal{E} \simeq \omega T r^2 \simeq \omega\pi r^2$.

Thus, it only remains now to find an analog of the crucial Eq. (6.5.19). To do so some physical input is required. For instance, since Eqs. (6.5.38)–(6.5.41) are looking exactly the same as the respective equations we already obtained for $\mathbf{H}_1$, the question arises: What then makes the sub-Riemannian geometry of S^3 different from that for $\mathbf{H}_1$? That is to say, can one expect different physics from these geometries? Theorem 5.5.1 casts some doubt about such difference. This time, the problem is not in mathematical differences, however, but rather in the major setup of physical experiments. Indeed, when treated classically, the problem about the motion of charged particle in the magnetic field can be formulated as follows. The centrifugal force must balance the Lorentz force

$$\frac{mv^2}{r} = \frac{|e|}{c} v |\mathbf{H}| \tag{6.5.42}$$

providing us with the Larmor radius (e.g. see Eq. (6.5.5b)):

$$r = \frac{mcv}{|e\mathbf{H}|} = \frac{2c}{m} \frac{\sqrt{E}}{|c\mathbf{H}|}.$$

Evidently, the radius is the same irrespective of the rotation — clockwise or counterclockwise — of the electron. Accordingly, the cyclotron frequency ω given by

$$\omega = 2\pi \frac{v}{2\pi r} = \frac{|e\mathbf{H}|}{mc}$$

is also non-negative. Using these results we obtain the energy

$$\mathcal{E} = \frac{mv^2}{2} = \frac{m}{2}(r\omega)^2. \tag{6.5.43}$$

Both the classical expression, Eq. (6.5.43), and the quantum, $\mathcal{E} = \hbar\omega(n+\frac{1}{2})$, do not recognize the direction of rotation of electron (or polarity of the magnetic field). That is these results are time-reversal invariant. Now, if we use the Heisenberg space $\mathbf{H}_1$ such invariance will no longer exist! From here we conclude that only contact geometry of S^3 is describing correctly (quantum) dynamics of charged particle in the magnetic field. We would now like to provide some extra details why this is so. First of all, we would like to formally identify the energy functional, Eq. (6.5.38), obtained for S^3 with that for $\mathbf{H}_1$, that is with Eq. (6.3.21), since they both look analytically the same. Next, we do the same for the C-C length. Then, we analyze the horizontally condition, Eq. (6.5.37), from perspective of physics we just described. To facilitate matters, we would like to use the dynamical Eqs. (5.6.5.4b). In which case we take into account that at least one of the oscillators is describing circular motion. When this observation is applied to Eq. (6.5.37), the already familiar result is obtained

$$x_1\dot{y}_1 - y_1\dot{x}_1 = r^2\omega \equiv t. \tag{6.5.44}$$

Clearly, we could have done the same for another oscillator with the same outcome. What is the physics of such a replacement? Well, when we discussed the superconducting ring in Secs. 3.4. and 3.5. we noticed that the superconducting current is linked (entangled) with the magnetic field (obeying div $\mathbf{H} = 0$ as usual). This means that the magnetic field lines topologically are circles. The $\mathbf{H}_1$ geometry used for description of particle in the magnetic field is unable to take into account this fact while the S^3 geometry is capable by design: Figs. 1(a) and (b) are exibiting two kinds of Hopf rings reflecting the time-reversal symmetry of the Hopf mapping problem. Since the equation analogous to Eq. (6.5.44) was already used in derivation of the patching condition, Eq. (6.5.20b), there is no need to repeat the same calculation once again. Notice only that the combination $2\pi r^2\omega$ requires $r^2\omega$ to be an integer. Then, the factor of 2π can be interpreted as half of the surface area of the sphere S^2 of unit radius in accord with Montgomery [189] and Pincall [206]. This concludes our discussion.

Chapter 7

Abrikosov Lattices, TGB Phases in Liquid Crystals and Heisenberg Group

We discussed liquid crystals in Secs. 3.2 and 3.3. Our discussion was based on the assumption that our readers are already familiar with basics of liquid crystals physics, e.g. see [75]. We shall continue our discussion in this subsection under the same assumptions. The existence of twist grain boundary phases (TGB) in liquid crystals was suggested by Lubensly and Renn in 1988 and was almost simultaneously discovered experimentally. The paper by Lubensky [78] capitalizes on the remarkable correspondence discovered by de Gennes in 1972 [77] between properties of superconductors and smectics A liquid crystals. This correspondence is summarized in Table 1 of [78] where it is extended by incorporating new result by Lubensky and Renn. These authors argue that the Abrikosov vortex lattice phase typical for superconductors of the second kind corresponds to the TGB phase in smectics A liquid crystals. Such a phase can be visualized by comparing Fig. 9 of [78] with Fig. 4 of this book depicting twisted arrangement of planes. Such an arrangement graphically illustrates the kernel $\alpha = 0$ for the standard contact 1-form $\alpha = dz + xdy$. A comparison yields complete isomorphism of patterns. Our discussion of contact structures has been in part motivated by the noticed correspondence. This observation allows us to bring in the results of Sec. 5.1. Use of contactomorphic transformation converts the standard 1-form α into $\alpha = -\frac{1}{4}dt + \frac{1}{2}(ydx - xdy)$ whose kernel is equivalent to a more familiar (to physicists) relationship, Eq. (3.4.13), as explained in Remark 6.5.1. Since Eq. (3.4.13) can be derived from the G-L functional, Eq. (3.3.7), describing conventional superconductivity, we almost succeeded in establishing the smectics A superconductivity correspondence. In order to obtain information about the Abrikosov vortex lattice from these results, some additional steps are needed. To put things into right perspective, we provide some background about such lattices in the following remark:

Remark 7.1. In 1957 Abrikosov [209], while analyzing possible solutions of the G-L model, Eq. (3.3.7), discovered new vortex lattice-type solutions of the G-L model. They will be briefly described below. Such new solutions were discovered experimentally only in 1967 by Essmann and Trauble [210]. Mathematically rigorous justification of Abrikosov's results was obtained only in 2010 in [211] even though there were many prior results, e.g. recall already cited [114, 152].

To describe both the Abrikosov lattices in superconductors and the TGB phases in liquid crystals it is sufficient, in principle, to use the G-L functional, Eq. (3.3.7). Such a treatment will not be too illuminating however. Thus, we follow [211] instead and begin with the G-L functional in the form

$$\mathcal{F}_{\text{G-L}}[\psi, \mathbf{A}, \mathbf{H}] = \int_{\mathbf{\Omega}} dV \left\{ |\text{curl}\, \mathbf{A} - \mathbf{H}|^2 + |\nabla_{\mathbf{A}}\psi|^2 + \frac{\kappa^2}{2}(1 - |\psi|^2)^2 \right\}, \tag{7.1}$$

where $\mathbf{H}$ is the external magnetic field, $\nabla_{\mathbf{A}} = \nabla - i\mathbf{A}$, $\{\mathbf{\Omega} \in \mathbf{R}^n, n = 2, 3$. In the case if $n = 3$, the minimization of $\mathcal{F}_{\text{G-L}}$ is made under the assumption that both ψ and $\mathbf{A}$ are independent of the z-coordinate. Accordingly, the infinite constant factor is absorbed into $\mathcal{F}_{\text{G-L}}$. Instead of keeping the external magnetic field explicitly in $\mathcal{F}_{\text{G-L}}$ it is more convenient to account for it by writing

$$\mathcal{F}_{\text{G-L}}[\psi, \mathbf{A}, \mathbf{H}] = \mathcal{F}_{\text{G-L}}[\psi, \mathbf{A}, \mathbf{0}] - 2\int_{\mathbf{\Omega}} dV \mathbf{H} \cdot \text{curl}\, \mathbf{A}. \tag{7.2}$$

Then $\mathcal{F}_{\text{G-L}}[\psi, \mathbf{A}, \mathbf{0}]$ has an obvious minimum for $\mathbf{A}= 0$, $\psi = 1$ corresponding to the superconducting solution. In addition, $\mathcal{F}_{\text{G-L}}[\psi, \mathbf{A}, \mathbf{H}]$ has a minimum for $\psi = 0$ and $\text{curl}\, \mathbf{A} = \mathbf{H}$ corresponding to the normal (that is non-superconducting) solution. The nature of superconducting state is controlled by the value of parameter κ determined by the metal/alloy composition. The critical value $\kappa_c = 1/\sqrt{2}$ divides superconductors into those of the first kind for which $\kappa < 1/\sqrt{2}$ and those of the second kind for which $\kappa > 1/\sqrt{2}$. For $\kappa_c = 1/\sqrt{2}$, it is possible to use the Bogomolny-type approach analogous to that discussed in Sec. 5.6.2. We shall exclude this single parameter point from consideration since [114, 148, 152] and

Sec. 5.6.2 contain exhaustive treatment of this case. Superconductors of the first kind undergo the first order phase transition when going from normal to superconducting state. The external magnetic field below the estimated temperature-dependent critical value H_c is expelled from the superconductor. This is the essence of the already discussed Meissner effect and such a superconducting state shall be called *the Meissner phase.* For the fixed temperature the magnetic field stronger than H_c destroys superconductivity. Superconductors of the second kind exhibit more complex behavior. For the fixed temperature above the upper critical magnetic field H_{2c} they are in the normal state, for magnetic fields $H_{1c} < H < H_{2c}$ they undergo the second order phase transition leading to the formation of the Abrikosov lattice (also known as the *mixed phase*). Below H_{1c} they act as if they are superconductors of the first kind by expelling magnetic field (that is superconductor is in the Messner phase in this regime). The analysis of these possibilities can be made using the G-L functional, Eq. (7.1), where it is permissible to put $\mathbf{H} = 0$ [211]. The nonzero external magnetic field is treated separately as explained below. Variation of zero field G-L functional leads to the G-L equations discussed in Sec. 3.3. Being interested in the regime $\kappa > \kappa_c$, these results must be supplemented by the following requirements.

In addition to gauge invariance of $\mathcal{F}_{\text{G-L}}[T_\eta\psi, T_\eta\mathbf{A}] = \mathcal{F}_{\text{G-L}}[\psi, \mathbf{A}]$ with respect to transformations of the type

$$T_\eta\psi = e^{i\eta}\psi \quad \text{and} \quad T_\eta\mathbf{A} = \mathbf{A} + \nabla\eta$$

the invariance with respect to translations

$$T_t\psi(x) = \psi(x+t) \quad \text{and} \quad T_t\mathbf{A}(x) = \mathbf{A}(x+t)$$

and rotations

$$T_R\psi(x) = \psi(Rx) \quad \text{and} \quad T_R\mathbf{A}(x) = R^{-1}\mathbf{A}(Rx)$$

is also required. In the case of $\Omega = \mathbf{R}^2$ it is possible to think about the vortex lattice $\mathcal{L}$. The lattice states are those for which both ψ and $\mathbf{A}(x)$ are doubly-periodic. Below we shall argue that such double periodicity is manifestation of the Heisenberg group action. Before doing so, the gauge translations and rotations generated by

$$\psi(x+t) = e^{ig_t(x)}\psi(x) \quad \text{and} \quad \mathbf{A}(x+t) = \mathbf{A}(\mathbf{x}) + \nabla g_t(x)$$

must respect both the G-L equations and double periodicity. In such a case, let $\Omega_{\mathcal{L}}$ denote any fundamental cell of doubly-periodic lattice. Consider the magnetic flux Φ through $\Omega_{\mathcal{L}}$. Define the magnetic flux $\Phi(\mathbf{A})$ through a lattice cell via

$$\Phi(\mathbf{A}) = \int_{\Omega_{\mathcal{L}}} d\mathbf{s} \cdot \operatorname{curl} \mathbf{A}. \tag{7.3}$$

By using Eq. (3.3.9) we can rewrite this result by analogy with Eq. (3.4.13) as follows:

$$\int_{\Omega_{\mathcal{L}}} d\mathbf{s} \cdot \operatorname{curl} \mathbf{A} = \int_{\partial\Omega_{\mathcal{L}}} d\mathbf{l} \cdot \mathbf{A} = \int_{\partial\Omega_{\mathcal{L}}} d\mathbf{l} \cdot \nabla\theta, \tag{7.4}$$

where it is assumed that the periodicity of $\psi = |\psi|e^{i\theta(\mathbf{x})}$ is assured by the phase factor $\theta(\mathbf{x})$. Since the function ψ is single valued, using Eq. (7.4), we obtain

$$\int_{\Omega_{\mathcal{L}}} d\mathbf{s} \cdot \operatorname{curl} \mathbf{A} = 2\pi n, \tag{7.5}$$

where $n = 0, \pm 1, \pm 2, \ldots$. The obtained result is in accord with that displayed in Eq. (3.4.13). This time, however, it can be interpreted slightly differently. Let $|\Omega_{\mathcal{L}}|$ be the area of the fundamental cell. Then, the external magnetic field $\mathbf{H}$ can be taken into account by the following requirement

$$|\mathbf{H}| \equiv b = \frac{1}{|\Omega_{\mathcal{L}}|} \int_{\Omega_{\mathcal{L}}} d\mathbf{s} \cdot \operatorname{curl} \mathbf{A} \equiv \langle \operatorname{curl} \mathbf{A} \rangle_{\mathcal{L}} = \frac{2\pi n}{|\Omega_{\mathcal{L}}|}. \tag{7.6a}$$

If the parameter b is assigned, it can be used as a condition determining the area of fundamental cell

$$|\Omega_{\mathcal{L}}| = \frac{2\pi n}{b}. \tag{7.6b}$$

The mathematical problem of minimization of the G-L functional, Eq. (7.1), can now be formulated as follows. For $d = 2$ define the energy $\mathcal{E}(\psi, \mathbf{A})$ via

$$\mathcal{E}(\psi, \mathbf{A}) = \frac{1}{|\Omega_{\mathcal{L}}|} \mathcal{F}_{\text{G-L}}[\psi, \mathbf{A}, \mathbf{0}] \tag{7.7}$$

and find the minimizers $\psi, \mathbf{A}$ for $\mathcal{E}$ under the condition that the average flux b is pre-assigned.

In view of Remark 6.5.1, the chain of Eqs. (7.4) determines the Heisenberg group. This observation is sufficient to establish the existence of Abrikosov lattice bypassing variational calculations of [211]. Before providing a sketch of arguments to establish such an existence, we would like to rewrite the obtained results in the language of liquid crystals where, in view of Fig. 4, uses of contact/Heisenberg geometry are natural.

Reformulation of the obtained results in the liquid crystalline language is nontrivial as can be seen from reading of [78]. Indeed, it requires discussion of the smectic-nematic-type transition which, thanks to de Gennes work, [77], can be described in superconductor-normal metal transition language. To describe the TGB phases it is necessary to bring into play also the cholesterics [75, 78]. Let us discuss this issue in some detail. The most optimal way to bring into play the cholesterics lies in modification of the distortional free energy density f_d defined by Eq. (3.2.1). Following [212] the modified f_d can be written as follows:

$$\begin{aligned} f_d[\mathbf{n}] &= K_1(\operatorname{div}\mathbf{n})^2 + K_2(\mathbf{n}\cdot\operatorname{curl}\mathbf{n}+\tau)^2 + K_3(\mathbf{n}\times\operatorname{curl}\mathbf{n})^2 \\ &\quad + (K_2+K_4)[\operatorname{tr}(\nabla\mathbf{n})^3 - (\operatorname{div}\mathbf{n})^2]. \end{aligned} \tag{7.8}$$

Terms in the second line take into account the boundary effects. They may or may not be included into free energy $f_d[\mathbf{n}]$ since if they are included, they can be reduced to the total divergence. Uses of Stoke's theorem makes them only important at the boundaries. But the same effects can be achieved by minimizing the total free energy functional (see below) and imposing the appropriate boundary conditions on the equations afterwards. In addition, the above expression for $f_d[\mathbf{n}]$ does not take into account the constraint $\mathbf{n}^2 = 1$ which is absent in superconductors. The presence of the parameter τ in Eq. (7.8) is responsible for twisting effects. These are absent in nematics but present in cholesterics [78]. Next, to make the correspondence with superconductivity mathematically well-defined, we have to assume, following [212], that $K_2 = K_3$. In such a case we obtain the result of central importance

$$(\mathbf{n}\cdot\operatorname{curl}\mathbf{n}+\tau)^2 + (\mathbf{n}\times\operatorname{curl}\mathbf{n})^2 = (\operatorname{curl}\mathbf{n}+\tau\mathbf{n})^2. \tag{7.9}$$

Substituting this into Eq. (7.8) and comparing with Eq. (3.2.11), we recognize that: (a) the director $\mathbf{n}$ has exactly the same meaning as the vector potential $\mathbf{A}$ for the static electromagnetic field since the Lagrange multiplier K_1 is enforcing the constraint $\nabla\cdot\mathbf{A} = 0$. If we write

$$f_d[\mathbf{n}] = K_1(\operatorname{div}\mathbf{n})^2 + K_2(\operatorname{curl}\mathbf{n}+\tau\mathbf{n})^2 \tag{7.10}$$

we recognize at once that the Lagrange multiplier K_2 enforces the already familiar force-free condition, Eq. (1.3.5). After these remarks we are ready to write the analog of the G-L functional, Eq. (7.1). In the appropriately chosen system of units it is given by

$$\mathcal{G}_{\text{G-L}}[\psi, \mathbf{A}, \tau] = \int_{\Omega} dV \Big\{ K_1 (\operatorname{div} \mathbf{A})^2 + K_2 |\operatorname{curl} \mathbf{A} + \tau \mathbf{A}|^2 + |\nabla_{\mathbf{A}} \psi|^2 + \frac{\kappa^2}{2}(1 - |\psi|^2)^2 \Big\}. \tag{7.11}$$

Evidently, such a form of the G-L-type functional takes into account the constraint $\mathbf{n}^2 = 1$ as well. Mathematically sound results are obtained in the limit $K_1, K_2 \to \infty$. In such a case the force-free equations

$$\operatorname{div} \mathbf{A} = \mathbf{0} \quad \text{and} \quad \operatorname{curl} \mathbf{A} + \tau \mathbf{A} = \mathbf{0}. \tag{7.12}$$

are enforced. Since the parameter τ is constant, the field $\mathbf{A}$ is Trkalian (Sec. 3.2.) implying equivalence of equations

$$\operatorname{curl} \mathbf{A} + \tau \mathbf{A} = \mathbf{0} \quad \text{and} \quad \operatorname{curlcurl} \mathbf{A} + \tau \operatorname{curl} \mathbf{A} = \mathbf{0}, \text{ etc.}$$

Without loss of generality we can always write $\tau \mathbf{A} = -\mathbf{H}$ so that the constraint

$$\operatorname{curl} \mathbf{A} - \mathbf{H} = \mathbf{0} \tag{7.13}$$

becomes exactly the same as in the G-L superconductivity functional, Eq. (7.1). If only the divergence-free vector fields are used, the equivalence between superconductors and liquid crystals becomes complete. The order parameter ψ for superconductors acquires the following meaning for liquid crystals.

When the external field $\tau = 0$, this order parameter describes the phase transition between smectics (A or C) phases and nematics [78]. When $\tau \neq 0$, the disordered phase is cholesteric and the ordered phase characterized by the nonzero ψ could be either smectic (A or C) — an analog of the Meissner phase in superconductors — or the TGB-analog of the mixed phase (or of Abrikosov lattice). Such an identification runs into some mathematical problems though. Indeed, in the case of superconductors, the results were obtained for two-dimensional domain Ω while in the present case the domain Ω is three-dimensional. To overcome this difficulty, the following observations are helpful. The force-free condition $\operatorname{curl} \mathbf{n} + \tau \mathbf{n} = \mathbf{0}$ can be equivalently rewritten as $\mathbf{n} \cdot \operatorname{curl} \mathbf{n} = -\tau$ since $\mathbf{n}^2 = 1$. In such a form it was introduced in Sec. 3.2 in connection with the description of cholesterics. At the same time, we know that the force-free equation is obtainable variationally, e.g. see Woltjer's theorem, Sec. 3.1. As discussed at length in Chap. 5, the force-free fields have their origin in contact geometry. How can this fact be explicitly used in the present case?

To inject some physics, we shall use [78, 213]. In particular, we notice that in both smectics A and smectics C the molecules segregate into layers (planes) forming a solid (that is periodic structure) in one direction (say, in z-direction) and liquid in other two (say, in $x-y$ plane). Let $\mathbf{N}$ be the normal to layers (that is positive direction in z-axis). Then the director $\mathbf{n}$ may or may not be collinear with $\mathbf{N}$. When it is collinear with $\mathbf{N}$, we are dealing with smectics A phase while if it is not collinear, the director has some nonzero projection onto $x - y$ plane. In this case we are dealing with the smectics C phase. Now take a look at Fig. 4 and complete it in your imagination in z-direction by planes. Add the director $\mathbf{n}$ to each plane and observe that the director for all planes is rotating around the x-axis. This is the TGB phase as depicted in Fig. 9.6.4(b) of the textbook by Chaikin and Lubensky, [213]. This is also the standard contact structure! Now we are having a periodicity in z-direction (because of the smectic order) and periodicity in x-direction (because of rotations around the x-axis). That is we are having double periodicity already encountered for the Abrikosov lattices! But in superconductors this periodicity was for two-dimensional layers while here, apparently, we are dealing with the three-dimensional pattern. Let us forget for the moment about the TGB phase, then we are dealing with smectics for which there is solid-type periodicity in z-direction. Following de Gennes [77] the (complex) order parameter $\psi(r)$ for the smectics can be written as $\psi(r) = |\psi|e^{iq_0 z}$ where $q_0 = \frac{2\pi}{d}$ with d denoting the layer spacing. In such notations the covariant derivative $\nabla_{\mathbf{A}}$ in Eq. (7.11) is given by $\nabla_{\mathbf{A}} = \nabla - iq_0\mathbf{A}$ since now $\mathbf{n} \rightleftarrows \mathbf{A}$. Thus, q_0 plays the role of a charge. In the case of double periodicity typical for the TGB phases the order parameter should be expressed in terms of elliptic functions. In the simplest case, following [78], we can write it as $\psi(r) = |\psi|e^{i(q_0 z \pm u(x))}$ where the phase $u(x)$ is expected to be periodic in x-direction. Such a presentation is too restrictive nevertheless. Instead, by analogy with superconductors, it is more advantageous to introduce another phase function $\theta(z, x)$ so that the order parameter can be written as $\psi(r) = |\psi|e^{i\theta(x,z)}$. By looking at the functional Eq. (7.11) it is convenient to choose $|\psi| = 1$. In such a case we end up with the already discussed London limit, Eq. (3.4.1), which in the present case allows us to use word-for-word arguments we used for 2D superconductors and with the same outcome. It only remains to demonstrate that the double periodicity is synonymous/indicative with/to the Heisenberg group. Fortunately, this was already done by Mumford, [193], so that interested readers are encouraged to look into this reference. Alternatively, more physically motivated is the introductory article by Cartier

[192]. Connections of double periodicity with Nil manifolds are examined in detail in [194] while the encyclopedic book by Thurston [183], explains the place of Nil manifolds in general (Thurston's) classification scheme of 3-manifolds. These results had been put into still broader and more geometrical context by Goldman [214] as follows.

Hyperbolic geometry is the most important of all geometries. Its physical significance stems from its immediate connection with the Lorentz group [183]. The simplest model of hyperbolic geometry is the upper Poincaré plane which via conformal transformation can be mapped into unit disc. Geodesics in both the upper plane and the disc are circular segments. The multivariable extension of this model results in either the half-space or the ball models of hyperbolic space. Isometries of hyperbolic space are groups of motions preserving geodesics. That is to say, their application is not moving the ends of geodesics located at the boundary of hyperbolic space. Such a boundary is the surface of a ball in the hyperbolic ball model. There are groups of motions acting at this surface as explained in our work [215]. In the case of three-dimensional hyperbolic space, these groups are known as Kleinian groups. If we enlarge the real hyperbolic space by making it complex, the ball model for the real hyperbolic space will survive. The boundary groups (acting on the ball surface) for such complex hyperbolic spaces are multidimensional Heisenberg groups. They replace the Kleinian groups. *Thus, quantum mechanics can be looked upon as sub-Riemannian dynamics taking place at the boundary of the complex hyperbolic space.*

Exercise 7.1. Use the above brief description of the complex hyperbolic space in order to connect it with the light-cone model of Minkowski space on which the Lorentz group is acting. Hint: Read Sec. 2.3 of Thurston's book [183]. Figure out how quantum mechanics (that is Heisenberg group) can be recovered from such a comparison. Hint: read [214, 215] in order to locate the exact place of sub-Riemannian geometry within the complex hyperbolic geometry.

Chapter 8

Sub-Riemannian Geometry, Spin Dynamics and Quantum-Classical Optimal Control

8.1 General Information

Efforts to make quantum computers are based on two assumptions/postulates. These can be formulated as follows. The first is originating from 1937 Schrödinger's paper, [216], written shortly after the famous Einstein, Podolsky and Rosen (EPR) paper of 1935 [217] was published. In [216], Schrödinger claimed that the *entanglement is the essence of quantum mechanics*. The problem lies, however, in the fact that the word "entanglement" is difficult to formalize mathematically.[a] [218, 219] provide seven definitions of entanglement given at different times by different authors. The most famous among these definitions is attributed to Bell [220]. Recent comments on Bell's work and on the status of hidden variables in quantum mechanics can be found in [221]. Experimental verification of quantum entanglement along with comprehensive theoretical treatment of this phenomenon is given, for example, in recently published [222]. To have entanglement requires having at least two quantum particles (e.g. spins). The ability of states to be entangled is deeply connected with the violation of Bell's inequalities. At the same time, the existence of entangled states leads to serious problems with causality and nonlocality in quantum mechanics and quantum field theory. According to [218–220] the causality and entanglement are mutually exclusive phenomena. Thus, the problem remains "to reconcile irreconcilable". This issue should be considered as totally independent of another issue which can be formulated as follows: How

[a]See, however, Appendix C.

important are entanglements in quantum computation?[b] Put differently: Since entanglements are believed to be purely quantum phenomena (actually, they are not, as explained below, especially in Appendix D), to what extent account of entanglements in quantum computations leads to significant speed-up of these computations as compared to classical? Exactly this issue was addressed in highly cited paper by Jozsa and Linden [223] where negative answer to the above question was obtained. Furthermore, the same authors argued that the quantum computational algorithm involving entanglements can be effectively simulated classically to within a prescribed in advance accuracy (tolerance). At about the same time Valliant came to the same conclusions by using different methods [224].

The second postulate concerns with the properties of coherence and decoherence. The easiest way to define decoherence is through statement that due to interactions with environment, quantum mechanical states are having finite lifetimes. If both the entanglement and the decoherence are purely quantum phenomena, then the interaction of quantum computer with the environment should be treated quantum mechanically as well. Evidently, at some level of such joint treatment, classical reality must emerge. This task seems to be very difficult (if not impossible!) to achieve as explained in the book by Zagoskin [225]. It is helpful at this point to bring the quotation from the paper [226] by Zagoskin and Blais.

"From physicist's standpoint, the most interesting part of quantum computing research may well be the possibility to probe the boundary between the quantum and the classical worlds. *The more macroscopic are the structures involved, the better. So far, the most "**macroscopic**" qubit prototypes that have been studied in the laboratory are certain kinds of superconducting qubits... Non-superconducting qubits realized so far are all essentially "microscopic"*.... The advantage of quantum superconducting cirquits (QSC) as implementation of qubits is, first and foremost, due to the *macroscopically quantum coherent ground state* of suprconductors, which (a) supports non-dissipative flow of electrical current and (b) suppresses or outright eliminates low energy elementary excitations... as solid state based device, a superconducting qubit and more complex QSC can be more easily scaled up, can have significant density, and can be produced using well-developed set of design and fabrication methods. *Moreover, since these cirquits are macroscopic, it can be simpler to manipulate and to read their state.* The necessity to operate at low temperatures is not a disadvantage

[b]E.g. read Appendix C.

of QSC, since almost all the quantum information processing requires low temperatures in order to suppress the effects of noise."

Thus if we accept the point of view advocated in [225, 226], we should first of all recall the existence of superconducting rings — macroscopic objects behaving quantum mechanically — discussed in Feynman's Lectures in Physics and in Sec. 3.4 of this book. The question arises: If we were able to treat the superconducting rings classically (that is using sub-Riemannian geometry), is there a way to describe particles with spin (that is qubits) using the formalism of sub-Riemannian geometry (that is of optimal control)? In this chapter we provide the affirmative answer to this question.

It is commonly believed [227] that known theory of optimal control is only applicable to classically behaving systems and that dynamics of quantum systems requires development of brand new quantum theory of optimal control. The sub-Riemannian dynamics of particle in the constant magnetic field yielding correct quantum mechanical spectrum as discussed in Secs. 6.4 and 6.5 provides a counterexample to such a belief. By capitalizing on this counterexample, we shall provide enough evidence making new theory of quantum optimal control unnecessary.

As the entanglement phenomena believed to represent the fundamental difference between classical and quantum mechanics, two recent papers, [228, 229], demonstrate the entanglement at the purely classical level. The phenomenon of classical entanglement, in fact, is known for some time (e.g. at least since 1954). It was noticed in correlated motion of two tagged particles undergoing Brownian dynamics in colloidal suspensions (e.g. read [11] listed in [9], on dynamics of colloidal suspensions). Examples just mentioned suggest that quantum computers can be somehow modelled classically in accord with rigorous mathematical results of [223, 224]. Surely, such classical description is not quite traditional still as discussed in the rest of this chapter and in Appendix D.

8.2 Quantum Computers Paradigm and Dynamics of 2-Level Quantum Systems

8.2.1 *What is Quantum Computation and Quantum Computer?*

We begin with formulating *Di Vincenzo criteria* by summarizing key features of all quantum computers [226, 230]. These are:

1. A qubit is a 2-level quantum system, which *can be controlled* and read out.

2. A quantum computer is a set of N qubits where, in addition, certain 2-qubit operations can be performed.
3. A decoherence time (i.e. the time during which the system maintains quantum coherence) must exceed the time necessary to perform a single or 2-qubit operation by a wide margin.

Actual/physical realizations of quantum computers as well as advantages and disadvantages of a partricular design (that is trends in the design of quantum computers) can be found at the following website: http://quist.lanl.gov. It should be noted, though, that even this database does not include all possible types of quantum computers. For this reason our readers may want to consult the systematically updated web links such as http://en.wikipedia.org/wiki/Quantum_computer (general information about quantum computers) or http://en.wikipedia.org/wiki/History_of_quantum_computing (history of quantum computing), etc.

8.2.2 *Connecting Quantum 2-Level Systems with Classical Reality by Using the Hopf Map*

The design of all quantum computers always begins with the detailed study of 2-level quantum systems. Such 2-level systems are associated with dynamics of spin-1/2 particle in the magnetic field **H** as discussed, for example, in Feynman's lectures on physics [99]. The quantum Hamiltonian $\hat{H}$ for this problem is known to be (up to unimportant for us constant) $\hat{H} = -\sigma\cdot\mathbf{H}$ (with the same notations as in Sec. 6.5) so that the Schrödinger (actually, the Pauli–Schrödinger) equation can be written as

$$i\frac{d}{dt}\chi = \hat{H}\chi. \tag{8.2.2.1a}$$

The two-component spinor $\chi^T = (a, b)$ is normalized in the usual way, that is as $(\chi^T)^*\chi = 1$. Explicitly written, Eq. (8.2.2.1a) acquires the following form:

$$i\frac{d}{dt}\begin{pmatrix} a \\ b \end{pmatrix} = -\frac{1}{2}\begin{pmatrix} H_z & H_x - iH_y \\ H_x + iH_y & -H_z \end{pmatrix}\begin{pmatrix} a \\ b \end{pmatrix}. \tag{8.2.2.1b}$$

By definition, a pure state φ, that is a qubit, is defined by $\varphi = a\binom{0}{1} + b\binom{1}{0} \equiv a\,|0\rangle + b\,|1\rangle$ with both a and b being some complex numbers subject to normalization condition $|a|^2 + |b|^2 = 1$. Mathematically, this condition determines a 3-sphere as discussed prior to Eq. (6.5.25). This time, however, following Feynman *et al.*, [231], it is convenient to introduce the combinations $x = ab^* + ba^*, y = i(ab^* - ba^*)$ and $z = aa^* - bb^*$. Use of these combinations along with Eq. (8.2.1b) produces

$$\frac{d}{dt}\begin{pmatrix} x \\ y \\ z \end{pmatrix} = \begin{pmatrix} 0 & H_z & -H_y \\ -H_z & 0 & H_x \\ H_y & -H_x & 0 \end{pmatrix}\begin{pmatrix} x \\ y \\ z \end{pmatrix}. \tag{8.2.2.1c}$$

Thus, by such Hopf (e.g. see Eq. (6.5.26)) mapping quantum dynamics on S^3 is mapped into classical dynamics on S^2. Since any ellipsoid can be conformally mapped into S^2, it is convenient to think about Eq. (8.2.2.1c) as describing the rigid body rotations. From comprehensive treatise on mechanics by Pars, [232], Chap. VII, we find that the above equation indeed describes the kinematics of the rotating rigid body with respect to (fixed) coordinate system whose origin coincides with the center-of-mass of the body. In such a setting it is possible to make the following identifications: $H_z = \omega_3, H_y = \omega_2$ and $H_x = \omega_1$, where ω_i, $i = 1$–3, are components of the vector $\boldsymbol{\omega}$ of angular velocities of the body with respect to the fixed frame, e.g. read [136], Chap. 6. This observation is sufficient for connecting dynamics of 2-level quantum system with the sub-Riemannian dynamics. In fact, it is possible to accomplish even more than just to establish such a connection! In order to explain why this is so, some facts from mechanics of rigid body rotations are helpful. For the sake of uninterrupted reading we provide a condensed summary of relevant results.

8.2.3 *Some Mathematical Facts about the Rigid Body Rotations and Their Applications to Physical Problems*

Let g be an element of some Lie group G. Select another element $h \in G$ and define the right and left translations respectively as $R_h(g) = gh$ and $L_h(g) = hg \, \forall \, g \in G$. These definitions allow us to introduce the inner automorphism of G, that is: $G \to G$, via $A_g = R_{g^{-1}} L_g$. It sends $h \in G$ to ghg^{-1}. This result is valid for any group (not just the Lie group!). To apply it to Lie groups we notice that each element of such a group is parameter-dependent (e.g. in the case of rotations we are having Euler's angles whose dynamics is parametrized by time t). Thus, let $t \to g(t)$ be a curve/trajectory issued from e (that is from the unit element) so that

$g(t) \simeq e + t\xi + o(t)$. Consider an inner automorphism

$$\begin{aligned} g(t)h(s)g(t)^{-1} &= (e + t\xi + \cdots)(e + s\omega + \cdots)(e + t\xi + \cdots)^{-1} \\ &= e + s[\omega + t(\xi\omega - \omega\xi) + o(t)] + o(s). \end{aligned} \tag{8.2.3.1}$$

In general, each element g may depend on several parameters. The tangent space to G at e is the vector space of the Lie algebra $\mathfrak{g}$ of G so that, for instance, $\xi \in \mathfrak{g}$, etc. Consider a special case of Eq. (8.2.3.1) by replacing $h(s)$ by ω. Then the combination $g\omega g^{-1}$ is the map $\mathfrak{g} \to \mathfrak{g} \, \forall g \in G$ and $\forall \omega \in \mathfrak{g}$. Such a combination is called $Ad_{g(t)}\ \omega$. In addition, we define another map $\mathfrak{g} \to \mathfrak{g}$ via

$$ad_\xi \omega = \left.\frac{d}{dt}\right|_{t=0} Ad_{g(t)}\omega = [\xi, \omega] \tag{8.2.3.2}$$

with $[\xi, \omega] = \xi\omega - \omega\xi$. The correspondence $g \to Ad_{g(t)}$ is called the *adjoint representation* of the Lie group G while the correspondence $\xi \to ad_\xi$ is called the *adjoint representation of the Lie algebra* $\mathfrak{g}$.

These results are sufficient for the first set of applications. In particular, in theory of rigid body rotations [68, 136, 232] one is using the skew symmetric matrices characteristic for the Lie algebra of SO(3) — the Lie group of rotations. This fact allows us to establish a remarkable isomorphism between the three-dimensional vectors $\boldsymbol{\xi}$ and 3×3 skew symmetric matrices $\boldsymbol{\xi}$ so that the commutator $[\xi, \omega] = \xi\omega - \omega\xi$ of such matrices is in one-to-one correspondence with the cross product of vectors $\boldsymbol{\xi} \times \boldsymbol{\omega}$.

Indeed, the basis of the Lie algebra $\mathfrak{so}(3)$ is made of the following three skew-symmetric matrices

$$X_1 = \begin{pmatrix} 0 & 0 & 0 \\ 0 & 0 & -1 \\ 0 & 1 & 0 \end{pmatrix}, \quad X_2 = \begin{pmatrix} 0 & 0 & 1 \\ 0 & 0 & 0 \\ -1 & 0 & 0 \end{pmatrix} \quad \text{and} \quad X_3 = \begin{pmatrix} 0 & -1 & 0 \\ 1 & 0 & 0 \\ 0 & 0 & 0 \end{pmatrix}. \tag{8.2.3.3}$$

Using them we obtain:

$$[X_1, X_2] = X_3, \quad [X_2, X_3] = X_1, \quad [X_3, X_1] = X_2. \tag{8.2.3.4a}$$

This result demonstrates that, indeed, the commutators of such matrices are equivalent to the cross products. Unfortunately, such an isomorphism between vectors and skew symmetric matrices exist only in three dimensions. It can be conveniently written as follows. Consider the map $\Psi \colon \bar{\mathbf{A}} \to \mathbf{X}$

which is explicitly given as

$$\bar{\mathbf{A}} \equiv \begin{pmatrix} 0 & -z & y \\ z & 0 & -x \\ -y & x & 0 \end{pmatrix} \to \mathbf{X} = \begin{pmatrix} x \\ y \\ z \end{pmatrix}.$$

Accordingly,

$$\bar{\mathbf{A}}\mathbf{X} \to \mathbf{\Psi}(\bar{\mathbf{A}}) \times \mathbf{X}. \tag{8.2.3.4b}$$

We had already encountered such type of mapping previously, in Sec. 6.5, e.g. see Eqs. (6.5.25)–(6.5.28). It will also be used below.

In the meantime, it is still possible to complicate matters by converting vectors, e.g. $\boldsymbol{\xi}$ and $\boldsymbol{\omega}$, into vector fields by making them $\mathbf{x}$-dependent, $\mathbf{x} \in \mathbf{R}^3$. In such a case it is possible to prove the following remarkable identity [68, 136]:

$$[\mathbf{a}, \mathbf{b}] = \mathrm{curl}(\mathbf{a} \times \mathbf{b}). \tag{8.2.3.5a}$$

This result follows from more general identity

$$\mathrm{curl}(\mathbf{a} \times \mathbf{b}) = [\mathbf{a}, \mathbf{b}] + \mathbf{a}\,\mathrm{div}\,\mathbf{b} - \mathbf{b}\,\mathrm{div}\,\mathbf{a} \tag{8.2.3.5b}$$

valid for any 3-manifolds. If we apply this result to Eq. (1.2.1) describing dynamics of Euler's incompressible fluid, then if $\mathbf{a}$ is $\tilde{\boldsymbol{\omega}}$ and $\mathbf{b}$ is $\mathbf{v}$, we obtain,

$$\frac{\partial}{\partial t}\tilde{\boldsymbol{\omega}} = [\tilde{\boldsymbol{\omega}}, \mathbf{v}]. \tag{8.2.3.6}$$

This equation is easily recognizable as Euler's equation describing dynamics of rotations of the rigid body. Since the vectors $\tilde{\boldsymbol{\omega}}$ and $\mathbf{v}$ are $\mathbf{x}$-dependent such an identification should be made with some caution. Details are explained in [68, 136]. Nevertheless, because of Arnold's inequality, Eq. (3.1.12), and its improved version, Eq. (3.2.21), valid even for non-Abelian gauge fields, it is possible to reduce the infinite-dimensional result, Eq. (8.2.3.6), to that used for description of dynamics of three-dimensional rotating rigid body. *Under such conditions the dynamics of 2-level quantum system is mapped into dynamics of classical Euler-type fluids. Thus, at least formally, we can think about the dynamics of such fluids in terms of quantum qubits!* Since, according to Feynman, [99], the dynamics of superconducting Josephson's junctions is described in terms of dynamics of 2-level system too — the result which is used essentially in quantum computers [225, 226] — we just re-obtained back the correspondence between the dynamics of ideal fluids and superconductivity discovered originally by Fröhlich [11]!

Once the adjoint representation of Lie groups and Lie algebras is defined, it becomes possible to define the coadjoint representation. It is needed, for example, for rigorous mathematical description of Euler's equations. A readable self-contained description of this mathematical concept can be found in our work, [233]. In this book it is presented in the Appendix D. It is easy to comprehend if we use it in the appropriate physical context. E.g. it can be thought of as some kind of a Fourier transform. Then, instead of the Lie algebra $\mathfrak{g}$, one can think of the vector space $\mathfrak{g}^*$ dual to $\mathfrak{g}$. As in Fourier analysis, one can think about the vectors of $\mathfrak{g}^*$ as being formed by the space of linear functionals defined on $\mathfrak{g}$. That is one needs to define some kind of an inner scalar product, just like it is done in quantum mechanics. Along with such a product it is possible then to define the coadjoint operators $Ad^*_{g(t)}$ and ad^*_ξ so that, for instance, $(Ad^*_g\xi)(\omega) = \xi(Ad_{g^{-1}}\omega)$ is the map $\mathfrak{g}^* \to \mathfrak{g}^*$ such that $g \in G$, $\omega \in \mathfrak{g}$ and $\xi \in \mathfrak{g}^*$. Analogously, $(ad^*_v\omega)(u) = \omega(ad_v u) = \omega([v,u])$ with both v and $u \in \mathfrak{g}$ while $\omega \in \mathfrak{g}^*$. These abstract concepts are mainly needed for multidimensional generalizations of the theory of rigid body rotations.[c] In three dimensions it is easy to rephrase these results in simple physical terms as it is done in the textbooks on mechanics for physicists. Basically, we are dealing with two coordinate systems. One is assumed to be fixed (we shall call it "*spatial*" or (s)) while another (we shall call it "*body*", or "*corps*" in French (c)) is rotating with respect to the first. Both coordinate systems have the same origin. The question arises: How are physical observables written in terms of coordinates of one of these systems related to those of the other? Specifically, if we are talking about the vector of the total angular momentum $\mathbf{M}_c(t)$ of the body in c-system, it is related to $\mathbf{M}_s(t)$ defined in s-frame via

$$\mathbf{M}_c(t) = Ad^*_{g(t)}\mathbf{M}_s(t). \qquad (8.2.3.7)$$

Any textbook on mechanics for physicists, e.g. see [234, 235], will have such a relation written explicitly. In particular, if the moments of inertia tensor is labelled as $\mathbf{I}$, then the angular velocity $\boldsymbol{\Omega} = \mathbf{I}^{-1}\mathbf{M}_c(t)$. Such a velocity in c-system has its analog $\boldsymbol{\omega}$ in s-system. Let us discuss this relationship in some detail.

Euler's equations in c-system can be written as

$$\frac{d}{dt}\mathbf{M}_c(t) = ad^*_\Omega\mathbf{M}_c(t). \qquad (8.2.3.8a)$$

[c]See, however, Appendix D, Sec. D.2.17. In it, these abstract notations and concepts become self-obvious in the context of selected physical examples.

Explicitly, following [136] we obtain,

$$\frac{d}{dt}\mathbf{M}_c(t) = [\mathbf{M}_c, \boldsymbol{\Omega}]. \tag{8.2.3.8b}$$

Since we already discussed the isomorphism between the cross products and commutators, the above equation can also be rewritten as

$$\frac{d}{dt}\mathbf{M}_c(t) = \mathbf{M}_c \times \boldsymbol{\Omega}. \tag{8.2.3.8c}$$

Following [234] we notice that up to an unimportant constant the same equation describes the motion of, say, an electron in the magnetic field, e.g. see Eq. (1.2.4). That is to say, the equation

$$\frac{d}{dt}\mathbf{L} = \mathbf{L} \times \mathbf{H} \tag{8.2.3.9}$$

in which $\mathbf{L} = \mathbf{r} \times \mathbf{v}$ and $\mathbf{H}$ is the magnetic field, is equivalent to Eq. (1.2.4) (or Eq. (6.2.2c)). Thus, the equation describing the motion of an electron in the magnetic field is exactly the same mathematically as Euler's equation for rotating body. But in Chap. 6 we demonstrated that such magnetic problem is the benchmark problem of sub-Riemannian geometry! Thus, the same must be true in the present case. This time, we shall treat this problem differently. Our treatment will be complementary to that discussed in Chap. 6.

8.2.4 *Poinsot versus Euler and Kirchhoff*

We have already learned that problems of sub-Riemannian geometry can be reformulated as problems of optimal control and that such problems typically originate from consideration of the Lagrangian/Hamiltonian dynamics in the presence of nonholonomic constraints. In this section we demonstrate that, even without knowing results of previous section, it is possible to rewrite dynamics of 2-level quantum spin system in terms of dynamics of rigid ball rolling on two-dimensional Euclidean plane without slippage. This dynamics provides a representative example of dynamical system with nonholonomic constraint. The connection between these two problems is important to put in some historical perspective.

Independently from Euler, Poinsot also developed theory of rotating bodies in the middle of 18th century. His theory is described in detail in the monograph by Felix Klein and Arnold Sommerfeld whose English edition was published only in 2010 [236], that is 110 years after its original German edition! The question arises: Why was there such a delay with

English edition and why it was necessary to publish this edition in 2010? Theoretical methods describing dynamics of rigid body discussed in [236] are as important today as they were more than 110 years ago. This concerns, in particular, the work of Poinsot on dynamics of rigid bodies. Although his contributions are briefly described in [136] and [234], the monograph by Klein and Sommerfeld provides much more detailed account. Instead of considering Poinsot contributions to dynamics of rigid body in detail, for our needs it is sufficient to discuss only the simplest case of a round sphere (instead of an ellipsoid of inertia treated by Poinsot) rolling over the plane without slippage. Such a problem was recently discussed in [237] in connection with dynamics of spin-1/2 quantum 2-level system. We would like to discuss some facts from this work.

Let a sphere of radius R roll over the curve Γ drawn on the plane $\mathbf{R}^2$. Let $\mathbf{t}$ be a unit vector tangent to Γ, and let $\mathbf{n}$ be a unit vector of the normal to the plane originating at the point of contact of the ball with the plane. Then, one can construct yet another vector $\mathbf{u} = \mathbf{n} \times \mathbf{t}$ so that if $\mathbf{X}$ is the coordinate on the surface of the sphere with respect to its center, then the equation of motion for $\mathbf{X}$ is given by

$$\frac{d}{dt}\mathbf{X} = \frac{\mathbf{u}}{R} \times \mathbf{X}. \tag{8.2.4.1a}$$

Denote the combination $\frac{\mathbf{u}}{R} = -\mathbf{H}$, so that the previous equation acquires the form

$$\frac{d}{dt}\mathbf{X} = \mathbf{X} \times \mathbf{H} \tag{8.2.4.1b}$$

identical with the already obtained Eq. (8.2.3.9). Using this correspondence one arrives at Eq. (8.2.2.1c). The obtained result lies at the foundation of the Poinsot approach to dynamics of the rotating rigid bodies even though the actual computations and arguments are somewhat different [136, 234]. Since mechanically we are now dealing with the dynamics of bodies with nonholonomic constraints, connections with sub-Riemannian geometry are evident. Indeed, for the rolling ball they are described in [174]. We would also like to mention connections between the obtained results and those for dynamics of rigid rods obtained by Kirchhoff. Such a connection was discussed in Sec. 2.5, in particular, in [45] by Singer. It is appropriate now to remind our readers some details about this connection having in mind physical applications.

We begin by comparing the Frenet–Serret Eq. (2.4.1) for curves with Eq. (8.2.2.1c). In fact, after simple relabelling of entries it can be used to describe dynamics of rigid rods. Detailed derivation of this correspondence is given in [238], Chaps. 18 and 19 and [237], pp. 67–70. In notations of [45] the generalized Serret–Frenet-type equations are given by

$$\frac{d}{dt}\begin{pmatrix} T \\ U \\ V \end{pmatrix} = \begin{pmatrix} 0 & k_2 & k_3 \\ -k_2 & 0 & k_1 \\ -k_3 & -k_1 & 0 \end{pmatrix}\begin{pmatrix} T \\ U \\ V \end{pmatrix}. \tag{8.2.4.2}$$

By comparing this equation with Eq. (2.4.2) we conclude that: $k_1 = \tau, k_2 = k$ and $k_3 = 0$ in the case of curves. This result can be easily understood physically. By definition, the rod has finite thickness/cross-section while the real (physical) thickness of the curve is ignored. From here, the rod has two curvatures while the curve has only one. The connection between rods and curves is established using the concept of centerline. If the rod's cross-section is a circle, the centerline is a union of centers of all these cross-sections. So, it can be visualized as some curve. The total elastic free energy $\mathcal{F}_e$ of the rod is given by

$$\mathcal{F}_e = \frac{1}{2}\int_0^T ds[\alpha_1 k_2^2 + \alpha_2 k_3^2 + \beta k_1^2]. \tag{8.2.4.3}$$

In this expression the first two terms represent bending while the third-twisting. In the symmetric case the constants $\alpha_1 = \alpha_2$ and if the torsion is constant along the centerline we are coming back to the case discussed in Remark 2.5.5.

Remark 8.2.4.1. Our readers should be aware of the fact that results just presented are also of use in the Faddeev–Skyrme model discussed in Sec. 3.2. Thus, if knotted/linked curves are associated with description of hydrodynamics of ideal fluids and/or Abelian electrodynamics, the knotted linked rods are appropriate for description of the abelianized QCD [83]. From here we obtain a formal connection between the dynamics of 2-level quantum systems and that of QCD. Incidentally, this observation formally solves the millennium gap problem.[d] More on this is discussed in our work, [181]. Connection with QCD upon development leads to the topological quantum computation. Some details are presented in Appendix C and the following chapter.

[d] E.g. Read http://www.claymath.org/millennium/Yang-Mills_Theory/

8.2.5 *Two-Level Quantum Systems. List of Applications*

The above remark indicates that both situations, that is rods and elastic curves, are physically realizable. It is also of importance to investigate mathematically under what conditions it is possible to replace the elastic rods by the elastica (that is to neglect torsion, and to consider $\alpha_1 = \alpha_2$ while keeping the length of the centerline fixed). A comprehensive mathematical study of this issue was undertaken by Jurdjevic [239, 240]. We refer our readers to his works for details. A simple minded reduction of Eq. (8.2.4.2) for rods to Eq. (2.4.1) for curves was studied in [241] where many physical applications are listed. In Appendix B we give mathematically rigorous derivation of dynamics of Euler's elastica based on uses of sub-Riemannian geometry.

[241] contains references to many physical applications of 2-level systems. It is appropriate at this point to comment at least on some of these applications. The story begins with the seminal work by Isidor Rabi [242] done in 1937. Not only it contains the formalism which is in use up to this day but also it discusses likely applications such as measurement of the magnetic moment of the neutron, the rotational moment of molecules and the nuclear moment of atoms. The last application had grown into the full fledged field of Nuclear Magnetic Resonance (NMR) — an indispensable tool in synthetic organic chemistry, medicine, quantum computers, etc. This happened, thanks to the efforts of Felix Bloch [243], whose work of 1946 has become the most read paper in the field of NMR. A good summary of this type of research is given in [244] with emphasis on the theory of quantum control. We shall discuss this topic below, in the sub-Riemannian setting. In the meantime, we would like to mention other applications. They began shortly after publication of Bloch's paper by applying the same ideology to optics (photon polarization), atomic physics (electron–electron scattering), nuclear physics (neutron polarization). Up to year 1961 these efforts are summarized in [245]. Further developments along these lines in optics and atomic spectroscopy up to 1975 are nicely summarized in the book, [246].

Applications of 2-level quantum mechanical systems to scattering theory of atoms and molecules had began in 1932 with works by Landau, Zener, Stueckelberg and Majorana.[e] Since readable derivation of the Landau–Zener formula is given in Landau and Lifshitz volume on quantum

[e]Exact references to original works of these authors can be found in Wikipedia under the title: "Landau–Zener formula."

mechanics, [247], paragraph 90, we recommend to our readers to consult this source for details first. The latest advancements associated with the ongoing research employing this formula are summarized in [248]. Use of the rolling ball model for description of the Landau–Zener (L-Z) transition has been done only in 2010 in [237] — the same year [248] was published. Various chemical applications and many technical details can be found in yet another review paper, [249]. Calculations in [249] were inspired by much earlier paper by Delos and Thorston [250]. Since the already cited paper, [241], is extending the paper by Delos and Thorston, the sub-Riemannian treatment of the Landau–Zener-type problems is assured.

In 1932, that is five years prior to publication of the paper by Rabi [242], Rosen and Zener [251] used for the first time the 2-level quantum system in the context of scattering (collision phenomena). Their mathematical treatment of 2-level systems had been reduced to the study of solutions of hypergeometric equations. The Weber-type equation was subsequently used by Zener for solution of the L-Z problem [247]. It is also of hypergeometric type. Years later, in 1985, Hioe and Carroll in [252] demonstrated that solutions of the majority of 2-level problems indeed involve uses of hypergeometric equations. We shall further comment on the mathematical meaning and significance of this fact later. In the meantime, by generalizing ideas of Rosen and Zener, Nikitin and Demkov developed, now very popular, 2-level models of scattering processes whose description involves hypergeometric functions. The present status of these models is reflected respectively in [253] and [254]. Furthermore, in [255] the Serret–Frenet formalism discussed in previous subsection was used for calculation of Berry's phase and in [237] it was claimed that the same methods can also be used for calculation of its classical counterpart, Hannay's angle.

8.2.6 *Contact Geometry on S^3 and Optimal Control of 2-Level Systems*

In Sec. 6.5 we noticed that the 3-sphere can be described with help of two complex numbers z_1 and z_2 such that they obey the equation $|z_1|^2 + |z_2|^2 = 1$. Thus, any pair $(z_1, z_2) \in S^3$ can be represented either as a vector (spinor) or as a matrix g, e.g. see Eq. (6.5.25), in a very much the same way as it is done for elements of the group SO(3). The question arises: Is such similarity descriptions exhausted by thus described mapping? The answer is: "No". This can be seen from Eq. (6.5.27). Equation (6.5.27) is remarkable in the sense that parametrization of the matrix g is made in terms of Euler's angles normally used for description of rotations in $\mathbf{R}^3$. If this is so, what can be said about the relationship between groups SU(2) and SO(3) in addition to what was said already? A very detailed answer to this question can be found in [256]. Thus, we shall follow this reference in providing the answer to the question just formulated. From this reference we find that two linear operators $\pm g$ (see Eq. (6.5.27)) correspond to two opposite points on S^3. If these points are identified, one obtains the real projective space $\mathbf{RP}^3$ on which the SO(3) group is acting. The mapping SU(2) $\to$ SO(3) is *epimorphism*. That is for each rotation described by SO(3) there are two unitary operators g and $-g$ from SU(2). Instead of giving an exact definition of epimorphism, we notice only that parametrization of SU(2) in terms of Euler's angles is somewhat unusual, that is: $0 \leq \varphi < 2\pi, 0 \leq \theta < \pi, -2\pi \leq \psi < 2\pi$. When going to SO(3) from SU(2) one should replace $-2\pi \leq \psi < 2\pi$ by more familiar $0 \leq \psi < 2\pi$. Loosely speaking, this fact implies that the group SO(3) is twice smaller than SU(2). Does it makes sense to think of the map SU(2) $\leftarrow$ SO(3)?[f] The answer is "Yes" if we think of SU(2) as of the fiber bundle (we already did so in Sec. 6.5). In such a case S^2 is the base space on which the group SO(3) is acting. Indeed, let P and Q be two points belonging to S^2. Let $G \in S^2$ so that $GP = Q$. However, $\forall P$ we can also find such points P for which $GP = P$. That is, there is always a rotation leaving the axis of rotation going through P and the center of a sphere O fixed. A moment of thought leaves us with the plane $\mathbf{R}^2$ — this is the plane of fixed points for the equation $GP = P$ and the group SO(2) is acting on this plane as a subgroup of SO(3). Since SO(2) is isomorphic to S^1 we can construct a coset SO(3)/S^1 on which the group action is free. Under such circumstances it is possible to talk about the fiber bundle with the base S^2 and the fiber S^1 so that SU(2) $\simeq S^2 \times S^1$ as discussed in Sec. 6.5.

With this information in our hands, we are ready to move forward. For instance, let us rewrite the kinematic Eq. (8.2.2.1c). We shall follow both [136] and [230] for this purpose. In particular, following Arnol'd, let $\mathbf{q}$ be the position of a point in s-system and, respectively, $\mathbf{Q}$ — its position in c-system. Then, there is some matrix B such that $\mathbf{q} = B\mathbf{Q}$. Accordingly, if $\mathbf{v} = \dot{\mathbf{q}}$, then $B\mathbf{V} = \mathbf{v}$. If $\boldsymbol{\omega}$ is the angular velocity in s-system, then $\boldsymbol{\omega} = B\boldsymbol{\Omega}$ with $\boldsymbol{\Omega}$ being the angular velocity in the c-system. From these identifications it is clear that the matrix B plays the central role in relating vectors in spatial frame (s) and rotating frame (c). In particular, choose

[f] The physical aspects of this question are discussed below, in Sec. 8.4.3.

once again $\mathbf{q} = B\mathbf{Q}$ so that $\dot{\mathbf{q}} = \dot{B}\mathbf{Q} + B\dot{\mathbf{Q}}$. In the case of pure rotation $\dot{\mathbf{Q}} = \mathbf{0}$ so that we obtain: $\dot{\mathbf{q}} = \dot{B}\mathbf{Q}$ or, equivalently,

$$\dot{\mathbf{q}} = \dot{B}B^{-1}\mathbf{q}. \tag{8.2.6.1a}$$

This equation is equivalent to Eq. (8.2.2.1.c) if we make an identification

$$\mathbf{A} \equiv \dot{B}B^{-1} = \begin{pmatrix} 0 & -\omega_3 & \omega_2 \\ \omega_3 & 0 & -\omega_1 \\ -\omega_2 & \omega_1 & 0 \end{pmatrix}. \tag{8.2.6.1b}$$

Such a form can be found in [232]. Using the isomorphism Ψ, (Eq. (8.2.3.4)), we can as well rewrite Eq. (8.2.6.1a) as

$$\dot{\mathbf{q}} = [\boldsymbol{\omega}, \mathbf{q}] \tag{8.2.6.1c}$$

that is

$$\mathbf{A}\mathbf{q} = [\boldsymbol{\omega}, \mathbf{q}] \tag{8.2.6.1d}$$

so that we again arrive at the equation analogous to Eq. (8.2.4.1a). The matrix $\mathbf{A} \in \mathrm{SO}(3)$. Evidently, the relationship analogous to Eq. (8.2.6.1c) should also exist in the c-system. That is to say, we expect that

$$\dot{\mathbf{Q}} = [\boldsymbol{\Omega}, \mathbf{Q}]. \tag{8.2.6.2}$$

So, we must have the following nontrivial consistency requirement

$$\frac{d}{dt}(B^{-1}\mathbf{q}) = [B^{-1}\omega, B^{-1}\mathbf{q}] \equiv B^{-1}[\boldsymbol{\omega}, \mathbf{q}]. \tag{8.2.6.3}$$

The last equality comes as requirement that the matrix B relates vectors in one system (say, c-system) to those in s-system. This requirement is very helpful in deriving Euler's Eq. (8.2.3.8b) as can be seen from [136]. Using these results, we obtain as well

$$B\mathbf{A}B^{-1}\mathbf{Q} = \tilde{\mathbf{A}}\mathbf{Q} = [\boldsymbol{\Omega}, \mathbf{Q}] = \frac{d}{dt}\mathbf{Q}. \tag{8.2.6.4}$$

By applying the Ψ-mapping, this can be rewritten as

$$\left(\frac{d}{dt}\mathbf{Q}\right)\mathbf{Q}^{-1} = \tilde{\mathbf{A}} \equiv \boldsymbol{\Omega}. \tag{8.2.6.5}$$

Explicitly (see [232, 235]),

$$\begin{aligned} \Omega_x &= \dot{\varphi}\sin\theta\sin\psi + \dot{\theta}\cos\psi, \\ \Omega_y &= \dot{\varphi}\sin\theta\cos\psi - \dot{\theta}\sin\psi, \\ \Omega_z &= \dot{\varphi}\cos\theta + \dot{\psi}. \end{aligned} \tag{8.2.6.6}$$

These equations, familiar to every physics major, will now be treated from the point of view of Hopf mapping and contact geometry of S^3.

The contact 1-form on S^3 was defined in Eq. (5.5.2). With help of parametrization defined in Eqs. (6.5.27) and (6.5.28) we obtain:

$$x_1 = \cos\left(\frac{\varphi+\psi}{2}\right)\cos\frac{\theta}{2},$$

$$x_2 = \sin\left(\frac{\varphi+\psi}{2}\right)\cos\frac{\theta}{2},$$

$$y_1 = \cos\left(\frac{\varphi-\psi}{2}\right)\sin\frac{\theta}{2},$$

$$y_2 = \cos\left(\frac{\varphi-\psi}{2}\right)\sin\frac{\theta}{2}.$$

Using such parametrization in 1-form, Eq. (5.5.2), after some calculation, [167], p. 236, we finally obtain[g]

$$\alpha_x = \frac{1}{2}(\sin\theta\sin\psi d\varphi + \cos\psi d\theta) \qquad (8.2.6.7a)$$

implying that $2\alpha_x = \Omega_x dt = \sin\theta\sin\psi d\varphi + \cos\psi d\theta$. The above result is certainly not unique in the sense that projections of $\mathbf{\Omega}$ onto x, y and z axes are defined with accuracy up to permutation. This can be seen from [104][h] where analogous calculations yielded different result:

$$\alpha_z = \frac{1}{2}(\cos\theta d\varphi + d\psi) \qquad (8.2.6.7b)$$

implying that $\Omega_z dt = 2\alpha_z$. Evidently, with some efforts it would be possible to obtain as well

$$\alpha_y = \frac{1}{2}(\sin\theta\cos\psi d\varphi - \sin\psi d\theta). \qquad (8.2.6.7c)$$

Our readers can find all needed details in [257], Chap. 3.

That all three 1-forms are contact can be seen from the fact that $d\alpha_i \neq 0$ identically, $i = x, y, z$. This observation allows us to develop the contact geometry for this problem. Unlike [167], where emphasis was made on purely mathematical aspects of such development, in this section we shall discuss the obtained results in the physical context.

[g] Uses of the subscript x will be justified momentarily.

[h] We used this reference in connection with our discussion of the Dirac monopole.

We begin with Euler's equations for rotating body in c-system in the absence of external torques. These are given by

$$\begin{aligned} &\frac{d}{dt}\Omega_x + \frac{I_z - I_y}{I_x}\Omega_y\Omega_z = 0, \\ &\frac{d}{dt}\Omega_y + \frac{I_x - I_z}{I_y}\Omega_z\Omega_x = 0, \\ &\frac{d}{dt}\Omega_z + \frac{I_y - I_x}{I_z}\Omega_y\Omega_x = 0, \end{aligned} \tag{8.2.6.8}$$

with $I_i, i = \{x, y, z\}$ being respective components of the inertia tensor (in the coordinate system in which such a tensor is diagonal). Surely, we are not going to discuss all possible solutions of these equations since not all solutions make sense physically. Instead, let us consider several simple but physically interesting cases. For instance, let $\Omega_x = 0$. This leaves us with the only possible solution: $\Omega_y = C_2, \Omega_z = C_3$, where C_2 and C_3 are some constants. Let us next consider the axisymmetric case $I_y = I_z$. In this case we obtain: $\Omega_x = C_1$, and two remaining equations are converted into

$$\begin{aligned} &\frac{d}{dt}\Omega_y = -\omega\Omega_z, \\ &\frac{d}{dt}\Omega_z = \omega\Omega_y, \end{aligned} \tag{8.2.6.9}$$

immediately yielding the solution $\Omega_y = A\cos\omega t$ and $\Omega_z = A\sin\omega t$ with $\omega = \frac{I_x - I_y}{I_y}\Omega_x$. Let us discuss now the physics behind these solutions. To do so, we notice that the kinetic energy K of the rotating body is given by

$$2K = I_x\Omega_x^2 + I_y\Omega_y^2 + I_z\Omega_z^2 \tag{8.2.6.10}$$

in which Ω_x, etc. are given by Eq. (8.2.6.6). In the absence of external forces the kinetic energy K coincides with the full Lagrangian $\mathcal{L}$ so that the Lagrangian equations (compatible with Euler's equations) will contain cyclic variables, e.g. φ and ψ which can be conveniently chosen as zero. In such a case we obtain:

$$2K = I_x\dot{\theta}^2 + I_y(\dot{\varphi}\sin\theta)^2 + I_z(\dot{\varphi}\cos\theta + \dot{\psi})^2. \tag{8.2.6.11}$$

In the case when $\Omega_x = 0$ and $\Omega_y = C_2, \Omega_z = C_3$ we obtain respectively

$$\dot{\theta}^2 = 0, \ \frac{M_y}{I_y} = \dot{\varphi}\sin\theta \quad \text{and} \quad \frac{M_z}{I_z} = \dot{\varphi}\cos\theta + \dot{\psi}. \tag{8.2.6.12}$$

To use these results we have to take into account the following. In the absence of external forces/torques the total momentum $\mathbf{M}$ is conserved. Without loss of generality, let its direction coincide with the z-axis of the s-system. Then, in c-system we obtain: $\mathbf{M}^2 = M_y^2 + M_z^2$, since $M_x = 0$ according to Eq. (8.2.6.12). In the case if $I_y = I_x = I$, we obtain, using Eq. (8.2.6.11),

$$\mathbf{M}^2 = I^2\dot{\varphi}^2 \sin^2\theta + I_z^2(\dot{\varphi}\cos\theta + \dot{\psi})^2. \tag{8.2.6.13}$$

From our study of Euler's equations we already know that the case $M_x = 0$ implies $M_y^2 = \text{const.}$ and $M_z^2 = \text{const.}'$. This means that the L.H.S. of the second and the third of Eq. (8.2.6.12) are constants. Next, since $M_x = 0$ the equation $\dot{\theta}^2 = 0$ implies $\theta \equiv \Theta = \text{const.}$ This means that the z-axis of the c-system is tilted with respect to the z-axis of the s-system by the fixed angle Θ that is the body in c-system is *precessing* around the z-axis of the s-system. In turn, this implies that $M_z = M\cos\Theta$ so that $\Omega_z = \frac{M}{I_z}\cos\Theta$. This means that the body rotates around the z-axis of s-system with angular velocity Ω_z. Accordingly, $\Omega_y = \frac{M}{I}\sin\Theta$ implying that $\dot{\varphi} = \frac{M}{I}$ according to the second of Eq. (8.2.6.12). This is the angular velocity of precession. Because $\frac{M}{I_z}\cos\Theta = \dot{\varphi}\cos\Theta + \dot{\psi}$ we obtain as well: $\dot{\psi} = \frac{M}{I_z}\cos\Theta - \frac{M}{I}\cos\Theta = \Omega_z\frac{I-I_z}{I} \equiv \tilde{\omega}$. This result is helpful to compare against previously obtained $\omega = \frac{I_x - I_y}{I_y}\Omega_x$. First we have to replace Ω_z by Ω_x. Second, in both cases we made an assumption about the axial symmetry. For $\tilde{\omega}$ we had $I = I_y = I_x \neq I_z$ while for ω we have $I_x \neq I_y = I_z = I$. The question arises: How legitimate is such a replacement physically? This question makes sense because in the first case we put $\Omega_x = 0$ while in the second we choose $\Omega_x = C_1$ where the constant C_1 is nonzero in general. The Landau and Lifshitz textbook, [235], provides an impression that such a replacement is legitimate. Jumping a bit ahead, we shall demonstrate that the first case does admit treatment by the methods of sub-Riemannian geometry (that is by the methods used in the theory of optimal control) while the second does not. Clearly, the second case could be realizable physically nevertheless.

To demonstrate that the above-mentioned replacement changes physics is easy. Indeed, using Euler's Eq. (8.2.6.8) we notice that the case $\Omega_x = 0$ is realizable only if $\Omega_y = C_2, \Omega_z = C_3$. When $\Omega_x \neq 0$, both Ω_y and Ω_z become time-dependent. While the condition $\theta = \Theta = \text{const.} \neq 0$ defines the precession cone in the first case, no such condition exist in the second case since the equation $\dot{\theta}^2 = \text{const.} \neq 0$ does not contain solution $\theta = \Theta = \text{const.}' \neq 0$. Thus, physically, these two cases are certainly not the same.

It is instructive to re-obtain these results by using the formalism of 2-level quantum systems. We shall do this by taking into account the isomorphism between Euler's Eq. (8.2.3.8c) and Lorentz equations for the charged particle in the magnetic field Eq. (8.2.3.9). We begin with Eq. (8.2.2.1) and consider the simplest case of a particle in the constant magnetic field which is oriented along the z-direction. In this case Eq. (8.2.2.1) acquires the form

$$i\frac{d}{dt}\begin{pmatrix} a \\ b \end{pmatrix} = -\frac{1}{2}\begin{pmatrix} H_z & 0 \\ 0 & -H_z \end{pmatrix}\begin{pmatrix} a \\ b \end{pmatrix} \tag{8.2.6.14}$$

immediately yielding the solution

$$\begin{pmatrix} a(t) \\ b(t) \end{pmatrix} = \begin{pmatrix} e^{\frac{itH_z}{2}} a(0) \\ e^{-\frac{itH_z}{2}} b(0) \end{pmatrix}. \tag{8.2.6.15}$$

Use of the Hopf map (defined after Eq. (8.2.2.1b) allows us to obtain the following solutions for the x, y and z:

$$\begin{aligned} x(t) &= x(0)\cos(H_z t) + y(0)\sin(H_z t), \\ y(t) &= y(0)\cos(H_z t) - x(0)\sin(H_z t), \\ z(t) &= z(0), \end{aligned} \tag{8.2.6.16}$$

in accord with Eq. (6.3.13). Incidentally, Eq. (8.2.2.1c) formally acquires the form of the Serret–Frenet Eqs. (2.4.1) and (2.4.2) for the moving triad frame. This is a special case of general reduction of the Kirchhoff rod equations discussed in Secs. 2.5 and 8.2.4.

The next level of complexity is associated with the case when we have both the constant nonzero magnetic field $H_z \equiv \alpha$ and the magnetic field H_{x-y} varying in the $x-y$ plane defined as follows: $H_{x-y} = B(\cos \alpha t, \sin \alpha t, 0)$. Precisely such a combination of fields is used in NMR experiments [258]. It was shown in [237] that this problem can be reduced to that for the static fields, that is

$$i\frac{d}{dt}\begin{pmatrix} A \\ B \end{pmatrix} = -\frac{1}{2}\begin{pmatrix} \alpha & B \\ B & -\alpha \end{pmatrix}\begin{pmatrix} A \\ B \end{pmatrix}, \tag{8.2.6.17}$$

where $A(t) = e^{\frac{it\alpha}{2}} a(t), B(t) = e^{\frac{-it\alpha}{2}} b(t)$. A quick look at Eqs. (8.2.2.1b) and (8.2.2.1c) reveals that the matrix

$$\begin{pmatrix} 0 & H_z & -H_y \\ -H_z & 0 & H_x \\ H_y & -H_x & 0 \end{pmatrix}$$

in Eq. (8.2.2.1c) acquires now the familiar Serret–Frenet form given by Eq. (2.4.2), that is

$$\begin{pmatrix} 0 & \alpha & 0 \\ -\alpha & 0 & B \\ 0 & -B & 0 \end{pmatrix}$$

so that the associated curvature κ of the elastica is α while the torsion τ is B. Thus, this is yet another example of the reduction of the Kirchhoff rigid rod equations to those of Euler's elastica. Furthermore, by comparing these results with those obtained by directly solving Euler's equations we obtain the following correspondence: $\alpha \rightleftarrows \Omega_z, B \rightleftarrows \Omega_x$. Evidently, since Euler's case was treated under the assumption of axial symmetry we can always relabel $\Omega_x \rightleftarrows \Omega_y$. This is permissible since in both cases we are having c-system precessing around s-system with constant angular velocity and constant cone angle. Details can be found in [235, 258].

With these observations in our hands we are ready to develop the sub-Riemannian/contact geometry for such simplified NMR-type process. Before doing so, we would like to notice that the full solution of the Pauli-like Eq. (8.2.6.17) along with its NMR-related interpretation is given in the Appendix to [258].

In our presentation of dynamics/optimal control of just described MNR-type process based on ideas of sub-Riemannian geometry we follow [167, 259, 260].

Exercise 8.2.6.1. Since these are purely mathematical works, we encourage our readers to reformulate results we are about to discuss in more traditional language (e.g. in the NMR language) used in physics literature.

Very fortunately, all ingredients of what is needed for such a reformulation were already obtained. Thus, we are left only with providing a summary. First, the dictionary of the sub-Riemannian — optimal control correspondence was presented at the beginning of Sec. 6.4. Second, in the context of dynamics on S^3 all needed elements were discussed in Sec. 6.5, e.g. see Eqs. (6.5.31)–(6.5.41). These results are either taken from [167, 259, 260] or are compatible with these references. This observation leaves us with the following questions. The optimal control problem on S^3 summarized in Eqs. (6.5.31)–(6.5.41) is designed for description of precession of the axi-symmetric body in the absence of external forces. Physically, it corresponds to the idealized situation which even in the context of NMR (let alone quantum computing!) is never realized in practice. Practice always involves some kinds of external forces as well as friction/dissipation which is typically temperature-dependent, etc. Is there a way to incorporate these effects into theory of optimal control, that is into the sub-Riemannian geometry? For recent efforts in this direction, read survey article by Khaneja [244]. Much more comprehensive analysis of dynamics of the 2-level systems with dissipation done with help of sub-Riemannian geometry was recently presented in [261, 262]. These papers provide solid foundation for development of sub-Riemannian (optimal control) treatment of decoherence problem for many-level systems.

Finally, we would like to comment on the aspects of optimal control having in mind other applications. For instance, in the book by Arnol'd and Khesin, [68], there is a mention of a remarkable correspondence found by Vishik and Dolzhanskii, [263], between the dynamics of heavy top described in any book on classical mechanics (that is axially symmetric rotating body in the field of gravity, e.g. Kovalevsky-type top) and that of magnetohydrodynamics of incompressible fluids discussed in Sec. 2.4. Presence of external forces of dissipative character causes the matrix in Eq. (8.2.4.2), to acquire diagonal terms as can be seen, for example, from results by Khaneja, [244]. *This means that development of the theory of optimal control for quantum 2-level systems (accounting for effects of dissipation)* [261, 262] *should also be useful for development of the theory of optimal control of magnetohydrodynamics devices.* It remains to be seen to what extent the developments in magnetohydrodynamics might affect the developments in quantum computers.

Remark 8.2.6.1. The above discussion is underscoring the usefulness of the sub-Riemannian point of view in what is considered as purely classical and as purely quantum mechanical type of problems. We already know that mathematics of sub-Riemannian geometry is essentially that of classical mechanical systems with nonholonomic constraints resulting in replacement of Minkowski-type space–time by some kind of fiber-bundle. It should be clear from the discussed examples that there is no need to additionally quantize dynamical systems with nonholonomic constraints since quantum mechanics emerges (without any *ad hoc* assumptions, that is without any need for quantization postulates!) directly from the sub-Riemannian formalism. Nevertheless, attempts at formal quantization of nonholonomic dynamical systems still continue up to the time of writing of this book, e.g. see [264]. It should also be noted that in spite of the fact that the systematic procedure of quantization of dynamical systems with constraints was initiated in physics by Dirac already in 1950 [265], it is still subject of intense research. The question arises: How are results of these efforts in physics related to those developed in mathematics in connection with the sub-Riemannian (that is contact) geometry? We shall discuss this and related topics in the next section.

8.2.7 *Dirac Quantization of Dynamical Systems with Constraints and Contact Geometry*

After Dirac's 1950 seminal paper, there had appeared many alternative ways to deal with dynamical systems with constraints, e.g. the BRST (Becci–Ruet–Stora–Tyutin) method of quantization of constrained dynamical systems [266], Batalin–Vilkovisky method [267], Faddeev–Jackiw method [268], etc. Although these methods are interrelated, no attempts were made to connect them with methods of sub-Riemannian geometry. Decisive steps in this direction were recently made by Arnol'd in [269]. Alternative method was proposed by Moser [270]. Unlike Arnold's approach, Moser's method is limited to the exactly integrable dynamical systems only. Connections with contact geometry and integrable systems were made more explicit in the work by Webster [271] (commissioned by Moser) where our readers will be able to recognize some of the results discussed in Sec. 3.1. of this book. Connection between Dirac's method of dealing with constraints and that involving the Lagrangian multipliers (used in sub-Riemannian geometry) is discussed employing Neumann's dynamical system as an example in [272, 273]. Since we studied such a system in the context of solutions semiflexible polymers [203], we believe that it is appropriate to introduce our readers to such type of problems in this chapter. We shall do it later. Now would like to discuss instead the dynamics of Josephson junctions in the light of noticed connections with dynamics of 2-level systems, of Kirchhoff rods, of Euler's elastica and, because of this, of conformational dynamics of semiflexible polymers.

8.3 Beyond the 2-Level Quantum Systems

8.3.1 *General Information*

In Sec. 8.2.1 we stated Di Vincenzo criteria applicable to all quantum computers. Nicely written book by Nakahara, [230], provides more details on this subject. Since in our book quantum computers are discussed only in the context of sub-Riemannian geometry (that is in the context of optimal control) with only one exception made in Appendix C, we shall focus attention of our readers on the most basic aspects of quantum computers related to the subject matters of this book. In particular, according to part two of Di Vicenzo criteria to build quantum computer requires coupling the 2-level systems so that at least two-qubit operations can be performed. This observation requires us to extend results of Sec. 8.2 in which we discussed only the dynamics of 2-level quantum systems. Extension of these results to multilevel systems is possible in a variety of ways. For instance, we can begin with [241, 252] or with much more comprehensive treatment of the same issues given in [274]. In these references the dynamics of 2-level systems is reformulated in terms of the associated with these systems hypergeometric equations. In our papers, [173, 275], our readers will be able to find a detailed study of hypergeometric equations of one and several variables. As results of our work, [177], indicate, all exactly solvable two-body-type problems of the standard quantum mechanics admit exact solutions by reduction of a specific problem to that involving solution of hypergeometric equations of one variable. In [173, 276] it is demonstrated that the hypergeometric equations of multiple arguments are linked with the Knizhnik–Zamolodchikov (K-Z) equations of conformal field theories. In turn, these equations are derivable from the Wess–Zumino–Witten-type models admitting re-interpretation in terms of quantum spin chains [276]. Since the K-Z equations are associated with braid groups [173], we obtain connections between quantum computers and knot/link theory (some basics on this topic are presented in Appendix C). Notice that the connection with knot/link theory was developed in great detail by Kitaev [277]. More recent and comprehensive update of his results can be found in [278].

Connections between quantum computers and spin chains can also be established using the entirely different arguments. In particular, based on information already presented it should be clear that the description of any quantum computer should involve the (pseudo) spin Hamiltonian of the following type [226]:

$$\hat{H} = \sum_a h_a \cdot \sigma_a + \sum_{a,b} \sigma_a \mathbf{J}_{ab} \sigma_b. \tag{8.3.1.1}$$

We wrote "pseudospin" for a reason. Even though theoretically it is always possible to operate with the Pauli spin operators $\boldsymbol{\sigma}_a$ so that the above Hamiltonian is that for some quantum spin chain, in practice use of quantum spin chains is typically not convenient. Surely, the above Hamiltonian by design satisfies the first and the second of Di Vincenzo criteria. However, the third requirement involving the decoherence might be difficult to fulfill in practice. Thus, in practice one has to begin with some kind of a device reproducing the dynamics of 2-level quantum spin system. Such a device should be readily accessible macroscopically and behave exactly like a quantum qubit. Next, such 2-level devices should somehow be coupled together so that the two-qubit operations should be easily executable. Finally, the reading of the outcome of execution of the sequence of two-qubit operations must be easily implementable. All three requirements can be executed with the help of interconnected Josephson junctions arrays working, naturally, at very low temperatures. Thus, first of all we need to discuss in some detail, that is better than it was done in Sec. 8.2, the types of Josephson junctions used in quantum computing. By doing so some related areas of research will be touched upon as well.

8.3.2 *Types of Josephson Junctions Used in Quantum Computing, Atomic Physics and Quantum Optics*

To begin our discussion, once again we remind our readers that this book is not about the design of quantum computers as such. That is we are not going to discuss the actual fabrication of Josephson junctions, of quantum spin chains, etc. We shall be working with devices which are already made and we shall demonstrate that their functioning admits description in terms of sub-Riemannian geometry (and/or classical theory of optimal control).

Let us recall that we had begun with our study of superconducting ring in Sec. 3.4. In Sec. 6.4 we re-interpreted results of Sec. 3.4 in the sub-Riemannian language. In Remark 6.5.1 we formally connected the superconducting ring description with the Heisenberg group while the full proof of this fact is given in Appendix A. In Chap. 7 we took full advantage of the obtained results by re-interpreting the TGB phases in the sub-Riemannian language. Finally, in Sec. 8.2, following Feynman, we argued that dynamics of Josephson junctions can be described in terms of quantum dynamics of 2-level systems which can also be recast into the sub-Riemannian language. Notice that the simplest Josephson junction can be made without using the superconducting rings. Such a device made of two superconductors divided by thin dielectric layer is known in the literature as current-driven "*phase qubit*" (since both superconductors are linked with the external electric circuit). Our readers are advised at this point to read recent review papers on uses of Josephson junctions in atomic physics, quantum optics and quantum computing [279–281]. Without going into details, from reading of these references it follows that the ring-shaped Josephson junctions can serve as: (a) artificial athoms, (b) as lasers, (c) as means of creating electromagnetically induced transparency (that is they can make objects invisible), (d) as means of communicating between qubits, etc. All these devices use superconducting rings placed in the magnetic field but modified by the extremely thin insulating surfaces (junctions) placed perpendicular to the ring's suprconducting current. Presence of insulating junctions in these rings causes some tunneling of supercurrents through such junctures.[i] This allows us to describe these rings also as Josephson junctions. In practice one can find one, two or three junctions along the ring.[j] In the first case these are known as rf SQUID's. Here "rf" stands for "radio frequency" and SQUID stands for "superconducting quantum interference device". In the case of two insulating junctions one has the dc SQUID (dc stands for the dc current) and, if such a ring is modified by attaching to it two leads connected to external battery (supplying the dc current), such a device is known as "*charge qubit*". Finally, if three junctions are installed into the ring and there are no leads, one is talking about the "*flux qubit*". It is essential that, while acting quantum mechanically, starting with superconducting ring discussed in Sec. 3.4, all these devices are macroscopic. This circumstance prompts us to inquire if the presence of junctions in just described ring-type Josephson junctions devices can destroy the sub-Riemannian description already established for the superconducting ring without such junctions. Based on the 2-level analysis of a single Josephson junction presented in Sec. 8.2 the answer to the above inquiry is negative. We shall discuss these and related topics in some detail in the next subsection.

Remark 8.3.2.1. If non-interacting qubits operate by the rules of sub-Riemannian geometry, can the dynamics of 3-level quantum system be also described in a sub-Riemannian way (that is by using methods of optimal control)? Positive answer to this question can be formulated starting from different initial positions. For instance, from purely experimental point of view, it is possible to impose on the Josephson junctions some working conditions so that they become 3-level quantum systems — e.g. macroscopic artificial atoms as discussed in detail in [281]. The positive answer is also assured from the theoretical side as explained in [282, 283]. The problem of

[i] Incidentally, similar effects occur at the juncture between the superconducting and normal metals. In such a case physics is even more complicated than that occurring in Josephson's junctions, e.g. read [225]. We are not going to discuss this case in our book.

[j] To insert more is technically possible but thus far has not been used in applications.

extension of optimal control results for these 3-level systems to the n-level systems using Pontryagin's theory of optimal control was discussed recently in [284]. Pontryagin's theory of optimal control is briefly dicussed in Appendices A and B.

8.3.3 *Josephson Junctions — Semiflexible Polymers Correspondence. A Gentle Introduction*

Already in Sec. 8.2 we related dynamics of a single Josephson junction to dynamics of Kirchhoff rods and elastic curves. In [239, 240] by Jurdjevic the one-to-one correspondence between dynamics of a rolling ball on flat surface and that of Euler's elastica (Sec. 2.5 and Appendix B) has been rigorously demonstrated. Since the sub-Riemannian dynamics of rolling ball is connected one-to-one with spin dynamics of 2-level systems, the correspondence between dynamics of a single Josephson junction and that of Euler's elastica can be considered as established. Nevertheless, it is of interest to re-establish such a connection without use of the rolling ball-Euler elastica correspondence for reasons which shall become obvious upon reading. This is accomplished in this subsection. In Appendix B the sub-Riemannian treatment of Euler's elastica is provided. Since Euler's elastica dynamics is exactly the same as conformational dynamics of semiflexible polymers, the discussed results could be useful for theoretical develpments of both areas of research — solutions of semiflexible polymers and quantum computers.

We begin our discussion by reproducing some known basic facts about dynamics of Josephson's junctions. Being armed with such results, we shall re-establish the Josephson junction — Euler's elastica correspondence. Since solutions of semiflexible polymers do not require ultralow temperatures for their experimental study, it is hoped that further development of the Josephson's junctions — Euler's elastica correspondence could be helpful for optimizing the decoherence problem.

Following Feynman's treatment, [99], of Josephson's junctions, suppose that we have two superconductors (1 and 2) of the same kind[k] separated by the insulating very thin cross-section. If ψ_1 and ψ_2 are the collectivized electron wave functions to the left and to the right of insulating barrier, the corresponding Schrödinger's equation in the absence of external magnetic

[k]For the superconductors to be of the same kind is not essential. However, this assumption causes some technical simplifications.

field is that for the 2-level quantum system, e.g. ($\hbar = 1$)

$$\begin{aligned} i\partial_t\psi_1 &= U\psi_1 + K\psi_2, \\ i\partial_t\psi_2 &= K\psi_1 + U\psi_2. \end{aligned} \tag{8.3.3.1a}$$

The constant K is characterizing the junction. In its absence, the above two equations would describe the lowest energy state for each superconductor. Since $K \neq 0$, some tunneling across the insulating barrier will take place. Suppose furthermore that both superconductors are connected to the dc battery. Then, the situation naturally becomes asymmetric causing U to split into U_1 and U_2 so that $U_1 - U_2 = qV$ where V is the potential difference across the junction and q is the charge (in superconductors $q = 2e$). It is convenient to count energy from the middle of the junction so that the above two equations acquire the following form

$$\begin{aligned} i\partial_t\psi_1 &= \frac{qV}{2}\psi_1 + K\psi_2, \\ i\partial_t\psi_2 &= K\psi_1 - \tfrac{qV}{2}\psi_2. \end{aligned} \tag{8.3.3.1b}$$

Since according to Sec. 3.4 the wave function ψ can be represented as $\psi = \sqrt{\rho}\exp(i\Phi)$, use of such an ansatz in Eq. (8.3.3.1b) leads to the following set of four equations:

$$\begin{aligned} \partial_t\rho_1 &= 2K\sqrt{\rho_1\rho_2}\sin\delta, \\ \partial_t\rho_2 &= -2K\sqrt{\rho_1\rho_2}\sin\delta, \\ \partial_t\Phi_1 &= K\sqrt{\frac{\rho_2}{\rho_1}}\cos\delta\Phi - \frac{qV}{2}, \\ \partial_t\Phi_2 &= K\sqrt{\frac{\rho_1}{\rho_2}}\cos\delta\Phi + \frac{qV}{2}. \end{aligned} \tag{8.3.3.2}$$

In these equations $\delta\Phi = \Phi_2 - \Phi_1$.

The first two equations imply that $\partial_t\rho_1 = -\partial_t\rho_2$ whose solution is $\rho_0 =$ const. Such a conclusion even though is formally correct mathematically, has some flaws physically. These are coming from the fact that the above four equations are not sufficient for physically correct description of a single Josephson's junctuon. The way out of the difficulty is sketched in [285]. This technical peculiarity makes 2-level description of Josephson's junction suspicious and puts connections with sub-Riemannian geometry in doubt. Fortunately, there is other way to establish such a connection to be described below. Leaving delicate details discussed in [285] to interested readers, we simply state that the current J through the junction can nevertheless be written as

$$\partial_t \rho_1 = J = J_0 \sin \delta\Phi, \tag{8.3.3.3}$$

where $J_0 = 2K\rho_0$ is the number characterizing the partricular Josephson's junction, provided that $\rho_1 = \rho_2 = \rho_0$, and ρ_0 on the R.H.S. of the above equation is treated as a constant.

The second two equations are simpler since they can be treated straightforwardly with the help of standard mathematical methods. Thus we obtain,

$$\partial_t \delta\Phi = 2\,\mathrm{eV}. \tag{8.3.3.4}$$

Equations (8.3.3.3) and (8.3.3.4) are the key equations of Josephson's theory. However, they still do not account for the magnetic field effects. Nevertheless, they are already interesting enough and deserve some comments.

First, following Josephson, one can describe *the dc Josephson effect.* It happens when $V = 0$. In such a case $\delta\Phi$ does not need to be zero. Let its value be, say, $\delta\Phi_0$. Then Eq. (8.3.3.3) describes some dc current through the junction whose maximum value is J_0. This effect has been confirmed experimentally. When the external current is applied whose magnitude is larger than J_0, a finite voltage suddenly reappears across the junction. Second, suppose we apply a *constant* voltage $V \neq 0$ to the junction. Then, Eq. (8.3.3.4) provides us with the relative phase $\delta\Phi = \delta\Phi_0 + \frac{2e}{\hbar}Vt$. In this expression we have restored the Planck constant and took into account that the charge $q = 2e$, that is equal to the charge of the Cooper pair. Using this result in Eq. (8.3.3.3) we obtain

$$J = J_0 \sin\left(\delta\Phi_0 + \frac{2e}{\hbar}Vt\right). \tag{8.3.3.5}$$

That is, even if we had applied the dc current and constant voltage to the junction, the current through the junction begins to oscillate with frequency $\omega = 2\pi\nu = \frac{2e}{\hbar}V$. This phenomen is known as the *ac Josephson effect.*

From the above description of the dc and ac Josephson effects it should be clear that the dc effect can be considered as equilibrium/stationary phenomenon. This was already recognized by Josephson [225]. This means that Eq. (8.3.3.3) can be obtained by minimization of the free energy functional $\mathcal{F}[\delta\Phi]$ of the type

$$\mathcal{F}[\delta\Phi]/(\phi_0/2\pi c) = -J_0\cos(\delta\Phi) - J\delta\Phi. \tag{8.3.3.6}$$

Here ϕ_0 was defined in Eq. (3.4.13). If now the finite voltage is applied to the junction, such a junction is going to act as a capacitor. It is known that the energy $\mathcal{E}$ stored in the capacitor is given by $\mathcal{E} = CV^2/2$, where C is the capacitance. Using Eq. (8.3.3.4) we obtain the following result for $\mathcal{E}$:

$$\mathcal{E} = \frac{C}{2}\left(\frac{1}{2e}\frac{d}{dt}\delta\Phi\right)^2. \tag{8.3.3.7}$$

By treating the free energy $\mathcal{F}[\delta\Phi]$ as mechanical potential energy while $\mathcal{E}$ as kinetic energy, we obtain the Lagrangian

$$\mathcal{L} = \mathcal{E} - \mathcal{F} \tag{8.3.3.8}$$

whose minimization produces the equation of motion

$$\frac{C\pi c}{2e^2\phi_0}\frac{d^2}{dt^2}\delta\Phi + J_0\sin(\delta\Phi) = J. \tag{8.3.3.9}$$

This equation can be obtained as well using Maxwell's equations for description of the dynamics of Josephson's junctions in electromagnetic field. Initially, it was demonstrated by Josephson [285]. Correctness of this equation was confirmed experimentally in [286]. In the limit $J = 0$ this equation becomes an equation for the pendulum whose characteristic frequency $\omega_j^2 = \frac{2e^2\phi_0 J_0}{C\pi c}$ was measured in [286]. Using the dc Josephson effect and knowing frequency ω_j^2 one can determine separately the constants C and K characterizing Josephson's junction. The *pendulum equation is also an equation for Euler's elastica.* Its sub-Riemannian origin is discussed in detail in Appendix B. *Thus, the connection with semiflexible polymers through Euler's elastica is established without use of the rolling-ball-elastica correspondence.* Numerous additional connections shall be discussed below, in this chapter. In particular, one of such additional connections emerges because of the following.

In theory of Josephson's junctions Eq. (8.3.3.9) is just the classical Lagrangian equation of motion obtainable by standard methods from the Lagrangian $\mathcal{L}$. Following the well-known prescriptions it is possible to quantize such a dynamical system. This idea is plausible since when the cosine term in $\mathcal{F}[\delta\Phi]$ is expanded and only the lowest order in such an expansion is being kept, we obtain the Lagrangian for standard harmonic oscillator for which the quantization procedure is routine. Such a simplification is possible but is accociated with some technical/conceptual complications described in detail in the book by Zagoskin, [225], pp. 67–70. These complications cannot be resolved rigorously by traditional methods known in physics literature. Use of such methods always requires assumptions which are impossible to justify mathematically. No such complications occur if dynamical problem described by the Lagrangian $\mathcal{L}$ is treated by methods of sub-Riemannian geometry! Appendix B provides detailed explanation. The most important outcome of these efforts lies in observation that both the classical Eq. (8.3.3.9) and its quantum analog, Eq. (B.27), are used in the theory of semiflexible polymers to be discussed below, in this chapter.

To make additional connections with polymers several steps are still required. First, the above derivation of Euler's elastica equation was made in the absence of magnetic field while from previous subsection we know that only phase qubit Josephson junction does not require magnetic field for its operation. In the qubit regime the dynamics of such a junction is described in terms of dynamics of 2-level quantum system, that is by Eq. (8.2.2.1). This is achieved by the Pauli-type Hamiltonian derived in [226] for such a junction. Thus, the sub-Riemannian description of the phase qubit dynamics is assured in this case. The charge and flux qubits require magnetic field for their operation as it was already explained in the previous subsection. Nevertheless, the dynamics of these Josephson's junctions is also described in terms of the Pauli-type Hamiltonians [226]. It is of methodological interest to uncover the sub-Riemannian nature of these devices by different means.

For this purpose we have to extend Josephson's basic Eqs. (8.3.3.3) and (8.3.3.4) to the case when they take into account the effects of magnetic field. The point of departure in this case is Eq. (3.3.5b) of Sec. 3.3. In the present case this equation should be rewritten as

$$\nabla\Phi = \frac{2e}{\hbar c}\left(\frac{mc}{2e^2 n_s}\mathbf{j}_s + \mathbf{A}\right). \tag{8.3.3.10}$$

Upon formal integration we obtain,

$$\Phi(\mathbf{x}) - \Phi(\mathbf{x}+\delta\mathbf{x}) = \frac{2e}{\hbar c}\int_C \left(\frac{mc}{2e^2 n_s}\mathbf{j}_s + \mathbf{A}\right)\cdot d\mathbf{l}, \tag{8.3.3.11}$$

where C is some contour (not nesessarily closed!), e.g. see [285] p. 15. Following Feynman [99], we would like to use the above result for the dc SQUID, that is for the superconducting ring with two junctions a and b and two leads which we shall denote as P and Q. Since the superconducting current is flowing only at the surface of such a ring (which we can imagine as having the donut shape), we can place the contour C inside the ring (the donut) thus making $\mathbf{j}_s = 0$. Then, following Feynman, we apply Eq. (8.3.3.11) to the upper (where the junction is a) and to the lower (where the junction is b)

halves of the superconducting donut with the result:

$$\delta\Phi(\frown)_{P\to Q} = \delta\Phi_a + \frac{2e}{\hbar c}\int_{C_a} \mathbf{A}\cdot d\mathbf{l} \tag{8.3.3.10a}$$

and

$$\delta\Phi(\smile)_{P\to Q} = \delta\Phi_b + \frac{2e}{\hbar c}\int_{C_b} \mathbf{A}\cdot d\mathbf{l}. \tag{8.3.3.10b}$$

By subtracting the first result from the second we obtain:

$$\delta\Phi_b - \delta\Phi_a = \frac{2e}{\hbar c}\oint \mathbf{A}\cdot d\mathbf{l}. \tag{8.3.3.12}$$

This is possible in view of the obvious symmetry requirement [99]: $\delta\,\Phi(\frown)_{P\to Q} = \delta\Phi(\smile)_{P\to Q}$. In the case of constant external magnetic field Eq. (8.3.3.11) allows us to re-obtain the sub-Riemannian result defining the Heisenberg group $\mathbf{H}_1$ as explained in Remark 6.5.1 and Appendix A. Thus, the dc SQUID can surely be studied by methods of sub-Riemannian geometry! Let us study now the rf SQUID. In this case the situation is simpler since we can use the 2-level Eq. (8.3.3.1b) modified by the magnetic field as explained in [285]. Since the resulting equations are still of 2-level-type, all results of Sec. 8.2 can be applied so that the sub-Riemannian nature of the rf SQUID is assured.

Exercise 8.3.3.1. Demonstrate that the flux qubit (that is the superconducting ring in the magnetic field with three Josephson junctions) can also be described using sub-Riemannian geometry. Hint: Use the locality of Josephson's equations and the fact that in the qubit regime such flux qubit is also described by the Pauli-like Hamiltonian [225, 226].

Remark 8.3.3.1. According to [287], it is permissible in Eq. (8.3.3.10) to put $\delta\Phi(\frown)_{P\to Q} = \delta\Phi(\smile)_{P\to Q} = 0$. In such a case, it is possible to use the phase differences $\delta\Phi$ in these equations in Euler's elastica, Eq. (8.3.3.9). Such an equation was indeed derived, e.g. see [285], p. 365, Eq. (12.2.15). This equation is a bit more general since it includes the effects of dissipation — negligible at ultralow temperatures — and effects of the iduced magnetic field originating from the effects of the induced current quantization (e.g. read comments immediately after Eq. (3.4.13)). Although we shall not go into these fine details of Josephson's junction problematics in this book, interested readers are encouraged to read about these phenomena, using either [287] or, more comprehensive, [285].

8.3.4 From Pauli–Schrödinger to Dirac Equation

We begin with the Pauli–Schrödinger-Dirac equation correspondence. Typically, it is obtainable in the nonrelativistic limit from the Dirac equation, e.g. see [288], paragraph 15. We shall not use this known path though for connecting these equations. Instead, we would like to argue that Eq. (8.2.2.1) can be interpreted as Dirac's equation. For this purpose we notice that Dirac's equation can be written in many equivalent forms [288]. For instance, in the two-component form ($\hbar = 1$, $\hat{\mathbf{p}} = -i\nabla$) it can be written as

$$
\begin{aligned}
i\frac{\partial \varphi}{\partial t} &= m\varphi + \boldsymbol{\sigma}\cdot\hat{\mathbf{p}}\chi, \\
i\frac{\partial \chi}{\partial t} &= -m\chi + \boldsymbol{\sigma}\cdot\hat{\mathbf{p}}\varphi,
\end{aligned}
\tag{8.3.4.1a}
$$

where $\boldsymbol{\sigma} = \{\sigma_1, \sigma_2, \sigma_3\}$ and m denoting respectively the Pauli matrices and the mass of, say, the electron. Alternatively, it can be written in the form of Schrödinger's equation

$$
\begin{aligned}
i\frac{\partial \psi}{\partial t} &= \hat{H}\psi, \\
\hat{H} &= \boldsymbol{\alpha}\cdot\hat{\mathbf{p}} + \beta m,
\end{aligned}
\tag{8.3.4.1b}
$$

where the matrices $\boldsymbol{\alpha}$ and β are defined by

$$
\boldsymbol{\alpha} = \begin{pmatrix} \mathbf{0} & \boldsymbol{\sigma} \\ \boldsymbol{\sigma} & \mathbf{0} \end{pmatrix}, \quad \beta = \begin{pmatrix} \mathbf{I} & \mathbf{0} \\ \mathbf{0} & -\mathbf{I} \end{pmatrix}
\tag{8.3.4.2}
$$

with $\mathbf{I}$ being the 2×2 unit matrix. Different but equivalent forms of Dirac's equation can be obtained from each other by unitary transformations of bispinors: $\psi' = U\psi$. Every such transformation should be accompanied by the transformation of Dirac matrices, e.g. $\boldsymbol{\alpha}' = U\boldsymbol{\alpha}U^\dagger, \beta' = U\beta U^\dagger$. By the appropriate choice of unitary transformation we may end up with the following form of the Dirac equation

$$
\begin{aligned}
i\frac{\partial \xi}{\partial t} &= m\eta + \boldsymbol{\sigma}\cdot\hat{\mathbf{p}}\xi, \\
i\frac{\partial \eta}{\partial t} &= m\xi - \boldsymbol{\sigma}\cdot\hat{\mathbf{p}}\eta.
\end{aligned}
\tag{8.3.4.1c}
$$

It reduces to

$$
i\frac{\partial \xi}{\partial t} = \pm\boldsymbol{\sigma}\cdot\hat{\mathbf{p}}\xi
\tag{8.3.4.3}
$$

in the limit of zero mass. In such a form the above equation can be identified with the equation for neutrino. In Sec. 3.1, we introduced the helicity and the helicity integral, e.g. see Eq. (3.1.13). At that time we noticed that the curl operator is playing the same role for the Laplacian as the Dirac operator for the Dalambertian, e.g. see Eqs. (1.3.2) and (1.3.3). At this point we realize that the force-free equation/condition $\operatorname{curl}\mathbf{A} = \pm\kappa\mathbf{A}$ must have its analog in the theory of neutrino.

Remark 8.3.4.1. Because of noticed correspondence, it should be possible to reformulate some neutrino problems in the language of hydrodynamics. Apparently, this still can be done even in the presence of neutrino mass. In such a case one should use the massive Majorana equation for neutral particle described by the Dirac-type equation [289]. In [277] Kitaev used Majorana fermions for quantum (topological) computation.

Remark 8.3.4.2. If we look for solution of Eq. (8.3.4.3) in the form of plane waves, e.g. $\xi = f(t)\exp(\pm i\mathbf{p}\cdot\mathbf{x})$, then it can be reduced to the Pauli–Schrödinger Eq. (8.2.2.1). The $\pm$ options correspond to two different choices for the direction of magnetic field. Using this observation, the sub-Riemannian treatment of neutrino dynamics is assured.

Remark 8.3.4.3. The analog of the force-free condition, $\operatorname{curl}\mathbf{A} = \pm\kappa\mathbf{A}$, in the case of neutrino dynamics can be formulated with help of the spin operator $\mathbf{\Sigma}$ [288] defined by

$$\mathbf{\Sigma} = \begin{pmatrix} \boldsymbol{\sigma} & \mathbf{0} \\ \mathbf{0} & \boldsymbol{\sigma} \end{pmatrix}. \tag{8.3.4.4}$$

It can be demonstrated [288] that the combination $\hat{\mathbf{p}}\cdot\mathbf{\Sigma}$ commutes with the Hamiltonian $\hat{H}$ defined by Eq. (8.3.4.1b) (with or without mass!). If this is so, then one can classify the eigenfunctions of $\hat{H}$ in terms of those satisfying the eigenvalue problem

$$\frac{1}{2}\hat{\mathbf{p}}\cdot\mathbf{\Sigma}\varphi_{\mu\mathbf{p}} = |\mathbf{p}|\mu\varphi_{\mu\mathbf{p}}; \quad \mu = \pm\frac{1}{2} \tag{8.3.4.5}$$

for the helicity operator. Here $\mathbf{p}$ is the Fourier momentum. Clearly, it is the field-theoretic analog of the force-free condition. Recall that we had already encountered such type of a condition in the field-theoretic context in Chap. 5.

Now we would like to connect just obtained results with those in our work, [290]. In doing so we shall also follow [291] — direct precursor of our work. This requires us to use the Fourier-transformed Eq. (8.3.4.1b) in which the matrices $\boldsymbol{\alpha}$ and β defined by Eq. (8.3.4.2) are replaced by $\tilde{\mathbf{\Sigma}}$ and

$\tilde{\beta}$ so that Eq. (36) of [291] reads

$$\begin{aligned} E\psi &= \hat{H}\psi, \\ \hat{H} &= \tilde{\boldsymbol{\Sigma}} \cdot \mathbf{p} + \tilde{\beta} m, \end{aligned} \tag{8.3.4.6}$$

with

$$\tilde{\boldsymbol{\Sigma}} = \begin{pmatrix} \boldsymbol{\sigma} & \mathbf{0} \\ \mathbf{0} & -\boldsymbol{\sigma} \end{pmatrix} \quad \text{and} \quad \tilde{\beta} = -\begin{pmatrix} \mathbf{0} & \mathbf{I} \\ \mathbf{I} & \mathbf{0} \end{pmatrix}. \tag{8.3.4.7}$$

The question arises: Is there a way to transform Eq. (8.3.4.1b) into Eq. (8.3.4.6)? This question is equivalent to the question about proving the existence of the matrix U such that

$$U\hat{H}U^{-1} = \boldsymbol{\alpha} \cdot \mathbf{p} + \beta m. \tag{8.3.4.8}$$

That is we have to find U such that

$$U\tilde{\boldsymbol{\Sigma}}U^{-1} = \boldsymbol{\alpha} \quad \text{and} \quad U\tilde{\beta}U^{-1} = \beta. \tag{8.3.4.9a}$$

It can be easily checked that if we choose

$$U = \begin{pmatrix} \mathbf{I} & \mathbf{I} \\ \mathbf{I} & -\mathbf{I} \end{pmatrix} \tag{8.3.4.9b}$$

then, the first of Eq. (8.3.4.9) is satisfied while the second is satisfied also if we replace $\tilde{\beta}$ by $-\tilde{\beta}$. Such a discrepancy is not essential however since the Dirac equation is capable of describing both particles (say, electrons) and antiparticles (say, positrons). It also describes solutions having space, time and space–time reversal symmetry. In each case there is some unitary transformation connecting such solutions to each other. In the case of polymers whose conformational properties are described by the Dirac equation (see below) the charge conjugation symmetry (relating particles to antiparticles) is not immediately applicable and other symmetries should be discussed on the case-by-case basis. In addition, we expect that the combination $\hat{\mathbf{p}} \cdot \boldsymbol{\Sigma}$ commuting with the Hamiltonian $\hat{H}$, Eq. (8.3.4.1b), should have its analog for the Hamiltonian defined by Eq. (8.3.4.6). This is indeed the case as shown in [291]. Following this reference, let us consider the special case when the vector $\mathbf{p}$ in Eq. (8.3.4.6) has only zth component. This is permissible since it is always possible to find the Lorentz transformation relating general and this special case. Then, by direct inspection, it can be established that the equation

$$\tilde{\boldsymbol{\Sigma}}_z \Psi_{\pm} = \pm \Psi_{\pm} \tag{8.3.4.10}$$

plays the same role in the present case as Eq. (8.3.4.5) for $\hat{H}$ defined by Eq. (8.3.4.1b). If we choose

$$\Psi_+ = \frac{1}{\sqrt{2}}\begin{pmatrix}1\\0\\0\\1\end{pmatrix} \quad \text{and} \quad \Psi_- = \frac{1}{\sqrt{2}}\begin{pmatrix}0\\-1\\-1\\0\end{pmatrix} \tag{8.3.4.11}$$

then, we also obtain:

$$\Psi_- = \tilde{\beta}\Psi_+. \tag{8.3.4.12}$$

The Schrödinger-like Eq. (8.3.4.1b) (or Eq. (8.3.4.6)) can be equivalently rewritten as follows:

$$\psi(x, t + \Delta t) = \exp(-i\hat{H}\Delta t)\psi(x, t) \tag{8.3.4.13}$$

or, more explicitly, either as

$$\begin{aligned}\Psi_+(p, \Delta t) &= [\mathbf{I} - i(p\tilde{\mathbf{\Sigma}}_z + m\tilde{\beta})\Delta t]\Psi_+(p, 0)\\ &= (\mathbf{I} - ip\Delta t])\Psi_+(p, 0) - mi\Delta t\Psi_-(p, 0)\end{aligned} \tag{8.3.4.14a}$$

or, as

$$\Psi_-(p, \Delta t) = [\mathbf{I} + ip\Delta t]\Psi_-(p, 0) - mi\Delta tm\Psi_+(p, 0). \tag{8.3.4.14b}$$

Equivalently, these equations can be compactly presented in matrix form as

$$\begin{pmatrix}\Psi_+(\varepsilon)\\ \Psi_-(\varepsilon)\end{pmatrix} = \begin{pmatrix}1 - ip\varepsilon & -i\varepsilon m\\ -i\varepsilon m & 1 + ip\varepsilon\end{pmatrix}\begin{pmatrix}\Psi_+(0)\\ \Psi_-(0)\end{pmatrix} \tag{8.3.4.15}$$

where we suppressed the argument p in wave functions and replaced Δt by ε. In such a form this equation can be iterated and the outcome of such an interaction admits nice statistical mechanics interpretation. Before supplying such an interpretation, we would like to replace m by $-m$ in Eq. (8.3.4.15) which is equivalent to replacement of $\tilde{\beta}$ by $-\tilde{\beta}$ thus making transformations given by Eq. (8.3.4.9) exact. This step will be justified below.

Remark 8.3.4.4. In view of Eq. (8.3.4.13) the recursion (or monodromy in mathematics jargon) Eq. (8.3.4.15) is equivalent to the Pauli–Schrödinger Eq. (8.2.2.1). Thus, both the neutrino and the massive Dirac equation can be treated with methods of sub-Riemannian geometry. In the case of individual semiflexible polymer chains whose conformational properties are described by the Euclidean version of the Dirac propagator [290], the sub-Riemannian connection allows us to relate (e.g. read Appendix B and the next section) the Dirac model of such type of polymer chain with Euler's elastica model for these chains.

8.3.5 *Quantum Random Walks — Semiflexible Polymers Correspondence*

In the famous book on path integrals by Feynman and Hibbs, [292], as an exercise, the Problem 2–6 is formulated. It requires us to reformulate the Dirac propagator in the path integral form. As it is always the case, we begin with some kind of discrete version of the path integral. In the case of $(1 + 1)$-dimensional Dirac propagator (discussed in Problem 2–6), it is convenient to develop results on a square lattice whose size of the link is ε. Furthermore, it is convenient to rotate the lattice counterclockwise by 45° with respect to some lattice corner to be considered as an origin of coordinate system. Then, for such a system the coordinate axes will go through the diagonals of lattice squares. Consider now the motion of a particle on such a lattice.

The rules of particle's motion are as follows. The particle can move straight ("trans") or change its direction (gauche) by 90° at the corners of squares of the lattice. The terms "trans" and "gauche" are adopted from polymer physics where polymers are modelled by some kind of a random walk of steps of equal length modelling monomer units. The successive units can follow each other without bending (trans) or may bend (gauche). In the last case this may cost some energy to do. The succsession of trans–gauche conformations can be encoded by 0's and 1's as discussed in Appendix C. As a result, a polymer conformation (for a specific polymer) can be identified with some number written in the binary system. *Already this feature makes polymers relevant for computation.* In such defined $X-Y$ plane, the X-axis represents space while the Y-axis time. The particle is allowed to move back and forth along the X-direction and only forth along the Y-direction. Let $\mathbf{a} = (x_1, y_1)$ and $\mathbf{b} = (x_2, y_2)$ be the initial and final locations of points for such a walk in $X-Y$ plane. Furthermore, let $N(R)$ be the total number of walks of N steps long, starting at $\mathbf{a}$, ending at $\mathbf{b}$, and having exactly R turns along their paths. According to Feynman, [292], the lattice propagator $K_{\alpha\beta}(\mathbf{a}, \mathbf{b})$ can be defined now as follows:

$$K_{\alpha\beta}(\mathbf{a}, \mathbf{b}) = \sum_{R=0}^{\infty} N_{\alpha\beta}(R)(i\varepsilon m)^R, \tag{8.3.5.1a}$$

where α and β correspond to the initial (final) orientations of the walk: to the right $(+)$ or to the left $(-)$. In the case of polymers the mass m contains information about the trans–gauche bending energy: to go straight costs nothing, but to change a direction costs m each time. Polymer's rigidity can therefore be defined (up to a constant) as the inverse of the mass. Massless particles, e.g. neutrinos, have rigid rods, or better, rigid coiled spirals as their polymer analogs. Clearly, for the rigid rods the rigidity is infinite. The imaginary i factor in the propagator, Eq. (8.3.5.1a), is written because such a propagator is meant to describe the motion of particles in Minkowski space. In the case of polymers, we shall eliminate the imaginary i at the end of our calculations and replace it by the real factor as it is done in the case of finite temperature quantum field theory. There, the time t is replaced by the inverse temperature $\tilde{\beta}$.

Typically, it is necessary to perform an average over initial orientations α and to sum over the final β. This operation has its analog in the continuum limit for the Dirac particle [288]. Because of this, the quantity of interest is actually

$$K(\mathbf{a},\mathbf{b}) = \sum_{\alpha,\beta} p(\alpha) K_{\alpha\beta}(\mathbf{a},\mathbf{b}), \tag{8.3.5.1b}$$

where $p(\alpha)$ represents the averaging over initial positions (and can be looked upon as the discrete version of polarization matrix [288]). In the discrete case it is sufficient to choose $p(+) = p(-) = \frac{1}{2}$. These factors can be safely absorbed into $K_{\alpha\beta}$. If $N_+(N_-)$ represents the total number of steps in the walk to the right (left), then we can define $N = N_+ + N_-$ and $M = N_+ - N_-$. In view of earlier made definitions, the time interval can be defined as

$$t_2 - t_1 = \frac{N\varepsilon}{\sqrt{2}} \tag{8.3.5.2a}$$

while the space interval is given by

$$x_2 - x_1 = \frac{M\varepsilon}{\sqrt{2}}. \tag{8.3.5.2b}$$

The factor $1/\sqrt{2}$ appears because our lattice has been rotated by 45° counterclockwise with respect to the original lattice. Introduce now a system of N classical Ising spins $\sigma_i, i = 1 - N$, and relate the ith step to σ_i. Then, for the ith spin $\sigma_i = +1(-1)$ we identify $(+)$ with the particle move to the right and $(-)$ to the left. With such a rule for spins the "magnetization" M can now be defined as

$$M = N_+ - N_- = \sum_{i=1}^{N} \sigma_i \tag{8.3.5.3}$$

while the number of turns R can be determined as

$$R = \frac{1}{2} \sum_{i=1}^{N-1} (1 - \sigma_i \sigma_{i+1}). \tag{8.3.5.4}$$

Then, the combination $(i\varepsilon m)^R$ acquires the following look

$$(i\varepsilon m)^R = \exp\left\{ j \sum_{i=1}^{N-1} (\sigma_i \sigma_{i+1} - 1) \right\}, \tag{8.3.5.5}$$

where $\exp(-2j) = i\varepsilon m$. The propagator (8.3.5.1a) can now be written in the form of a partition function $Z_{\sigma_1 \sigma_N}(N, \mu; j)$ for one-dimensional classical Ising model in constant magnetic field μ, i.e.

$$\begin{aligned} K_{\alpha\beta}(\mathbf{a}, \mathbf{b}) &= Z_{\sigma_1 \sigma_N}(N, \mu; j) \\ &= \sum_{\{\sigma\}} \exp\left\{ \mu \sum_{i=1}^{N} \sigma_i + j \sum_{i=1}^{N-1} (\sigma_i \sigma_{i+1} - 1) \right\}. \end{aligned} \tag{8.3.5.6}$$

In this expression $\{\sigma\}$ represents the spin summation: from σ_2 to σ_{N-1}. By appropriately rescaling space and time variables we define respectively $t = t_b - t_a$, $x = x_b - x_a$ and $M\mu = -ipx$ which is the same as $\mu = -ip\varepsilon$. Then, the partition function for the 1D Ising model can be calculated with the help of standard transfer matrix method. Thus, we obtain:

$$Z_{\sigma_1 \sigma_N}(N, \mu; j) = \lambda_+^N \varphi_+(\sigma_1)\varphi_+(\sigma_N) + \lambda_-^N \varphi_-(\sigma_1)\varphi_-(\sigma_N), \tag{8.3.5.7}$$

where $\lambda_\pm$ and $\varphi_\pm$ are respectively the eigenvalues and the eigenfunctions of the transfer matrix $H(\sigma_i, \sigma_{i+1})$ whose explicit form is given by

$$H(\sigma_i, \sigma_{i+1}) = \exp\left[\frac{\mu}{2}(\sigma_i + \sigma_{i+1}) + j((\sigma_i \sigma_{i+1} - 1) \right]. \tag{8.3.5.8}$$

Since $\exp(-2j) = i\varepsilon m$ and $\mu = -ip\varepsilon$, to order $O(\varepsilon)$ the above transfer matrix can be written as

$$\|H\| = \begin{pmatrix} 1 - ip\varepsilon & i\varepsilon m \\ i\varepsilon m & 1 + ip\varepsilon \end{pmatrix}. \tag{8.3.5.9}$$

This result coincides with the already obtained transfer matrix defined in Eq. (8.3.4.15) where we need to change m to $-m$ (guaranteeing correctness of the unitary transformation, Eq. (8.3.4.9a), with the unitary matrix defined by Eq. (8.3.4.b)). The eigenvalues of the matrix $\|H\|$ are given by

$$\lambda_\pm = 1 \mp i\varepsilon m E \quad \text{with } E^2 = 1 + \frac{p^2}{m^2}. \tag{8.3.5.10}$$

Accordingly, the eigenfunctions $\varphi_{\pm}$ are given by

$$\begin{aligned}
\varphi_{+}(+) &= \left[\frac{1}{2}\left(1-\frac{p}{mE}\right)\right]^{\frac{1}{2}}, \\
\varphi_{+}(-) &= -\left[\frac{1}{2}\left(1+\frac{p}{mE}\right)\right]^{\frac{1}{2}}, \\
\varphi_{-}(+) &= \left[\frac{1}{2}\left(1+\frac{p}{mE}\right)\right]^{\frac{1}{2}}, \\
\varphi_{-}(-) &= \left[\frac{1}{2}\left(1-\frac{p}{mE}\right)\right]^{\frac{1}{2}}.
\end{aligned} \tag{8.3.5.11}$$

In view of Eq. (8.3.5.1b), the final result for the Dirac propagator $G(p,t)$ is given by

$$\begin{aligned}
G(p,t) &= \sum_{\alpha,\beta} p(\alpha)K_{\alpha\beta}(p,t) \\
&= \left(1-\frac{1}{E}\right)e^{-imEt} + \left(1+\frac{1}{E}\right)e^{+imEt}.
\end{aligned} \tag{8.3.5.12}$$

In arriving at this result it was taken into account that $e = \lim_{\varepsilon\to 0^+}(1+\varepsilon)^{\frac{1}{\varepsilon}}$. From the above derivation, it should be clear that the restriction to $p = p_z$ is just a matter of convenience. Use of unitary (Lorentz-type) transformation can restore the full p-dependence. Therefore, the momentum p in the above propagator can be safely replaced by $\mathbf{p}$. Although the result Eq. (8.3.5.12) for the Dirac propagator is exactly the same as can be obtained by iteration of Eq. (8.3.4.15), its appearance is not self-obvious.

While we shall explain this in the context of semiflexibe polymers in the next section, in this section we would like to discuss the obtained results from the point of view of quantum random walks. For reader's convenience, a glossary of needed terminology is provided in Appendix C. We strongly advise the readers to read this glossary before reading the following text.

To re-obtain Eq. (8.3.4.15) in the context of quantum random walks (QRW) we must use Eq. (C.36) in which we have to replace the parameter θ by εm. In such a case we obtain the following result for the quantum coin C operator:

$$\begin{aligned}
C = \exp(i\sigma_x \varepsilon m) &= \begin{pmatrix} \cos\varepsilon m & i\sin\varepsilon m \\ i\sin\varepsilon m & \cos\varepsilon m \end{pmatrix} \\
&= \begin{pmatrix} 1 & i\varepsilon m \\ i\varepsilon m & 1 \end{pmatrix} + O((\varepsilon m)^2).
\end{aligned} \tag{8.3.5.13}$$

At the same time, to get the explicit *form* for the shift operator S it is convenient to use the discrete Fourier transform for the wave functions defined on the regular lattice. In such a case, the shift operator S reads

$$S = \exp\left(-i\sigma_z p\varepsilon\right) = \begin{pmatrix} \exp(-ip\varepsilon) & 0 \\ 0 & \exp(ip\varepsilon) \end{pmatrix}$$
$$= \begin{pmatrix} 1 - ip\varepsilon & 0 \\ 0 & 1 + ip\varepsilon \end{pmatrix} + O(\varepsilon^2). \quad (8.3.5.14)$$

Using these results, the combination SC discussed in Appendix C reads

$$SC = \begin{pmatrix} 1 - ip\varepsilon & i\varepsilon m \\ i\varepsilon m & 1 + ip\varepsilon \end{pmatrix} + O(\varepsilon^2). \quad (8.3.5.15)$$

Evidently, it is coinciding with the transfer matrix in Eq. (8.3.4.15) (or (8.3.5.9)), provided that we replace m by $-m$. The physical meaning of such a replacement was discussed immediately after Eq. (8.3.4.15). Thus, Eq. (8.3.4.15) has exactly the same meaning as Eq. (C.35) (Appendix C) valid for both continuous and discrete-time quantum random walks. In Appendix C we mentioned the work of Childs, [293], who considered the relationship between the discrete-time and continuous-time quantum random walks. Equations (8.3.5.13)–(8.3.5.15) allow us to investigate this relationship further using the Dirac propagator as an example. For this purpose, taking into account that $\varepsilon = \Delta t \equiv t/N$ we write $C = C(t/N)$ and $S = S(t/N)$ so that by using the Trotter product formula[1] we obtain,

$$\lim_{N\to\infty} U(t/N)^N = \lim_{N\to\infty} [S(t/N)C(t/N)]^N = \exp\{\pm i(-\sigma_z p + \sigma_x m)t\}$$
$$\equiv \exp(\pm i\hat{H}t). \quad (8.3.5.16)$$

This result is in agreement with both Eqs. (8.3.4.13) and (C.31) derived for continuous-time quantum random walks. It should be noted that these results were obtained without the use of 2-qubit gates needed for execution of logical operations. Such a deficiency happens to be not essential. It can be easily corrected as explained in [294]. The correction involves use of the discretized version of the full $(3+1)$-dimensional Dirac propagator. It should also be noted that exactly the same quantum algoritm is used for quantum cellular automata (QCA) modelling quantum lattice gases [295].

[1]That is $\exp(t(A+B)) = \lim_{n\to\infty}[\exp(tA/n)\exp(tB/n)]^n$ for any two self-adjoint operators A and B.

In fact, what we just described can be identified with the two-component QCA. E.g. read Sec. 5 of [295].

8.4 Semiflexible Polymers and Quantum Computers

8.4.1 *General Information*

The connections between semiflexible polymers and quantum computers can be anticipated based on observation that the succession of trans–gauche conformations along polymer's backbone can be encoded by 0's and 1's. As a result, the specific polymer conformation (for a given polymer) can be identified with some number written in the binary system. Details are discussed in Appendix C. To do computation requires changing conformation. Development of such an idea (to some extent) can be found in the paper by Lloyd [296].[m] Even though much more details related to Lloyd's paper can be found in [297], to our knowledge, the design of quantum computers based on Lloyd's ideas still remains as a project waiting for completion. The difficulty lies in our ability/inability to manipulate the individual trans–gauche conformations in a reliable way. Next evidence of likely connection between semiflexible polymers and quantum computers was presented in Sec. 8.3.3 and Appendix B when we discussed the correspondence between the equation describing dynamics of Josephson's junctions and that describing Euler's elastica. This connection will be further discussed below in this section. In the meantime, we would like to mention yet another relation between polymers and quantum computers. This time, in the context of adiabatic quantum computation. The idea of adiabatic quantum computation was presented in the highly cited paper by Farhi *et al.* [298] in 2000. This idea was utilized in [299] in the context of protein folding. Since the discussion of the adiabatic quantum computation would take us far away from topics discussed in this book, we shall refrain from making further comments, especially, because we have at our disposal more immediately relevant examples of uses of semiflexible polymers for quantum computation. These are based on results of Sec. 8.3.5

[m]Incidentally, in Lloyd's paper the protocol for quantum of computing is based on results obtained for classical reversible computation for which uses of Fredkin's gates are essential. Because such computation is topological in nature it can be used for quantum algorithms as well. This is demonstrated in Lloyd's paper. Subsequently, other topological methods were developed for quantum computation as explained in Appendix C.

that is on uses of discrete and continuous-time QRW's. This is justifiable since in [300] it was demonstrated that the adiabatic quantum computation is equivalent to the standard quantum computation described in Appendix C. In the same appendix we explained the intimate connection between random walks (RW) and computations done either using classical or quantum computer. Indeed, in both cases computation can be modelled by some kind of Turing's machine — a finite state automaton. The computational process can be visualized as some kind of a walk — from one vertex to the next (determined by the state of a given vertex) — on the directed graph, e.g. see Eq. (C.34) of Appendix C. Thus, descriptions of QRW's and quantum computers are interrelated. Surprisingly, this idea was brought to completion only very recently. In [301] Childs developed fine details of this correspondence by describing in detail how the universal quantum computation can be achieved by using the continous-time QRW's. That the same goal can be achieved using the discrete-time QRW's was demonstrated a bit later in [302]. In Appendix C we noticed that QRW becomes classical RW when decohered. In real life such a decoherence is realized either by sequential measurements of the path of the walker or, in polymer's language, by the loss of rigidity for such a walk (e.g. caused by increase in temperature or changes in ionic strength, e.g. for DNA's and proteins in solution, etc.). As explained in Appendix C, the degree of intrusion into quantumness is characterized by the parameter p: $0 \leq p \leq 1$. Since the nonzero p destroys unitarity, this makes QRW-quantum computers correspondence questionable. Fortunately, such type of interpolation between quantum and classical RW's was recently elevated to general principle of how quantum computations should be done [303]. The latest efforts in this direction are summarized in [304]. The nonunitary of QRW's is believed to be essential in biosynthetic energy transfer [305, 306] working perfectly at room temperatures. This fact provides us with a compelling reason for looking at possibilities of utilizing semiflexible polymers in quantum computation (or, at least in solving some computer-related problems) since quantum computers working at room temperatures are certainly more attractive than those requiring ultra low temperatures.

Exercise 8.4.1.1. Investigate to what extent ideas outlined in [306] are compatible with those presented in [303, 304].

Fortunately, there are even stronger reasons for looking into this direction. These are based on observation that besides the digital-type quantum computers described in Appendix C there are also analog quantum computers. These are called "quantum simulators" [307]. Since typically these devices do not require error corrections and even quantum gates, the requirements for building them are much less demanding than those for quantum computers. For this reduction in requirements one should pay a price. By design, these controllable quantum systems can simulate only one pre-selected quantum system. For instance, quantum behavior of the Dirac propagator was recently quantum simulated by using a single trapped ion [308–309]. Other uses of quantum simulators employing cold trapped ions can be found in [310].

In connection with these results, it should be noted the following. In 1969, Symanzik reformulated theory of interacting quantum fields in terms of interacting polymers (in solutions[n]) [311]. This reformulation was subsequently used for obtaining the most rigorous results for quantum field theories [312]. Thus, when it comes to quantum simulators of quantum field theories, polymer solutions are the most appropriate candidates. The authors of [308–309] claimed that their quantum simulator (trapped ion) can reproduce such features of the Dirac equation (Dirac propagator) as Zitterbewegung and Klein's paradox. As we demonstrated in [39], in polymer's language the Zitterbewegung is just a spontaneous coiling of the backbone of semiflexible polymer. The detailed study of Klein's paradox is presented in [313]. It is associated with behavior of the Dirac propagator in the presence of boundaries. In the language of polymers such a task was completed in our work, [314], without actual mention of Klein's paradox. This deficiency was corrected in [315] quoting results of our work.

The above results provide us with sufficient reasons to develop semiflexible polymers — quantum computers correspondence further. To do so we still need to provide some information about semiflexible polymers, especially in regard to their connection with the Dirac propagator. This is accomplished in the next subsection.

8.4.2 *Some Helpful Facts from Theory of Semiflexible Polymers: Dirac versus Kratky–Porod*

Fully flexible polymers can be well-modelled by the well-known standard random walk. That is to say, in the continuum limit their conformational characteristics are described by the Gaussian distribution. Such a

[n]He did not had in mind actual physical solutions. But, physical polymers needed for modelling of quantum fields do exist in solutions only.

distribution is formally obtainable as solution of the diffusion equation. It is convenient also to associate with it the Green's function — a propagator.[o] Such a propagator can be used for calculation of statistical properties of fully flexible (Gaussian) polymer chains. For instance, let our polymer be of total contour length N and let $\mathbf{r}(\tau)$ describe a position of a polymer segment having contour position τ. By this we mean that we use natural parametrization for the polymer's backbone selecting $\mathbf{r}(0)$ as the origin and $\mathbf{r}(N)$ as the end with respect to some chosen frame of reference.[p] Let $\mathbf{r}(0) \equiv \mathbf{r}_0$ and $\mathbf{r}(N) \equiv \mathbf{r}_1$ so that we can construct the end-to-end vector $\mathbf{R} = \mathbf{r}_1 - \mathbf{r}_0$. The retarded Green's function $G_0(\mathbf{R}, N)$ for the diffusion equation then serves as moment-generating function. Explicitly, following our work, [290], we introduce the Fourier-transformed propagator via

$$G_0(\mathbf{p}, N) = \int d\mathbf{R} G_0(\mathbf{R}, N) \exp(i\mathbf{p} \cdot \mathbf{R}) \tag{8.4.2.1}$$

so that for the Gaussian-type propagator we obtain: $G_0(\mathbf{p}, N) = \exp(-A\mathbf{p}^2 N)$, where A is some (known) constant. The first physically interesting quantity to calculate is $\langle \mathbf{R}^2 \rangle$ which is the average (over orientations) square of the end-to-end distance for a single polymer chain. It is given by

$$\langle \mathbf{R}^2 \rangle = -\frac{d}{d\mathbf{p}^2} \ln G_0(\mathbf{p}, N) \mid_{\mathbf{p}=0} = AN. \tag{8.4.2.2}$$

The same result for semiflexible polymers formally requires us to replace $G_0(\mathbf{p}, N)$ by Green's function for semiflexible polymers. Such a function is not so easy to obtain. If we are interested only in calculations of $\langle \mathbf{R}^2 \rangle$ there is a way to obtain the result without explicit knowledge of Green's function. It is described in the Landau–Lifshitz textbook, [317]. We would like to put their result into a broader context following our work, [318]. We begin by rewriting the Gaussian-type propagator $G_0(\mathbf{R}, N)$ in the path integral form

$$G_0(\mathbf{R}, N) = \int_{\mathbf{r}(0)=\mathbf{0}}^{\mathbf{r}(N)=\mathbf{R}} \mathfrak{D}[\mathbf{r}(\tau)] \exp\{-S_N[\mathbf{v}(\tau)]\}, \tag{8.4.2.3}$$

where the action functional $S_N[\mathbf{v}(\tau)]$ is given by

$$S_N[\mathbf{v}(\tau)] = \frac{3}{2l} \int_0^N d\tau \langle \mathbf{v} \cdot \mathbf{v} \rangle \tag{8.4.2.4}$$

---

[o] Typically, in polymer literature it is common to use the retarded-type propagators, e.g. consult our work, [316].

[p] Such a choice makes our Green's functions retarded.

and $\mathbf{v}(\tau) = d\mathbf{r}/d\tau$, $\langle \mathbf{v} \cdot \mathbf{v} \rangle = \sum_{i=1}^{3} v_i v^i$. Here the parameter l describes the scale (e.g. the size of the monomer of which the polymer chain is made) below which one cannot treat polymer as the Gaussian-type walk. Alternatively, if one begins with a walk on some kind of a lattice (e.g. read Appendix C), then l represents the lattice spacing. To describe semiflexible polymers one has to replace the functional $S_N[\mathbf{v}(\tau)]$ by

$$\tilde{S}_N[\mathbf{v}(\tau)] = \frac{3}{2l} \int_0^N d\tau \langle \mathbf{v} \cdot \mathbf{v} \rangle + \frac{\eta}{2} \int_0^N d\tau \langle \mathbf{t} \cdot \mathbf{t} \rangle. \tag{8.4.2.5}$$

Here η is some phenomenological parameter responsible for rigidity of polymer's backbone. Its meaning will be further clarified below. Since the tangent vector $\mathbf{t}$ is defined by $\mathbf{t} = d\mathbf{v}/d\tau$ and since using natural parametrization we obtain $\langle \mathbf{v} \cdot \mathbf{v} \rangle = 1$, our readers are asked to compare the functional $\tilde{S}_N[\mathbf{v}(\tau)]$ against the functional $\mathcal{F}(\gamma)$ defined in Theorem 2.5.2, Chap. 2. Such a comparison allows us to make the identification $\langle \mathbf{t} \cdot \mathbf{t} \rangle = \kappa(s)^2$ where $\kappa(s)$ is the curvature of the curve. With these definitions the path integral for the Kratky–Porod-type model of semiflexible polymer chain can be written by analogy with that given by Eq. (8.4.2.3)

$$G(\mathbf{v}(N), \mathbf{v}(0)) = \int_{\mathbf{v}(0)}^{\mathbf{v}(N)} \mathfrak{D}[\mathbf{v}(\tau)] \prod_{\tau} \delta(\mathbf{v}^2 - 1) \exp\left\{ -\frac{\eta}{2} \int_0^N d\tau \kappa^2(\tau) \right\}. \tag{8.4.2.6}$$

Remark 8.4.2.1. It should be noted that the traditional definition of the Kratky–Porod (K-P) model (also known as the worm-like chain model) is not based on uses of path integrals, e.g. see http://en.wikipedia.org/wiki/Worm-like_chain. If the constraint in the path integral is disregarded, we shall obtain the path integral Eq. (8.4.2.3) in which it is required to make obvious replacement $\mathbf{r}(\tau) \rightarrow \mathbf{v}(\tau)$. The integral would be of Gaussian-type and is straightforwardly calculable. The presence of a constraint results in the Brownian motion on the sphere S^2 instead of $\mathbf{E}^3$. Although the calculation of the path integral for this case can be done [318], it is associated with some technical difficulties as compared to the flat (that is $\mathbf{E}^3$) Gaussian case. Nevertheless, if we are interested only in calculations of averages of the type $\langle [\mathbf{v}(\tau) - \mathbf{v}(0)]^2 \rangle_G$ physical considerations are sufficient. Indeed, since geometrically the constraint $\mathbf{v}^2 = 1$ represents S^2, for times $0 < \tau \ll N$ the Brownian motion on S^2 can be safely replaced by that on the plane. In such a case, the result for $\langle[\mathbf{v}(\tau)-\mathbf{v}(0)]^2\rangle_G$ can be obtained immediately

$$\langle[\mathbf{v}(\tau)-\mathbf{v}(0)]^2\rangle_G = \frac{2}{\eta}\tau. \tag{8.4.2.7a}$$

Noting that $\mathbf{v}^2 = 1$, the above result can be rewritten as

$$\langle 2 - 2\mathbf{v}(\tau)\cdot\mathbf{v}(0)\rangle_G = 2 - 2\langle\mathbf{v}(\tau)\cdot\mathbf{v}(0)\rangle_G = \frac{2}{\eta}\tau. \tag{8.4.2.7b}$$

Upon exponentiation we obtain,

$$\langle\mathbf{v}(\tau)\cdot\mathbf{v}(0)\rangle_G = \exp\left(-\frac{\tau}{\eta}\right). \tag{8.4.2.8}$$

This result allows us to obtain $\langle\mathbf{R}^2\rangle$ via

$$\begin{aligned}\langle\mathbf{R}^2\rangle &= \int_0^N d\tau \int_0^N d\tau' \langle\mathbf{v}(\tau)\cdot\mathbf{v}(0)\rangle_G \\ &= 2N\eta - 2\eta^2\left[1-\exp\left(-\frac{N}{\eta}\right)\right].\end{aligned} \tag{8.4.2.9}$$

In such form it is obtained in the Landau–Lifshitz textbook, [316], by different arguments which are not readily generalizable.

Consider now just obtained Eq. (8.4.2.9) in two limits: (a) low rigidity ($\eta \to 0$) and (b) high rigidity ($\eta \to \infty$). In the first case we obtain: $\langle\mathbf{R}^2\rangle \simeq 2N\eta$. This is known as the Gaussian result for flexible polymers. In the second case we obtain: $\langle\mathbf{R}^2\rangle = N^2$. This is exactly the result for QRW's as discussed in Appendix C! Since this result was obtained without use of the Dirac propagator, we have to explain at this point why uses of the Dirac propagator for purposes of such calculations are more advantageous. To do so, we need to go back to Eqs. (8.3.5.10)–(8.3.5.12). Transition to the Euclidean space requires some care. Since we are having a combination $\pm imEt$, we can, for instance, perform the usual rotation: $\pm it \to \hat{N}$, where the "+" or "−" sign is determined by physical considerations (e.g. whether we are dealing with retarded or advanced Green's function). Alternatively, we can replace m by $\pm im$. Again the choice of sign will be fixed by physics. Thus, we end up with $E^2 = 1 - \frac{p^2}{m^2}$ so that there is a possibility that $E^2 < 0$. Such a possibility does have physical meaning as we would like to explain now. To do so we shall be guided by the result Eq. (8.4.2.2).

Specifically, using Eq. (8.3.5.12) and replacing $G_0(\mathbf{p}, N)$ by

$$G(\mathbf{p}, \hat{N}) = 2\cosh(mE\hat{N}) + \frac{2}{E}\sinh(mE\hat{N}), \qquad (8.4.2.10)$$

where $E^2 = 1 - \frac{\mathbf{p}^2}{m^2}$, and the parameter $\hat{N}$ differs from N by some constant to be determined. The combined use of Eqs. (8.4.2.2) and (8.4.2.10) produces

$$\langle \mathbf{R}^2 \rangle = -\frac{1}{G}\frac{\partial G}{\partial E}\frac{\partial E}{\partial \mathbf{p}^2}\bigg|_{\mathbf{p}=0} = \frac{\hat{N}}{2m} - \frac{1}{4m^2}(1 - \exp(-2m\hat{N})). \qquad (8.4.2.11)$$

This result should be compared with that in Eq. (8.4.2.9). Such a comparison requires us to choose $m = (2\sqrt{2}\eta)^{-1}$ and $\hat{N} = N\sqrt{2}$. The factor $\sqrt{2}$ is caused by the chosen geometry of our lattice (counterclockwise rotated with respect to the initial one). Thus, the replacement $m \to \pm im$ happens to be sufficient for reproduction of Eq. (8.4.2.9). Although physical meaning of the mass is clarified now by its (just obtained) relation to phenomenological rigidity parameter η, it is possible to do even better study for m in order to determine the sign "+" or "−" in substitution $m \to \pm im$. In doing so we follow our work, [318]. In it, we used some results of the Nobel laureate, Paul Flory.

The story begins with modelling of a monodisperse ensemble of semiflexible polymers as some kind of semiflexible random walks on a regular lattice with coordination number z. Fixing our attention at particular polymer chain first we are confronted with the following situation. After a particular step by the walker was already made, the walker is confronted with two options: (a) to continue going in the same direction (trans), or (b) to change the direction (gauche), provided that the immediate self-reversals are excluded. The first option is weighted with the conditional probability one, while the second with the factor $(z-2)\exp(-E)$, where the dimensionless (trans–gauche) energy E can be computed quantum mechanically, in principle. Next, we need to take into account the presence of other polymers. After the first walker was placed on the lattice, other walkers "should adjust their walking behavior" in order not to run into the first walker. This presents some combinatorial problem which was solved by Flory with the result for the average number of bends (turns) R per polymer molecule per effective link given by

$$R = \frac{(z-2)\exp(-E)}{1 + (z-2)\exp(-E)}. \qquad (8.4.2.12)$$

It is of interest to relate R just obtained with that given by Eq. (8.3.5.1). In [318] it is shown that the average for R defined by Eq. (8.3.5.1) is given by

$$R = -\frac{1}{2}\frac{\partial}{\partial j}\frac{1}{N}\ln Z = \frac{1}{2}(1 - \tanh j), \tag{8.4.2.13}$$

where $Z = G(\mathbf{p} = \mathbf{0}, \hat{N})$. Notice that we previously defined j via $j = -\frac{1}{2}\ln(\varepsilon im)$. Now, we can make a replacement $m \to -im$ causing selection of "$-$" sign in $m \to \pm im$. At the same time, if we choose $j = -\frac{1}{2}\ln[(z-2)\exp(-E)]$ in Eq. (8.4.2.13) we shall re-obtain the result Eq. (8.4.2.12) implying: $\varepsilon m = (z-2)\exp(-E)$.

The obtained results demonstrate the advantage of the Dirac model over the K-P since they connect much better with the lattice-type polymer models. This advantage is amplified many times when one is interested in calculations of averages like $\langle \mathbf{R}^{2n} \rangle$, $n = 2, 3, 4, \ldots$ as it was done in our work, [319]. Subsequently, theoretical results of [319] demonstrated a remarkable agreement between the theory, the experiment and the numerical simulations. They are summarized in the Wikipedia, e.g. read http://sasfit.ingobressler.net/manual/ WormLikeChainEXV.

For the purposes of what follows, it is instructive to discuss some results from [319]. Specifically, comparison between different models and with numerical results causes us to consider instead of $G(\mathbf{p}, \hat{N})$ in Eq. (8.4.2.10) its "truncated" version

$$\tilde{G}(\mathbf{p}, \hat{N}) = \frac{2}{\mathcal{E}}\sin m\mathcal{E}\tilde{N}, \tag{8.4.2.14}$$

where $\mathcal{E} = \frac{\mathbf{p}^2}{m^2} - 1$. Such truncated version is not an artefact. It appears as result of different averaging prescription (over the initial and the final orientations of the chain) as compared with that in Eq. (8.3.5.1b). By repeating the same calculations as in Eq. (8.4.2.11) we obtain:

$$\langle \mathbf{R}^2 \rangle = -\frac{1}{G}\frac{\partial G}{\partial E}\frac{\partial E}{\partial \mathbf{p}^2}\bigg|_{\mathbf{p}=0} = \frac{4}{3}a^2 x\mathcal{L}(x). \tag{8.4.2.15}$$

Here we introduced the Langevin function $\mathcal{L}(x) = \coth x - 1/x, x = 3N/2a, 2a = l$ and l was defined in Eq. (8.4.2.4). At the first sight, expressions Eqs. (8.4.2.11) and (8.4.2.15) look different. But explicit plotting of both expressions for $\langle \mathbf{R}^2 \rangle$ shows that they are practically coinciding. Furthermore, computation of higher moments using $\tilde{G}(\mathbf{p}, \hat{N})$ demonstrates obvious superiority over that by using $G(\mathbf{p}, \hat{N})$ in Eq. (8.4.2.10). The emerging Langevin function reveals additional physics built into $\tilde{G}(\mathbf{p}, \hat{N})$ revealing its close ties with the 2-level quantum system described by Eq. (8.2.2.1). For the sake of what follows, it is helpful to provide some details (even though they are expected to be well-known) about this connection.

Originally, the Langevin function was used for description of paramagnetism. It should be considered as limiting case of the Brillouin function described as an exercise in the ninth volume of the Landau and Lifshitz course, [13]. We reformulate this exercise as follows.

Exercise 8.4.2.1. The quantum Hamiltonian $\mathcal{H}$ for spin $\mathbf{S}$ (not necessarily 1/2!) interacting with the external magnetic field $\mathbf{B}$ is given by $\mathcal{H} = -\mathbf{S}\cdot\mathbf{B}$. Find the partition function (L. Brillouin, 1927) for such a system. Hint: Michael Stone, [320], re-obtained Brillouin's result using supersymmetry in 1989. A quick introduction to supersymmetry can be found in [321].

We would like to make some comments on Brillouin's solution. Since the energies of spin $\mathbf{S}$ of a particle in a field $\mathbf{B}$ are $-S|\mathbf{B}|, \dots, +S|\mathbf{B}|$, the partition function $Z(\beta)$ is obtained straightforwardly as

$$Z(\beta) = e^{\beta SB} + \cdots + e^{-\beta SB} = \frac{\sinh[(2S+1)\frac{\beta B}{2}]}{\sinh\left(\frac{\beta B}{2}\right)}. \tag{8.4.2.16a}$$

From here, the average magnetization $\langle S \rangle$ is calculated as follows:

$$\langle S \rangle = \beta^{-1}\frac{\partial}{\partial B}\ln Z(\beta) = \frac{2S+1}{2}\coth\left[(2S+1)\frac{\beta B}{2}\right] - \frac{1}{2}\coth\left(\frac{\beta B}{2}\right). \tag{8.4.2.16b}$$

It is convenient to rescale the magnetic field: $\mathbf{B} \to \tilde{B}/S$ and simultaneously divide both sides of the previous equation by S. Then, the Brillouin function $B(x)$ can be redefined as follows:

$$\frac{\langle S \rangle}{S} = B(x) = \frac{2S+1}{2S}\coth\left[\frac{(2S+1)}{2S}x\right] - \frac{1}{2S}\coth\left(\frac{1}{2S}x\right). \tag{8.4.2.16c}$$

In the limit $S \to \infty$ we obtain

$$B(x) \simeq \coth x - \frac{1}{x} = \mathcal{L}(x), \tag{8.4.2.16d}$$

where $x = \beta\tilde{B}$. It is illuminating to re-obtain this result in an old fashioned way using classical statistical mechanics. In this case the partition function is given by

$$Z(\beta) = \int d\Omega \exp(\beta SB\cos\theta) = \frac{4\pi\sinh(\beta SB)}{\beta SB}, \tag{8.4.2.17a}$$

where $d\Omega = d\varphi d\theta \sin\theta$. The average magnetization is obtained now as

$$\langle S \rangle = \beta^{-1} \frac{\partial}{\partial B} \ln Z(\beta) = S\mathcal{L}(\beta S B)$$

or, equivalently, as

$$\frac{\langle S \rangle}{S} = \mathcal{L}(x) \tag{8.4.2.17b}$$

to be compared with Eq. (8.4.2.15). The above result can be generalized if we take into account that the partition function for the K-P model can be equivalently presented as

$$Z_N = \int \prod_{i=0}^{N} \frac{d\Omega_i}{4\pi} \exp\left(K \sum_{i=1}^{N} \mathbf{s}_i \cdot \mathbf{s}_{i-1}\right) \tag{8.4.2.18}$$

as demonstrated in our works, [203, 318]. Here the unit vectors $\mathbf{s}_i$ are classical spins obeying the constraint: $\mathbf{s}_i^2 = 1$. With such defined partition function, it becomes possible to consider a model whose Hamiltonian

$$-\mathcal{H} = K \sum_{i=1}^{N} \mathbf{s}_i \cdot \mathbf{s}_{i-1} + \sum_{i=0}^{N} \mathbf{f} \cdot \mathbf{s}_i \tag{8.4.2.19}$$

describes one-dimensional three-component classical Heisenberg ferromagnet in the constant magnetic field. We shall take the full advantage of just described polymer-magnet correspondence beginning from the next subsection.

8.4.3 *Qubits from Hairpins*

In our work, [322], we noticed that as soon as we leave the domain of fully flexible (Gaussian) polymers and enter the domain of semiflexible polymers, there exists a countable infinity of models describing such polymers. Between these models, those employing the Dirac propagator are unquestionably superior to others. However, physical reality, e.g. solutions of liquid crystalline-like polymers, may require uses of, say, the K-P-type models as explained in [39]. The purpose of this subsection is to provide new interpretation of the already obtained results. For this task, we follow de Gennes, [323], who introduced the notion of hairpins for semiflexible polymers. He considered the kink-type solutions of the equation obtained by minimization of the following elastic energy functional

$$\mathcal{F}[\theta] = \mathcal{F}_0 + \frac{1}{2}\int_0^N ds \left[Q \sin^2\theta(s) + B\left(\frac{d\theta}{ds}\right)^2\right]. \tag{8.4.3.1}$$

In view of known identity $\sin^2 x = \frac{1}{2}(1 - \cos 2x)$, the Lagrangian density $\mathcal{L}[\theta]$ for $\mathcal{F}[\theta]$ coincides with that given in Eq. (8.3.3.8) for Josephson's junctions. It is also compatible with just obtained Hamiltonian, Eq. (8.4.2.19). In Appendix B we explain why the equation of "motion" obtained by minimization of $\mathcal{F}[\theta]$ coincides with that for Euler's elastica. The minimization problem for Euler's elastica is treated by methods of sub-Riemannian geometry. The functional $\mathcal{F}[\theta]$ can be naturally extended to $(1+1)$ dimensions. Such a generalization is known as the sine–Gordon model. Comprehensive classical and quantum treatment of the sine–Gordon model is given in the monograph [324] by Faddeev and Takhtajan. By adapting the same methodology as in [114], the Lagrangian density $\mathcal{L}$ for the sine–Gordon model in the dimensionless rescaled variables is given by

$$\mathcal{L}[\theta] = \frac{1}{2}\partial_\mu \theta \partial^\mu \theta + (\cos\theta - 1), \tag{8.4.3.2}$$

where the signature of (1+1)-dimensional space–time is $(1, -1)$ with $+1$ for time and -1 for space. The above Lagrangian density is invariant with respect to shifts $\theta \to \theta \pm 2\pi$. Because of this, consider now the kink-type solutions. To make a connection with [114] we notice that the contour variable s in $\mathcal{F}[\theta]$ plays a role of "time" (but it can also be considered as one-dimensional space variable which we can relabel as x). In the limit of long polymer chains (that is when $N \to \infty$) it is possible to completely adopt results of [114]. Then, the topological charge $\hat{Q}$ (the winding number) can be defined as

$$\hat{Q} = \frac{1}{2\pi}\int_{-\infty}^{\infty} dx \left(\frac{d\theta}{dx}\right) \tag{8.4.3.3}$$

while the Bogomolny bound in the present case is given by

$$E \geq \int_0^{2\pi\hat{Q}} 2\left|\sin\frac{\theta}{2}\right| d\theta = 4|\hat{Q}|\left[-\cos\frac{\theta}{2}\right]_0^{2\pi} = 8|\hat{Q}|. \tag{8.4.3.4}$$

The bound is attained when

$$\frac{d\theta}{dx} = \pm 2\sin\frac{\theta}{2}. \tag{8.4.3.5a}$$

By selecting, for instance, sign + we obtain

$$\theta(x) = 4\tan^{-1}(x - a). \tag{8.4.3.5b}$$

In this expression a is some constant. Such a solution describes the kink if the boundary conditions $\theta(-\infty) = 0, \theta(+\infty) = 2\pi$ are imposed.

By applying now de Gennes polymer interpretation of the results just obtained we need to restore the dimensionalities, that is we write $x \rightleftarrows s, a = 0,\ \theta \to 2\theta$. By introducing the characteristic (contour) length of the hairpin $2l$ (which is de Gennes terminology for the kink), that is the "thickness" of the kink along the s-axis, the hairpin solution obtained by de Gennes reads as follows:

$$\theta(s) = 2\tan^{-1}(s/l). \tag{8.4.3.5c}$$

In such an interpretation the hairpin is some local sharp bend (by 180°) of the polymer chain. As in the theory of kinks, there could be many hairpins simultaneously.

Remark 8.4.3.1. Since $\cos\theta \simeq 1 - \frac{\theta^2}{2!} + \frac{\theta^4}{4!} + \cdots \to (\frac{\theta^2}{2} - 1)^2$, it is possible to use such an expansion in the one-dimensional version of the Lagrangian Eq. (8.4.3.2) in order to write the path integral for one-dimensional Ising model. Details are given in the book by Polyakov [41]. Based on this observation, a connection with the Euclidean version of Dirac propagator, Eq. (8.4.2.10), is established at once![q]

Typically, the concept of hairpins is used in interpreting experimental results involving stretching of individual semiflexible polymer chains. In addition to the theoretical results presented in the already cited [203, 325] discuss both theory and experimental setup of stretching experiments. There are two kinds of stretching: one, of nematic-type, is caused by the presence of other semiflexible polymers interacting with a particular polymer. Another is caused by the stretching of the anchored (to some surface) polymer chain by pooling it by another end with help of hydrodynamic, magnetic, electric, etc. stretching forces. For example, in [326] the dielectric response of semiflexible polymer was considered within the framework of the de Gennes hairpin model. The authors obtained the following generating/partition function $Z(\mathbf{f})$

$$Z(\mathbf{f}) = 2\left[\cosh(AN) + \frac{\varphi}{A}\sinh(\varphi A)\right], \tag{8.4.3.6}$$

where $A^2 = \nu^2 + \varphi^2$, $\nu = \tilde{\mathbf{f}} \cdot \vec{l}, \varphi = \exp(-\frac{h}{k_B T})$. Here $\vec{f}$ is the stretching force, h is the hairpin energy, while the parameter $\vec{l}$ is indirectly related to the mean hairpin density $\bar{n}$ defined by

$$\bar{n} = -k_B T \frac{\partial}{\partial h} \ln Z(\mathbf{f}) \tag{8.4.3.7}$$

[q]E.g. see Eq. (8.3.5.6).

so that $N/\bar{n} = 1/\varphi$. The bending energy E defined in Eq. (8.4.2.12) is related to $|\tilde{\mathbf{l}}|$ via

$$|\tilde{\mathbf{l}}| = \frac{1}{2}\frac{E}{h}\frac{k_BT}{h}.$$

Equation (8.4.2.10) for the Dirac propagator can be identified now with Eq. (8.4.3.6) for hairpins. This requires us to make the following identifications

$$A^2 \Leftrightarrow m^2E^2, \nu^2 \Leftrightarrow \frac{\mathbf{p}^2}{2}, \varphi^2 \Leftrightarrow m^2.$$

To complete such an identification, following de Gennes [203], we need to notice that

$$Z(\mathbf{f}) = \int d\mathbf{R}G(\mathbf{R}, N)\exp(\mathbf{f}\cdot\mathbf{R}). \tag{8.4.3.8}$$

The partition function $Z(\mathbf{f})$ would become the Fourier transform of $G(\mathbf{R}, N)$ if we make a replacement $\mathbf{f} \Leftrightarrow \pm i\mathbf{p}$. Conversely, we can use the Fourier-transformed Green's function and make a substitution $\mathbf{p} \Longrightarrow \pm i\mathbf{p}$ at the end of calculations.

Remark 8.4.3.2. In the hairpin model just described, the presence of other chains was not assumed. That is to say, the hairpins in this model occurred because permanent electric dipoles were attached to the individual links of the backbone and the constant electric field was applied. In such a case the situation is identical to that described by Eq. (8.4.2.17) with obvious replacement of magnetic field by electric and magnetic dipole moment by electric dipole moment.

We introduced all these details about hairpins, etc. having in mind to connect hairpins with qubits. To do so is easy at this point.[r] Indeed, using Eq. (8.3.5.16) and Eqs. (C.32) and (C.33) of Appendix C, we obtain the following Pauli–Schrödinger-type equation for (1 + 1)-dimensional Dirac propagator

$$i\frac{\partial}{\partial t}\chi = (\sigma_z p + \sigma_x m)\chi. \tag{8.4.3.9a}$$

Notice that this quantum result becomes identical to the "classical" one for semiflexible polymers in the presence of stretching forces if we make replacements: $\mathbf{p} \Rightarrow \pm i\mathbf{p}$ and $m \Rightarrow \pm im$. *Because of this, we shall use the above quantum version. This is the key observation allowing us to connect qubits with hairpins.*

[r]Our readers are encouraged to try doing so before seeing the answer on the next page.

Remark 8.4.3.3. There is some delicate place in making such an identification. In quantum mechanics time variable is the same for all particles while in the present case the role of time is played by the polymer contour length so that different polymers formally live in different times. In quantum mechanics such many-worlds interpretation, e.g. see http://en.wikipedia.org/wiki/Many-worlds_interpretation, does exists too. It removes many paradoxes inherent in traditional formulations of quantum mechanics but is still debatable by quantum mechanics/quantum computers community. Nevertheless, it is fully embraced in the theory of Brownian motion, e.g. read [327]. To keep focused on the issues of this book, we shall refrain from further comments on this subject

With parameters p and m physically interpreted in polymer context, the equivalent of Eq. (8.2.2.1b) can now be written as

$$i\frac{\partial}{\partial t}\begin{pmatrix} a \\ b \end{pmatrix} = -\frac{1}{2}\begin{pmatrix} p & m \\ m & -p \end{pmatrix}\begin{pmatrix} a \\ b \end{pmatrix}. \tag{8.4.3.9b}$$

Since this equation coincides with Eq. (8.2.6.17), we can use the Hopf mapping in order to rewrite it in accord with Eqs. (8.2.2.1c). Thus, we obtain:

$$\frac{d}{dt}\begin{pmatrix} x \\ y \\ z \end{pmatrix} = \begin{pmatrix} 0 & p & 0 \\ -p & 0 & m \\ 0 & -m & 0 \end{pmatrix}\begin{pmatrix} x \\ y \\ z \end{pmatrix}. \tag{8.4.3.9c}$$

These are just either the Serret–Frenet-like Eq. (2.4.1) or Eq. (8.2.6.1) describing the kinematics of rotations. Clearly, the hairpins were used in order to establish the meaning of parameters p and m in the above equations. From the discussion around Eq. (8.2.6.17) it is already known that exactly the same equations are used in NMR and, hence, in quantum computers [225, 230]. This observation provides us with a chance to talk about uses of quantum mechanics in quantum computers in the context of solutions of semiflexible polymers.

We begin with the following question:

If we can solve Eq. (8.4.3.9c), *will the obtained solution contain the same information as can be obtained from* Eq. (8.4.3.9b)?

To provide an answer, we need more than the already known results from mathematics. Indeed, for both equations obtaining a solution requires assigning of the initial condition. Since Eq. (8.4.3.9c) is macroscopic, there should be no problem in preparing the initial condition, provided that we find a meaningful physical interpretation for the triplet (x, y, z). In the quantum case described by Eq. (8.4.3.9b), the situation is different. It boils down to our ability/inability to prepare the initial state, that is the initial wave function. Although this issue was considered in great detail by Feynman [99] in early '60s, apparently, his explanations were not sufficient in view of still ongoing debates on this topic, e.g. read [328, 329] as examples. We shall not take sides in these debates.[s] Instead, using results for the Hopf mapping presented in Sec. 6.5, especially, Eqs. (6.5.26) and (6.5.28), elementary trigonometry provides us with the following:

$$x = \sin\theta\cos\phi, \quad y = \sin\theta\sin\phi, \quad z = \cos\theta. \tag{8.4.3.10}$$

These relations define the *Bloch vector* whose end is moving on the *Bloch sphere* [230] defined by the constraint: $x^2 + y^2 + z^2 = 1$. *Because of this constraint it is possible to think about the vector* $(x, y, z)^T$ *in probabilistic terms as it is done for* $(a, b)^T$. Notice that the angle ψ is not present in coordinates of the Bloch vector. In Sec. 8.2.6, we noticed that the mapping $\mathrm{SU}(2) \to \mathrm{SO}(3)$ is epimorphism. This simply means that some information is lost under mapping. This observation provides the formal answer to the question posed above. However, this answer is based exclusively on mathematics. When physics kicks in, the situation improves considerably. According to the standard textbook on quantum computing by Nielsen and Chuang, [C.4], the angle ψ should be attributed to the overall phase of the wave function. In quantum mechanics such a phase is normally ignored (unless we study the effects of gauge fields, gravity, etc.). To our knowledge, it is ignored in all applications of quantum mechanics to quantum computers. And, exactly *this fact, allows us to think about quantum mechanics of spin, Eq.* (8.4.3.9b) *in terms of classical dynamics associated with Eq.* (8.4.3.9c). This dynamics can be interpreted in terms of Euler's equations, Serret–Frenet equations, Bloch equations, etc. as discussed in this chapter and in other places in this book. Very interesting interpretation of these equations exists in the context of computer vision [47].

Remark 8.4.3.4. Already from the paper by Bloch [243], it follows that Bloch's equations contain dissipative terms. In the language of quantum computing these are the terms causing decoherence. This fact prompts the following problems.

Exercise 8.4.3.1. 1. Can the Hopf mapping survive if the Bloch equations contain dissipative terms? Hint: Read [230] as the starting point,

[s]See however, Sec. 9.2 below.

and then Appendix D. It can be found in every book on quantum computation that 1-qubit contains much more information than the classical bit. This statement implies that instead of just two outcomes in classical computations, at least in principle, we are having the entire Bloch sphere in the quantum case. Formally, this information is encoded in the parameters characterizing the Bloch vector, Eq. (8.4.3.10). This is only true if we can indiscriminately perform measurements of the spinor wave function for the qubit. This is not a simple task, however, as results of [330], p. 28, and Appendix D indicate, even in the absence of dissipation. Since the Bloch vector is just an image of the evolving spinor, the Bloch sphere happens not to be fully accessible for the evolving Bloch vector. Investigate what happens to these results/observations if dissipation effects are present. Is there the meaning of dissipation effects in the context of computer vision?

Remark 8.4.3.5. Study of questions formulated in the above exercise is essential for the design of any quantum computer as explained in Sec. 8.2.1 and Appendix D.

Remark 8.4.3.6. If we consider the phenomenological Bloch equation with dissipation axiomatically, then methods of sub-Riemannian geometry (that is of optimal control) can be applied to this problem, e.g. read [244, 261, 262].

Remark 8.4.3.7. Since the phenomenological Bloch equations (with or without dissipation) are macroscopic in nature, they can be used without any connection with quantum mechanics, as we just explained using examples from different fields of study. This observation is essential in our search for macroscopic systems whose dynamics obeys these equations. It is causing us to look at conformations of semiflexible polymers as alternative to qubits. Incidentally, in the pattern/computer vision theory connections between the Serret–Frenet equations and Euler's elastica (Appendix B) had already been established for some time [331].

The connection with the Serret–Frenet equations also causes us to inquire about relations of these equations to Euler's elastica. These are formally given by the entries of the matrix in Eq. (8.4.3.9c). But what is the meaning of the Bloch vector in such a context?

Following [332] and using results of Sec. 6.5, it becomes possible to rewrite the action functional, Eqs. (8.4.2.5) and (8.4.2.6), for the K-P model. Specifically, using Eqs. (6.5.26) and (8.4.3.10) and taking into account that in the context of the K-P model we have to identify $\mathbf{v} \Leftrightarrow \mathbf{x}$ we obtain,

$$\frac{1}{2}\left(\frac{d\mathbf{x}}{d\tau}\right)^2 = 2[\dot{\mathbf{z}}^\dagger\dot{\mathbf{z}} + (\mathbf{z}^\dagger\dot{\mathbf{z}})^2]. \tag{8.4.3.11}$$

It is convenient to rewrite the second term in the above equation as follows. Let $a = i\mathbf{z}^\dagger\dot{\mathbf{z}}$ so that

$$\frac{1}{2}\left(\frac{d\mathbf{x}}{d\tau}\right)^2 = 2[\dot{\mathbf{z}}^\dagger\dot{\mathbf{z}} - a^2]. \tag{8.4.3.12}$$

The rationale for introducing a lies in its direct relationship with the Berry phase. We shall return to this topic shortly below. In the meantime, the same results can be re-obtained with the help of matrix g defined by Eq. (6.5.27). Indeed, it is shown in [332] that

$$\frac{1}{2}\left(\frac{d\mathbf{x}}{d\tau}\right)^2 = \mathrm{tr}(\dot{g}^\dagger\dot{g}) + \frac{1}{2}\left[\mathrm{tr}(\sigma_3 g^{-1}\dot{g})\right]^2. \tag{8.4.3.13}$$

If now we replace the matrix g by $g\exp(i\alpha(\tau)\sigma_3)$, this will lead to

$$\mathrm{tr}(\dot{g}^\dagger\dot{g}) \rightarrow \mathrm{tr}(\dot{g}^\dagger\dot{g}) - 4\alpha\dot{\alpha} + 2\dot{\alpha}^2 \tag{8.4.3.14}$$

while $-2a^2$ term is replaced now by $-2a^2 + 4\alpha\dot{\alpha} - 2\dot{\alpha}^2$ so that the combination in the R.H.S. of Eq. (8.4.3.13) stays U(1) invariant. This invariance brings us to the conclusion that the coordinates of the Bloch vector are exactly the same as those used for the K-P model of semiflexible polymers so that we expect the dynamics of the Bloch vector to obey the same equations of motion as those for the vectors in the K-P model. The questions still remain: (a) How do these equations look like explicitly? (b) What is the relevance of Eq. (8.4.3.9c) to the K-P model?

Evidently, questions (a) and (b) are connected. To answer both of these questions we notice that to obtain these equations we used the fact that the euclideanized version of the Dirac propagator nicely describes conformational properties of semiflexible polymers and these polymers are also described with the help of the K-P model. Can K-P model be restored from Eq. (8.4.3.9c)? Fortunately, the answer to this question was already obtained by Kirchhoff long time ago as it was discussed in Sec. 2.5 and, later, in Sec. 8.2.4. Equation (8.2.4.2) is the same as Eq. (8.4.3.9c), provided that $k_3 = 0$. But such situation happens in the limit when the thickness of the rod can be ignored. If we do not ignore the thickness, can Eq. (8.2.4.2) be connected with the K-P model? We shall demonstrate that this is still possible in the next section.

8.4.4 *Semiflexible Polymers, Dirac Monopole and Heisenberg Ferromagnets*

By combining Eqs. (8.4.2.5), (8.4.2.6) and (8.4.3.13) the path integral for the K-P polymer chain can be formally written as

$$\int D[g(\tau)]\exp(-S[g(\tau)]), \tag{8.4.4.1a}$$

where

$$S[g(\tau)] = \int_0^N d\tau \left\{ \text{tr}(\dot{g}^\dagger \dot{g}) + \frac{1}{2}[\text{tr}(\sigma_3 g^{-1}\dot{g})]^2 \right\}. \tag{8.4.4.1b}$$

Using the (Hubbard–Stratonovich) identity

$$\int_{-\infty}^{\infty} dx \exp\{-ax^2 \pm ibx\} = \sqrt{\frac{\pi}{a}} \exp\left\{-\frac{b^2}{4a}\right\} \tag{8.4.4.2}$$

generalized to the path integrals, we can rewrite the action $S[g(\tau)]$ in the exponent of the path integral Eq. (8.4.4.1a) as follows:

$$S[g(\tau)] = \int_0^N d\tau \left\{ \text{tr}(\dot{g}^\dagger \dot{g}) \pm \frac{ix(\tau)}{2}\text{tr}(\sigma_3 g^{-1}\dot{g}) \right\} + \frac{1}{2}\int_0^N d\tau\{x(\tau)\}^2, \tag{8.4.4.3}$$

provided that we select $a = 1/2, b^2 = [\text{tr}(\sigma_3 g^{-1}\dot{g})]^2$. To obtain physically meaningful results, we shall assume that $x(\tau)$ in the above equation is independent of τ. Then, the last term can be dropped, while the first term can be transformed with the help of Eq. (6.5.27). Calculations performed in [332] allow us to write

$$\text{tr}(\dot{g}^\dagger \dot{g}) = \frac{1}{2}[\dot{\theta}^2 + (\dot{\psi} + \dot{\varphi}\cos\theta)^2 + \dot{\varphi}^2 \sin^2\theta] \tag{8.4.4.4}$$

easily recognizable as the kinetic energy of the fully symmetric top, Eq. (8.2.6.11) (that is the top for which $I_x = I_y = I_z$). From Arnol'd, [136], we know that variables ψ and φ are cyclical. This means that $\dot{\psi} = c_1$ and $\dot{\varphi} = c_2$. By selecting $c_1 = -c_2$ we re-obtain de Gennes hairpin Lagrangian, Eq. (8.4.3.1)! Surely, there are other options too. For instance, if we suppress ψ, then, upon implementation of standard quantization protocol, we shall obtain the Schrödinger-like (Euclidean-rotated) equation for the rigid rotator. This type of equation allows us to obtain the retarded Green's function used for calculation of averages for the K-P model [318]. Typically, in polymer physics other options, e.g. keeping ψ nonzero [332], are not considered. This fact is in accord with what was said above about the

Bloch vector: it also does not contain information about ψ. What is then the physical meaning of the second term in Eq. (8.4.4.3)? As results of [332] indicate, it is possible to rewrite it as

$$\bar{g}\int_0^N \frac{i}{2}\text{tr}(\sigma_3 g^{-1}\dot{g})d\tau = \pm\int_0^N d\tau\mathbf{A}(\tau)\cdot d\mathbf{n}, \tag{8.4.4.5}$$

where $\bar{g} \equiv x$ and the vector potential 1-form is defined by the Dirac monopole result, Eqs. (5.6.3.9) and (5.6.3.10). The same expression also admits an interpretation in terms of Berry's phase. For closed trajectories use of Stoke's theorem makes this expression an integer times 2π and, since it is located in the exponent of the path integral, Eq. (8.4.4.1a), it can *formally* be dropped. To drop this (Berry phase) term from the path integral is sometimes permissible and sometimes not as explained in our works, [64, 203]. Basically, this term is associated with the torsion. If the rod-like curve is not closed and is not anchored, there is no need to worry about this term.

It is very instructive to re-obtain these results using alternative methods. We begin by rewriting the action functional in Eq. (8.4.2.6) as follows:

$$S[\mathbf{n}(\tau)] = \frac{\eta}{2}\int_0^N d\tau\left(\frac{d\mathbf{n}}{d\tau}\right)^2 + \frac{1}{2}\int_0^N d\tau\lambda(\tau)(\mathbf{n}^2 - 1). \tag{8.4.4.6}$$

Minimization of $S[\mathbf{n}]$ produces the following equations of "motion"

$$\ddot{n}_i = \tilde{\lambda}n_i, \quad i = 1\text{–}3. \tag{8.4.4.7a}$$

The Lagrangian multiplier $\tilde{\lambda} = \lambda/\eta$ is used in the above action to enforce the differential geometric constraint $\mathbf{n}^2 = 1$ well known from the differential geometry of curves. Because of this constraint, it follows that

$$\mathbf{n}\cdot\dot{\mathbf{n}} = 0 \tag{8.4.4.8}$$

implying $\dot{\mathbf{n}}\cdot\dot{\mathbf{u}} + \mathbf{n}\cdot\ddot{\mathbf{n}} = 0$. Accordingly, in view of Eq. (8.4.4.7), $\tilde{\lambda} = -\dot{\mathbf{n}}\cdot\dot{\mathbf{n}}$. Thus, we have the following dynamical problem on S^2 :

$$\frac{d^2\mathbf{n}}{dt^2} + |\dot{\mathbf{n}}|^2\mathbf{n} = 0, \quad |\mathbf{n}|^2 = 1, \quad \mathbf{n}\cdot\dot{\mathbf{n}} = 0. \tag{8.4.4.7b}$$

Let us introduce the velocity $\dot{\mathbf{n}} = \mathbf{y}$, so that the above system of equations can be rewritten as

$$\dot{\mathbf{n}} = \mathbf{y}, \quad \dot{\mathbf{y}} = -|\mathbf{y}|^2\mathbf{n}, \quad |\mathbf{n}|^2 = 1, \quad \mathbf{n}\cdot\mathbf{y} = 0. \tag{8.4.4.9}$$

Elements of the phase space are vectors $\mathbf{y}$ attached to the points $\mathbf{n}\in S^2$. The set of vectors

$$\{(\mathbf{n},\mathbf{y}) \in \mathbf{R}^6; |\mathbf{n}|^2 = 1, \mathbf{n}\cdot\mathbf{y} = 0\} \tag{8.4.4.10a}$$

forms the tangent bundle $\mathbf{T}(S^2)$ on the sphere S^2. $\mathbf{T}(S^2)$ can be identified with the phase space of this dynamical system. The velocity $|\dot{\mathbf{n}}| = |\mathbf{y}|$ is constant along any solution for such a system. Indeed,

$$\frac{d}{dt}|\mathbf{y}|^2 = 2\mathbf{y}\cdot\dot{\mathbf{y}} = -2|\mathbf{y}|^2(\mathbf{n}\cdot\mathbf{y}) = 0. \tag{8.4.4.11}$$

This allows us to normalize $\mathbf{y}$ as well, that is to require $|\mathbf{n}| = 1 = |\mathbf{y}|$. The unit tangent bundle $\mathbf{T}_1(S^2)$ can now be defined as

$$\{(\mathbf{n},\mathbf{y}) \in \mathbf{R}^6; |\mathbf{n}| = 1, \mathbf{n}\cdot\mathbf{y} = 0, |\mathbf{y}| = 1\}. \tag{8.4.4.10b}$$

Thus $\mathbf{T}_1(S^2)$ is going to be mapped one-to-one onto rotation group SO(3). The infinitesimal generators $X_i, i = 1$–3, for this group had already been defined by Eq. (8.2.3.3). In particular, consider the following matrix differential equation

$$\frac{d}{dt}U = UX_3, \tag{8.4.4.12}$$

where $U \in \mathrm{SO}(3)$. Its solution is $U(t) = U(0)\exp(tX_3)$. By straightforward computation it is easy to demonstrate that, in fact,

$$U(t) = U(0)\begin{pmatrix} \cos t & -\sin t & 0 \\ \sin t & \cos t & 0 \\ 0 & 0 & 1 \end{pmatrix}. \tag{8.4.4.13}$$

Now, let us use the Ψ mapping defined by Eq. (8.2.3.4b). In the present case, let us make the matrix U out of column vectors $\mathbf{n},\mathbf{y},\mathbf{z}$. Then, $\frac{d}{dt}U = (\dot{\mathbf{n}},\dot{\mathbf{y}},\dot{\mathbf{z}}) = (\mathbf{y},-\mathbf{x},\mathbf{0})$. The last result follows if we make an identification $\mathbf{n} = [\cos t, \sin t]$, $\mathbf{y} = [-\sin t, \cos t]$ and take into account Eqs. (8.4.4.9) and (8.4.4.10b).

Evidently, our use of the matrix X_3 is a bit restrictive because it makes the z-direction to look special. To restore the democracy we notice that *any* SO(3) matrix $\mathbf{A}$ can be represented as

$$\mathbf{A} = \sum_{i=1}^{3}\omega_i X_i = \begin{pmatrix} 0 & -\omega_3 & \omega_2 \\ \omega_3 & 0 & -\omega_1 \\ -\omega_2 & \omega_1 & 0 \end{pmatrix}. \tag{8.4.4.14}$$

Such restoration of democracy is not only plausible but essential in view of the rotational invariance of the original Lagrangian for the action $S[\mathbf{n}(\tau)]$. Therefore, the matrix equation

$$\frac{d}{dt}U = U\mathbf{A} \tag{8.4.4.15}$$

is equivalent to the already known Eq. (8.2.6.1a) so that the one-to-one correspondence between the rotations of rigid body and classical dynamics of the K-P model is established. This means that previously obtained Eq. (8.4.3.9c) can indeed be associated with the "dynamics" of K-P model.

Remark 8.4.4.1. In Sec. 8.2.4 we discussed the nonholonomic kinematics of the ball rolling on $\mathbf{R}^2$, e.g. see Eq. (8.2.4.1a) and comments around this equation. In Appendix B we discussed dynamics of Euler's elastica from the sub-Riemannian point of view. In a very detailed study using methods of sub-Riemannian geometry, Jurdjevic [239, 240] demonstrated that the trajectory of the point of contact for the rolling ball is the same as can be obtained by solving equations of motion for Euler's elastica. The results just presented provide independent proof that this is indeed the case.

The question still remains: How these results are related to the Dirac monopole? To answer this question we need to recall the Landau–Lifshitz Eq. (2.3.12b). Using known facts about this equation [324], it can be rewritten in notations we just used as follows:

$$\frac{\partial}{\partial t}\mathbf{n}(\tau, t) = \mathbf{n} \times \frac{d^2}{d\tau^2}\mathbf{n}, \quad \mathbf{n}^2 = 1. \tag{8.4.4.16}$$

Such an equation describes dynamics of the classical spin in the one-dimensional (classical analog of) Heisenberg ferromagnetic chain. In view of the results of Chap. 2, not surprisingly, the K-P model is isomorphic to the static version of the classical Heisenberg ferromagnet. This fact was discussed in detail in our works [203, 318]. It follows from Eq. (8.4.4.16) that the static version of the L-L equation

$$\frac{d^2}{d\tau^2}\mathbf{n} = \tilde{\lambda}\mathbf{n} \tag{8.4.4.17}$$

is exactly the same as already obtained Eq. (8.4.4.7). Next, following Veselov, [333], we consider once again the time-dependent version of the L-L equation, Eq. (8.4.4.16), and look for solutions of the traveling-wave type, that is for solutions in the form: $\mathbf{n} = \mathbf{q}(x - at)$. Substitution of such an ansatz into the L-L equation produces:

$$-a\dot{\mathbf{n}} = \mathbf{n} \times \ddot{\mathbf{n}}, \quad \mathbf{n}^2 = 1. \tag{8.4.4.18}$$

Or, equivalently,

$$\ddot{\mathbf{n}} = \tilde{\lambda}\mathbf{n} + a\dot{\mathbf{n}} \times \mathbf{n}. \tag{8.4.4.19}$$

This result coincides with the equation of motion for the Dirac monopole, Eq. (6.5.6). Thus, the connection with the Dirac monopole is established, as promised. To generalize the obtained results, say, to multiqubits more work is needed. The first step in this direction is done in the next section.

8.4.5 *From the L-L Equation to Neumann's Model. Back to Contact Geometry*

We would like to remind to our readers that the K-P model defined by the path integral, Eq. (8.4.2.6), is just one of a *countable infinity* of models of semiflexible polymer chains as explained in [322]. Our readers may be puzzled by the fact that one and the same physical reality-semiflexible polymers can be represented by so many different models. The resolution to this puzzle is simple: going from real physical polymer chain to the model representing this chain is always associated with some loss of information (very much like in the case of Hopf mapping). This is understandable since it always happens when switching from mechanical to statistical mechanical description of just about anything. Sacrificing chemical individuality of a given polymer chain can be done up to some extent only. This extent depends upon physical properties we are interested in. For instance, if we are interested in conformational properties only and only at the level of a single polymer chain, its individuality is squeezed into the chain rigidity parameter η defined in Eq. (8.4.2.6). Conformational properties of semiflexible polymers can be modelled on a computer. The most successful computer-generated model of semiflexible polymers reproducing experimentally observed form factors for single polymer chain was proposed some time ago by Pedersen and Shurtenberger. The computer-generated plots for the form factor for semiflexible polymers for various parameter domains were compared against both the experimental data and our analytical result for the static formafor [319] obtained with the help of the Dirac propagator. The comparison yielded an unusually complete agreement between theory, on one hand, and numerical and experimental data, on the other, in the wide range of physical parameters. The results are documented in the Wikipedia, e.g. read Sec. 8.4.2. The obtained excellent agreement selects both the Dirac and the K-P models as most relevant for description of conformational problems of semiflexible polymers. These results were developed for individual polymer chains. That is for individual qubits if we use the quantum computer terminology. In polymer terminology such situation makes physical sense only for extremely low polymer concentrations. When concentrations are not low, that is in the case of multiqubits, the situation with polymers resembles that encountered in atomic physics of multielectron atoms. For such atoms the self-consistent Hartree and Hartree–Fock methods are typically used for description of motions of individual electrons in the effective field of other electrons and nuclei. The simplest self-consistent theory for the nematic-type polymers is the theory proposed by Maier and Saupe [75, 213]. It is of exactly the same nature as the Weiss molecular field approximation for ferromagnets which is also discussed in [75, 213]. When this type of approximation is applied to solutions of semiflexible polymers, some caution should be exercised since interactions between polymer segments are more complicated than those between Ising-type spins in ferromagnets. The detailed study of the validity of the Maier–Saupe-type interaction approximation was made in [334]. But even in this work only interactions in the nematic liquid crystals were studied. The individual molecules in such crystals are rod-like so that the flexibility is not taken into account at all. In our work, [322], it is demonstrated that such a neglect is not essential and can be accounted for by the appropriate interpretation of the coupling constant responsible for the polymer–polymer interactions. Under such assumptions theory of concentrated solutions of semiflexible polymers was developed in [318]. However, the full use of the Maier–Saupe-type approximation compatible with already cited [334] was made much more recently, in [335]. This work was inspired by much earlier work, [336] by Warner *et al.* Since in [335] the results of [336] were rederived anyway starting from more refined microscopic model, by leaving these derivations aside, [336] can be safely taken as starting point of our discussion. Following [203] and [336] we have to (effectively) replace the K-P model described by Eq. (8.4.4.6) by

$$S[\mathbf{n}(\tau)] = \frac{\eta}{2}\int_0^N d\tau \left(\frac{d\mathbf{n}}{d\tau}\right)^2 - \frac{g}{2}\int_0^N d\tau(3\mathbf{n}_z^2 - 1) + \frac{1}{2}\int_0^N d\tau\lambda(\tau)(\mathbf{n}^2 - 1), \tag{8.4.5.1a}$$

or, more generally, by

$$S[\mathbf{n}(\tau)] = \frac{\eta}{2}\int_0^N d\tau \left(\frac{d\mathbf{n}}{d\tau}\right)^2 - \frac{1}{2}\int_0^N d\tau(\mathbf{n}, \mathbf{B}\cdot\mathbf{n}) + \frac{1}{2}\int_0^N d\tau\lambda(\tau)(\mathbf{n}^2 - 1), \tag{8.4.5.1b}$$

where the matrix $\mathbf{B}$ is assumed to be symmetric. Presence of an extra term in the action functional, Eq. (8.4.5.1), (as compared with the original K-P functional, Eq. (8.4.4.6)) is caused by the effects of interaction of a given polymer chain with other chains at the mean field (Mayer–Saupe) level. By analogy with Eq. (8.4.4.7a) minimization of $S[\mathbf{n}(\tau)]$ produces the equation of "motion"

$$\ddot{n}_i = -B_{ij}n_j + \lambda n_i, \quad i = 1\text{–}3. \tag{8.4.5.2}$$

Since our "particle" is constrained to the sphere as before, we should take into account the restrictions $\mathbf{n} \cdot \dot{\mathbf{n}} = 0$ and $\dot{\mathbf{n}} \cdot \dot{\mathbf{n}} + \mathbf{n} \cdot \ddot{\mathbf{n}} = 0$ on $\mathbf{n}$. Using the last constraint and the equation of motion, Eq. (8.4.5.2), we obtain

$$-\mathbf{n} \cdot \mathbf{Bn} + \lambda + \dot{\mathbf{n}}^2 = 0$$

implying

$$\lambda = \mathbf{n} \cdot \mathbf{Bn} - \dot{\mathbf{n}}^2. \tag{8.4.5.3}$$

By inserting this result back into Eq. (8.4.5.2) we obtain Neumann's system of equations which is manifestly nonlinear. Nevertheless, it is exactly integrable! The Neumann dynamical problem (and the model associated with it) is rather easy to formulate. If the action $S[\mathbf{n}(\tau)]$, Eq. (8.4.4.6), describes the geodesic motion of the particle of unit mass on the sphere S^2, one can think of the geodesic motion on the ellipsoid. This was discussed in 1884 in the monograph by Jacobi [178, 236]. Carl Neumann, knowing about Jacobi work, solved similar problem in 1859 [270–273]. Instead of looking at the geodesics on the ellipsoids, he looked at the motion of the particle constrained to the sphere in the presence of the potential whose analytical form originates from the equation for ellipsoid. This is conceptually similar to the already mentioned problem about the motion of a charged particle in the presence of the Dirac monopole (Sec. 6.5) and, as we shall demonstrate, indeed these problems are interrelated. Following [337], we consider the Hamiltonian of the type

$$H = \frac{1}{2}\{|\mathbf{x}|^2|\mathbf{y}|^2 - (\mathbf{x} \cdot \mathbf{y})^2 + \mathbf{x} \cdot \mathbf{Bx}\} \tag{8.4.5.4}$$

constrained to the tangent bundle $\mathbf{T}(S^2)$, e.g. see Eq. (8.4.4.10a). Set $\mathcal{F}_1 = \frac{1}{2}(|\mathbf{x}|^2 - 1)$ and $\mathcal{F}_2 = \mathbf{x} \cdot \mathbf{y}$, and consider the extended Hamiltonian

$$H^* = H - \lambda_1\mathcal{F}_1 - \lambda_2\mathcal{F}_2 \tag{8.4.5.5}$$

for which we need to calculate the Poisson brackets on $\mathbf{T}(S^2)$. Specifically, such calculation yields:

$$\{H, \mathcal{F}_1\} = 0, \{H, \mathcal{F}_2\} = -\mathbf{x} \cdot \mathbf{Bx}, \quad \{\mathcal{F}_1, \mathcal{F}_2\} = |\mathbf{x}|^2 = 1. \tag{8.4.5.6}$$

If we require that $\{H^*, H\} = 0$, this will bring us the following results:

$$\lambda_1 = \mathbf{x} \cdot \mathbf{B}\mathbf{x} \quad \text{and} \quad \lambda_2 = 0. \tag{8.4.5.7}$$

The Hamiltonian equations for the constrained system can now be read as follows:

$$\dot{\mathbf{x}} = H^*_{\mathbf{y}} = \mathbf{y} \quad \text{and} \quad \dot{\mathbf{y}} = -H^*_{\mathbf{x}} = -\mathbf{B}\mathbf{x} + (-|\mathbf{y}|^2 + \lambda_1)\mathbf{x}. \tag{8.4.5.8}$$

They coincide with Eq. (8.4.5.2) as required. Now, however, we know that if the Lagrangian multiplier λ_1 is a constant, then the equation $\lambda_1 = \mathbf{x} \cdot \mathbf{B}\mathbf{x}$ indeed defines an ellipsoid. Notice that the connection of this method of dealing with constraints with Dirac's method, [265], was established only very recently in the already cited references [272, 273]. In this section we are not going to discuss solutions of Neumann problem. Our readers can find this information in, e.g. [178, 272, 273, 337]. The same problem was solved recently by the loop group method in [338]. Instead, we shall discuss how the Neumann problem is related to the properties of concentrated solutions of semiflexible polymers, to the Dirac monopole and, ultimately, to quantum computers.

We begin with the discussion of the Landau–Lifshitz (L-L) equation in Secs. 2.3–2.5 of Chap. 2. Although in these sections it was defined in $(1+1)$ space–time dimensions, originally, it was derived for $(1+3)$ dimensions. The full $(1+3)$-dimensional version of this equation reads

$$\frac{\partial}{\partial t}\mathbf{S} = \mathbf{S} \times (\boldsymbol{\nabla}^2\mathbf{S} + \mathbf{J} \cdot \mathbf{S}) \tag{8.4.5.9}$$

where, in addition to the already familiar diffusion term, the term $\mathbf{J} \cdot \mathbf{S}$ is formally added. Evidently, this term is not participating in spin dynamics but contributes to static solutions. For example, consider the static version of the simplified $(1+1)$-dimensional L-L model. It is given by

$$\frac{d^2}{dx^2}\mathbf{S} + \mathbf{J} \cdot \mathbf{S} = \boldsymbol{\lambda}\mathbf{S}, \quad \mathbf{S}^2 = 1. \tag{8.4.5.10}$$

Comparison between Eqs. (8.4.5.2) and (8.4.5.10) suggests, that the static $(1+1)$-dimensional version of the L-L equation coincides with the Neumann equation. This means that all results discussed in Secs. 2.3–2.5 are preserved! But now we can accomplish more. For instance, as in the K-P case, again following Veselov, [333], we consider a dynamical version of the L-L equation and look for solutions of the traveling-wave type. Then, the substitution of such an ansatz into the above equation produces

$$-a\dot{\mathbf{q}} = \mathbf{q} \times (\ddot{\mathbf{q}} + \mathbf{J} \cdot \mathbf{q}), \quad \mathbf{q}^2 = 1. \tag{8.4.5.11a}$$

Equivalently,

$$\ddot{\mathbf{q}} + \mathbf{J} \cdot \mathbf{q} = \lambda \mathbf{q} + \mathbf{a}\dot{\mathbf{q}} \times \mathbf{q} \tag{8.4.5.11b}$$

where $\lambda = (\mathbf{q}, \mathbf{J} \cdot \mathbf{q}) - \dot{\mathbf{q}}^2$. This equation coincides with Eq. (6.5.6) for the Dirac monoplole when $\mathbf{J} = 0$. Evidently, in the case if $\mathbf{J}$ has only one component this conclusion will still remain correct. Thus, in such circumstances all previously obtained results for the Dirac monopole will apply to the present case. This will be independently demonstrated in the next subsection.

Since the dynamics of a "charged particle" in the presence of the Dirac monopole is of sub-Riemannian-type (contact-type) problem as discussed in Chap. 6, Sec. 6.5, the same is true for the Neumann problem. This is indeed the case as shown in detail in [272, 273]. The alternative treatment of such type of problems is discussed in [339]. In [340] the same nonholonomic systems were considered by converting dynamical problems with nonholonomic constraints to those describing dynamics of the non-Abelian gauge fields. Such a connection was suggested some time ago by Shapere and Wilczek [341] in the context of swimming of microorganisms. In [272, 273] it was demonstrated that the Dirac method of treating the constrained dynamical systems [265] can be substituted by that developed for nonholonomic systems, that is by the methods developed in contact geometry. Following Arnol'd, [269], p. 1126, it remains to check how "diracians" can be replaced by "contactians", say, in the context of Neumann's problem.

8.4.6 *Delicate Inter-relationship between Classical and Quantum Neumann's Problems in the Context of Semiflexible Polymers, Euler's Elastica and Ion Traps*

In 1989, Wolfgang Paul was awarded the Nobel Prize in Physics for his invention of the quadrupole ion trap (http://en.wikipedia.org/wiki/Paul_trap). Such traps are being used in quantum information science and in many other related fields dealing with quantum processes at the atomic scale, e.g. read [342]. Remarkably enough, the mathematical description of these traps smoothly merges with that for semiflexible polymers we just discussed. To explain why this is so we begin with some excerpts from our work, [177].

In the second half of 19th century while studying the three-body celestial mechanics problem, especially the stability of Lunar orbits (the Sun, Moon and Earth system), Hill came up with his, now famous, Hill's equation. To write down this equation in dimensionless form it is convenient to introduce the Hill operator $Q(q) = -\frac{d^2}{dt^2} + q(t)$ where the potential $q(t)$ is periodic: $q(t) = q(t+1)$. In such notations the Hill equation can be written as

$$Q(q)x = \lambda x. \tag{8.4.6.1}$$

Remark 8.4.6.1. By comparing Eq. (8.4.6.1) with the eigenvalue problem for the Euler's elastica operator, Eq. (B.27), we establish a correspondence between the Hill's equation and that for quantized Euler's elastica. More accurately, in the case of elastica we are dealing with the so-called Mathieu's equation (Eq. (8.4.6.11) below) which is just a special case of Hill's equation. Both equations are studied by the same mathematical methods developed by Floquet in the 19th century. These methods were substantially upgraded by the end of 20th century as explained in [177] and will be discussed further, in Chap. 9. Mathieu's equation is relevant for dynamics of Josephson's junctions, e.g. see [225], p. 102, and in exactly same form it is also being used in the context of quadrupole ion traps [343, 344].

Since this is the second order differential equation, formally it should have two solutions. These solutions depend upon the boundary conditions. For instance, for periodic solutions of the type $x(t) = x(t+2)$ the "spectrum" of Eq. (8.4.6.1) is discrete and is given by

$$-\infty < \lambda_0 < \lambda_1 \leq \lambda_2 < \lambda_3 \leq \lambda_4 < \cdots \uparrow +\infty.$$

The word "spectrum was put in quotation marks because of the following. Equation (8.4.6.1) does have a normalizable solutions only if λ belongs to the pre-assigned intervals $(\lambda_0, \lambda_1), (\lambda_2, \lambda_3), \ldots, (\lambda_{2i}, \lambda_{2i+1}), \ldots$. In such a case the eigenfunctions x_i are normalizable in the usual sense of quantum mechanics and form an orthogonal set. From the Lunar the ory it follows that it is possible to assign yet another boundary conditions, e.g. $x(0) = x(1) = 0$. In this case the discrete spectrum also exists but it lies exactly in the gaps between the intervals just described, i.e. $\lambda_1 \leq \mu_1 \leq \lambda_2 < \lambda_3 \leq \mu_2 \leq \lambda_4 \cdots$ (e.g. read references to mathematical works in [177]). For such a set there also exists the set of normalized mutually orthogonal eigenfunctions. Since the set of pre-assigned intervals should always be finite, this means that, beginning with some pre-assigned i, we are left with $\lambda_{2i} = \lambda_{2i+1}\ \forall\, i > i$. These double eigenvalues do not have independent physical significance since they can be determined by the set of single eigenvalues (for which $\lambda_{2i} \neq \lambda_{2i+1}$). Because of this, the potentials $q(t)$ in the Hill's equation are called *finite gap* potentials.[t] Hence, it is sufficient to discuss only such potentials which possess finite single spectrum. The auxiliary μ-spectrum is then determined by the gaps of the single spectrum. With this information defined, we are ready to connect the Hill and Neumann equations. We begin by rewriting Eqs. (8.4.5.2), (8.4.5.3) in the following equivalent form

$$\ddot{\boldsymbol{\xi}} = -\mathbf{A}\boldsymbol{\xi} + u(\boldsymbol{\xi})\boldsymbol{\xi}, \quad u(\boldsymbol{\xi}) = \phi(\boldsymbol{\xi}) - \langle \dot{\boldsymbol{\xi}}, \dot{\boldsymbol{\xi}} \rangle, \quad \phi(\boldsymbol{\xi}) = \langle \boldsymbol{\xi}, \mathbf{A}\boldsymbol{\xi} \rangle, \quad \langle \boldsymbol{\xi}, \boldsymbol{\xi} \rangle = 1. \tag{8.4.6.2}$$

Without loss of generality, we can assume that the matrix $\mathbf{A}$ is already in the diagonal form: $\mathbf{A} = \mathrm{diag}(\alpha_1, \ldots, \alpha_n)$. With such an assumption, Eq. (8.4.6.2) can be rewritten in the following suggestive form

$$\left(-\frac{d^2}{dt^2} + u(\boldsymbol{\xi}(t))\right)\xi_k = \alpha_k \xi_k, \quad k = 1, \ldots, n. \tag{8.4.6.3}$$

Thus, if we can prove that $u(\boldsymbol{\xi}(t)) = q(t)$ defined in Eq. (8.4.6.1) the promised connection between the Hill and Neumann equations will be established. Fortunately, this is indeed the case. Following [177] let us consider the Fuchsian-type equation

$$y'' + \frac{1}{2}\phi y = 0, \tag{8.4.6.4a}$$

where the potential ϕ is determined by the Schwarzian derivative $[f]$, i.e.

$$\phi = [f] = \frac{f' f''' - \frac{3}{2}(f'')^2}{(f')^2}, \tag{8.4.6.4b}$$

where $f = y_1/y_2$ with y_1 and y_2 being two independent solutions of Eq. (8.4.6.4a) subject to the normalization condition: $y_1' y_2 - y_2' y_1 = 1$. For the future use, we would like to consider solutions of Eq. (8.4.6.4a) on the circle S^1. Also, let us consider a map of the circle of the type: $F(t+1) = F(t) + 1$. Furthermore, let $t = F(\xi)$ so that substitution $y(t) = Y(\xi)\sqrt{F'(\xi)}$ leaves Eq. (8.4.6.4a) form-invariant, i.e.

$$Y'' + \frac{1}{2}\Phi Y = 0$$

with the potential Φ being defined now as $\Phi(\xi) = \phi(F(\xi)[F'(\xi)]^2 + [F(\xi)]$. The infinitesimal transformation $F(\xi) = \xi + \delta\varphi(\xi)$ with δ being some

[t] Since there is only finite number of gaps $[\lambda_1, \lambda_2], [\lambda_3, \lambda_4], \ldots$ where the spectrum is forbidden.

small parameter and $\varphi(\xi)$ being some function to be determined causes the potential $\Phi(\xi)$ to change as follows: $\Phi(\xi+\delta\varphi)=\phi(\xi)+\delta(T\varphi)(\xi)+O(\delta^2)$. Here $(T\varphi)(\xi)=\phi(\xi)\varphi'(\xi)+\frac{1}{2}\varphi'''(\xi)+2\phi'(\xi)\varphi(\xi)$. Next, we relabel $\delta\rightleftarrows t$

$$\lim_{t\to 0}\frac{\Phi-\phi}{t}=\frac{\partial\phi}{\partial t}=\phi(\xi)\varphi'(\xi)+\frac{1}{2}\varphi'''(\xi)+2\phi'(\xi)\varphi(\xi). \tag{8.4.6.5}$$

Thus far the perturbing function was left undefined. If now we choose $\varphi(\xi)=\phi(\xi)$, then the above equation acquires the form

$$\frac{\partial\phi}{\partial t}=\frac{1}{2}\phi'''(\xi)+3\phi'(\xi)\phi(\xi) \tag{8.4.6.6a}$$

immediately recognizable as the *Korteweg–de Vries equation* (KdV). For our purposes, it is sufficient to consider only the static version of the KdV equation, that is

$$\phi'''(\xi)+6\phi'(\xi)\phi(\xi)=0. \tag{8.4.6.6b}$$

Let us take a look now at $u(\boldsymbol{\xi})=\phi(\boldsymbol{\xi})-\langle\dot{\xi},\dot{\xi}\rangle$ defined by Eq. (8.4.6.2). Consider an auxiliary functional $\varphi(\xi)=\langle\xi,\mathbf{A}^{-1}\xi\rangle$ for which we shall assume that $\varphi(\xi)=u(\boldsymbol{\xi})$. If this is so, then

$$\frac{du}{dt}=2\langle\dot{\boldsymbol{\xi}},A\boldsymbol{\xi}\rangle-2\langle\ddot{\boldsymbol{\xi}},\dot{\boldsymbol{\xi}}\rangle. \tag{8.4.6.7}$$

Because of the normalization constraint $\langle\boldsymbol{\xi},\boldsymbol{\xi}\rangle=1$, we conclude that $\langle\ddot{\boldsymbol{\xi}},\dot{\boldsymbol{\xi}}\rangle=0$. Thus, $\frac{du}{dt}=2\langle\dot{\boldsymbol{\xi}},A\boldsymbol{\xi}\rangle$. We would like to compare this result against $\frac{d\varphi}{dt}$. To calculate this derivative we need to use Eq. (8.4.6.2) resulting in $\frac{d\varphi}{dt}=2\langle\dot{\boldsymbol{\xi}},A^{-1}\boldsymbol{\xi}\rangle$. Because of the assumption $\varphi(\xi)=u(\xi)$ we have to demand now that $\langle\dot{\boldsymbol{\xi}},A^{-1}\boldsymbol{\xi}\rangle=\langle\dot{\boldsymbol{\xi}},A\boldsymbol{\xi}\rangle$. Suppose that this is the case and consider furthermore

$$\frac{d^2u}{dt^2}=2\langle\ddot{\boldsymbol{\xi}},A^{-1}\boldsymbol{\xi}\rangle+2\langle\dot{\boldsymbol{\xi}},A^{-1}\dot{\boldsymbol{\xi}}\rangle. \tag{8.4.6.8}$$

By using Eq. (8.4.6.2) again we obtain,

$$\frac{d^2u}{dt^2}=-2+2u\varphi+2\langle\dot{\boldsymbol{\xi}},A^{-1}\dot{\boldsymbol{\xi}}\rangle. \tag{8.4.6.9}$$

Finally, we calculate as well $\frac{d^3u}{dt^3}$ using the results just obtained

$$\frac{d^3u}{dt^3}=2\frac{du}{dt}\varphi+4u\frac{du}{dt}=6u\frac{du}{dt}. \tag{8.4.6.10}$$

By noticing that in Eq. (8.4.6.6b) we can always make a rescaling $\phi(\xi) \to \lambda\phi(\xi)$ we select $\lambda = -1$. In this case Eqs. (8.4.6.6b) and (8.4.6.10) coincide.

Thus, we just established connections between the classical Neumann and "quantum" Hill (or Mathieu) equations. It is of interest to arrive at the same result differently by quantizing the Neumann problem. Such a task was accomplished in great detail in [345] with the main result (of relevance to our discussion) presented on p. 5378 stating that the quantum version of the classical Neumann problem is indeed given in terms of the eigenvalue problem for the Hill equation, in accord with Eq. (8.4.6.3). By looking at Eq. (8.4.5.1) we notice that physically, it is sufficient to treat only one component (say, z) via use of Neumann's equations. In such a case we end up with the Mathieu equation ([345], p. 5378)

$$\frac{d^2\psi}{dx^2} + [\delta - 2\varepsilon\cos 2x]\psi = 0 \qquad (8.4.6.11a)$$

which is Eq. (5.7) of our work, [203] on hairpins in semiflexible polymers. Obtained result is in complete accord with Remark 8.4.6.1. Following our work, [203], it is instructive to make several important comments regarding this equation. First, after making a substitution $\psi(x) = z(x)\sqrt{\sin x}$, Eq. (8.4.6.11a) is converted into

$$\left[\frac{d^2}{dx^2} + \cot x\frac{d}{dx} - 2\varepsilon\cos^2 x - \frac{1}{4} - \frac{1}{4\sin^2 x}\right] z(x) = Ez(x), \qquad (8.4.6.11b)$$

where the constant E is related to δ. Apart from the extra factors $-\frac{1}{4} - \frac{1}{4\sin^2 x}$ this equation coincides with Eq. (4.5) of [336]. Equation (8.4.6.11b) is also known as the *associated Mathieu equation* and is the special case of a spheroidal wave equation. As it is demonstrated in [203] presence of extra factors permit us to interpret Eq. (8.6.4.11b) as the Schrödinger-type equation for the motion of charged particle in the presence of the Dirac monopole. These results are in complete accord with those established in the previous subsection.

Remark 8.4.6.2. In [346, 347] Kuznetsov was able to find a connection between equation like Eq. (8.6.4.11b) and XXX-like Heisenberg spin chain using the inverse scattering method. His results are in accord with those in previous section since classical dynamics of $(1+1)$-dimensional L-L chain is the same as classical dynamics of XXX spin chains.

Remark 8.4.6.3. Should the polymer chain be closed (circular or knotted) as in Chap. 2, Sec. 2.5, then we could use the Hasimoto transform converting the L-L dynamics into that for the nonlinear Schrödinger equation (NSE), see also Appendix B of our work, [181], for a quick introduction to this topic. As is explained in [181], the NLS is the mean field approximation to the one-dimensional nonideal Bose gas model. This means that elementary excitations in the XXX 1D Heisenberg chain ferromagnet are the same as in the nonideal Bose gas. In Appendix C (Sec. C.5) we discussed the possibility of making the entangling gate using excitations in the nonideal Bose gas. In the light of results presented in this chapter as well as in Appendix D, Sec. D.2.14, it is becoming clear that not only qubits can be made of hairpins but also the entangling gates. When superimposed with ideas of Lloyd [296], it is being hoped that results presented in this chapter may help in designing polymer-type "quantum" computer or, at least, to use polymers for modelling processes taking in place in quantum computers. In fact, such polymer-quantum computer symbiosis had already began, e.g. read again [298, 299].

Chapter 9

From Contact Geometry to Contact Topology

9.1 General Information

There were numerous attempts to include topological considerations into practical problems involving contact geometry. Most noticeable among them are summarized in the monograph by Arnol'd "Singularities of Caustics and Wave Fronts" [348]. Since his book is ideologically related to the book by Maslov and Fedoriuk [349] on "Semiclassical Approximation in Quantum Mechanics" — the subject matter seemingly more familiar to every physicist — it makes sense to illustrate some topological ideas of contact geometry in the context of quantum mechanics, at least at the beginning. This will be accomplished in the next subsection. In the meantime, it is appropriate to mention several other publications which use topological ideas of contact geometry for solving practical problems. These include (but not limited to) [350, 351]. Among the latest works in this direction we would like to mention, for example, [352, 353]. The first of them deals with contact geometry and topology of the restricted three-body problem. The restricted three-body problem has long and interesting history [232]. Among other researchers it has attracted attention of Henry Poincaré whose contributions to this subject are discussed in Appendix 9 of [136]. By commenting on the works by Poincaré, Birkhoff and, more recent authors such as Conley and Zehnder, Arnol'd came up with his famous Arnol'd conjecture which was ultimately solved in early '80s by Andreas Floer. Floer's contributions as well as works of Arnol'd and others are nicely summarized in the monograph by Hofer and Zehnder [354]. Since in this book the task is to emphasize physical aspects of contact geometry and topology, we shall largely avoid discussion of purely mathematical advancements made in the area of contact geometry and topology. We shall only comment on those which show some promise of potential applications in physics.

9.2 Mathematics and Physics of the Cauchy Problem in Quantum Mechanics: Viktor Maslov versus David Bohm

The Cauchy problem for Schrödinger's equation [349] provides us with an opportunity to compare physical and mathematical approaches to quantum mechanics utilizing mathematical and physical results discussed across this book.

We begin with Schrödinger's equation

$$i\hbar\frac{\partial}{\partial t}\psi = -\frac{\hbar^2}{2m}\nabla^2\psi + V(x)\psi \tag{9.2.1a}$$

supplemented with the initial condition

$$\psi\mid_{t=0} = \psi_0(x)\exp\left\{\frac{i}{\hbar}S_0(x)\right\}. \tag{9.2.1b}$$

Since Planck's constant plays role of a small parameter, it is customary to look for an asymptotic solution in the WKB form, that is in the form

$$\psi(t,x;\lambda) = \exp\{i\lambda S(x)\}\sum_{j=0}^{\infty}(i\lambda)^{-j}\varphi_j(x), \tag{9.2.2}$$

where $\lambda = \hbar^{-1} \to \infty$ (when $\hbar \to 0$) is small parameter and $S(x)$ and $\varphi_j(x)$ are determined iteratively upon substitution into Eq. (9.2.1a) with help of the initial condition, Eq. (9.2.1b). As Maslov demonstrates [349], it is impossible to construct a solution in the form given by Eq. (9.2.2) existing for $0 \leq t \leq T$ and for all $x \in \mathbf{R}^n_x$. This means that known from every textbook for physicists WKB method can sometimes fail to describe the transition from quantum to classical mechanics correctly! Very recently failure of the WKB method (but for a very different reason!) was noticed in physics literature as well [355]. Since the WKB can be looked upon as mathematical reformulation of the Bohr correspondence principle, it follows then that this principle cannot be uniformly confirmed/supported by the conventional WKB method. A survey of various formulations of this principle was recently collected in [356] with the outcome: "In general, there is no unique prescription (for the correspondence principle) of how to obtain classical mechanics from quantum mechanics covering all possible cases and situations met in practice. *In our opinion, all the efforts made so far support the thesis that quantum mechanics as we know it, is substantially different from classical mechanics*". Written in 2006, this paper contains extensive quotations from various sources, with one major exception. It is not mentioning the results of David Bohm, [357], whose prophetic insight into the problem of quantum-classical correspondence cannot be ignored. Because of many similarities, we are going to discuss Bohmian approach to quantum mechanics along with that by Maslov side-by-side.

In the light of results of Appendix D we notice that, although mathematically the Cauchy problem stated in Eqs. (9.2.1) is formulated correctly (that is in the standard way accepted in mathematics), there is a serious problem of constructing the initial wave function experimentally. As if anticipating future troubles with this issue, Maslov redefines the Cauchy problem in a much more satisfactory way. To our knowledge, this (other) satisfactory way came not as result of reading of Bohm's papers. Nevertheless, Maslov's treatment was developed almost in parallel to that by Bohm although 15–17 years later. Discussing both approaches in parallel will allow us to illuminate the central role of contact geometry and topology in these issues.

Following the standard WKB protocol Maslov begins with the Hamilton–Jacobi Eq. (1.2.19) of Chap. 1. For reader's convenience we rewrite it here in more convenient form

$$\frac{\partial S}{\partial t} + \frac{1}{2m}(\nabla S)^2 + V(x) = 0. \tag{9.2.3a}$$

At the same time, Bohm using the same initial data, Eq. (9.2.1b), obtains (without any approximations!)

$$\frac{\partial S}{\partial t} + \frac{1}{2m}(\nabla S)^2 + V(x) - \frac{\hbar^2}{4m}\left[\frac{\nabla^2 P}{P} - \frac{1}{2}\frac{(\nabla P)^2}{P^2}\right] = 0. \tag{9.2.4a}$$

Here $P(x) = [\psi_0(x)]^2$. Bohm obtains yet another equation

$$\frac{\partial}{\partial t}P + \nabla \cdot \left(P\frac{\nabla S}{m}\right) = 0 \tag{9.2.4b}$$

which becomes an equation of continuity if the following arguments (by Bohm) are applied.

"Now, in the classical limit ($\hbar \to 0$) the above equations are subject to a very simple interpretation. The function $S(\mathbf{x})$ is a solution of the Hamilton–Jacobi equation.[a] If we consider an ensemble of particle trajectories which are solutions of the equations of motion, then it is a well-known theorem in mechanics that if all of these trajectories are normal to any given surface of constant S, then they are normal to all surfaces of constant S and $\nabla S(\mathbf{x})/m$ will be equal to the velocity vector, $\mathbf{v}(x)$, for any particle passing the point $\mathbf{x}$. Equation (9.2.4b) can therefore be re-expressed as

$$\frac{\partial}{\partial t}P + \nabla \cdot (P\mathbf{v}) = 0. \tag{9.2.4c}$$

This equation indicates that it is consistent to regard $P(\mathbf{x})$ as the probability density for particles in our ensemble. For in that case, we can regard $P\mathbf{v}$ as the mean current of particles in our ensemble, and Eq. (9.2.4c) then simply expresses the conservation of probability."

In Maslov's asymptotic treatment there is also a kind of continuity equation which also plays an important role in the whole development but without any physical interpretation. For Eq. (9.2.3a) one can unambiguously formulate the Cauchy problem. For this it is sufficient to define the $(n-1)$-dimensional domain $U \in \mathbf{R}_y^{n-1}$ so that the initial value for $S(x)$ is

$$S(x) = S_0(y) \tag{9.2.3b}$$

since it is assumed that $x = f(y)$.[b] Clearly, instead of the Hamilton–Jacobi equation it is possible to set up the same Cauchy-type problem for the Hamiltonian equations

$$\frac{dx}{dt} = \frac{\partial H}{\partial p}, \quad \frac{dp}{dt} = -\frac{\partial H}{\partial x} \tag{9.2.3c}$$

supplemented with the initial Cauchy data

$$x(0) = x^0(y), \quad p(0) = p^0(y), \;\; y \in U. \tag{9.2.3d}$$

Projections of the phase space trajectories $\{x(t,y), p(t,y)\}$ onto $\mathbf{R}_x^n$ will be called "*rays*". The action S is being restored with the help of known connection: $dS = pdx$, as discussed in Chap. 5 so that

$$S(x(t,y)) = S_0(y) + \int_0^t p \cdot \frac{dx}{d\tau} d\tau. \tag{9.2.5a}$$

[a]That is Eq. (9.2.3a) (our comment).
[b]Here, following Maslov, we do not write the subscripts explicitly as well as we do not use the boldface.

The integral is taken along the phase trajectory, as required. If the correspondence $x \rightleftarrows x(t,y)$ is one-to-one it becomes possible to replace the coordinates $\{x_1, \dots, x_n\}$ by $\{t, y_1, \dots, y_{n-1}\}$. The one-to-one correspondence is possible if and only if the Jacobian

$$J(t,y) = \det\left[\frac{\partial x(t,y)}{\partial(t,y)}\right] \tag{9.2.6}$$

is nonzero. At $t = 0$ this Jacobian is nonzero by design (as can be proven) but for $t > 0$ this may or may not be the case. Following optical analogy, we shall call the manifolds for which $J(t,y) = 0$ as manifold of *caustics*.

Remark 9.2.1. Evidently, the mathematical description of caustics in quantum mechanics is the same as that in optics. Thus, all results of Arnol'd book, [348], are applicable to quantum mechanics as well. This means, in particular, *that all topological problems discussed in* [348], *and, more recently, in* [358] *involving contact geometry are relevant to quantum mechanics as well.* A condensed summary of these results and ideas is given in another review paper by Arnol'd, [359].

Remark 9.2.2. *Existence of caustics is the main obstacle* in designing the asymptotic solutions, Eq. (9.2.2), in the whole space–time domain $0 \leq t \leq T$, $x \in \mathbf{R}^n_x$. This circumstance makes the WKB method and, with it — the Bohr correspondence principle — ill-defined.

Remark 9.2.3. While the family of rays might have caustics, the trajectories themselves are nonsingular! That is there are no caustics for trajectories!

The Cauchy data, Eqs. (9.2.3d), define $(n-1)$-dimensional manifold Λ_0^{n-1} in the phase space. Such a manifold is not at all arbitrary! It is determined by the *contact condition* $\alpha = 0$, where the 1-form $\alpha = dS - pdx$ was defined by Eq. (5.2.7). Explicitly, the equation for the initial Lagrangian plane is given by

$$dS_0(y) = p^0(y)dx^0(y). \tag{9.2.7a}$$

Remark 9.2.4. Notice that the equation $dS/dx = p$ can be rewritten (in view of Eq. (9.2.1b)) as

$$\frac{\hbar}{m}\nabla S = \mathbf{v} \tag{9.2.7b}$$

which is an equation for the velocity of superconducting current as discussed in Sec. 3.3. of Chap. 3. From Remark 6.5.1, we deduce then that such an interpretation is sufficient for recovery of the Heisenberg group $\mathbf{H}_1$ (see also Appendix A) and, hence, of quantum mechanics! From here it follows that contact geometry — a branch of classical mechanics — captures the essence of quantum mechanics. This result we encountered repeatedly in various places in the book.

From Eq. (9.2.7a), it follows that on Λ_0^{n-1}

$$ddS_0(y) = 0 = d(p^0(y)dx^0(y)) = \sum_{j=1}^{n} dp_j \wedge dx_j \equiv \omega^2. \tag{9.2.8}$$

That is, in accord with Definition D.2.17.2 (Appendix D), we have to conclude that Λ_0^{n-1} is the Lagrangian manifold! As in thermodynamics, for Lagrangian manifolds the integral $\int_C pdx$ is path-independent. The Liouville theorem is telling us that the time shift(flow) $g^t(x^0(y), p^0(y)) = \{x(t,y), p(t,y)\}$ will leave the 2-form ω^2 unchanged implying that the manifold $g^t(\Lambda_0^{n-1}) = \Lambda_t^n$ will remain Lagrangian during the time evolution. Based on this result it is convenient to introduce a jet of trajectories

$$\Lambda^n = \bigcup_{-\infty<t<\infty} g^t(\Lambda_0^{n-1})$$

which is $(n+1)$-dimensional Lagrangian manifold.

Remark 9.2.5. Because of this observation it becomes possible to replace the Cauchy problem for the Schrödinger Eq. (9.2.1) by that describing the evolution of Λ_t^n. This conclusion was reached by Maslov in mid '60s while Bohm reached the same conclusion in 1951 but interpreted it physically (read below).

Once accepted, this observation allows all quantum mechanical calculations to be performed and to be interpreted in terms of classical mechanics. The initial data on the Lagrangian manifold Λ_0^{n-1} could be assigned with some effort as explained below.

Let $r, r^0 \in \Lambda^n$, then

$$S(r) = S(r^0) + \int_{r^0}^{r} p \cdot dx. \tag{9.2.5b}$$

For short times the point $r = (x,p) \in \Lambda^n$ is located in the vicinity of Λ_0^{n-1} so that the relationship $p = p(x)$ is assured. This means that in such a case we can write $r = (x, p(x))$ so that $S(r) = S(x)$. Set $\Lambda_T^n = \Lambda^n \cap \{0 \le t \le T\}$ and look for a projection onto $\mathbf{R}_x^n$ $(x \in \mathbf{R}_x^n)$. When such a projection is diffeomorphic, $S(r)$ can be considered as dependent upon x.

This diffeomorphism may not survive time evolution. During this evolution there could be at least points $r \in \Lambda^n$ which do not admit diffeomorphic projection to $\mathbf{R}_x^n$. Such a situation can be visualized as follows. Suppose that our Lagrangian manifold is two-dimensional — just like the ocean's surface. In stormy weather the moving incoming waves are making folds (which subsequently break). On such folds, the diffeomorphic projection is being lost.

Remark 9.2.6. The classification of singularities (that is of all possible folds) of Lagrangian manifolds is a topological problem. It is discussed in [348, 358, 359].

The question arises: Can violation of the relationship $p = \frac{\partial S}{\partial x} = p(x)$ be caused by the caustics $J(t, y) = 0$? The answer to this key question leads to new (as compared to WKB) way of producing the asymptotic expansions. This was done by Maslov.

Going back to Bohm's paper, [357], we read the following:

(p. 170, bottom) "For in order to obtain results that are equivalent to those of the usual interpretation of the quantum theory, we are required to restrict the value of the initial particle momentum to $\mathbf{p} = \nabla S(\mathbf{x})$. From the application of Hamilton–Jacobi theory to Eq. (9.2.4a), it follows that this restriction is consistent, in the sense that *if it holds initially, it will hold for all time. Our suggested new interpretation of the quantum theory implies, however, that this restriction is not inherent in the conceptual structure.* We shall see in Sec. 9, for example, that it is quite consistent in our interpretation *to contemplate modifications in the theory*, which permit an arbitrary relation between $\mathbf{p}$ and $\nabla S(\mathbf{x})$.[c] The law of force on the particle can, however, be so chosen that in the atomic domain, $\mathbf{p}$ turns out to be very nearly equal to $\nabla S(\mathbf{x})$, while in processes involving very small distances, these two quantities may be very different.[d]"

(p. 171 top-right) "... with the aid of such modifications in the theory, we could in principle measure the particle positions and momenta precisely, and thus violate the uncertainty principle. As long as we restrict ourselves to conditions in which Schrödinger's equation is satisfied, and in which $m\mathbf{v} = \nabla S(\mathbf{x})$,[e] however, the uncertainty principle will remain an effective practical

[c]Indeed such modifications are needed if caustics are present (our observation).

[d]Surely, in Bohm's paper these are just intuitive comments.

[e]In Remark 9.2.4 we explained that use of the relation $m\mathbf{v} = \nabla S(\mathbf{x})$ is in fact implying the existence of Schrödinger's equation so that there is no need to use them both.

limitation on precision of measurements. This means that at present, the particle *positions and momenta should be regarded as hidden variables*. . . [f]"

Remark 9.2.7. In Appendix C as well as in Chap. 8 we used conformational statistics of polymers for modelling of all kinds of quantum mechanical phenomena. In polymer physics Bohm's hidden variables are used routinely and have physical meaning.

Remark 9.2.8. Bohm's predictions had been checked and solidly confirmed experimentally in 2011 and 2012, e.g. read [360, 361]. These works use some refinements of quantum measurement theory described in [362]. More comprehensive up-to-date review of quantum measurement theory is given in [363]. As a result of these experiments, a flurry of activity around Bohmian quantum mechanics had been initiated in these two years. The monographs [364] and [365] published in this time period provide good background of Bohmian mechanics and contain many up-to-date applications of general principles. They do not discuss though works by Maslov, Arnol'd and other mathematicians.

Following Maslov, to deal with the problem of designing new asymptotic expansions, it is useful to keep in mind that on the Lagrangian manifold Λ^n there is some freedom of choice of selection of local coordinates. They can be made either of $\{x_1, \ldots, x_n\}$ or of $\{p_1, \ldots, p_n\}$ or of sets $\{p_{(\alpha)}, x_{(\beta)}\}$, $(\alpha) = (\alpha_{i_1}, \ldots, \alpha_{i_k}\}, (\beta) = (\beta_{j_1}, \ldots, \beta_{j_{n-k}}\}$ which *do not* contain conjugate coordinate and momentum, that is they have no pairs of the type (x_j, p_j).[g] In the vicinity of a "bad" point $r \in \Lambda^n$ which cannot be projected diffeomorphically onto $\mathbf{R}^n_x$ it becomes necessary to choose some (or all) p's from the above set (p-representation) and to complement them with x's so that the total number of coordinates is n as required. As soon as we replaced some of x's by p's we need to redo the WKB expansion method. For this purpose we have to introduce the multidimensional Fourier transform $\mathcal{F}$ and its inverse $\mathcal{F}^{-1}$ via

$$(\mathcal{F}_{\lambda, x\to p} u(x))(p) = \left(\frac{\lambda}{2\pi i}\right)^{\frac{n}{2}} \int \exp[-i\lambda\langle x, p\rangle] u(x) dp, \tag{9.2.9a}$$

$$(\mathcal{F}^{-1}_{\lambda, p\to x} u(p))(x) = \left(\frac{\lambda}{-2\pi i}\right)^{\frac{n}{2}} \int \exp[i\lambda\langle x, p\rangle] u(p) dp. \tag{9.2.9b}$$

---

[f]This conclusion of Bohm is in accord with conclusions of Maslov summarized in Remark 9.2.5.

[g]With such a construction, in accord with Bohm, the Heisenberg's uncertaity principle is respected.

In complete analogy with expansion Eq. (9.2.2) it is possible to have an expansion in p-representation of the same kind and to write

$$\psi(t,x;\lambda) = (\mathcal{F}^{-1}_{\lambda,p\to x}u(p))(x)$$
$$= \left(\mathcal{F}^{-1}_{\lambda,p\to x}\left[\exp\{i\lambda\tilde{S}(p)\}\sum_{j=0}^{\infty}(i\lambda)^{-j}\tilde{\varphi}_j(p)\right]\right)(x). \qquad (9.2.10)$$

Surely, such a Fourier transform can (and should) be done just with part of available variables. New asymptotic solutions can now be built as follows.

Choose some domain $\Omega \in \mathbf{R}^n_x$, then there could be two options for Λ^n: (a) It may have points of Λ^n which project onto Ω diffeomorphically. There could be several sub-domains of Λ^n with such a property. Then, the asymptotic solution Eq. (9.2.2) should be replaced by the sum of asymptotic solutions in accord with the number of domains of Λ^n; (b) If Ω contains a caustic, then in the vicinity of caustic the solution is chosen in the form given by Eq. (9.2.10), perhaps with part of variables in p-representation. Both solutions of (a) and (b)-types should be glued together in the same way as we did when we constructed a fiber bundle in Chap. 6, Sec. 6.5.[h] That is we should deal with the cocycle conditions, etc. Such gluing operations are responsible for emergence of the Maslov indices and for replacement of the familiar Bohr–Sommerfel'd quantization conditions occurring in the WKB calculations by the Maslov quantization conditions (see below).[i] In the case if the underlying classical system is exactly integrable, e.g. that for the harmonic oscillator, the Maslov quantization condition produces the exact spectrum.

For completeness of our presentation, we provide a formal definition of the Maslov index following Arnol'd, [136].

Definition 9.2.1. The Maslov index of an *oriented curve* on a Lagrangian manifold is defined as the number of passages from the negative side of the manifold of singularities to the positive side, minus the number of passages in the other direction.[j] It is being assumed that the ends of the curve are nonsingular and that the curve intersects only the manifold of simple singular points and only with nonzero angles.

---

[h] Recall Exercise 3.5.1 and Sec. 6.5.

[i] In the present case the situation is complicated by the fact that we have to take into account orientations of the Lagrangian surfaces when gluing them together.

[j] Words "negative" and "positive" are associated with orientations of the Lagrangian (sub)manifolds.

Remark 9.2.9. It can be demonstrated [366] that the Morse index, e.g. see Definition D.2.13.2. (Appendix D), is a special case of the Maslov index.

Remark 9.2.10. More recently, the Conley–Zehnder index, [D.52], p. 293, was introduced. It is a kind of Maslov index, playing an important role in Floer's theory mentioned in Sec. 9.1. In Floer's theory the Maslov index is also known as the *relative Morse index*. A very accessible description of Floer's theory (at the level of presentation of this book) can be found in the book by Jost [17]. A detailed relationship between Morse theory, Conley–Zehnder index and Floer homology is developed in the paper by Salamon [367]. More on connections between Morse theory, Hamiltonian systems and Maslov indices can be found in the long and detailed paper by Salamon and Zehnder [368].

The Maslov indices of closed curves enter also asymptotic formulas for stationary problems of quantum mechanics. Specifically, consider the one-dimensional Hamiltonian $H = p^2/2 + U(q)$ and suppose that on the level manifold, that is for $E = H$, there lies also the Lagrangian submanifold L. If a sequence of numbers $\mu_N \to \infty$ satisfies the conditions

$$\frac{2\mu_N}{\pi}\oint_\gamma pdq \equiv ind\gamma \ (\text{mod}4) \ (\text{Maslov quantization condition}) \qquad (9.2.11\text{a})$$

for all closed contours γ on L, then the equation $\frac{1}{2}\nabla^2\psi = \lambda^2(U(q) - E))\psi$ has a series of eigenvalues $\lambda_N = \mu_N + O(\mu_N^{-1})$. In the case if $U(q) = q^2/2$ the following exact result is obtained [349, 366]:

$$\mu \oint_{H=E} pdq = 2\pi\left(N + \frac{1}{2}\right), \quad N = 0, 1, 2, \ldots, \qquad (9.2.11\text{b})$$

$\mu = h^{-1}$ with h being the Planck constant. From here, the exact result (in the appropriate system of units) for the harmonic oscillator $E_N = h(N + \frac{1}{2})$ follows.

Remark 9.2.11. Since every exactly integrable system in the action-angle variables can be reduced to the set of harmonic oscillators [136, 235], this means that the fiber bundle construction of asymptotic results by Maslov which was sketched above is sufficient for quantization of any integrable system. These results are fully compatible with those presented in Chap. 6 (Sec. 6.5), in Appendix A and in Remark 9.2.4 — all obtained by methods of sub-Riemannian geometry.

Remark 9.2.12. Although mathematically correct, Maslov's results for asymptotics of the wave functions are hard to obtain both mathematically and physically. The most difficult case mathematically is when the eigenvalues have multiplicities (that is in the case when the entanglement is taking place), e.g. read Arnol'd, [136], Appendix 10 (on multiplicities) and p. 445 (on asymptotics).

9.3 From Maslov and Bohm to Bell and Beyond

The asymptotic analysis developed by Maslov can be transferred to many-body exactly integrable systems. Such a transfer is typically associated with what is known in literature as Duistermaat–Heckman formula. Because its derivation and its meaning are directly connected with the results of Appendix D, it makes sense to discuss it in some detail. This will be done in the next section. In this section we would like to discuss instead the following.

Unlike Maslov's, Bohm's papers [357] already contain many-body extension of the formalism and, with it, the entanglement issues related to the EPR paper, [217]. On p. 186, we find the following statement by Bohm:

"Thus, the "quantum mechanical" forces may be said to transmit uncontrollable disturbances *instantaneously from one particle to another through the medium of the* ψ-field."

To understand what Bohm had in mind we need to go back slightly in his paper and to bring the following quotation (p. 171, bottom-right):

"We conclude that our suggested interpretation of the quantum theory provides a much broader conceptual framework than that provided by the usual interpretation,[k] for all of the results of usual interpretation are obtained from our interpretation if we make the following three special assumptions *which are mutually consistent*:

(1) That the ψ-field satisfies Schrödinger's equation.
(2) That the particle momentum is restricted to $\mathbf{p} = \nabla S(\mathbf{x})$.[l]
(3) That we do not predict or control the precise location of the particle, but have in practice, a statistical ensemble with probability density $P(\mathbf{x}) = |\psi(\mathbf{x})|^2$. The use of statistics is, however, *not inherent in the*

[k] By this, he had in mind the Copenhagen interpretation.

[l] In view of Remark 9.2.4 the requirement (1) is a corollary of the requirement (2). Surely, they are mutually consistent as stated by Bohm.

conceptual structure, but merely a consequence of our ignorance of the precise initial conditions of the particle."

The fact that use of statistics is "not inherent in the conceptual structure" lies as well at the heart of Maslov's approach described in the previous section. There were many attempts to reduce quantum mechanics to what is effectively the classical statistical mechanics, e.g. read [369]. There were also numerous attempts to make quantum mechanics as part of classical mechanics, e.g. read [370], and [6–22] therein. The closest to our Remark 9.2.4 is the recent paper by Gosson, [371], emphasizing the role of Lagrangian manifolds. However, the Heisenberg group mentioned in his paper was not looked upon through the glasses of sub-Riemannian geometry (Appendix A). At the same time, complex hyperbolic geometry discussed in great detail in [214], not only contains the Heisenberg group, the Cauchy–Riemann and sub-Riemannian geometries but also connects them with the hyperbolic structure of our Minkowski space–time. Furthermore, since the complex hyperbolic geometry includes real as a special case, this means that one can think about the hyperbolicity of our space–time as result of existence of some kind of *hyperbolic knots* in the otherwise flat space–time. Space complementary to such knots is hyperbolic [372]. This topological result is relatively easy to understand. Indeed, following our work, [373], let us consider a sphere S^2 (that is $\mathbf{R}^2$ + point at ∞). If we make three holes in such a sphere (in fact, two is sufficient since the third is the point at ∞) the motion on such trice punctured sphere will be the same as that on the hyperbolic (Lobachevsky/Poincaré) plane. Imagine that these three holes were made by some (hyperbolic) knot piercing through this plane. Then, since the positioning of such a plane can be arbitrary with respect to some fixed plane, the whole 3D space acquires hyperbolicity. By analogy, we can think about taking some planes, or even infinite lines (strings) out of 3-space, then their complement could become hyperbolic.

As an example of this kind we consider the complex projective space $\mathbf{CP}^n$ discussed repeatedly in Appendix D. Kiernan [374] proved the following remarkable theorem.

Theorem 9.3.1. *Let $H_1, \ldots, H_{2n+1}$ be $2n+1$ hyperplanes in general position in $\boldsymbol{CP}^n$, then $M = \boldsymbol{CP}^n \backslash \{H_1, \ldots, H_{2n+1}\}$ is complete hyperbolic space.*

Remark 9.3.1. The meaning of "complete hyperbolic" can be found in [215].

Remark 9.3.2. The existence of knots in the sky was indeed checked very recently [375] with somewhat strange (for us) outcome: "We rule out at 95% confidence models[m] that predict more than six detectable cosmic textures[n] on the full sky". From what has been said above, it follows that even one cosmic texture could be enough to make our space hyperbolic. Six textures are more than sufficient for such a purpose.

With these comments, it is still essential not to disentangle the dynamics of quantum mechanics from that of the measuring devices [363]. The most insightful approach to this problem (other than that by Bohm) was made by John Bell [220]. To discuss his input in the context of Bohmian mechanics we follow recent paper by Braverman and Simon, [376]. These authors begin it by making comments on Bhom's result $P(\mathbf{x}) = |\psi(\mathbf{x})|^2$. While the probability distribution of particle coordinates-hidden variables — is given only probabilistically via $P(\mathbf{x})$, the velocity $\mathbf{v}$ at a given position $\mathbf{x}$ could be obtained deterministically.

Remark 9.3.3. This experimental result (see immediately below) provides realization of the concept of Lagrangian manifold used by Maslov instead of the initial wave function. It also provides support to Remark 9.2.4 in which we stated that the essence of quantum mechanics (that is of the Heisenberg group $\mathbf{H}_1$) is encoded in such information.

Based on theoretical work by Wiseman, [377], (see also [378]) experiments described in [360] demonstrate that, indeed, the individual Bohmian velocity can be measured. This has become possible due to use of the theory of weak measurements, [362]. The result by Wiseman is extendable to multiparticle case along the lines of Bohm's papers [357]. In such a case the topological (entanglement) issues emerge already at the level of just two particles so that Bell's input into this issue becomes immediately relevant. His theoretical analysis provides affirmative support to the conclusion reached by Bohm that "quantum mechanical forces may transmit uncontrollable disturbances *instantaneously from one particle to another through the medium of the* ψ-field". It should be noted though that papers by Bohm do not take statistics of particles explicitly into account. An attempt at correction of this deficiency and at checking the instantaneity in Bohmian mechanics was recently made in the paper by Braverman and Simon [376]. It is instructive to sketch the idea behind their experimental set up so that

[m] That is the cosmological models.
[n] That is knots, as explained in the paper.

theoretically minded readers can get some feeling of the price to be paid for confirmation of each theoretical conclusion.

Begin by combining Eqs. (9.2.1b) and (9.2.4b) in order to obtain the one-particle velocity

$$\mathbf{v}(\mathbf{x}) = \frac{\mathbf{j}(\mathbf{x})}{|\psi(\mathbf{x})|^2}. \tag{9.3.1a}$$

Here

$$\mathbf{j}(\mathbf{x}) = \frac{\hbar}{m}\mathrm{Im}[\psi(\mathbf{x})\nabla\psi(\mathbf{x})]. \tag{9.3.1b}$$

In the two-particle case the above one-particle wave function is replaced by $\psi(\mathbf{x}_1, \mathbf{x}_2)$ (initially without specification of statistics). In such a case, the velocity of the first (second) particle $\mathbf{v}_1(\mathbf{x}_1, \mathbf{x}_2)$ is given by the same formula as above in which $\psi(\mathbf{x}) \to \mathbf{v}_1(\mathbf{x}_1, \mathbf{x}_2)$ (or $\mathbf{v}_2(\mathbf{x}_1, \mathbf{x}_2)$) and $\mathrm{Im}[\psi(\mathbf{x})\nabla\psi(\mathbf{x})] \to \mathrm{Im}[\psi(\mathbf{x}_1, \mathbf{x}_2)\nabla_1\psi(\mathbf{x}_1, \mathbf{x}_2)]$ (or $\mathrm{Im}[\psi(\mathbf{x}_1, \mathbf{x}_2)\nabla_2\psi(\mathbf{x}_1, \mathbf{x}_2)$). The velocity field $\mathbf{v}_1(\mathbf{x}_1, \mathbf{x}_2)$ is being interpreted as the velocity for the particle 1 at position $\mathbf{x}_1$, provided that particle 2 is located at position $\mathbf{x}_2$. If statistics is ignored and particles are not entangled, then $\psi(\mathbf{x}_1, \mathbf{x}_2) = \psi(\mathbf{x}_1)\psi(\mathbf{x}_2)$, so that $\mathbf{v}_1$ in this case is independent of location of particle 2. To account for statistics, entanglement and instantaneity in a double slit-type experiment Braverman and Simon suggest to choose two double slits so that particle 1 is emitted towards the left (double) slit while particle 2 towards the right. The emitting source is constructed in such a way that at the time when each particle is in the plane of its respective double slit, the wave function of the total system is symmetric[o]

$$\frac{1}{\sqrt{2}}(f_u(x_1)f_u(x_2) + f_d(x_1)f_d(x_2)). \tag{9.3.2a}$$

The subscripts "u" and "d" denote respectively the "up" and "down" so that, for instance, $f_u(x_1)$ is the wave function of "particle" 1 going to the upper slit. It has zero overlap with $f_d(x_1)$ for the "particle" 1 going through the "down" slit. To complicate matters a phase shifter is introduced and placed just behind the lover slit for the "particle" 1 so that the wave function looks now as

$$\frac{1}{\sqrt{2}}(f_u(x_1)f_u(x_2) + e^{i\phi}f_d(x_1)f_d(x_2)). \tag{9.3.2b}$$

[o]This is justified since the experiments are expected to be made with photons, that is with bosons. And the authors provide arguments, based on already published research papers, how to make photons to act as if they are particles with some prescribed masses (interested readers should read the original paper for details).

Suppose that the first "particle" is going toward the left double slit and the second toward the right. If particles are entangled, then by changing ϕ on the left-hand side should also affect the interference pattern on the right. The double slits could be at any distance apart from each other and the local phase shift ϕ acting on the "particle" 1 only must affect the measurable velocities of both particles thus demonstrating the instantaneity. It remains to be seen if such a setup indeed will produce the desired result.

9.4 Harmonious Coexistence of Classical and Quantum Mechanics: All About and Around the Duistermaat–Heckman Formula

Although what is discussed in this section is connected with results of previous section and of Appendix D, our exposition will only point at topological methods required for obtaining these results. Thus, this section does not contain contact geometry/topology results as such but is logically essential since it is linked with what was just said above.

Method of coadjoint orbits of Kirillov, Kostant and Soriau (Appendix D.2.17), the Atiyah–Guillemin–Sternberg theorem (Appendix D.2.2) and the basics of Morse theory (Appendix D.2.13) — all make the Duistermaat–Heckman (D-H) theorem look very natural. The crux of the matter lies in demonstrating that the method of stationary phase used, say, in Maslov's calculations, when applied to multidimensional integrals whose integration measure is torus action invariant allows us to obtain the exact results instead of just asymptotics as in the WKB or Maslov-type calculations.

Following our works, [233, 379], where our readers will be able to find all needed references, we begin with the simplest example based on results discussed in Appendix D.2.3. In Appendix D we described symplectic dynamics on the sphere S^2 of unit radius. The symplectic volume Ω for this case is $\Omega = d\theta \wedge dz$ while the reduced volume $\tilde{\Omega} = dz$. Since the Hamiltonian $\mathrm{H} = z$ is the Morse–Bott function on S^2, it has two extrema ± 1. Consider the integral

$$I(\beta) = \int_{\mathcal{M}_{\mathrm{red}}} \tilde{\Omega} \exp(-\beta \mathrm{H}) = \frac{1}{\beta}(\exp(\beta) - \exp(-\beta)). \tag{9.4.1}$$

In this integral the reduced phase space is $\mathcal{M}_{\mathrm{red}} = S^2/S^1$ so that $\tilde{\Omega} = dz$, $z \in [-1, 1]$, and β is some parameter/number. The above integral is already the D-H formula! The rest of discussion of this section will be devoted to proving that it is indeed the case and to various kinds of generalizations of this result so that we shall be able to compare this generalization with what is known in literature as the D-H formula. The most efficient way of proving these results is the following.

Consider the limit $\beta \to 0^+$ in Eq. (9.4.1). Naturally, $I(\beta \to 0^+) = 2$. But 2 is the Euclidean volume of the simplest cross-polytope — the line segment $[-1,1]$.[p] To understand what we just obtained better, we need to introduce the following related integral

$$I(z) = \int_s^t dx e^{zx} = \frac{e^{tz}}{z} - \frac{e^{sz}}{z} \tag{9.4.2}$$

and to compare it against the analogous sum

$$S(z) = \sum_{k=s}^{t} e^{kz} = \frac{e^{tz}}{1-e^{-z}} + \frac{e^{sz}}{1-e^{z}}. \tag{9.4.3}$$

The sum can be interpreted as the partition function for two quantum harmonic oscillators whose frequencies are having opposite signs. This observation suggests interpretation of this sum in terms of some kind of hidden supersymmetry. Although this is indeed the case, we are not yet ready to take advantage of this observation. What we need to do first is to ask a question: Is there a way to transform $I(z)$ into $S(z)$? This is the question raised in the paper by Khovanskii and Poukhlikov. Why should we care about this? Well, we just obtained the "classical" result $I(\beta \to 0^+) = 2$, and it is of interest to find its "quantum" analog. The quantum analog is obtained if we are going to ask the following question: If the cross polytope is put on lattice (just like all rational integral polytopes discussed in Appendix D) then, what is the discrete volume of such a polytope? In the case of a segment $[-1,1]$ the discrete volume is 3-two extremal points and the point zero-center of symmetry of such polytope. If we would blow up a segment $[-1,1]$, e.g. to $[-m,m]$ then, the number of points inside the polytope, including boundaries, is $2m+1$. Obviously, for large m this result is very much the same as that we would obtain should we blow up the unit sphere S^2 accordingly. Thus, going back to $S(z)$ and taking limits of summation as m and $-m$ and taking into account that for $s \to 0$

$$\frac{1}{1-e^{-s}} = \frac{1}{s} + \frac{1}{2} + \frac{s}{12} + O(s^2) \tag{9.4.4}$$

and

$$\frac{1}{1-e^{s}} = -\frac{1}{s} + \frac{1}{2} - \frac{s}{12} + O(s^2) \tag{9.4.5}$$

[p] E.g. read Remark D.2.3.1.

we obtain $S(z \to 0^+) = 2m + 1$ as required allowing us to rewrite the sum $S(z)$ for the segment $[-m, m]$ in the equivalent form

$$S(z) = \sum_{k=-m}^{m} e^{kz} = \exp(-zm) \left[\frac{\exp(z(2m+1)) - 1}{\exp z - 1} \right]. \tag{9.4.6}$$

From Sec. D.2.15 of Appendix D it follows that any rational convex polytope can be made out of balloon sculpture whose basic elements are segments-deformation retracts of S^2. To construct such a sculpture (and the polytope) and to associate with it the multidimensional version of Eq. (9.4.6) we have to use the concept of zonotope introduced in Appendix D.2.2. Thus, we begin with the cubic lattice of dimension d. Let $\mathrm{Vert}\,\mathcal{P}$ denote the vertex set of such a (hyper)cube and let $\{u_1^v, \dots, u_d^v\}$ be the orthogonal basis made of vectors (not necessarily of unit length) made of highest weight vectors (D.2.12) of the Weyl–Coxeter reflection group B_d appropriate for the cubic symmetry. These vectors are oriented along the positive semi-axes with respect to the center of symmetry of (hyper) cube. When parallel translated to the edges ending at particular hypercube vertex $\mathbf{v}$, they can point either in or out of this vertex. In terms of just introduced notations the d-dimensional version of Eq. (9.4.6) acquires the following form.

$$\sum_{\mathbf{x} \in \mathcal{P} \cap \mathbf{Z}^d} \exp\{\langle \mathbf{z}, \mathbf{x} \rangle\} = \sum_{\mathbf{v} \in \mathrm{Vert}\mathcal{P}} \exp\{\langle \mathbf{z}, \mathbf{v} \rangle\} \left[\prod_{i=1}^{d} (1 - \exp\{-z_i u_i^v\}) \right]^{-1}. \tag{9.4.7}$$

Because of zonotope construction (Appendix D.2.2) the obtained results remain correct for any centrally symmetric polytope (perhaps with replacement of the highest weight set $u_1^v, \dots, u_d^v$ by another set of the appropriate symmetry).

Remark 9.4.1. For an arbitrary convex polytope Eq. (9.4.7) (appropriately amended as discussed) was obtained (seemingly independently) by many authors and in different contexts — all listed in [233, 379]. The most important for us is the fact that in the theory of representations of Lie groups/algebras the above relation is known as the Weyl character formula.

Consider formally the case when the parameter z is small. Then by writing $1 - \exp\{-z_i u_i^v\} \simeq z_i u_i^v$ and replacing summation by integration on the L.H.S. of Eq. (9.4.7) we obtain the D-H formula

$$\int_{\mathcal{M}} \tilde{\Omega} \exp\{\langle \mathbf{z}, \mathbf{x} \rangle\} = \sum_{p \in v} \frac{e^{\langle \mathbf{z}, \mathbf{p} \rangle}}{\prod_i z_i u_i^v}. \tag{9.4.8}$$

We would like to connect this result with results of previous sections. To do so, following book Guillemin and Sternberg, [380], we introduce the

Gaussian integral

$$\frac{1}{\sqrt{2\pi}}\int e^{irx^2/2}e^{-i\xi x}dx = e^{\frac{i\pi}{4}}r^{-\frac{1}{2}}e^{-i\frac{\xi^2}{2r}} \tag{9.4.9a}$$

and its multidimensional analog

$$\left(\frac{1}{\sqrt{2\pi}}\right)^n\int e^{i\langle\mathbf{x}Q\mathbf{x}\rangle/2}e^{-i\xi\cdot\mathbf{x}}d\mathbf{x} = e^{\frac{i\pi}{4}\mathrm{sign}Q}\left|\det Q\right|^{-\frac{1}{2}}e^{-i\frac{\langle\xi Q^{-1}\xi\rangle}{2}}. \tag{9.4.9b}$$

Consider next the saddle point calculation of the multidimensional integral of the type $\int_X e^{itf}\mu$ where X is compact oriented n-dimensional manifold with volume form μ. When $t\to\infty$, we obtain (taking into account Eq. (9.4.9b))

$$\left(\frac{t}{2\pi}\right)^{\frac{n}{2}}\int_X e^{itf}\mu = \sum_p c(p)e^{itf(p)} + O(t^{-1}). \tag{9.4.10}$$

Here the summation is done over all critical points and

$$c(p) = \exp\left(i\frac{\pi}{4}\mathrm{sign}d^2f(p)\right)\left[\det d^2f(p)\right]^{-\frac{1}{2}}. \tag{9.4.11a}$$

This result can be improved by noticing that in the vicinity of p we can always expand (see Appendix D, Eq. (D.87a))

$$f(x) = f(p) + \frac{1}{2}\sum_i \alpha_i(p)x_i^2 + \cdots$$

so that, in fact, for symplectic manifolds (that is when $n\to 2m$)

$$c(p) = \exp\left(-i\frac{\pi}{2}\right)\left[\prod_{i=1}^m \alpha_i(p)\right]^{-1}. \tag{9.4.11b}$$

In those cases when it can be proven that the term $O(t^{-1})$ in Eq. (9.4.10) is exactly zero,[q] we can put $t=1$ in this expansion thus resulting in the D-H formula

$$\int_X e^{if}\tilde{\mu} = (i)^m\sum_p \frac{e^{if(p)}}{\prod_{i=1}^m \alpha_i(p)}. \tag{9.4.12}$$

Obtained result is in accord with Eq. (9.4.8).

[q]This happens exactly in the D-H case!

Exercise 9.4.1. By noticing that for the unit sphere $S^2 = X$ we can write $z = \pm\sqrt{1 - x^2 - y^2}$, use results of Morse theory (Appendix D.2.15) in order to demonstrate that Eq. (9.4.1) is indeed the D-H formula, Eq. (9.4.8). Hint: Consult the review article by Vergne on "Applications of equivariant cohomology", [D.65].

By comparing Eqs. (9.4.2a) and (9.4.2b) with Eqs. (9.4.7) and (9.4.8) it should become obvious that if there is a way to relate the "classical" result Eq. (9.4.2a) with "quantum" Eq. (9.4.2b), then the same method should work in relating the D-H classical result, Eq. (9.4.8) and "quantum" result, Eq. (9.4.7). Such a method was indeed found by Khovanskii and Poukhlikov.[r] They introduced the Todd operator

$$Td(z) = \frac{z}{1 - e^{-z}}. \tag{9.4.13}$$

Using it, we obtain

$$Td\left(\frac{\partial}{\partial h_1}\right) Td\left(\frac{\partial}{\partial h_2}\right) \left(\int_{s-h_1}^{t+h_2} e^{zx} dx\right)\Bigg|_{h_1=h_2=0} = \sum_{k=s}^{t} e^{kz}. \tag{9.4.14}$$

Also, the result

$$Td\left(\frac{\partial}{\partial \mathbf{z}}\right) \exp\left(\sum_{i=1}^{n} p_i z_i\right) = Td(p_1, \ldots, p_n) \exp\left(\sum_{i=1}^{n} p_i z_i\right) \tag{9.4.15}$$

happens to be the most useful. By applying it to

$$i(x_1, \ldots, x_k; \xi_1, \ldots, \xi_k) = \frac{1}{\xi_1 \cdots \xi_k} \exp\left(\sum_{i=1}^{k} \xi_i x_i\right) \tag{9.4.16}$$

we obtain,

$$Td\left(\frac{\partial}{\partial \mathbf{x}}\right) i(\mathbf{x}, \xi) = \frac{1}{\prod_{i=1}^{k}(1 - \exp(-\xi_i))} \exp\left(\sum_{i=1}^{k} \xi_i x_i\right). \tag{9.4.17}$$

While the final result on the R.H.S. looks the same as the individual (vertex) term entering into the sum over vertices on the R.H.S. of Eq. (9.4.7), Eq. (9.4.16) has the same structure as each term in the sum over vertices on the R.H.S. of Eq. (9.4.8). From here it follows that the (quantum) Weyl

[r] E.g. read [233, 379] and references therein.

character formula can be obtained from the classical D-H formula with the help of the Todd transform.

When commenting on the result, Eq. (9.4.2b) we noticed that it is hinting at supersymmetry which might be involved relating the "classical" integral $I(z)$ to "quantum" sum $S(z)$. Said differently, the Todd operator is typically associated with the index theory for elliptic operators [381] and, hence, with cohomology. Connections between cohomology and supersymmetry for the D-H-type problems had been worked out in great detail in the monograph, [382], by Guillemin and Sternberg who attribute the ideas of equivariant (co)homology to Henri Cartan. He is considered the founder of this field. To avoid duplication, we would only outline the key ideas. To facilitate our readers' understanding, we strongly recommend prior reading of Appendix D.

Let us begin with Definition D.2.5.3. If G is a compact Lie group acting on a topological space X freely, then topologically the quotient X/G is behaving in a very much the same way as X itself. Thus, if G acts freely on X, the equivariant cohomology groups of X are the same as of X/G. That is,

$$H_G^*(X) = H^*(X/G). \tag{9.4.18a}$$

Let, for instance, $G = X$, then we obtain

$$H_G^*(X) = H^*(\text{pt.}), \tag{9.4.18b}$$

where pt. means "point". But in the simplest case of the torus action of S^1 on S^2 discussed in Appendix D.2.3, we encountered two fixed points so that the torus action is not free! Assume that we can replace X by X^* on which the torus action (or action of other Lie group G) is free. Then X^*/G is a topological space on which G acts freely by design so that, again, we can use the ordinary cohomology. Suppose, furthermore, that the space X^* is homotopy equivalent to X. Such a space is possible to construct as a product $X^* = X \times E$ where E is contractible space on which G acts freely. Thus, we are coming to the standard way of defining the equivariant cohomology via prescription

$$H_G^*(X) = H^*((X \times E)/G). \tag{9.4.19}$$

The above result depends upon existence of space E. That is to say, given a compact Lie group G, is there a contractible space E on which G acts freely? It can be demonstrated that every compact group G can be embedded as a subgroup of $\mathrm{U}(n)$ for large enough n. In Appendix D we already studied

U(n) — the Lie group playing the central role in quantum mechanics. It happens that, as it is happening in quantum mechanics, n could become infinite. For any finite n the space E on which U(n) is acting without fixed points is contractible. For demonstration of the fact that it is contractible even when $n \to \infty$ we refer to [382]. In addition to this reference, we strongly recommend our readers to read classical papers by Atiyah and Bott, [383], and Witten, [384]. Incidentally, both papers begin with discussion of the D-H formula. While [383] can be considered as condensed summary of the book, [382], in [384] Witten extends the D-H formula to the non-Abelian case needed for treatment of two-dimensional free non-Abelian gauge field theories.

Exercise 9.4.2. The non-Abelian generalization of the D-H formula still makes it a classical object. To make it quantum object requires use of the Todd transform. As we already know, it converts the D-H formula into the Weyl character formula. How such a transform should look like in the non-Abelian setting? What is the group-theoretic meaning of the new Todd transformed result which is replacing the standard Weyl character formula?

To proceed, we need, following [D.91], to introduce the "tautological 1-form" on the cotangent bundle T^*M of symplectic manifold M. Such a bundle was defined in Sec. D.2.17 (Appendix D). In physics literature [234, 235] the cotangent bundle is known as phase space. Thus, let $\lambda \in T^*M$ and let w be a tangent vector at λ to T^*M. Evidently, a point $\lambda \in T^*M$ is defined by $\lambda := (q, w)$, $q \in M$, $w \in T_\lambda(T^*M)$. T^*M is smooth manifold of dimension $2n$. The canonical projection π is defined as usual by $\pi(q, w) = q$[s] or, equivalently, as $\pi(\lambda) = q$. The canonical projection π induces mapping $\pi_* : T_\lambda(T^*M) \to T_qM$. The tautological 1-form $s_\lambda \in \Lambda^1(T^*M)$ at λ should look the same independent of the choice of coordinate system. Let, as usual, $s_\lambda = p_i dq^i$ and let $\xi \in T_\lambda(T^*M)$. Then, if

$$\xi = \xi^i(q)\frac{\partial}{\partial q^i} + \xi^i(w)\frac{\partial}{\partial w^i}, \quad \text{then } \pi_*(\xi) = \xi^i(q)\frac{\partial}{\partial q^i}. \tag{9.4.20}$$

For the tautological 1-form s, by definition, we obtain

$$s(\xi) = \eta(\pi_*(\xi)) = \eta\left(\xi^i(q)\frac{\partial}{\partial q^i}\right) = \xi^i(q)\eta_i \tag{9.4.21}$$

[s]Recall definition of the Lagrangian manifold.

which is equivalent to $s(\xi) = \eta(\pi_*(\xi)) = \eta_i dq^i(\xi)$. This result allows us to clarify Definition D.2.3.1 of the moment map (Appendix D, Eq. (D.39)).

Definition 9.4.1. Let the symplectic 2-form ω be given by $\omega = -ds$, then the function $\Phi^\xi : M \to \mathfrak{g}^*$ defined by $s(\xi)$ is the moment map. Furthermore, $\xi \in \mathfrak{g}$.

To demonstrate that this is indeed the case, we first apply the Cartan formula, Eq. (5.5.9), to ω. Thus we have

$$\mathcal{L}_\xi \omega = (di_\xi + i_\xi d)\omega.$$

But $d\omega = 0$. If, in addition, Eq. (D.39) holds, we get as well: $\mathcal{L}_\xi \omega = di_\xi \omega = dd\Phi^\xi = 0$. Thus, the invariance of 2-form ω is contingent upon the existence of the moment map Φ^ξ. Next, if $\mathcal{L}_\xi \omega = 0$, what can be said about $\mathcal{L}_\xi s$? We have

$$(d\imath_\xi + \imath_\xi d)s_\lambda = d[\xi^i(q)\eta_i] - i_\xi \omega. \tag{9.4.22}$$

If we demand that $\mathcal{L}_\xi s_\lambda = 0$, we re-obtain Eq. (D.39), that is $d\Phi^\xi = i_\xi \omega$ as required. Being armed with these results we shall use them now in equivariant setting. For this purpose, we also need to notice that, in view of Eq. (D.36) Φ^ξ should be identified with the Hamiltonian. Next, following Atiyah and Bott, [383], we introduce an equivariant 2-form $\tilde{\omega}$ via

$$\tilde{\omega} = \omega - \mathbf{u} \cdot \Phi^\xi \quad \text{along with the exterior differential,} \tag{9.4.23a}$$

where $\mathbf{u} \cdot \Phi^\xi \equiv u^i \xi^i(q)\eta_i$. The exterior differential d should now be replaced by d_ξ given by

$$d_\xi = d + \sum_i u^i \xi^i \imath(\xi_i). \tag{9.4.23b}$$

Atiyah and Bott connect this differential with that introduced by Witten in his influential work "On supersymmetry and Morse theory", [385]. For this purpose it is sufficient to replace all u^i's by just one parameter u.[t] Consider then a product $d_\xi \tilde{\omega}$. Restricting ourself by Witten's case and restricting (without loss of generality) to treatment of just one u-component we obtain

$$(d + u\imath(\xi))(\omega - u\Phi^\xi) = d\omega + u\imath(\xi)\omega - ud\Phi^\xi = (\imath(\xi)\omega - d\Phi^\xi)u = 0. \tag{9.4.24}$$

[t] This fact is just one of many connecting equivariant cohomology with supersymmetry as discussed in detail in [383]. For a quick introduction into this topic in the physics language, read [233, 380].

Thus if initially the 2-form ω was closed, now the corrected 2-form $\tilde{\omega}$ is equivariantly closed (if Eq. (D.39) holds). This fact allows us to introduce the equivariant exponential $\alpha = \exp(i\tilde{\omega})$ for which $d_\xi\alpha = 0$. This observation happens to play the decisive role in the proof of the Berline–Vergne localization theorem [D.64] which can be stated as follows.

Theorem 9.4.1. (Beline–Vergne) *Let G be a compact Lie group acting on a compact, oriented manifold M. Let $\xi \in \mathfrak{g}$ be such that the corresponding vector field ξ_M has isolated zeros, and suppose that $d_\xi\alpha = 0$ holds. Then the D-H formula, Eq.* (9.4.12), *holds.*

The idea of the proof is based on two main ingredients. These are: (a) The careful use of the Stoke's theorem; (b) Use of the Morse theory, especially of Eqs. (D.88d) *and* (D.92). *In this case by comparing these equations with Eq.* (D.26c) *we recognize that we are dealing with the set of independent harmonic oscillators and, thus, at each critical point the problem is reduced to that leading to the D-H result, Eq.* (9.4.1). *By analogy with Morse theory we have to prove that in some coordinate system quadratic expansion around critical point is exact. For just one oscillator this problem is solved in Exercise* 9.4.1. *Multidimensional case is elegantly treated in* [D.64].

Remark 9.4.2. For those who are interested in various mathematical generalizations and alternative proofs of just obtained results, the book [D.24] by Audin is a good reference.

Remark 9.4.3. With all elegance of the D-H formula we still have to recognize that this nevertheless is a classical result. As such, it is based on known fact from mechanics that all exactly integrable systems in the appropriately chosen variables (e.g. action-angle) are reducible to the set of harmonic oscillators as explained by Arnol'd [136]. Use of the Todd transform allows us to obtain the quantum results (Weyl's character formula). Alternatively, it is possible to use either Maslov's results (Sec. 9.2) or Witten's supersymmetric approach, [385].

All these methods do not include many-body interacting fermion and boson systems. Thus, they all can be considered as some improvements of the old Bohr–Sommerfel'd method of doing quantum mechanics. Yet, to obtain corrections to Bohr–Sommerfel'd results making them exact requires lots of mathematical ingenuity. Still, it is appropriate to recall at this point that even though Niels Bohr received the Nobel Prize in Physics in 1922 for explanation of the atomic spectra of periodic system of elements while treating multielectron atoms as hydrogen-like, he was the one who remained unhappy with his own results already for He. The task of improving the results for He using Bohr's methods was formulated by Max Born. He gave an assignment to his assistants, Pauli, Jordan and Heisenberg, to make improvements in Bohr's theory in accordance with Bohr's wishes. The situation in pre-quantum mechanics, just prior to 1925, the year when new quantum mechanics was born, is described in great detail in the book by Born, [386]. These incredibly difficult calculations were abruptly put aside in 1925. Still, when it comes to He, new quantum mechanics did not solve the He problem exactly even though everybody familiar with the spectroscopy knows that the He spectrum is as sharp as that for H. Thus, if we would follow Heisenberg's original recipe [173], there should be a way to solve the He atom exactly. At this moment, as is well known, the quantum mechanics of He atom is "solved" with the help of the variational Hartree–Fock method.

Remarkably, use of methods of classical mechanics for solving problems of atomic and nuclear physics with full account of Pauli principle etc. had already begun in 1980 with works by Rowe, Ryman and Rosensteel [387]. For the sake of our readers we are bringing the full quotation of the Abstract to the paper by Rosensteel [388] published in 2011.

"Algebraic mean field theory (AMFT) is a many-body physics modeling tool which firstly, is a generalization of Hartree–Fock mean field theory, and secondly, an application of the orbit method from Lie representation theory. The AMFT ansatz is that the physical system enjoys a dynamical group, which may either be a strong or a weak dynamical Lie group G. When G is a strong dynamical group, the quantum states are, by definition, vectors in one irreducible unitary representation (irrep) space, and AMFT is equivalent to the Kirillov orbit method for deducing properties of a representation from a direct geometrical analysis of the associated integral co-adjoint orbit. AMFT can be the only tractable method for analyzing some complex many-body systems when the dimension of the irrep space of the strong dynamical group is very large or infinite. When G is a weak dynamical group, the quantum states are not vectors in one irrep space, but AMFT applies if the densities of the states lie on one non-integral co-adjoint orbit. The computational simplicity of AMFT is the same for both strong and weak dynamical groups. This paper formulates AMFT explicitly for unitary Lie algebras, and applies the general method to the Lipkin–Meshkov–Glick su(2) model and the Elliott su(3) model. When the energy in the su(3) theory is a rotational scalar function, Marsden–Weinstein reduction simplifies AMFT dynamics to a two-dimensional phase space."

Our readers are not only encouraged to read this paper but also many publications listed in references to it.

The Weyl's character formula in mathematics literature is known as partition function of statistical mechanics in physics literature. If such a calculation is done by the path integral methods, it requires us to use closed paths/orbits [389]. The mathematical problem of existence of such orbits on various manifolds is an outstanding problem in mathematics as it was mentioned in Sec. 6.5 of Chap. 6. Now we would like to discuss the same issue from the point of view of contact geometry and topology. This is the subject of the following section.

9.5 Mathematics and Physics of Weinstein's Conjecture: From Classical Statistical Mechanics to Seiberg–Witten Monopoles

Following [D.52, 390, 391], we would like to discuss the essence of Weinstein's conjecture mathematically first. For this purpose, suppose that we are given some classical Hamiltonian H and suppose that we are working with the microcanonical ensemble of statistical physics, that is with an ensemble in which the constraint $H = E$ is enforced. With such a constraint we associate the energy surface $S(E) = H^{-1}(E)$. Naturally, $S(E)$ is invariant with respect to Hamiltonian flow *which is expected to have no zeros.* The *basic question* about this flow is:

Are any of the orbits of this flow closed?

To provide an answer, we begin by recalling the Poincaré recurrence theorem described in the book by Arnol'd, [136], pp. 71–74. As is well known, this theorem is a corollary of the Liouville's theorem. The Poincaré theorem lies at the foundation of ergodic theory and allows us, when it is valid, to replace the microcanonical ensemble by the canonical one. If it can be proven that all orbits on $S(E)$ are closed, then the system is expected to behave nonergodically so that transition to the canonical ensemble cannot be performed. Thus the existence or nonexistence of periodic orbits is of major importance for statistical mechanics. It is also of major importance for mechanics, quantum mechanics (as discussed at the end of previous section), electrodynamics, hydrodynamics and even for the non-Abelian gauge field theories as it was discussed in Chaps. 1–5. From mechanics we know that the dynamical system will behave ergodically on hyperbolic surfaces.

This observation can be elevated to the general case. Indeed, the answer to the basic question is independent of the choice of the Hamiltionian function and depends only on the surface $S(E)$.

Let α be 1-form "living" on an odd-dimensional manifold M^{2n-1}. This is a contact form allowing us to construct the volume form $\alpha \wedge (d\alpha)^{n-1}$ and the Reeb vector field X_R defined by Eqs. (5.5.6a). The *contact structure* $\xi = \ker \alpha$ is the set of hyperplanes, e.g. see Fig. 4. In 1978, Alan Weinstein formulated the following:

Weinstein Conjecture: A Reeb vector field on a closed manifold M^{2n-1} admits a periodic orbit.

In 1987, Viterbo was able to prove that compact energy surfaces in $\mathbf{R}^{2n}$ of contact-type do have periodic orbits. It happens that Viterbo's n-dimensional result is not readily extendable already to the arbitrary 3-manifold. This extension was made much later by Taubes in 2007–2009 [392, 393] using results of Seiberg–Witten (S-W) theory. The action functional in the S-W theory looks exactly the same as that in Eq. (5.6.3.1). This equation was discussed in Chap. 5 in connection with the non-Abelian monopoles. Although the functional Eq. (5.6.3.1) looks the same as in the S-W theory the actual mathematical treatment of this functional in S-W theory is different. Pedagogically accessible treatment of the S-W theory at the level of this book is given in the lecture notes by Moore [394]. Initially the S-W theory was designed for description of 4-manifolds. Taubes solution of the Weinstein conjecture was facilitated by his previous work (from '90s) in which he demonstrated that if the 4-manifold M is symplectic, computation of the S-W invariants of M can be reduced to a certain counting of (pseudo)holomorphic curves. For a quick introduction to this topic, read an introductory paper by Donaldson [395]. This observation connects the S-W invariants with those by Gromov. "The basic idea is that one deforms the S-W equations using a large multiple of the symplectic form and shows that solutions of the deformed S-W equations concentrate along holomorphic curves" [391]. The next step involves the reduction from 4-manifold M to 3-manifold $\tilde{M}$. Since the 4-manifolds are symplectic while 3-manifolds $\tilde{M}$ are contact, this reduction can be made using standard methods which we would like to describe now.

We begin with the symplectic 4-manifold M which by design is made as the product $M = \mathbf{R} \times \tilde{M}$. Let α be a contact form on $\tilde{M}$. Consider a 1-form $\lambda = e^t\alpha$, $t \in \mathbf{R}$. We would like to demonstrate that such 1-form λ is playing the same role for M as more familiar 1-form $\beta = pdq$ defined on M in the sense that if the 2-form ω on M is obtained as $d\beta$ then, in our case, we obtain instead $d\lambda = e^t(d\alpha + dt \wedge \alpha) = \omega$. Using this result the symplectic volume form ω^n is obtained as follows:

$$\omega^n = ne^t dt \wedge \alpha \wedge (d\alpha)^{n-1} \neq 0.$$

From Eq. (D.36) of Appendix D we know that $dH = i_{X_H}\omega$. In the present case let the role of the vector field X_H be played by $X_t = \frac{\partial}{\partial t}$. Then, we obtain

$$i_{X_t}\omega = i_{X_t}e^t(d\alpha + dt \wedge \alpha) = e^t\alpha = dH = \lambda \tag{9.5.1}$$

as required.

Definition 9.5.1. The manifold $M = \mathbf{R} \times \tilde{M}$ (with 2-form $d\lambda = \omega$) is called *symplectization* of $\tilde{M}$ (with 1-form α).

Next, in Chap. 5, Sec. 5.5, we introduced the Liouville vector field Y. It is defined by $\mathcal{L}_Y\omega = \omega$. Since $\mathcal{L}_Y = d\imath_Y + \imath_Y d$ and since ω is the symplectic 2-form, the Liouville equation should be understood as $d\imath_Y\omega = \omega$. In the present case, we easily obtain: $Y = X_t$. Using Eq. (9.5.1) we obtain as well

$$di_{X_t}\omega = d\lambda = \omega. \tag{9.5.2}$$

Thus, in accord with results of Sec. 5.5, the Liouville vector field X_t can be used in order to obtain the contact 1-form from the symplectic two-form $i_{X_t}\omega = \lambda$. This result holds irrespective of the construction (symplectization) we just used. The obtained results can be formulated now as

Theorem 9.5.1. ([D.52], p. 113) *Let M be symplectic manifold equipped with the 2-form ω. Suppose that $\tilde{M} \subset M$ is compact hypersurface (in fact, a contact manifold). Then the following are equivalent.*

(i) *There is a contact form α on $\tilde{M}$ such that $d\alpha = \omega\,|_{\tilde{M}}$.*
(ii) *There exists Liouville vector field $X_t : U \to TM$, defined in the neighborhood U of $\tilde{M}$, which is transverse to $\tilde{M}$.*

Proof.[u] (i) Suppose that (ii) is proven then, according to the results of Sec. 5.5, just quoted above, we obtain the contact form λ in a standard way: $i_{X_t}\omega = \lambda$. Since X_t is the Liouville vector field, we re-obtain Eq. (9.5.2) as required.

(ii) Again, we have already designed our Liouville vector field X_t explicitly as described above. So, it does exist. What remains to be demonstrated

[u]Our proof is somewhat different from that given in [D.52].

is that it is transverse to $\tilde{M}$. This can be understood best by considering some example. For instance, we select that discussed in Sec. 5.5. In that section we had a symplectic manifold $\mathbf{R}^4$ and a contact hypersurface S^3 living in $\mathbf{R}^4$. In Appendix D we discussed such a hypersurface from the point of view of dynamical systems. Evidently, the results discussed in Appendix D.2.1 are sufficient to make a connection with the relation $S(E) = H^{-1}(E)$ defined above. To make this conclusion even stronger, we would like to add the results of Sec. 5.6.5, especially Eqs. (5.6.5.4a,b), and those of Sec. 6.5. From all these examples it follows that: (a) the Reeb vector field is made of vectors tangent to S^3 and (b) the Liouville vector field $Y = \frac{1}{2}r\partial_r$ is the radial vector field which is perpendicular (that is transverse) to the surface of the sphere S^3. □

Exercise 9.5.1. Try to prove the transversality of the Reeb and Liouville vector fields for any contact manifold, not just S^3. Hint: Learn Moser's trick [139] and use it for solving this problem.

Following [139], let us now return to the equation $dH = i_{X_H}\omega$ and compare it with $i_R d\lambda = i_R\omega$. For S^3, it was demonstrated in Sec. 5.5 that $i_R\omega = 0$ for surfaces of constant energy $dH = 0$. For this case we obtain:

$$i_R\omega = 0 \quad \text{and} \quad i_{X_H}\omega = 0. \tag{9.5.3}$$

From here it follows that the Reeb and the Hamiltonian vector fields (up to scaling factor) represent *the same* vector field implying that the constant energy (hyper)surfaces are being foliated by the Reeb contact vector fields!

Obtained results clarify the meaning of Weinstein's conjecture. They can be sharpened by the use of the following definitions.

Definition 9.5.2. ([D.52]) Compact hypersurface $\tilde{M}$ in a symplectic manifold M is said to be of *restricted contact type* if it is transverse to a Liouville vector field X defined on all of M.

Definition 9.5.3. ([D.52]) A contact manifold $\tilde{M}$ is said to be *symplectically fillable* if it is embedded as a hypersurface of contact type into a symplectic manifold M in such a way that $\tilde{M} = \partial M$.

Evidently, understanding (and appreciation) of the proof of Weinstein's conjecture by Taubes for 3-manifolds cannot be made without using the results we just have described. What makes Weinstein's conjecture so difficult to prove?

When the manifold $\tilde{M}$ is three-dimensional, its Euler's characteristic $\chi(\tilde{M}) = 0$. Euler's characteristics can be expressed with help of the

Poincaré–Hopf index theorem. Many examples of practical uses of this theorem can be found in our works, [397, 398]. These works bring into play theory of foliations.[v] Although this theory was initiated in physics by Faraday, was developed by Maxwell and by many others in physics, e.g. read http://www-spof.gsfc.nasa.gov/Education/whfldlns.html, mathematicians typically disregard these contributions and present theory of foliations without any mention of its origins. Typical foliations used in physics are known in mathematics as "oriented foliations". These are associated with the fields of force. In mathematics sometimes foliations do have directions (as in the Poincaré–Benedixson qualitative theory of differential equations in the plane, e.g. read http://minitorn.tlu.ee/~jaagup/uk/dynsys/ds2/limit/Poincare/Poincare.html) and sometimes they do not, as in the case of textures in liquid crystals discussed in [396, 397]. For such a case it is not possible to introduce the field of force while the indices for such *line* fields can still be introduced/defined and computed. This is exactly the case for both the dynamics of textures in liquid crystals and that of particles in $2+1$ gravity. In such force-free situations the dynamics is described in terms of the formalism of train tracks invented by Thurston [183]. Examples are provided in the already cited [396, 397] as well as in [398]. Using theory of foliations, Helmut Hofer was able to prove the Weinstein conjecture in 1993 for all overtwisted contact forms on any closed orientable 3-manifolds $\tilde{M}$ [399]. Recall that the overtwisted contact form was defined in Eq. (5.5.19a). Extension of these results to contact manifolds of dimensions higher than three was done by Albers and Hofer in 2009 [400]. A nice review of these and related results with emphasis on the theory of foliations and pseudo-holomorphic curves is given in [401]. Our readers may also benefit from reading the Ph.D. thesis by Gabriele Benedetti entitled "An approach to the Weinstein conjecture via J-holomorphic curves".[w] Finally, it is impossible to master the field of contact geometry and topology of 3-manifolds without reading Eliashberg's review paper, [402].

The fact that Euler's characteristics of 3-manifolds is zero is a source of the following difficulty. The Poincaré–Hopf index theorem *does* allow foliation fields on 3-manifolds to have singularities, that is to have zeros. What really matters is the *signed sum* of these zeros (these signed zeros are just like positive and negative electric charges in electrostatics, e.g. recall Sec. 3.5). The Weinstein conjecture is formulated under the assumption that

[v] We discussed some examples of foliations in Sec. 2.4.
[w] E.g. see http://www.dm.unipi.it/~abbondandolo/teaching/tesi/benedetti.pdf

the foliation field does not have zeros. From [391] we find that "Although for the 3-torus it is easy to construct vector fields with no closed orbit, already for S^3 the question of whether all vector fields have closed orbits is difficult. It turns out that the answer is 'no'" because S^3 can have both closed and open orbits. In the case of closed orbits they might be knotted/linked [137, 403]. This dichotomy: open-closed, already for S^3, is the major source of difficulty of proving the conjecture.

Since the proof by Taubes (seemingly) is "more physical", we would like to make few additional comments about his proof. We warn our readers that, thus far, Taubes proof is limited to 3-manifolds only while the method of J-holomorphic curves is not restricted to three dimensions.

Following [391], on a larger scale the proof involves:

(a) defining some kind of count of Reeb orbits with appropriate signs;
(b) demonstration that this count is a topological invariant;
(c) explicit calculation of this invariant and demonstration that it is nonzero.

Slightly refined version of these steps involve:

(a) defining of some kind of a chain complex generated by Reeb orbits;
(b) demonstration that the homology of this chain complex is a topological invariant;
(c) computation of this homology and demonstration that it is nontrivial.

For a prototype of the chain complex one can consider the Morse homology. We mention this fact only because we already have the results of our Appendix D.2.13 from which we can freely borrow. In particular, if $c_i(f)$ counts the number of critical points with index i, then Euler's characteristic $\chi(X)$ of the manifold X is defined by

$$\sum_{i=0}^{n}(-1)^i c_i(f) = \chi(X). \tag{9.5.4}$$

From here it follows that, provided that $\chi(X) \neq 0$, the number of critical points is bound by $|\chi(X)|$. But for 3-manifolds $\chi(X) = 0$ and such an estimate becomes useless. One way to go around this difficulty is to consider the equations describing the gradient flow.

This fact provides us with an opportunity to introduce to our readers ideas of gradient flow analysis in the physical context. These can be found in our papers [181, D.35]. In physics literature the gradient flow is known under many names, e.g. under the name "stochastic quantization" [404] or "critical dynamics", etc. Not surprisingly, exactly these methods are being used in the theory of spin glasses [405] and in adiabatic quantum computing (associated with/related to protein folding) [298, 299]. Again, not surprisingly, these methods can be reformulated in the supersymmetric language as it was first recognized by Witten in his seminal work [385] connecting supersymmetry and Morse theory. Deep down, the idea of all these methods can be traced back to the work by Ludwig Boltzmann on H-theorem as explained in our paper, [D.35].

Going back to mathematics, choose a metric g on X and use it to define the gradient vector field ∇f. If p and q are critical points, *a (downward) gradient flow from p to q, is a map $\gamma : \mathbf{R} \to X$* such that

$$\frac{d}{ds}\gamma(s) = \nabla f(\gamma(s)), \tag{9.5.5}$$

provided that $\lim_{s\to\infty}\gamma(s) = p$ and $\lim_{s\to-\infty}\gamma(s) = q$. Let $\mathcal{M}(p,q)$ denote the set of gradient lines: from p to q. For generic metric g, $\mathcal{M}(p,q)$ is a manifold of dimension

$$\dim \mathcal{M}(p,q) = ind(p) - ind(q). \tag{9.5.6}$$

This is so because the descending manifold $\mathcal{D}(p)$ is the set of points which can be reached by the downward flow starting at p and the ascending manifold $\mathcal{A}(q)$ is defined analogously. Thus, $\mathcal{M}(p,q) = \mathcal{D}(p) \cap \mathcal{A}(q)$. If $\mathcal{D}(p)$ is transverse to $\mathcal{A}(q)$, then Eq. (9.5.6) follows by dimension counting [17]. Under such circumstances it is convenient to introduce the *Morse–Smale pair* (f,g). For a given Morse function f and for generic metrics g, the pair (f,g) is Morse–Smale. It is convenient to introduce a related manifold $\mathcal{M}(p,q)/\mathbf{R}$ for which

$$\dim\{\mathcal{M}(p,q)/\mathbf{R}\} = ind(p) - ind(q) - 1.$$

In such a case if $ind(q) = ind(p) - 1$, then $\mathcal{M}(p,q)/\mathbf{R}$ is discrete. This fact is of central importance since it allows to introduce the Morse chain complex $\mathrm{C}_i^{\mathrm{Morse}}(X,f,g)$ as follows:

$$C_i^{\mathrm{Morse}}(X,f,g) := \mathbf{Z}\{p \in X \mid df_p = 0, ind(p) = i\}.$$

With help of this result the differential $\partial : C_i^{\mathrm{Morse}}(X,f,g) \to C_{i-1}^{\mathrm{Morse}}(X,f,g)$ is defined by counting gradient flow lines as follows: if p is an index i critical point, then

$$\partial p := \sum_{ind(q)=ind(p)-1} \#\frac{\mathcal{M}(p,q)}{\mathbf{R}}.q. \tag{9.5.7}$$

In this expression # denotes "signed count". To develop (Morse) homology it is essential to prove that $\partial^2 = 0$. This can indeed be done. For the pedagogical treatment of this issue at the level of this book, please consult [17, D.59]. To proceed, we need to introduce the Reeb orbits/flow and to repeat the above calculations. Since technically this is becoming rather involved, we only summarize the major achievements. It was demonstrated that solutions to the four-dimensional S-W equations on $\mathbf{R} \times \tilde{M}$ are obtained as gradient flow lines of the S-W functional $\mathcal{F}_{\mathrm{S-W}}$. The Morse homology (just outlined) for $\mathcal{F}_{\mathrm{S-W}}$ should now be replaced by the Seiberg–Witten–Floer homology (recall Sec. 9.1). Roughly speaking, the S-W-Floer homology is the homology of a chain complex which is generated by solutions of the S-W equations on $\tilde{M}$, and whose differential $\partial_{\mathrm{S-W-F}}$ counts number of solutions to the S-W equations on $\mathbf{R} \times \tilde{M}$. Details of this construction are extremely nontrivial and are discussed in detail in the monograph by Kronheimer and Mrowka, [19], entitled "Monopoles and Three-Manifolds". Thus, one way or another, we just have returned to the monopoles. Recall that we already discussed them in Chap. 5, [19] elevates this discussion to the levels accessible only to mathematicians at this moment. Furthermore, it connects monopoles with knots (Legendrian knots) in the way not seen in physics literature [406]. To correct this deficiency, we would like to introduce our readers to the Legendrian and transverse knots/links since they are by-products of contact geometry and topology. This is accomplished in the next section.

9.6 Quantum Money, Lagrangian, Legendrian and Transverse Knots and Links and the Associated Grid Diagrams

In the past 10–15 years mathematics of contact geometry and topology had been in the state of rapid and substantial growth. This growth thus far had not affected physical sciences, except the theoretical high energy physics where uses of contact geometry and topology have not been systematic. Since our book is about applications of contact geometry and topology in physics, it is unreasonable to describe mathematical developments in this field detached from physical context. In our opinion, a middle ground is reached by discussing the topic of quantum money. Such a selection enable us to bridge the gap between practical physics and abstract mathematics.

The issue of quantum money have been discussed recently by Farhi *et al.*, [407]. Popular exposition of these results along with the history of the topic can be found in [408]. From this reference we find that "Any digital good can be perfectly copied. This is a major headache for software companies (and for the entertainment industry), and is the reason why digital cash does not exist. The quantum mechanical "no cloning theorem" (e.g. read [C.4]) suggests that, at least in principle, it is possible to design quantum systems that cannot be copied. Several recent works proposed to use such type of systems for digital money". Among these the proposal by Farhi *et al.*, [407], seems the most promising. We would like to describe it briefly now making an emphasis on subject matters the most relevant to our book.

The mint (the facility for manufacturing money, say, coins) begins his work by generating a uniform superposition

$$|\text{initial}\rangle = \frac{1}{\sqrt{|\mathcal{B}|}} \sum_{e \in \mathcal{B}} |e\rangle|0\rangle, \tag{9.6.1}$$

where $\mathcal{B}$ is some pre-assigned large set. It then applies some function (that is operator) to e, causing the initial state to go to

$$\frac{1}{\sqrt{|\mathcal{B}|}} \sum_{e \in \mathcal{B}} |e\rangle|f(e)\rangle.$$

After that the mint measures the value of f. If the result of the measurement is v, then the residual state is going to be

$$|\$_v\rangle|v\rangle = \frac{1}{\sqrt{|\mathcal{N}_v|}} \sum_{\substack{e \in \mathcal{B} \\ f(e)-v}} |e\rangle|v\rangle. \tag{9.6.2}$$

Definition 9.6.1. The quantum money is the classical serial number v and the quantum state $|\$_v\rangle$.

To verify the money state, the verifier first have to measure the value of f on the given state $|\$_v\rangle$ to confirm that it indeed has value v. Then, it is required that the state is in a uniform superposition of states with the same value of v. The verification that the state is in uniform superposition is the main challenge for this process.

To implement this scheme, the authors use some facts from knot theory. Specifically, for $|\text{initial}\rangle$ they choose a uniform superposition over planar grid diagrams (to be defined below). The function f which the mint measures is the Alexander polynomial of the oriented link represented by a given planar grid diagram. The verification procedure is based on the

Reidemeister moves, which should not change the value of the polynomial since it is a topological invariant. In what follows, we shall assume that our readers are familiar with the basics of knot and link theory at least at the level of our review article, [64].

It is convenient to represent knots and links by planar grid diagrams. The same diagrams will be used for Legendrian and transverse knots/links (to be discussed below). A planar grid diagram G is $d \times d$ grid on which we place dX's and dO's. According to the rules, there should be exactly one X and one O in each row and in each column. No cell of the diagram can contain both X and O simultaneously. In each row a horizontal line between O and X is being drawn while a vertical line is being drawn in each column between the X and O. Whenever the horizontal and vertical lines intersect, the vertical line always crosses the horizontal line from above. Knots/links in the grid diagram can be assigned an orientation. E.g. each vertical edge goes from X to O, and each horizontal edge goes from O to X. The grid diagram G can then be specified by two disjoint permutations π_X and π_O so that the X's have coordinates $\{(i, \pi_X(i))\}$ and the O's have $\{(i, \pi_O(i))\}\, \forall\, i \in \{1, \ldots, d\}$.

Definition 9.6.2. Two permutations are said to be *disjoint* if $\forall\, i, \pi_X(i) \neq \pi_O(i)$. Any two permutations π_X and $\pi_O \in S_d$ define a planar grid diagram $G = (\pi_X, \pi_O)$.

With such rules, every knot/link can be represented by many different grid diagrams. Thus, the basis vectors in the Hilbert space can be encoded by the grid diagrams via identification: $|G\rangle = |\pi_X, \pi_O\rangle$. By construction, each diagram G will be encoded by the specific disjoint permutation as required. The dimension of the Hilbert space $d(G)$ is the number of elements in permutation. From combinatorics it follows that the number of disjoint pairs of permutations of d elements is equal to $d!$ times the number of permutations that have no fixed points. It is given by $\left[\frac{d!}{e}\right]$ where the symbol $[\ ]$ denotes the largest integer of the ratio $\frac{d!}{e}$.

To generate quantum money, the mint chooses a security parameter $\bar{D}$ which enters into (non-normalized) distribution

$$y(d) = \begin{cases} \frac{1}{d!\left[\frac{d!}{e}\right]} \exp\left\{-\frac{(d-\bar{D})^2}{2\bar{D}}\right\} & \text{if } 2 \leq d \leq 2\bar{D}, \\ 0 & \text{otherwise.} \end{cases} \tag{9.6.3}$$

Based on this definition, one defines an integer-valued function $q(d) = \left[\frac{y(d)}{y_{\min}}\right]$, where $y_{\min}$ is the minimum value of $y(d)$ for $d \in 2, \ldots, 2\bar{D}$. The mint then prepares the state (up to normalization)

$$\sum_{d=2}^{2\bar{D}} d!\sqrt{q(d)}|d\rangle$$

and uses an unitary transformation and the additional register to produce

$$\sum_{d=2}^{2\bar{D}} d!\sqrt{q(d)}|d\rangle \left(\frac{1}{d!} \sum_{\pi_X, \pi_O \in S_d} |\pi_X, \pi_O\rangle \right).$$

The next step is to measure whether or not the two permutations are disjoint. If the outcome is "disjoint" (it happens with probability e^{-1}), the mint computes the dimension d in the first register so that the initial state can now be explicitly obtained as

$$|\text{initial}\rangle = \frac{1}{\sqrt{N}} \sum_G \sqrt{q(d(G))}|G\rangle \tag{9.6.4}$$

with $(\sqrt{N})^{-1}$ being a normalization factor. If the measurement tells that permutations are "not disjoint", the mint starts the process over. The distribution $q(d)$ is selected in such a way that if one were to measure $d(G)$ on the $|\text{initial}\rangle$, then the distribution of results is going to be very close to the Gaussian with mean $\bar{D}$ restricted to integers $2 \leq d \leq 2\bar{D}$. Using the $|\text{initial}\rangle$ the mint computes the Alexander polynomial $A(G)$ into another register and then measures this register, thus obtaining the polynomial p. The resulting state $|\$_p\rangle$ is given by

$$|\$_p\rangle = \frac{1}{\sqrt{N}} \sum_{\substack{A(G)=p \\ G}} \sqrt{q(d(G))}|G\rangle. \tag{9.6.5}$$

Using Definition 9.6.1 we recognize that the quantum money now consists of $|\$_p\rangle$ and the polynomial p represented by the set of its coefficients. If the polynomial p happens to be zero (e.g. for splitable links), the mint should reject that state and start over. By omitting the rest of computational details which can be found in [407] we would like to focus on the main issue: What is it in the above scheme that makes quantum money so hard to forge? The main task for the potential attacker is to be able to solve the grid diagram equivalence problem, that is the problem of recognizing whether or not two grid diagrams represent the same link. Even for the quantum computers the authors believe that it is going to be difficult. In fact, the problem of deciding whether or not a grid diagram represents the unknot is already believed to be hard.[x]

Calculation of the Alexander polynomial for knots and links is nicely described in the books by Dale Rolfsen, [409] and Lou Kauffman, [410].[y] However, these books do not include information on Legendre and transverse knots/links or on other knot/link polynomials subsequently discovered. This gives us an opportunity to discuss in this section the Lagrangian, Legendre and transverse knots in the context of quantum money and then in the context of other physics-related problems. Clearly, such a discussion is aimed at physics educated readers as is our review, [64].

Thus, we begin with some facts about the Lagrangian knots. Typically, these are not discussed in mathematics literature on contact geometry and topology. They are discussed, nevertheless, in the review paper by Arnol'd, [411], which we would like to follow. In Sec. I he writes: "... definition of a knot starts from standard straight line in 3-space. A knot is represented by an embedding which differs from the standard one only in a finite ball". From here one obtains the following (recall results of Appendix D.2.17).

Definition 9.6.3. A Lagrangian knot type is a connected component of the space of Lagrangian embedding of $\mathbf{R}^n$ into standard symplectic space $\mathbf{R}^{2n}$, differing from the embedding of a fixed Lagrangian plane only inside finite balls.

Then, Arnol'd is raising a question:

Do there exist nontrivial Lagrangian knot types? The trivial type is that containing the standard Lagrangian plane. He then gives a negative answer referring to works by Eliashberg, Polterovich and Luttinger.

Theorem 9.6.1. *All Lagrangian knots in the standard symplectic 4-space are trivial in the sense of differential topology.*

Remark 9.6.1. For higher dimensions the answer to Arnold's question is yet unknown. It is appropriate to recall that all traditionally defined knots are trivial in four dimensions [199].

Having finished with Lagrangian knots we are ready to discuss the Legendrian and transverse knots.

[x] This problem was in fact recently solved by Kronheimer and Mrowka. See Sec. 9.7.4 below.

[y] Less conventional computation of the Alexander polynomial can be easily computerized is given in our work, [398].

Let $\tilde{M}^3$ be a contact 3-manifold with α being a contact one-form so that the contact structure ξ on $\tilde{M}^3$ is given by $\xi = \ker \alpha$. It is given by the set of hyperplanes, as in Fig. 4.

Definition 9.6.4. A Legendrian knot K_L in a contact manifold $(\tilde{M}^3, \xi)$ is an embedded S^1 that is always tangent to ξ.

Definition 9.6.5. A transverse knot K_T in a contact manifold $(\tilde{M}^3, \xi)$ is an embedded S^1 that is always transverse to ξ.

Recall that the usual knots known in physics literature, which from now on we shall call as "*topological knots*", are embeddings of S^1 into S^3 (that is into contact manifold!) without specification of tangency or transversality. Naturally, the Legendrian and transvere knots are related to topological knots. We would like now to make this relationship more precise.

Before doing so, it is instructive to make a digression and to look at the higher-dimensional analogs of knots we just defined following Etnyre, [412], whose treatment of such knots is restricted by description of contact structures on $\mathbf{R}^{2n+1}$ only. Such a restriction is not too severe in view of the Darboux theorem for contact manifolds [139] which is locally establishing that the 1-form α on $(2n+1)$-dimensional contact manifold $\tilde{M}$ can be brought (using contactomorphisms) to the standard form

$$\alpha = dz - \sum_{j=1}^{n} p_j dx_j$$

appropriate for $\mathbf{R}^{2n+1}$.

As in dimension three, where there are two standard projections onto $\mathbf{R}^2$, in higher dimensions there are also two projections that are useful for study of Legendrian knots. The first one is the *Lagrangian projection*:

$$\pi : \mathbf{R}^{2n+1} \to \mathbf{C}^n : (x_1, p_1, \ldots, x_n, p_n, z) \to (x_1, p_1, \ldots, x_n, p_n) \qquad (9.6.6a)$$

and the second one is the *front projection*

$$\Pi : \mathbf{R}^{2n+1} \to \mathbf{R}^{n+1} : (x_1, p_1, \ldots, x_n, p_n, z) \to (x_1, \ldots, x_n, z). \qquad (9.6.6b)$$

From Appendix D.2.18, we recognize at once that Eqs. (D.137b) and (9.6.6b) are the same. This explains why such knots are called Legendrian. Clearly, they can be looked upon as fiber bundles with $(x_1, \ldots, x_n, z)$ being bases. The projection which is given by Eq. (9.6.6a) is called *Lagrangian* since exactly such projection was ingeniously used by Maslov (Sec. 9.2) instead of the standard Lagrangian projection given by $(x_1, \ldots, x_n)$, e.g. see Eq. (D.137a).

Thus, if K_L is the embedded Legendrian knot, then $\pi(K_L)$ is the immersed Lagrangian submanifold of $\mathbf{C}^n$. The embedding of K_L into $\mathbf{R}^{2n+1}$ can be recovered (up to translation in the z-direction) from $\pi(K_L)$ as shown in [412]. Specifically, pick a point $p \in \pi(K_L)$ and choose any z-coordinate for p. Then, z-coordinate for any other point p' is determined by

$$\sum_{j=1}^{n} \int_{\gamma} p_j dx_j, \tag{9.6.7}$$

where γ is any path in $\pi(K_L)$ from p to p'. If the integral $\int_\gamma$ is independent of the path γ, then the embedding becomes an immersion into $\mathbf{R}^{2n+1}$. But the integral $\int_\gamma$ is independent of the path γ when the Lagrangian immersion $i : K_L \to \mathbf{R}^{2n}$ is exact. The obtained results demonstrate the sub-Riemannian flavor of Legendrian knots.

After this digression, we are ready to discuss what the Legendrian and transverse knots have in common with topological knots in S^3. In doing so we follow [412, 413]. We highly recommend to our readers to read both of these references for more details.

We begin with the standard 1-form α given by Eq. (5.2.7). Let the knot K_L (Legendrian) be parametrized by the parameter $\theta : 0 \leq \theta \leq 2\pi$. Thus if $\alpha = dz - pdx$ and if

$$\theta \to (x(\theta), p(\theta), z(\theta))$$

is the map $S^1 \to \mathbf{R}^3$, then the Legendrian knot K_L is given by the equation

$$\frac{dz}{d\theta} - p(\theta)\frac{dx}{d\theta} = 0. \tag{9.6.8}$$

As it is done with topological knots, we have to project K_L into $\mathbf{R}^2$. This can be done either with the Lagrangian or with the front projection. Let us describe the front projection first. In this case we should have $\Pi : (x, p, z) \to (x, z)$. In view of Eq. (9.6.8) we can recover $p(\theta)$ by using the equation $p(\theta) = \frac{dz}{dx}$. This is possible as long as $\frac{dx}{d\theta} \neq 0$. However, instances when $\frac{dx}{d\theta} = 0$ should not be excluded. They give rise to *cusps* in the projection. This feature is absent for topological knots. To cope with the cusps, the following rules for drawing knot diagrams are implemented:

1. the diagram should have no vertical tangencies;
2. the only non-smooth points are horizontal cusps;
3. at each crossing the slope of the overcrossing strand is smaller than undercrossing.

If for the topological knots there are Reidemeister moves, for the Legendrian knots there are their analogs as well.

Definition 9.6.6. For any topological knot, there are Legendrian knots representing it. The projection of the Legendrian knot (that is its front diagram) can be obtained from that for the topological knot by adding cusps and zig-zags according to the above three rules.

Each topological knot belongs to exactly one knot type. These are being described as C_i where C is the crossing number and i is an index within the class of knots with the same crossing number. The mirror knots (e.g. see Fig. 1) are accounted by writing instead $m(C_i)$. Classification of topological knots is well described in Rolfsen's book, [409]. Since for each topological knot there is more than one Legendrian knot, the topological knot type C_i is invariant of Legendrian knots associated with the given topological knot. We shall write such an invariant as $k(K_L)$.

Definition 9.6.7. With $k(K_L)$ being defined, let $\mathcal{L}(K_L)$ denote the set of all Legendrian knots having the same topological invariant $k(K_L)$.

This number is just one out of three classical Legendrian invariants making the Legendrian knots/links much more likely candidates for quantum money projects. Two other topological invariants are: the Thurston–Bennequin number $tb(K_L)$ and the rotation number $r(K_L)$. These topological invariants could be valuable for description of circular DNA and proteins [64, 414]. This is so because of the following.

Let D_{K_L} denote the front projection diagram for K_L. Then, the $tb(K_L)$ can be defined as

$$tb(K_L) = wr(D_{K_L}) - c(D_{K_L}), \tag{9.6.9}$$

where $wr(D_{K_L})$ is the writhe for the diagram D_{K_L} for oriented knot, e.g. see [64, 409, 410], while $c(D_{K_L})$ is half of the cusps in D_{K_L}. The rotation number $r(K_L)$ is defined analogously as

$$r(K_L) = \frac{1}{2}(c_{\downarrow}(D_{K_L}) - c_{\uparrow}(D_{K_L})), \tag{9.6.10}$$

where $c_{\downarrow}(D_{K_L})$ is the total number of downward cusps (these are oriented cusps in which the orientation (of the arrow) is down) and, accordingly, $c_{\uparrow}(D_{K_L})$ the same for the upward cusps. These results can be made physically more relevant if, following [412], we shall use the Lagrangian projection instead of the front projection.

Definition 9.6.8. The Lagrangian projection π of a Legendrian knot is given by

$$\pi(K_L) : \mathbf{R}^3 \to \mathbf{R}^2 : (x, p, z) \to (x, p).$$

For $\pi(K_L)$, we obtain: $d\alpha \mid_{x,y\in\mathbf{R}^2\in\pi(K_L)} = 0$ in accord with Eq. (D.129).

Two Lagrangian projections of the same knot can, in principle, be brought in correspondence with the help of sequence of Reidemeister moves. This can be done only in principle since in practice some other moves need to be introduced which will be discussed in the sequel.

In the meantime, we notice that it can be proven that using Lagrangian projection $\pi(K_L)$, instead of Eq. (9.6.9) we obtain,

$$tb(K_L) = wr(\pi(K_L)) \tag{9.6.11}$$

while for the rotational number Eq. (9.6.10) is being replaced by

$$r(K_L) = win(\pi(K_L)). \tag{9.6.12}$$

Evidently, win = "winding" means the same as "twisting".

Remark 9.6.2. In such a case we can think about the relevance in this context of Pohl's formula frequently used in calculations involving DNA and proteins [64, 410]. We would like to warn our readers against hasty uses of this formula in the present case.

Let us go back now to Definition 9.6.5 of the transverse knot. As before, let $\theta \to (x(\theta), p(\theta), z(\theta))$ be parametrization for the transverse knot K_T. By analogy with Eq. (9.6.8), we have now (for the positively oriented K_T)

$$\frac{dz}{d\theta} - p(\theta)\frac{dx}{d\theta} > 0. \tag{9.6.13}$$

The front projection is defined as before: $\Pi : (x, p, z) \to (x, z)$. The same is true for the Lagrangian projection. These results allow us to introduce yet another

Definition 9.6.9. The self-linking number sl of the transverse knot K_T is given by

$$sl = wr(\Pi(K_T)). \tag{9.6.14}$$

This result is helpful to compare with that given by Eq. (9.6.11).

Definition 9.6.6. can be extended now as follows.

Definition 9.6.10. (Stabilization) Given a Legendrian knot K_L there is a way to obtain another Legendrian knot which belongs to $k(K_L)$ using *stabilization*. This is achieved by removing a line segment and replacing it with the line with zig-zag. The zig-zag can be oriented: going from top-to-bottom (down) or going from bottom-to-top (up). Analogous procedure exists for Lagrangian projection. In this case it visually reminds that for the topological knots. For instance, one can replace an ordered segment with the up(down) curl. The stabilization is called "*positive*" and is denoted as $S_+(K_L)$ if down cusps (curls) are added and "*negative*" $S_-(K_L)$ in the opposite case. From here it can be demonstrated that

$$tb(S_\pm(K_L)) = tb(K_L) - 1 \tag{9.6.15a}$$

and

$$r(S_\pm(K_L)) = r(K_L) \pm 1. \tag{9.6.15b}$$

Theorem 9.6.2. *The topological classification of knots is equivalent to stable classification of Legendrian knots*

Proof. Read [412].

Remark 9.6.3. For transversal knots there is analogous result. The very existence and use of these knots should remind physically educated reader about existence of dual lattices in the solid state physics or the high and low temperature expansions on dual lattices, e.g. for the Ising model, etc. Sometimes it is much easier to prove some result for the transverse knots than for Legendrian ones and vice versa.

Remark 9.6.4. In the previous section we noticed the role of foliations in topological arguments. The same is true in the present case as well. Read [139, 412]. Uses of foliations can enlarge the number of potential physical applications considerably. We leave this task for our readers to develop (going into Farady's footsteps).

While comparing and classifying the Legendrian knots it is useful to know about the Bennequin inequality (sometimes it is also known as the Thurston–Bennequin inequality). It is being phrased with the help of Definition 9.6.7 as follows:

$$\overline{tb}(k(K_L)) = \max\{tb(K_L) \mid K_L \in \mathcal{L}(K_L)\}. \tag{9.6.16a}$$

By definition $\overline{tb}(k(K_L))$ is the knot invariant. The analogous result exists for the transverse self-linking number. For topological knots the analogous estimate was made in our work, [415]. Somewhat similar arguments were recently used by Ng, [416], who by using the Khovanov homology obtained an estimate for tb analogous to that by Bennequin. Self-contained exposition of the Khovanov homology is given in an excellent review paper by Lou Kauffman, [417].

The Khovanov homology associates to any oriented knot/link K a bigraded Abelian group $\mathrm{HKh}^{*,*}(K)$, where the first grading is called the quantum (or Jones) and the second is the homological grading. If the torsion is disregarded, the two-variable Poincaré polynomial (Khovanov polynomial) is given by

$$Kh_K(q,t) = \sum \dim(Hkh^{i,j}(k) \otimes \mathbf{Q})q^i t^j. \tag{9.6.17}$$

This result is remarkably similar to the bigraded Poincaré polynomial, Eq. (6.1), obtained in our work, [418], in connection with classification of Veneziano amplitudes.

Exercise 9.6.1. Demonstrate that such a coincidence is not completely accidental. Hint: Read [379], especially those sections which discuss Veneziano amplitudes in the knot-theoretic context. Also notice that connections between the Veneziano amplitudes and topological field theories was established in our work, [419].

Using the above Poincaré polynomial, Ng obtained the following estimate

$$\overline{tb}(K) \leq \min\{\deg_q Kh_K(q,t/q)\} \tag{9.6.16b}$$

which compares well with previously obtained Eq. (9.6.16a).

Remark 9.6.5. For a given knot/link K the polynomial $Kh_K(q,t)$ can be calculated by the Mathematica package, [420].

Remark 9.6.6. In connection with just discussed results and the content of Theorem 9.6.2 we have to notice the following. From [412] we find that the process stabilization and destabilization of knots/links is not symmetric. It is much easier to stabilize knots/links than to destabilize them. The process of (de)stabilization can be checked by the above inequality in view of Eq. (9.6.15a). They say that the knot K_L destabilizes if there is another knot K'_L such that $K_L = S_{\pm}(K'_L)$.

The above results can be complicated by systematic use of the connected sum operation # as it was also mentioned in [64]. For instance,

$$tb(K_{L_1} \# K_{L2}) = tb(K_{L_1}) + tb(K_{L_2}) + 1$$

and

$$r(K_{L_1}\#K_{L2}) = r(K_{L_1}) + r(K_{L_2}).$$

Etnyre and Honda proved the following

Theorem 9.6.3. *Let $K = K_1\#\cdots\#K_n$ be connected sum of knots in $\boldsymbol{R}^3$ with K_i being prime type of knot. The map*

$$[\mathcal{L}(K_1) \times \cdots \times \mathcal{L}(K_n)]_{\sim} \to \mathcal{L}(K_1\#\cdots\#K_n),$$

given by connected sum is a one-to-one correspondence where $\sim$ is generated by either (1) *permutations or* (2) *destabilizations (for the associated Legendrian knots) of the type*

$$(\ldots, S_{\pm}(K_{Li}), \ldots, K_{L_j}, \ldots) \to (\ldots, K_{Li}, \ldots, S_{\pm}(K_{Lj}), \ldots).$$

Corollary 9.6.3. *Given any integers n and m there is a topological knot type $k(K_L)$ and Legendrian knots $K_{L1}, \ldots, K_{Ln} \in \mathcal{L}(K_L)$ with the same tb and r that are not Legendre isotopic even after m stabilization of any type.*

Such Legendrian knots are impossible to distinguish with help of the knot invariants already listed. Therefore, in addition to those invariants that are listed, a number of new invariants was invented. We shall discuss them in the next section even though this problem is still not completely solved. In the meantime, we would like to discuss how all these results are related to the grid diagrams.

For the grid diagram G of size $d \times d$ there is a minimal number d, called the *arc index* $\alpha(K)$, which is the minimal grid number over all grid diagrams that represents a given knot/link K. By construction, $\alpha(K)$ is a topological invariant of K. Grid diagrams can be used both for topological knots as well as for Legendrian and transverse knots. *Since this feature could considerably enhance the security of quantum money scheme outlined above, it is worth discussing the computational details of the knot/link recognition scheme applied to the Legendrian knots/links.*

There are three types of moves on the grid diagrams known as *Cromwell moves.* These are lattice equivalents of the Reidemeister moves. Thus, we have

Definition 9.6.11. (Cromwell moves [421, 422]) There are three types of moves on the grid diagram. These are: translation, commutation and stabilization/destabilization.

Remark 9.6.7. Evidently, use of stabilization/destabilization is needed only if, instead of topological knots/links, we are going to study Legendrian knots/links.

Translation consists of vertical and horizontal translations. Vertical translation moves the top most row of the grid diagram to the bottom of the diagram or vice versa while leaving the rest unchanged. Horizontal translation, by analogy, moves the leftmost column of the diagram to the rightmost and vice versa.

Remark 9.6.8. Translations just defined are topologically equivalent to the conversion of the planar grid diagram into that for a torus.

Commutation interchanges two adjacent rows or two adjacent columns, provided the two conditions are satisfied. These are:

1. To commute two rows (or columns), four X's and O's in those rows (columns) must lie in different columns (rows).
2. Line segments connecting X's and O's in the rows (or columns) must be either nested, or disjoint.

Stabilization increases the grid number by 1 and replaces a single $X(O)$ in the diagram by 2×2 square with two X's (O's) and one $O(X)$. *Destabilization* is just the process inverse to stabilization. There are eight types of stabilization. These are: X:NW, X:NE, X:SW, X:SE, O:NW, O:NE, O:SW and O:SE. The first part corresponds to the letter in 1×1 subgrid while the second to the position in the 2×2 box. For instance, the NW means "North–West", etc. It is possible to write any O (de)stabilization as a composition of translations, commutations and exactly one X (de)stabilization located at the diagonally opposite position (e.g. NE is diagonally opposite to SW). Thus, it is sufficient to consider (de)stabilization by X's.

Thus defined Cromwell moves allow to be narrowed down to the Legendrian or transverse knots/links. Specifically,

Legendrian links are grid diagrams modulo torus translation, commutation, and X:NE and X:SW stabilization and destabilization moves.

Transverse links are grid diagrams modulo torus translation, commutation and X:NE, X:SW and X:SE stabilization and destabilization moves.

In such a setting, an enumeration of the Legendrian or transverse links becomes a problem of enumeration of the grid diagrams up to the appropriate equivalence relation.

Definition 9.6.12. Two grid diagrams represent isotopic topological links (knots) if and only if they are related by some sequence of Cromwell moves.

Under these circumstances the equivalence classes for Legendrian knots are in one-to-one correspondence with the equivalence classes for transverse knots (modulo negative stabilization). This allows us to look at, say, only the transverse knots/links which are in a certain sense the *transverse pushoffs* of the Legendrian knots/links. More specifically, following [424] we apply X:NE and X:SW stabilization and destabilization moves respectively to the Legendrian knots K_L. Since such moves change (tb, r) by $(-1, 1)$ and $(-1, -1)$ respectively, we obtain: $S_+S_-(K_L) = S_-S_+(K_L)$. Therefore, *the transverse knots can be seen as Legendrian knots modulo negative Legendrian stabilization.* The X:NW stabilization descends to the transverse knots, where it is called *transverse stabilization*. It decreases the sl by 2. The destabilization increases tb by 1 (for Legendrian knots) and sl by 2 (for transverse knots). Since these invariants are bounded from above (for any topological type), there are knots/links which cannot be further destabilized! All these properties are useful when making programs of knot recognition. Here are a few additional properties worth mentioning.

Given an oriented Legendrian knot K_L by reversing an orientation, one obtains another knot/link $-K_L$, e.g. see Fig. 1. Such an operation corresponds to the switching between X's and O's in a grid diagram. It also replaces (tb, r) by $(tb, -r)$. The orientation-reversal affects stabilization: $S_+(-K_L) = -S_-(K_L)$. The *Legendrian mirror* $\mu(K_L)$ is the end result of application of contactomorphism $(x, p, z) \to (-x, p, -z)$ to K_L. Such an operation is equivalent to rotating the grid diagram by 180° followed by replacing (tb, r) by $(tb, -r)$. The combination of these two symmetries descends to transverse knots where it is called the *transverse mirror*.

The description of computer programs implementing the above rules can be found at

http://www.math.duke.edu/˜ng/atlas/

Our readers should be warned that with all sophistications present in the above programs they are unable to unambiguously recognize some knots/links. This is so because the processes of stabilization–destabilization changes the size of the grid diagram so that the existing algorithms are sometimes unable to prove in finite time that two grid diagrams represent the same Legendrian or transverse knot/link. The recognition process involves uses of Cromwell moves, and of "classical" invariants such as Thurston–Bennequin tb, rotation r and self-linking sl numbers. However, as it is demonstrated in the review by Etnyre, [412], sometimes it becomes necessary to use other "non-classical" invariants. These are going to be discussed briefly in the next section. Before doing so, it is of interest to further discuss the design of the recognition program a little bit for reasons which will become obvious upon reading.

The simplest computer knot recognition program using grid diagrams is described in [424]. The computer program presented in the weblink given above is refinement of that given in [424]. The program in [424] ignores knot/link orientation. In the grid diagram this is reflected by the fact that X's and O's now lose their individuality. They can both be replaced by 1's while the rest of places on the same row/column is occupied now by 0's. Such an action produces the *Cromwell matrix* (made of 1's and 0's). It has exactly two 1's in each row or column. Our readers are encouraged to take a look at such a matrix (Fig. 2 of [424]). Evidently, there is a one-to-one correspondence between the nonoriented knots/links and Cromwell matrices just defined. Once this is recognized, it is possible to implement Cromwell moves using such matrices, even including stabilization.[z] Use of these operations on Cromwell matrices could result in the situation when two 1's are next to each other either in the same row or in the same column. In such a case the simplification is possible: a Cromwell $n \times n$ matrix can then be reduced to $(n-1) \times (n-1)$ as depicted in Fig. 3 of [424]. The rows/columns with such adjacent 1's are being deleted. This process can be continued until it stops at some step so that the resulting Cromwell's matrix becomes *irreducible*. Surely, it could be irreducible at the very beginning of this process. *Thus, there is one-to-one correspondence between the irreducible grid diagrams and irreducible Cromwell matrices.* This fact is of central importance. It can be further amplified by introducing the norm of the Cromwell matrix as follows. Say, let the Cromwell matrix have the following look

$$\begin{bmatrix} 1 & 0 & 0 & 1 & 0 \\ 0 & 1 & 0 & 0 & 1 \\ 1 & 0 & 1 & 0 & 0 \\ 0 & 1 & 0 & 1 & 0 \\ 0 & 0 & 1 & 0 & 1 \end{bmatrix}.$$

Then, the norm is given by 1001001001101000101000101. That is to create this norm we just placed next to each other the rows of the above matrix. Thus constructed norm is reminiscent to that used in analysis of

[z]E.g. read [413], p. 18.

genetic sequences in biology or sequences of Dehn twists in topological dynamics, e.g. as discussed in [398]. From Appendix C we know how it can be used in classical computers. In the case of genetic sequences a sequence in question is being compared against some known (reference) one. Something of this sort was implemented in [413]. In it, the path-searching algorithm was implemented. The idea can be described as follows. Each grid diagram is represented by a vertex of some graph. Two vertices are connected if they represent the grid diagrams that related to each other by just one (Cromwell) move. For nondirected grid diagrams, each such move is reversible. Thus, to determine whether two grid diagrams represent the same (nonoriented) Legendrian knot/link is equivalent to determining whether there is a path between two given vertices on a graph just described. It is actually having infinite number of vertices. Use of Grassmanians (Appendix D.2.8–D.2.10) could be appropriate for such a problem. Till now they had been used in problems related to computer vision/image recognition, e.g. read [47, 425]. Clearly, these methods had not been used in [413]. They should be taken into account in the future developments of programs of knot recognition. It could also be that knot recognition programs could be helpful for recognition of images/computer vision.

9.7 Latest Developments in Contact Geometry and Topology. A Guided Tour with Physics Applications in the Perspective

9.7.1 *General Information*

In the previous section we mentioned that three "classical" knot invariants: tb, r and sl are unable to describe/recognize all types of Legendrian (or transverse) knots/ links. As a result, a number of new knot/link invariants was suggested and the process of finding new invariants is far from being complete. Since this is a book about applications of contact geometry and topology in physics, it is unwise to duplicate mathematical papers describing these new invariants in detail. Instead, we are offering a "guided tour" accompanied by our comments about potential usefulness of selected mathematical works to physics.

We begin our tour by making some quotations from Arnol'd review paper, [411], p. 3331. "The existence of symplectic topology[aa] was first

[aa]And contact topology, as it is described in the paper.

appreciated by the mathematical community when the characteristic class entering into the quantization conditions was discovered. I christened it the *Maslov class* in [30]"[bb] Arnol'd continues by saying that the terminology "*Lagrangian manifold*" came also from Maslov's work. Arnol'd writes that "The Maslov class associates to any closed curve on a Lagrangian submanifold an integer (its Maslov index), which depends only on the homology class of the curve. The Maslov index of a curve is the intersection index of the curve with the caustic of the projection of the Lagrangian submanifold to the base space of the fibration (to the configuration space)".[cc]

Remark 9.7.1. Thus, all theories of Legendrian knots/links have their origin in quantum mechanics and theory of wave propagation. [348, 358] contain many examples of these interrelations. In particular, [358], written in 1995, contains four conjectures which were subsequently proven by Chekanov and Pushkar [426] in 2005. As a by-product of their work, a couple of new knot invariants were presented. These are worth discussing. We shall explain why this is so momentarily. Below, we follow [412, 413, 427] summarizing results of [426].

9.7.2 *Solution of the Planar Ising Model on the Grid Diagram and Chekanov–Pushkar–Eliashberg Invariants*

Although the exact solution of two-dimensional Ising model was originally obtained by Onsager in 1947, we follow Vdovichenko whose solution of the Ising model is described in Landau and Lifshitz volume on statistical physics, [317]. For completeness of our presentation, we reproduce here some facts about the two-dimensional Ising model.

We begin with the $n \times n = N$ grid as in previous section. Unlike the previous section, where X's and O's were selectively placed on the grid, this time the classical spins $\sigma_{kl} = \pm 1$ are being placed on each lattice site. The total energy $E(\sigma)$ of spin–spin interactions is given by

$$E(\sigma) = -J \sum_{k,l=1}^{n} (\sigma_{kl}\sigma_{kl+1} + \sigma_{kl}\sigma_{k+1l}), \tag{9.7.2.1}$$

[bb] In [30] Arnol'd introduced the Lagrangian Grassmannian (Appendix D.2.17) needed for computation of the Maslov index.
[cc] E.g. see Eq. (9.6.6b).

where $J > 0$ is some parameter. If the inverse temperature is β, the partition function $Z(\beta)$ can be defined as

$$Z(\beta) = \sum_{\{\sigma\}} \exp(-\beta E(\sigma)). \tag{9.7.2.2}$$

In this expression the summation $\sum_{\{\sigma\}}$ is taken over all 2^N possible spin configurations. To facilitate calculations, it is useful to notice that

$$\exp(\theta\sigma_{kl}\sigma_{k'l'}) = \cosh\theta(1 + \sigma_{kl}\sigma_{k'l'}\tanh\theta).$$

With help of this result we can rewrite $Z(\beta)$ as follows:

$$Z(\beta) = (1 - x^2)^{-N}\Sigma(\beta), \tag{9.7.2.3a}$$

where $x = \tanh\theta$, $\theta = \beta J$,

$$\Sigma(\beta) = \sum_{\{\sigma\}} \prod_{k,l}^{n} (1 + x\sigma_{kl}\sigma_{kl+1})(1 + x\sigma_{kl}\sigma_{k+1l}). \tag{9.7.2.3b}$$

By expanding this product and summing individual terms over lattice spins we obtain the collection of nonoriented grid diagrams, exactly like those described in [424]. Vdovichenko found ingenious way to organize these diagrams and, it happens, that Chekanov and Pushkar came up independently with exactly the same ideas. We shall discuss these ideas from several directions.

First, we follow Chekanov's paper, [427], in order to present statement of the problem from the mathematical perspective. Chekanov writes: "It is easy to show that every smooth (that is topological) knot admits a Legendrian realization." This means that a generic set of front diagrams $\pi(K_L)$ is going to be made out of cusps and transverse double points required to be linked with each other without loose ends.

Remark 9.7.2.1. This construction is completely analogous to that one encounters when one has to construct all Feynman's diagrams out of a given set of vertices [389, 428].

Next Chekanov writes: "A natural question to ask is whether there exists a pair of Legendrian knots which have the same classical invariants but are not Legendrian isotopic. The answer is positive. In [426] such a pair of Legendrian knots (smooth knot type 5_2) was found explicitly. Their topological and classical invariants are the same but they are not Legendrian isotopic (recall Corollary 9.6.3). To deal with this situation, new invariants were suggested. We shall discuss one of them in this subsection and another in the next.

Before doing so, we would like to place these results into the proper perspective. Colin, Giroux and Honda discovered [427] that the number of Legendrian knots with a given classical invariants is finite. Such knots/links are called *Legendrian simple* since their Legendrian isotopy types are determined by the classical invariants. These results could be of great help for quantum money project, provided that the atlas of Legendrian simple knots is compiled.

Discussion of the first invariant requires us to introduce the concept of *ruling*. Given front diagram of a given Legendrian knot *the ruling is defined as one-to-one correspondence of left and right cusps, together with the decomposition of the front diagram into a union of pairs of paths, each connecting the corresponding pair of left and right cusps, and satisfying the following conditions*:

1. All paths are smooth, except possibly at crossings. The paths are always going from left to right or from right to left (that is orientation is unimportant).
2. The two paths corresponding to the same pair of cusps never intersect, except at the cusps.
3. At each crossing where two paths intersect there could (or could not) be a *switch*.

A switch is defined by the rules (to be defined momentarily) by which each crossing is being resolved (very much like those used for converting a topological knot diagram into collection of Seifert circles). To visualize rulings, we suggest our readers to go to

http://www.haverford.edu/math/jsabloff/Josh_Sabloff/Research_files/aim-intro.pdf

and then to p. 12 (rulings of front diagrams). Using the same reference, we recognize that the cusps can always be smoothened so that we can always recover a projection of the topological knot into $\mathbf{R}^2$.

Remark 9.7.2.2. In the case of two-dimensional Ising model, the expansion and summation over spins at each lattice site will create the nonoriented grid diagrams. Cromwell moves put these diagrams on the torus (as explained in previous section) and the same torus topology is also used for the Ising model.

Remark 9.7.2.3. The rules by which the switches at crossings are recognized are exactly the same as those established by Vdovichenko for the

Ising model and lead to the same results. Indeed, correctly resolved set of crossings, called *Maslov switches*, leads to the decomposition of the nonoriented knot diagram into the set of nonoverlapping circles/discs [412, 427].

Finally, we define switches as those crossings which are resolved as bends touching each other at one point. There are two ways of making such bends out of a given crossing. Then the two remaining options — to go straight at the crossing — cannot be qualified as switches.

Remark 9.7.2.4. Thus, Maslov's switches are, in fact, Vdovichenko's switches since for the Ising model the rules are *exactly the same* [317] but were obtained in 1967!

Mathematically rigorous presentation of Vdovichenko's work is given in [429, 430]. Usefulness of statistical mechanics for designing knot/link invariants (for *topological knots*) was recognized some time ago [64, 431]. Extension of ideas and methods of statistical mechanics and topological field theory for designing invariants of Legendrian knots/links was initiated in 1999 in seminal paper by Eliashberg, Givental and Hofer [432]. This paper could be considered as substantial extension of earlier work by Witten, [433], on topological field theory since it describes the Legendrian and transverse knots/links and their invariants — feature non-existent in previous field-theoretic treatments.

From the above definitions, it follows that a particular ruling is determined by a particular set of Maslov switches. It can be demonstrated that the set of all rulings is invariant of the Legendrian knot. This result can be refined as follows. Let F_{K_L} be the front diagram of a Legendrian knot. Let R be the ruling for F_{K_L}, then $\#(R)$ is the total number of left (or right) cusps and $S(R)$ is the total number of Maslov switches. Define the number $\gamma(R) = \#(R) - S(R)$. It is a *complete ruling invariant* of the Legendrian knot K_L.

Definition 9.7.2.1. Following Chekanov, the Maslov number $m(K_L)$ of the Legendrian knot is twice the rotation number $r(K_L)$ (e.g. see Eq. (9.6.10)).

These results can still be refined with the help of just defined $m(K_L)$. Details are given in [412, 413].

9.7.3 *Several Remarks about Chekanov–Eliashberg Differential Graded Algebra (DGA) and Symplectic Field Theory*

The differential graded algebra (DGA) was announced by Chekanov in [427]. It was also independently developed by Eliahberg [434]. In [435]

Chekanov provided full details of DGA allowing us to make connections with results of Sec. 9. Indeed, from [435] we find that: "Differential graded algebra of a Legendrian link in $\mathbf{R}^3$ is a particular case of more general Morse-theoretic construction[dd] which associates with every generic Legendrian submanifold L in a contact manifold $(\tilde{M}, \alpha) \ldots$ a differential graded algebra whose generators are trajectories of the Reeb flow starting and ending on L. The differential is defined in terms of J-holomorphic curves in the symplectization of $(\tilde{M}, \alpha)$.[ee] In terms of J-holomorphic curves DGA is developed in the already cited [432] so that Chekanov's work provide some kind of a simplification of the formalism of symplectic field theory developed in [432]. As it is explained by Etnyre [412], DGA by Chekanov is part of more general theory [432] which is useful in its own right since it is capable of producing new topological invariants (e.g. Chekanov–Poincaré polynomial, etc.) which could be obtained in much harder way employing symplectic field theory and Heegaard–Floer homology to be discussed in the next section. Because the DGA is having its origins in the "Morse-theoretic construction" not surprisingly that it uses elements of Floer's theory [354] (Sec. 9.1) for crucial proofs (e.g. read [427]). Thus, one can find in DGA the homological expressions similar to Eq. (9.5.7) originating in S-W–Floer homology.

Since Sec. 4 of [412] provides a nice summary of Chekanov's paper [435], it allows us to skip details altogether. For readers interested in these details, we recommend Sec. 2 of [418], for algebraic background needed to understand Sec. 4 of [412]. The fact that a noncommutative associative graded $\mathbf{Z}_2$ algebra is used in these calculations is a simple reflection of the fact that all developments in [427] involve non oriented Legendrian knots (just like those described at the end of the previous section).

9.7.4 *Some Thoughts about the Heegaard–Floer Homology*

Professionals working in this field are well aware that this subject is so huge that requires at least a book as thick as ours just to cover the basics. Thus, to avoid falling into a trap of going into some specific mathematical details, we decided mainly to provide a large scale overview of the field.

The story begins with two review papers by Michael Atiyah [436, 437] which are not optional but essential. Having read these papers, our readers

[dd]Exactly of the same type as discussed in Sec. 9.5.
[ee]E.g. read again Sec. 9.5.

are invited then to read [438] where general ideas are illustrated by detailed examples done in the style and at the level of this book. To make reading easier, we recommend to have [148, 181] as supplementary. Across the book we mentioned several times about the Seiberg–Witten theory and, in Sec. 9.5, about the Seiberg–Witten–Floer (S-W-F) homology. As it will be explained below, the S-W-F homology and the Heegaard–Floer (HF) homology are equivalent in three dimensions. This quite recently established mathematical fact allows us to replace extremely sophisticated machinery of the S-W-F homology summarized in [19] by the much simpler combinatorial theory associated with the already familiar grid diagrams. This means that whatever was said above about the grid diagrams survives in its entirety in the HF homology theory. This also means that all topological invariants we discussed will also survive. It is quite another story if this new HF theory is capable of producing new invariants of Legendrian knots. Having set the stage, we proceed with some needed historical background.

The achievements (that is discoveries of new topological invariants) in topology of 3- and 4- manifolds discussed in papers by Atiyah are concerned with the events of '80s. In the '90s the Seiberg–Witten equations were discovered leading to construction of Seiberg–Witten invariants. Although we mentioned these invariants several times, e.g. see [17–19, 394], we still recommend reading the review paper by Donaldson, [439], nicely reflecting the state of the art at the mid of '90s. By 2000 the efforts in designing new invariants of 3- and 4-manifolds had been split between two groups of mathematicians. The efforts (till 2007) of the first group (Kronheimer, Mrowka, etc.) are summarized in the monograph, [19], while the efforts (till 2005) of the second group (Ozsvath, Szabo, Rasmussen, etc.) are summarized in the review paper by McDuff, [440]. This reference is also not optional. It should be read by everyone who is serious about studying this field.

Remark 9.7.4.1. The Seiberg–Witten theory is a supersymmetric non-Abelian version of the Ginzburg–Landau (G-L) theory of superconductivity discussed in Chap. 3 (see also [17]). In Chap. 5 we treated the "bosonic" non-Abelian version of this theory and obtained monopoles. The monopoles were also obtained in the S-W theory [19]. The Bogomol'nyi monopole equations of Sec. 5.6.2 we interpreted as the non-Abelian version of the force-free condition. As it is demonstrated in Chap. 5, the existence of the force-free condition is the signature of contact geometry. Therefore, not entirely surprisingly, Mrowka and Rollin in [406] established a connection between the Legendrian knots and monopoles. For the latest efforts in this direction, read [441].

Remark 9.7.4.2. Connection between the Legendrian knots and monopoles just mentioned bring us back all the way to Chap. 3 where we discussed the hydrodynamics of ideal fluids and the Faddeev–Skyrme model for pure Yang–Mills fields. In Chap. 5 we established Arnol'd-type inequality, Eq. (5.6.7), for the Y-M fields which is the non-Abelian version of Arnol'd inequality, Eq. (3.1.12), for ideal fluids. It is the basis of the Vakulenko–Kapitanskii result, Eq. (3.2.24), for the pure Y-M fields. In Sec. 5.6.4 in agreement with authors of [157], who claimed that "the linked magnetic tubes in the Faddeev–Skyrme model actually playing the role of monopoles" we made an attempt, in Sec. 5.6.5, at demonstrating that this is indeed the case by providing a sketch of contact treatment of monopoles. In fact, much more sophisticated analysis by Mrowka and Rollin, [406], based on earlier work, [442], by Kronheimer and Mrowka indicates that the $(3+1)$ decomposition of 4-manifolds described above, in Sec. 9.5, can be used in the present case as well so that the contact results of Sec. 5.6.5 remain valid. And if this is so, we can invoke the results of Manton for monopeles discussed in detail in Sec. 4.3 of [181]. As Manton demonstrated, the monopole functional, Eq. (5.6.3.1), is obtainable from the pure four-dimensional Y-M functional. If this is so, then it is possible to build the whole Standard Model of particle physics topologically, out of knots and links as it was argued in [181]. Recently, Finkelstein, [443], and, independently, Billson–Thompson *et al.*, [444] provided construction of the Standard Model based on knots/links which must be postulated in their works.

Furthermore, by commenting on Vakulenko–Kapitanskii result, Eq. (3.2.24), for pure Y-M fields in their book, [68], Arnol'd and Khesin on p. 128 write:

"... some recent computer experiments for the relaxation process of an initial mapping with nonzero Hopf invariant exhibit the following phenomenon. In the equivariant case (of S^1 acting by rotations on $\mathbf{R}^3$ and S^2)[ff] one observes an energy gap over the poles, where the rotation axis intersects the sphere".

We remind our readers that [181] is about existence of the gap in the excitation spectrum of pure Y-M fields. At the same time Eqs. (9.4.2a) and (9.4.2b), related to Arnol'd–Khesin comments, provide classical and

[ff] E.g. read Appendix D.2.3 and look at Fig. D.3.

quantum statistical mechanical treatments of this dynamical problem. The same problem can be treated with help of the rolling sphere discussed in Chap. 8. In the same chapter the rolling sphere problem was mapped into that for dynamics of elastic curves and elastic rods. Chapter 8 and Appendix B are devoted to just this type of a problem. In [181], we arrived at exactly the same conclusions by using different and more powerful arguments. These arguments allowed us to include the dynamics of gravity into the Standard Model and to treat both the Standard Model and gravity topologically in terms of knots and links. Now, based on what was just explained, it turns out that such knots and links are Legendrian.

After this long digression, we are back on track to discuss the achievements of Ozsvath–Szabo and the associates in the field of HF homology after 2005.

In addition to [440], the clearest explanation of the HF homology done in the "old" style (that is between 2000 and 2006) can be found in the work by its major proponents: Ozsvath and Szabo in [445]. The Ph.D. thesis by Sarkar, [446], is another excellent source. In particular, in this source we found the clearest explanation of the connection between the original Floer homology done in symplectic context [17, 354] and the HF homology. In 2006, Sukharit Sarkar along with Peter Ozsvath and Ciprian Manolescu submitted to Annals of Mathematics their ground breaking paper, [447], published only in 2009, in which the Heegaard diagrams used in knot theory for quite some time [409] were replaced by the grid diagrams. A complete, clear and readable presentation of this new way of looking at the HF homology is given (in addition to [447] which is also quite readable), in the paper by Manolescu, Ozsvath, Szabo and (Dylan) Thurston [448].

Since the original Floer's theory was designed for description of 3-manifolds, it is not totally surprising that the three-dimensional adaptation of the S-W-F theory (e.g. read Sec. 9.5) coincides with that by Floer. This can already be seen by comparing the homological Eq. (9.5.7) with the equation at the top of p. 649 of the Annals paper, [447]. Nevertheless, the rigorous proof of this result came only in 2010. [449] is the first in the sequence of five papers. Another group of researches using different methods had proven the same at about the same time, [450] (plus another three papers by the same authors). Why it is taking that much to prove the equivalence? It is taking that much because the traditional (that is S-W-F) way of doing things involves computation of gauge-theoretic or symplectic invariants via counting of solutions of nonlinear elliptic partial differential equations. As such, this counting is extremely difficult to develop analytically. Because of this, it was also difficult to establish the equivalence rigorously even though both theories were applied to the same objects of study with about the same outcome. However, once the correspondence was established, the count of invariants on the grid diagrams is becoming routine. It already had admitted a computerization, e.g. use this link

http://katlas.org/wiki/Heegaard_Floer_Knot_Homology

The logic behind the computer program done by Jean-Marie Droz is explained in detail in his Ph.D. thesis, [451]. A condensed but readable summary of his thesis is given in [452]. Still more and up-to-date details can be found in the paper by his advisor, Anna Belyakova [453].

Remark 9.7.4.3. Why our physically educated readers should care and get excited about these, seemingly pure mathematical results?! For several reasons. *First*, these results indicate, (and not for the first time) that results obtainable with uses of supersymmetric quantum field theories can be obtained without any uses of supersymmetry. The experimental data of particle physics strongly suggest that the supersymmetry is very unlikely to be discovered experimentally [454]. However, as it was already mentioned, Witten's supersymmetric contributions to the Morse theory, [385], had received high acclaim in mathematics [383] so that supersymmetry does exist in Nature, e.g. in stochastic quantization, in critical dynamics, etc. This means that even though there are no supersymmetry between the particles in Nature, the supersymmetry is still lingering behind some *real* physical processes without any trace of the doubt. Established equivalence could be of value therefore, even for particle physics. *Second*, such discovered equivalence could be of value for statistical physics and for quantum computing. In Sec. 5 of our review paper, [64] we discussed in detail the connection between the Jones polynomial and the Potts model (which is just the extension of the Ising model, Eqs. (9.7.2.1) and (9.7.2.2) in which individual spins can take not two values ± 1 but, say, n values, e.g. roots of unity, etc.) first noticed by Lou Kauffman and developed by him and others as discussed in Sec. 5 of [64]. Recently, in [455], Kauffman investigated the possibility of extension of the noticed correspondence to cover the case of Khovanov polynomial and, more broadly, to Khovanov homology. He also discussed likely applications to quantum computing in both [417] and [455]. Recently, Olga Plamenevskaya defined an invariant of transverse links (in S^3 with standard contact structure $\xi_{st} = \ker\alpha$) as distinguished element of the Khovanov homology of the link. In [456], she discussed how her results are connected with the HF invariants. To this one should add the result by Ng, [416] which we already discussed. From another direction two papers by Ozsvath–Szabo group [457, 458] established: (a) how transverse knots are being distinguished by the HF homology and (b) how the Legendrian (and transverse) knots are being distinguished by the HF homology. Thus, the connections with statistical mechanics and usefulness of these results to quantum computing definitely are worth of investigating.

To bring this section to conclusion we would like to list some of the latest developments. Among these we especially noticed the following.

1. Recently, Kronheimer and Mrowka demonstrated [459] that Khovanov homology is capable of recognizing the unknot. In their proof they are using Floer's instanton homology of 3-manifolds.[gg] This result could be useful for quantum money project discussed in Sec. 9.6.
2. J-M. Droz and E.Wagner had demonstrated how grid diagrams can be used for description of Khovanov homology [460].
3. In [461] Shumakovich following earlier paper by Ozsvath *et al.* [462] used the odd Khovanov homology to improve the result by Ng, [416], for the upper bound on the Thurston–Bennequin invariant.
4. Shumakovich had also recently written a self-contained review on Khovanov homology [463] which nicely complements the already recommended review by Kauffman, [417]. The following three papers [464–466] supplementing the above two are also highly recommended.

Finally, the summary of accomplishments and open problems in HF homology theory is presented in [467], by one of the active developers of this field, Ciprian Manolescu.

The following quotation from the abstract to the latest Ph.D. thesis entitled "Knot Floer Homology and Categorification" by Allison Gilmore, [468], we believe, is very appropriate for closing this subsection and for the Closing Remarks for the entire book which follow:

"With the goal of better understanding the connections between knot homology theories arising from categorification and from Heegaard–Floer homology, we present a self-contained construction of knot Floer homology in the language of HOMFLY-PT homology. Using the cube of resolutions for knot Floer homology defined by Ozsvath and Szabo, we first give a purely algebraic proof of invariance that does not depend on Heegaard

[gg] E.g. read [181, 438] for the basic facts on this topic.

diagrams, holomorphic disks, or grid diagrams. Then, taking Khovanov's HOMFLY-PT homology as our model, we define a category of twisted Soergel bimodules and construct a braid group action on the homotopy category of complexes of twisted Soergel bimodules. We prove that the category of twisted Soergel bimodules categorifies the Hecke algebra[hh] with an extra indeterminate and its inverse adjoined...."

[hh]E.g. read Appendix C of this book.

Closing Remarks: The Unreasonable Effectiveness of Contact Geometry and Topology in Physical Sciences

In 1960, Eugene Wigner wrote a paper entitled "The unreasonable effectiveness of mathematics in the natural sciences" [469] which created a multitude of responses, e.g. read http://en.wikipedia.org/wiki/The_Unreasonable_Effectiveness_of_ Mathematics_in_the_Natural_Sciences

These responses are worth reading along with Wigner's original paper. It is being hoped that after reading this book our readers may begin to think about the unreasonable effectiveness of contact geometry and topology, an example of some kind of paradigm — the master key, allowing to unlock just about any door in the natural sciences. Since mechanics is the oldest of physical disciplines and since Vladimir Arnol'd is most well-known for his contributions into mechanics, not surprising that he believed that mathematics as a whole should be considered as some branch of physics [470]. His opinion to a large extent is being shared by Jaffe and Quinn [471]. However, their article had received a multitude of responses from the top mathematicians such as Atiyah, Borel, Mandelbrot, Thurston and others [472, 473] with opinions ranging from strongly supportive to strongly objective.

Recently, physical sciences had become saturated with all kinds of emergent/emergence motives/interpretations/themes some of which can be found in the nicely written book by Robert Carroll [474] where among other things Bohmian quantum mechanics is described in detail. If we follow the logic of Arnol'd in [470] as some kind of a guiding principle in the stormy seas of emergence trends, we will end up with a situation typically encountered when reading the encyclopedic dictionary in which every word is explained through multitude of other words ad infimum. Fortunately, under different occasion, in [411], Arnol'd added some extra details to opinions expressed in his article [470]. We feel, that these are the most appropriate at this place of the book. Indeed, he writes: "The power of the symplectic and contact geometries depends on unification of many apparently disjoint branches of mathematics that these theories provide. It is comparable to the unification of most branches of mathematics provided by linear algebra or, more geometrically, by projective geometry. Some 19th century mathematicians objected to the tendency toward projectivization of affine and Euclidean geometries. Cayley finally settled the problem, proclaiming "projective geometry is all geometry". In the same sense one might now say 'symplectic geometry is all geometry,' but I prefer to formulate it in a more geometrical form: *contact geometry is all geometry*".

Since contact geometry came from physics, and since our book is meant to demonstrate that, in addition, "*contact geometry is all physics*", from such a standpoint Arnol'd's final comments made in [411] can indeed be taken as a guiding principle.

Surely, because "nothing is inscribed in stone", we feel, that the comments made by Richard Hamming, a well-known computer science mathematician, on Wigner's article reflect the essence of the existing conceptual problems (including all kinds of visionary remarks) in natural sciences the most eloquently and are as valid for Wigner's article as they are for Arnol'd's. We would like to finish this book by spelling out the Humming's principles and by illustrating these principles by examples. Thus, here they are:

1. Humans see what they look for.
2. Humans create and select the mathematics that fit a situation.[a]
3. Mathematics addresses only a part of human experience.
4. Evolution has primed humans to think mathematically.

To illustrate these principles, we made a selection of examples based on their relevance to the results of this book.

We begin with the ground breaking experimental results by Rutherford done in 1911 [475] in which he studied scattering of electrons from hydrogen atoms. Using formalism of classical mechanics he calculated the differential cross-section for scattering of electrons by hydrogen's nuclei, that is

[a]Whatever word "situation" could mean (our remark).

by protons. Detailed calculations of the cross-section leading to this result are given in the Landau–Lifshitz volume on mechanics [235]. Everybody working in physics knows that the same formula can be obtained quantum mechanically with the help of the first Born approximation [476]. A remarkable agreement between Rutherford's theoretical predictions based on laws of classical mechanics and experimental data nicely illustrates Humming principles 1 and 2. Indeed, every professional working in physics knows that in spite of excellent agreement between theory and experiment classical mechanics was dismissed as flawed when applied to objects at atomic scales.

Subsequent discovery/formulation of quantum mechanics created the fundamental problem: *How one can relate/connect microscopic and macroscopic realities to each other?* The best answer (at the time of birth of quantum mechanics) was given by Heisenberg whose results we analyzed in detail in [173]. They are also summarized in Appendices C and D of this book. The modern answer is given in Appendices A and B though and have practically nothing in common with the original Heisenberg's ideas. This is so in spite of the fact that in these appendices Heisenberg's group in fact is derived, but macroscopically, that is without any reference to subatomic phenomena!

Already in this macroscopic reality, when we are making a very accurate measurements, say, of length, temperature, pressure, etc. we encounter fluctuations causing us to deal with averages. First we have to collect the data for a given observable. We shall call these data a *spectrum.* Second, we have to worry about the statistical description of these data. Quantum mechanically, these issues are discussed at the beginning of Appendix D. Suppose that we are given the data and we are asked to restore/to describe the observable to which we do not have the direct access. Heisenberg was confronted exactly with this problem which he successfully solved in 1925 by inventing what is currently known as the matrix quantum mechanics. In it, each macroscopic observable has its own quantum spectrum created by eigenvalues of some matrix (finite or infinite). The spectrum is assumed not to have multiplicities (that is there should be no multiple eigenvalues of the same sign and magnitude) since such a redundancy apparently cannot be classically recognized. As we know from Appendix D, such multiplicities are in fact responsible for the entanglement processes. And, we are now aware that these are macroscopically recognized!

The results by Heisenberg were further improved and expanded by Max Born and his assistant, Pascual Jordan, to such an extent that already in 1926 a detailed set of lecture notes on quantum mechanics written by M. Born and translated into English was published by the MIT Press. An excellent collection of original papers on quantum mechanics, containing the MIT lectures, was recently assembled and published by Stephen Hawking [477]. In this publication on p. 924 the following statements can be found:

"Consider for instance the hydrogen atom. The classical theory not only gives the orbits of the electron but seems to assign a meaning to the position of the electron at each instant.[b] In the new theory the energy and moment of momentum[c] of a state can be given, but it appears to be impossible to give any further description of this state as a geometrical orbit and even more impossible to fix the position of the electron at any instant.[d] *Space points and time points in the ordinary sense do not exist. These concepts can only be introduced subsequently in limiting cases.*

On the other hand it seems to me that we have a right to use the terms "orbit" or even "ellipse", "hyperbola", etc. in the new theory,[e] *if we agree to interpret them rationally and to understand by them the quantum processes which go over in the limit to the orbits, ellipses, hyperbolas, etc. of the classical theory...* ".

Thus, as it follows from Born's lecture notes, the correspondence issue is resolved by Born in a somewhat metaphysical way providing an example of the third Humming's principle.

The correspondence principle by Bohr was discussed from the modern perspective in Chap. 9. As it was noticed by Maslov, the WKB method, the mathematical restatement of Bohr's correspondence principle, is ill-defined. When corrected, the known quantum mechanical spectrum is obtained by using classical mechanics involving sophisticated topological methods, e.g. those of contact and symplectic geometry. In the case of exactly integrable systems the obtained spectrum is exact! The same results can be re-obtained more efficiently with the help of the Duistermaat–Heckmann formula. Furthermore, the many-body quantum mechanical systems normally treated by the Hartree–Fock (that is by the conventional mean field)-type methods

---

[b] These, of course, were the assumptions of Rutherford when he was deriving his formula for the cross-section.

[c] This phrase looks a bit strange but it is reproduced exactly as it is.

[d] Recall, however, the latest double slit and the Heisenberg uncertainty principle checking experiments mentioned in Chap. 9.

[e] In fact, this was *exactly* the case described in Heisenberg's very first paper on quantum mechanics written in 1925 and analyzed in detail in [173].

can be efficiently treated by the method of coadjoint orbits of classical mechanics (second Humming's principle).

Exactly 101 years after Rutherford's scattering experiments were performed, the article in Physics Today was published [476] based on the development of ideas of pioneering article by Wannier [478] written in 1953. In it the quantum mechanical description of near-threshold fragmentation of atomic/molecular systems was replaced by that using methods of classical mechanics. These findings are consistent with results of our paper, [275], where it was demonstrated that mathematical theory of coagulation-fragmentation processes can be successfully used for development of quantum mechanics, quantum field and string theories. Thus, 101 year later scattering experiments in atomic and molecular physics are again based on ideas and methods envisioned by Rutherford, perhaps adjusted to the needs of fragmentation dynamics experiments (second Humming's principle).

Next, although it is considered to be the most convincing experiment illustrating principles of quantum mechanics (http://en.wikipedia.org/wiki/Stern-Gerlach_experiment) which played crucial role in development of quantum mechanics, the Stern–Herlach experiment was explained recently in purely classical terms in two papers by Franca, [479, 480] and in the detailed work by Stenson [481] (Humming's principles 1 and 2).

Although we explained the essence of entanglement classically in Appendix D one can find in literature other settings demonstrating entanglement classically. For instance, an example of entanglement is provided by Khrennikov [482] in the context of classical stochastic process, by Krauter and collaborators [483] in the context of correlated motion (with dissipation) of two macroscopic bodies, by Lee and collaborators [484] in the context of vibrations of millimeter sized diamonds at room temperature, etc. (second Humming's principle)

Einstein was unhappy with Copenhagen formulation of quantum mechanics. He felt that "God is not playing dice". Deep down such a feeling of Einstein could be explained by irreconcilable differences between then "new" quantum mechanics and "old" general relativity. In Appendix D it is demonstrated how for any mechanical system whose motion is described by the set of Lagrangian equations it is possible to rewrite one-to-one these equations in the form of geodesic equations (Eq. (D.131)) which are looking exactly the same as in Einstein's relativity. Lesser known is the fact stated in the book by Kirillov [485], p. 246, that all eigenvalue equations of quantum mechanics are also equations for geodesics. His terse description of this fact was made absolutely clear in lecture notes by Sternberg [486]. In accord with Arnol'd, these results fit nicely into formalism of contact geometry and topology (second Humming principle).

Moreover, as we already noticed, there is still a way to develop quantum mechanics combinatorially (e.g. see [275] and [173]). Since combinatorics can be looked upon as part of group theory and group theory can be linked with contact geometry, we are back to prophetic statement by Arnol'd. In addition, though, the combinatorial origins of quantum mechanics can also be explained from the point of view of theory of games, e.g. see the paper by Landsburg [487]. By looking from this direction, we are coming to the conclusion that "God does play dice but... in accordance with strict rules of combinatorics". Experimental validation of quantum game theory can be found, for instance, in the paper by Hannauske with collaborators [488]. The most spectacular verification of quantum game theory was recently obtained in connection with correlated behavior of insects [489]. (Humming's principles 2, 3 and 4.) The quantum game theory can be explained in the language of quantum random walks and, hence, semiflexible polymers. This opens the door toward thinking about functioning of living organisms using terminology of quantum computers. (Humming's principles 2, 3 and 4.) As it was mentioned in Chap. 8, the problems of such kind already begin at the level of description of folding of individual proteins (for which the adiabatic quantum computing methodology is used [490]). However, some progress had been achieved as well at the level of a single cell, etc. [491] (third Humming's principle).

Since the results of Appendix A indicate that quantumness is not intrinsic for microcosm, one can think about uses of quantum mechanics (and, hence, contact geometry and topology) at the scales of, say, our Solar system and, may be, even larger. Although research in this direction was initiated in our work, [177], the results of recent monograph [492] indicate that much more could be done for development of astronomical-quantum mechanical correspondences. Incidentally, such correspondences were systematically used by Max Born in his search for new quantum mechanics [386][f] (second Humming principle).

---

[f]The original version of this book, the lecture notes in German, was published in Göttingen in 1924.

We are finishing this book with yet another quotation from Arnol'd [493] which, in turn, he borrowed from the letter by Lagrange to D'Alambert written in 1781:

"The shaft becomes too deep... and if new veins would not be discovered, places for geometry in Academy will become what are already Arabic chairs at the universities."

We highly recommend the reader to read the rest of this remarkable publication.

Appendix A

Heisenberg Group in the Context of Sub-Riemannian Geometry and Optimal Control

A.1 The Laplace–Young Equation in the Presence of Surfaces and Its Connection with Dido Problem and Its Dual

In [A.1] we discussed the following well-known Laplace–Young problem (in three dimensions). Consider a droplet of new phase characterized by the free energy $\mathcal{F}_3$ given by

$$\mathcal{F}_3 = \sigma A - \Delta p V. \tag{A.1}$$

Here σ is the surface tension, A and V are respectively the area and the volume of the droplet while Δp is the difference between the internal and external pressures. For a spherical droplet of radius R the above functional acquires the following form

$$\mathcal{F}_3(R) = \sigma 4\pi R^2 - \Delta p \frac{4}{3}\pi R^3. \tag{A.2}$$

Minimizing $\frac{d\mathcal{F}_3(R)}{dR} = 0$, we obtain the familiar Laplace–Young equation known from the theory of phase transitions

$$\frac{2\sigma}{R^*} = \Delta p. \tag{A.3}$$

Evidently, the same problem can be studied in two dimensions (recall droplets of oil on the surface of your chicken soup). For such a case we obtain

$$\mathcal{F}_2(R) = \sigma 2\pi R - \Delta p \pi R^2. \tag{A.4}$$

This time, however, we would like to complicate matters a little bit. We shall assume that there is a boundary and that the oily droplet is capable of wetting the boundary. This requires us to rewrite $\mathcal{F}_2(R)$ accordingly. In particular, if the Monge parametrization is used, we obtain

$$\mathcal{F}_2 = \sigma \int dx \sqrt{1 + \left(\frac{dy}{dx}\right)^2} - \Delta p \int dx y(x). \tag{A.5}$$

The same expression can be written in the reparametrization-invariant form

$$\begin{aligned}\mathcal{F}_2 &= \sigma \int_0^N d\tau \sqrt{\left(\frac{dx}{d\tau}\right)^2 + \left(\frac{dy}{d\tau}\right)^2} - \frac{\Delta p}{2} \int_0^N d\tau \left(x\frac{dy}{d\tau} - y\frac{dx}{d\tau}\right) \\ &\equiv \sigma L - \Delta p A. \end{aligned} \tag{A.6}$$

The minimization problem can now be formulated as follows.

For the fixed length L of the "string" whose ends are anchored at the boundary determine the shape of the figure bounded by the string and the boundary so that such a figure will have the maximum possible area. In such a form this optimization problem is known as queen Dido problem.[a] This problem comes straight from the ancient Greek literature and, as such, it is about 2,800 years old. There is a *dual formulation* which is closer in spirit to the Laplace–Young problem [A.2]. Minimize the length L of the closed (in the sense just described) curve subject to the constraint that the signed area

$$A = \frac{1}{2} \int_0^N d\tau \left(x\frac{dy}{d\tau} - y\frac{dx}{d\tau}\right)$$

enclosed by a curve is a fixed constant. In such a dual form our readers had already a chance to encounter this problem in Sec. 6.3 while in Sec. 6.5 in Exercise 6.5.2 we were asking our readers to reconstruct the Heisenberg group using such dual formulation. In the style of L-L course we are now in the position to make such a reconstruction. To do so, requires us only to collect some results from the main text. In particular, by rewriting an inequality

$$l(\gamma(\tau)) \leq \sqrt{N}\sqrt{2\mathcal{E}(\gamma(\tau))},$$

obtained in Sec. 6.3 as

$$N \int_0^N d\tau \left[\left(\frac{dx}{d\tau}\right)^2 + \left(\frac{dy}{d\tau}\right)^2\right] \geq \left(\int_0^N d\tau \sqrt{\left(\frac{dx}{d\tau}\right)^2 + \left(\frac{dy}{d\tau}\right)^2}\right)^2 \tag{A.7}$$

[a] http://www.math.uiuc.edu/~laugesen/dido-isoperimetry-history.pdf

we observe that by employing natural parametrization, that is by requiring $(\frac{dx}{d\tau})^2 + (\frac{dy}{d\tau})^2 = c$, where c is some constant, the above inequality becomes an equality. This means that instead of the Carnot–Caratheodory distance, see Eq. (6.3.24), we can minimize the energy functional $\mathcal{E}(\gamma(\tau))$ as described by Eqs. (6.4.1) and (6.4.2). That is, by switching from sub-Riemannian to optimal control language, we have to look for the minimum of the functional [A.3]

$$\mathcal{F}[u,v] = \int_0^N d\tau(u^2 + v^2), \tag{A.8}$$

under conditions that the time-dependent functions u and v control the motion described by the equations

$$\dot{x} = u, \dot{y} = v, \quad \dot{t} = yu - xv. \tag{A.9a}$$

Using Sec. 6.1, we take advantage of the fact that the standard 1-form $\alpha = dz + xdy$ is contactomorphic to the 1-form $\alpha = -\frac{1}{4}dt + \frac{1}{2}\,(ydx - xdy)$. Because of this and, by comparing Eqs. (A.5) and (A.6), we can replace the optimal control dynamical problem described by Eqs. (A.9a) by the equivalent problem given by

$$\dot{x} = u, \quad \dot{y} = v, \quad \dot{z} = xv. \tag{A.9b}$$

Following [A.4], it is instructive to rewrite the above system of equations in the matrix form, e.g. as follows:

$$\begin{pmatrix} 0 & \dot{x} & \dot{z} \\ 0 & 0 & \dot{y} \\ 0 & 0 & 0 \end{pmatrix} = \begin{pmatrix} 1 & x & z \\ 0 & 1 & y \\ 0 & 0 & 1 \end{pmatrix} \begin{pmatrix} 0 & u & 0 \\ 0 & 0 & v \\ 0 & 0 & 0 \end{pmatrix} = \begin{pmatrix} 0 & u & xv \\ 0 & 0 & v \\ 0 & 0 & 0 \end{pmatrix}. \tag{A.10}$$

Such a rewriting brings immediately the Heisenberg group into play.

A.2 Some Facts about the Heisenberg Group

Following [A.5] we would like to describe the properties of this group in some detail. Thus, we begin with noncommutative group G. The noncommutativity is measured by the commutator. That is, if both h and k belong to G, their commutator $[h, k]$ is defined as $hk(kh)^{-1}$. In particular, if $[h, k] = e$ where e is unit of group G, then the elements h and k commute, otherwise, they do not commute. The set of elements commuting with all other elements is called *center of the group*. Now let $[K, G]$ be the subgroup of G generated by all commutators $[k, g]$ such that $k \in K$ and $g \in G$ where K is a subgroup of G. In particular, if $K = G$, then $[G, G]$ is called *commutator subgroup* of G.

Next, if G is some group, then define a sequence $\Gamma_n(G), n \geq 1$, as follows: $\Gamma_0(G) = G, \Gamma_{n+1}(G) = [\Gamma_n(G), G]$. This allows us to introduce the concept of nilpotency. G is said to be *nilpotent* if there is some finite n such that $\Gamma_n(G) = e$. The smallest integer n with the above property is called the *class of nilpotency* of G. Specifically, let

$$\mathbf{G} = \left[\begin{pmatrix} 1 & a & c \\ 0 & 1 & b \\ 0 & 0 & 1 \end{pmatrix};\ a, b, c \in \mathbf{R} \right] \tag{A.11}$$

and consider two representatives of $\mathbf{G}$, e.g.

$$A = \begin{pmatrix} 1 & a_1 & a_3 \\ 0 & 1 & a_2 \\ 0 & 0 & 1 \end{pmatrix}, \quad B = \begin{pmatrix} 1 & b_1 & b_3 \\ 0 & 1 & b_2 \\ 0 & 0 & 1 \end{pmatrix},$$

then

$$AB = \begin{pmatrix} 1 & a_1 + b_1 & a_3 + b_3 + a_1 b_2 \\ 0 & 1 & a_2 + b_2 \\ 0 & 0 & 1 \end{pmatrix}. \tag{A.12}$$

Accordingly,

$$A^{-1} = \begin{pmatrix} 1 & -a_1 & -a_3 + a_1 a_2 \\ 0 & 1 & -a_2 \\ 0 & 0 & 1 \end{pmatrix}, \quad B^{-1} = \begin{pmatrix} 1 & -b_1 & -b_3 + b_1 b_2 \\ 0 & 1 & -b_2 \\ 0 & 0 & 1 \end{pmatrix}. \tag{A.13}$$

Using just obtained results, we construct a commutator

$$[A, B] = \begin{pmatrix} 1 & 0 & a_1 b_2 - b_1 a_2 \\ 0 & 1 & 0 \\ 0 & 0 & 1 \end{pmatrix}. \tag{A.14}$$

so that the commutator subgroup of $\mathbf{G}$ is

$$\Gamma_1(\mathbf{G}) = \left[\begin{pmatrix} 1 & 0 & l \\ 0 & 1 & 0 \\ 0 & 0 & 1 \end{pmatrix};\ l \in \mathbf{R} \right] \tag{A.15}$$

and, therefore, by direct and easy computation we obtain

$$\Gamma_2(\mathbf{G}) = [\Gamma_1(\mathbf{G}), \mathbf{G}] = e. \tag{A.16}$$

So that the class of nilpotency of the Heisenberg group **G** is 2. *The nilpotency is measure of noncommutativity of the group.* Evidently, there are plenty of noncommutative groups whose nilpotency class is larger than 2. Whether or not they are of physical significance we leave it to our readers to figure out.

Exercise A.1. Lately in physics there had been a great deal of activity in the area of the so-called *deformation quantization*[b] as well in the area of *Nambu mechanics.*[c] These developments raise many questions. One of them: How quantization emerging from use of nilpotent groups of class larger than 2 (sometimes used in the theory of optimal control) is related to deformation quantization? More accurately, if there are physical phenomena requiring (identified with), say, the use of nilpotent groups of class larger than 2, can the same phenomena be described in terms of the deformation quantization? Is there any relationship of Nambu mechanics to such phenomena?

Clearly, though, that already class 2 is of central physical significance. It is of such importance because the group **G** is in fact the Heisenberg group. Since it is a Lie group, it is useful to compare it with the SO(3) group of rotations discussed in Chap. 8. In both cases we are dealing with 3×3 matrices. While the basis of the Lie algebra $\mathfrak{so}(3)$ is given by Eqs. (8.2.3.3) the basis of the Lie algebra for the Heisenberg group can be obtained analogously and is given explicitly by the following three matrices:

$$\mathbf{e}_1 = \begin{pmatrix} 0 & 1 & 0 \\ 0 & 0 & 0 \\ 0 & 0 & 0 \end{pmatrix}, \quad \mathbf{e}_2 = \begin{pmatrix} 0 & 0 & 0 \\ 0 & 0 & 1 \\ 0 & 0 & 0 \end{pmatrix}, \quad \mathbf{e}_3 = \begin{pmatrix} 0 & 0 & 1 \\ 0 & 0 & 0 \\ 0 & 0 & 0 \end{pmatrix} \tag{A.17}$$

whose commutators are[d]

$$[\mathbf{e}_1, \mathbf{e}_2] = \mathbf{e}_3, \quad [\mathbf{e}_1, \mathbf{e}_3] = 0, \quad [\mathbf{e}_2, \mathbf{e}_3] = 0. \tag{A.18}$$

That is these are *exactly of the same type* as for operators X_1, X_2 and R defined respectively by the Eqs. (6.1.1) and (6.1.2)! Because of this, it is possible to bring the system of Eqs. (A.8)–(A.10) in correspondence with those used for the particle in magnetic field discussed in detail in Chap. 6. This will be done further below. In the meantime, we notice that by analogy with

---

[b] E.g. see http://en.wikipedia.org/wiki/Weyl_quantization#Deformation_quantization
[c] http://en.wikipedia.org/wiki/Nambu_dynamics
[d] Note that for the Lie algebras the commutators are defined as $[A, B] = AB - BA$, as usual.

groups SU(2) and SO(3) (e.g. see respectively Eqs. (6.5.25)–(6.5.28), for SU(2), and Eqs. (8.2.3.3), (8.2.3.4), for SO(3)) we can now also establish the correspondence (a bijection) between $\mathbf{R}^3$ and $\mathbf{G}$. This can be done with help of Eqs. (A.12) and (A.13). Indeed, consider a formal mapping ϕ of the type

$$\phi(x,y,z) \to \begin{pmatrix} 1 & x & z \\ 0 & 1 & y \\ 0 & 0 & 1 \end{pmatrix}. \tag{A.19}$$

In view of Eq. (A.12), we obtain the noncommutative group law structure on $\mathbf{R}^3$ defined by the following obvious composition law

$$(x,y,z) \circ (x',y',z') = (x+x', y+y', z+z'+xy'). \tag{A.20}$$

This composition law differs somewhat from that given in the main text, e.g. see Eq. (6.3.25), but can be brought to exactly the same form by coordinate changes. Indeed, the difference between just obtained result and that given by Eq. (6.3.25) is very much of the same type as the difference between two contactomorphic 1-forms $\alpha = dz + xdy$ and $\alpha = -\frac{1}{4}dt + \frac{1}{2}\,(ydx - xdy)$. The second is obtainable from the first by coordinate changes. Details can be found in [A.5], p. 4.

Definition A.1. The space $\mathbf{R}^3$ together with the composition law, Eq. (A.20), (or Eq. (6.3.25)) is called respectively *nonsymmetric* (*or symmetric*) Heisenberg group and is denoted as $\mathbf{H}_1$.

Corollary A.1. *In view of the mapping ϕ defined by Eq.* (A.19) *we can as well call the Lie group $\mathbf{G}$ as Heisenberg group and, therefore, we shall relabel it as $\mathbf{H}_1$ from now on.*

A.3 Connections with Sub-Riemannian Geometry and Pontryagin's Theory of Optimal Control

At this point we are ready to make some connections between (Heisenberg) equations of motion, Eq. (A.10), and results of Sec. A.2. This will eventually bring us back to results already obtained in Chap. 6. Thus, taking into account Eq. (A.17) we can rewrite Eq. (A.10) as follows:

$$\dot{\mathbf{G}} = \mathbf{G}(u\mathbf{e}_1 + v\mathbf{e}_2)$$

or, which is the same,

$$\mathbf{G}^{-1}\dot{\mathbf{G}} = u\mathbf{e}_1 + v\mathbf{e}_2 = \dot{x}\mathbf{e}_1 + \dot{y}\mathbf{e}_2 \tag{A.21}$$

Now, let us compare this result with Eq. (6.3.10a)) of the main text. To do so, we also have to notice the isomorphism between Lie algebras given by Eqs. (6.1.1), (6.1.2) on one hand and by Eqs. (A.18) on the other. This allows us to identify $\mathbf{G}^{-1}\dot{\mathbf{G}}$ in Eq. (A.21) with the operator $\dot{\mathbf{x}}$ in Eq. (6.3.10a). After this, with help of the metric tensor defined by Eq. (6.3.18) we obtain finally the sub-Riemannian Hamiltonian Eq. (6.3.16) which is just a rewritten version of the Hamiltinian defined by Eq. (6.2.4)! This observation allows us to explain already obtained results in the language of Pontryagin's theory of optimal control. In doing so, we follow [A.3].

Having in mind the functional $\mathcal{F}[u,v]$ defined by Eq. (A.8), let us consider its most general analog

$$F[\mathbf{x},\mathbf{u};T] = \int_0^T dtg(\mathbf{x}(t),\mathbf{u}(t)) \tag{A.22}$$

for which we have to find a minimum under the conditions that $\mathbf{x}$ is subject to the differential equation constraint: $\dot{\mathbf{x}} = \mathbf{f}(\mathbf{x},\mathbf{u})$, provided that the initial $\mathbf{x}(t=0) = \mathbf{x}_0$ and the final $\mathbf{x}(T) = \mathbf{x}_T$ points are pre-assigned. Furthermore, variables $\mathbf{x}$ belong to M, where M is smooth manifold of dimension n (configuration space) and the control variables $\mathbf{u} \in \Omega$, where Ω is *closed* subset of k-dimensional Euclidean space $\mathbf{R}^k$. Both functions $\mathbf{f}$ and the *cost function* g are assumed to be smooth and non-negative. To state the necessary conditions dictated by the Pontryagin maximum principle, one need to introduce an auxiliary parametrized Hamiltonian (defined on phase space T^*M)

$$\hat{H}(\mathbf{x},\mathbf{p},\mathbf{u}) = \langle \mathbf{p}\cdot\mathbf{f}(\mathbf{x},\mathbf{u})\rangle - p_0 g(\mathbf{x},\mathbf{u}), \tag{A.23}$$

where $p_0 \geq 0$ is a fixed constant while the momenta $\mathbf{p} \in T^*M$. Evidently, both $\mathbf{p}$ and p_0 are just the Lagrangian multipliers. Now let $t \to u^*(t)$ be a curve that satisfies the following relationship along the trajectory $t \to (\mathbf{x}(t),\mathbf{p}(t))$ in T^*M:

$$H(\mathbf{x}(t),\mathbf{p}(t),\mathbf{u}^*(t)) = \max_{\mathbf{u}\in\Omega} \hat{H}(\mathbf{x}(t),\mathbf{p}(t),\mathbf{u}). \tag{A.24a}$$

Equation (A.25) (below) defines $\mathbf{u}^*$ implicitly as a function (or set of functions) of $\mathbf{x}$ and $\mathbf{p}$. This observation allows us to write as well

$$H^*(\mathbf{x}(t),\mathbf{p}(t),t) = H(\mathbf{x}(t),\mathbf{p}(t),\mathbf{u}^*(t)). \tag{A.24b}$$

Since $\mathbf{u}^*$ maximizes the functional $\hat{H}$, its partial derivatives (in $\mathbf{u}$) evaluated at $\mathbf{u} = \mathbf{u}^*$ must vanish, that is we have to require

$$\frac{\partial H^*}{\partial u_i} = 0, \quad i = 1,\ldots,k. \tag{A.25}$$

This equation provides (at least theoretically) $\mathbf{u}^*(t) = \mathbf{l}(\mathbf{x}(t), \mathbf{p}(t))$ implying that

$$H^*(\mathbf{x}(t), \mathbf{p}(t), t) = H(\mathbf{x}(t), \mathbf{p}(t)).$$

The above results provide the essence of the Pontryagin maximum principle. Following [A.3], we would like to adopt these general statements to the optimization problem given by Eqs. (A.8) and (A.9a). In such a case the Hamiltonian $\hat{H}(\mathbf{x}, \mathbf{p}, \mathbf{u})$ acquires the following explicit form:

$$\hat{H} = p_1 u + p_2 v + p_3(yu - xv) - \frac{1}{2}(u^2 + v^2). \tag{A.26}$$

The choice of the factor 1/2 in the last term will be justified momentarily. Minimization of Eq. (A.25) produces:

$$\frac{\partial H^*}{\partial u} = \frac{\partial H^*}{\partial v} = 0,$$

that is

$$u = p_1 + p_3 y \quad \text{and} \quad v = p_2 - p_3 x. \tag{A.27}$$

If we rescale $p_3 : p_3 \to 2p_3$ and, using Eq. (A.9b), we replace u by $\dot{x}$ and v by $\dot{y}$, then Eq. (A.27) coincide with the first two Eq. (6.3.11). Furthermore, if we use Eq. (A.27) in Eq. (A.26) we re-obtain the benchmark Hamiltonian, Eq. (6.2.4), in accord with [A.3], p. 355. Thus, we just demonstrated that Pontryagin's theory of optimal control is equivalent to the sub-Riemannian dynamics discussed in Chap. 6.

Appendix B

Sub-Riemannian Dynamics of Josephson Junctions

B.1 Nilpotency, Bracket-Generation and Solvability

In Appendix A we demonstrated that the Heisenberg group $\mathbf{H}_1$ is nilpotent. It happens that for development of sub-Riemannian geometry it is possible to use slightly wider class of groups which include $\mathbf{H}_1$ as a special case. We would like now to explain such a necessity of extension in simple physical terms. For this purpose, let us recall the map Ψ introduced in Sec. 8.2.3. Using Ψ the triple of basis elements X_i, $i = 1$–3, of the Lie algebra $\mathfrak{so}(3)$ defined by Eq. (8.2.3.3) can be identified with the triple x, y, z of $\mathbf{R}^3$. In such a case the vector $c_1X_1 + c_2X_2 + c_3X_3 \in \mathbf{R}^3$ if $c_i \in \mathbf{R}$. In view of this, it is possible to say that vectors X_i, $i = 1$–3 span $\mathbf{R}^3$. Since any n-dimensional manifold M locally (that is in the vicinity of some point p) can be modelled by $\mathbf{R}^n$ it is possible to write as well $\mathcal{D}_p - \text{span}\{X_1, \ldots, X_k\} \in T_pM$, that is $\mathcal{D}_p$ is a k-dimensional subspace of $T_pM, k \leq n$, called a *distribution*.[a] In the light of these definitions, it is obvious that the Heisenberg group can also be mapped onto $\mathbf{R}^3$ so that $c_1\mathbf{e}_1 + c_2\mathbf{e}_2 + c_3[\mathbf{e}_1, \mathbf{e}_2] \in \mathbf{R}^3$. Consider now in full generality the following ascending sequence of linear subspaces of T_pM

$$\mathcal{D}_p^1 = \mathcal{D}_p,$$
$$\mathcal{D}_p^2 = \mathcal{D}_p^1 + [\mathcal{D}_p, \mathcal{D}_p^1],$$

[a]Normally, distributions are associated with 1-forms. Since there is 1–1 correspondence between these 1-forms (forming a cotangent bundle in mechanics) and vector fields (forming a tangent bundle in mechanics) we apply term "distribution" to vector fields following [167].

$$\mathcal{D}_p^3 = \mathcal{D}_p^2 + [\mathcal{D}_p, \mathcal{D}_p^2], \tag{B.1}$$
$$\vdots$$
$$\mathcal{D}_p^{n+1} = \mathcal{D}_p^n + [\mathcal{D}_p, \mathcal{D}_p^n].$$

The distribution D_p is said to be *bracket-generating* at $p \in M$ if there is an integer $r \geq 1$ for which $\mathcal{D}_p^r \in T_pM$. The integer is called *step*. The distribution associated with the Lie algebra of the Heisenberg group $\mathbf{H}_1$ is of step 2 in view of the relationship $c_1\mathbf{e}_1 + c_2\mathbf{e}_2 + c_3[\mathbf{e}_1, \mathbf{e}_2] \in \mathbf{R}^3$. Since *in sub-Riemannian geometry only the bracket-generating distributions are used*, it is convenient to introduce further ramifications. For instance, let $\Gamma(\mathcal{D}_p)$ denote the *horizontal distribution*, e.g. in the case of $\mathbf{H}_1$ it is $\Gamma(\mathcal{D}_p) = \text{span}\{\mathbf{e}_1, \mathbf{e}_2\}$. Then, define recursively the following set of commutators

$$\begin{aligned} C^{(1)} &= \{[X, Y]; X, Y \in \Gamma(\mathcal{D}_p)\}, \\ C^{(2)} &= \{[C^{(1)}, Z]; Z \in \Gamma(\mathcal{D}_p)\}, \\ &\vdots \\ C^{(n+1)} &= \{[C^{(n)}, Z]; Z \in \Gamma(\mathcal{D}_p)\}. \end{aligned} \tag{B.2}$$

This sequence allows us to introduce the notion of nilpotency in a way different from that given in Appendix A. In Appendix A the nilpotency was defined at the level of Lie groups and without any reference to horizontality. Now, however, Eqs. (B.2), allow us to define the nilpotency at the level of the Lie algebra. Specifically, if $C^{(n)} = 0, n \geq 1$, then such a distribution is called nilpotent with *nilpotency class* n. The distributions which are bracket-generating, that is suitable for development of sub-Riemannian geometry are not necessarily nilpotent. E.g. consider a distribution made of two vectors

$$X_1 = \partial_{x_1} + e^{x_2}\partial_t \quad \text{and} \quad X_1 = \partial_{x_2}. \tag{B.3}$$

Since

$$[X_1, X_2] = -e^{x_2}\partial_t, \tag{B.4}$$

vectors X_1, X_2 and $[X_1, X_2]$ are linearly-independent and span $\mathbf{R}^3$. Such distribution is bracket-generating and is of step 2. However, it is not nilpotent since the iterated commutators, Eqs. (B.2), never vanish because of the presence of the factor e^{x_2}. We shall encounter somewhat similar situation in the case of dynamics of Euler's elastica as it will be explained below. At the same time there are some nilpotent distributions which are not bracket-generating (at least at some points in the parameter space). For example, let

$$X = \partial_x, \quad Y = \partial_y + tx\partial_t. \tag{B.5}$$

Such a distribution is nilpotent with nilpotency class 2 but is not bracket-generating for $t = 0$. The above concepts are closely related to the concept of solvable groups. At the level of the Lie algebras such a concept can be defined as follows [A.4]. If L is the Lie algebra of some Lie group, define recursively $L^{(1)} = [L, L], L^{(2)} = [L^{(1)}, L^{(1)}], \ldots, L^{(n+1)} = [L^{(n)}, L^{(n)}], \ldots$ so that there is an inclusion sequence

$$L \supset L^{(1)} \supset L^{(2)} \supset \cdots \supset L^{(N)} = 0$$

truncated at some $N \geq 1$. In such a case the Lie algebra is called *solvable.* If the Lie algebra is solvable, the associated with its Lie group is also solvable. It can be proven (see below) that every nilpotent group is solvable. Indeed, for the nilpotent Lie algebra using Eqs. (B.2) we can equivalently recursively define $L_{(1)} = [L, L], L_{(2)} = [L, L_{(1)}], \ldots, L_{(n+1)} = [L, L_{(n)}], \ldots$ so that we shall again have an inclusion chain

$$L \supset L_{(1)} \supset L_{(2)} \supset \cdots \supset L_{(N)} = 0$$

for some $N \geq 1$ and since $L_{(n)} \supset L^{(n)}$ [A.4] the proof follows. If this is so the question arises: Is every solvable group bracket-generating? If the answer is "yes" this will imply that only solvable groups can be used in sub-Riemannian geometry and/or in the theory of optimal control. This indeed happens to be the case [B.1].

B.2 Euler's Elastica. The Sub-Riemannian Treatment

In Sec. 8.3.3 we obtained Eq. (8.3.3.9) describing dynamics of a single Josephson junction. For $J = 0$ this equation is known as equation describing Euler's elastica. In the main text we explained why such an equation should be linked with Josephson's function dynamics. In this Appendix we shall discuss the sub-Riemannian nature of this equation.

We begin with the formulation of variational problem associated with Euler's elastica following [A.4]. Let $\mathbf{a}_0 = (x_0, y_0)$ and $\mathbf{a}_1 = (x_1, y_1)$ be two points (not necessarily distinct if the elastica is closed [B.2]) in the plane $\mathbf{R}^2$ and let $\mathbf{v}_0$ and $\mathbf{v}_1$ be two unit vectors (that is $|\mathbf{v}_0| = |\mathbf{v}_1| = 1$) attached respectively to these points. Based on this information, it is required to determine a shape of the elastic curve whose ends are located at the above points and whose directions at these ends is given respectively by $\mathbf{v}_0$ and $\mathbf{v}_1$.[b]

Let the elastic curve $\mathbf{r}(\tau)$ be naturally parametrized. That is $\mathbf{r}(\tau) = \{x(\tau), y(\tau)\}, \tau \in [0, L]$ where L is the length of the curve. Introduce an angle $\theta(\tau)$ between the velocity vector $\dot{\mathbf{r}}(\tau) \equiv \mathbf{v}$ and the positive direction of the x-axis. Then the elastica problem can be formulated as follows. Consider the system of equations

$$\begin{aligned} \dot{x} &= u_1 \cos\theta, \\ \dot{y} &= u_1 \sin\theta, \\ \dot{\theta} &= k(\tau) \equiv u_2 \end{aligned} \tag{B.6}$$

so that $\mathbf{q}(0) = (x, y, \theta)(0) = (0, 0, 0), \mathbf{q}(t) = (x_1, y_1, \theta_1)$. Because the curve is naturally parametrized, $\mathbf{v}_0 = (1, 0)$ and, accordingly, $\mathbf{v}_1 = (\cos\theta_1, \sin\theta_1)$. The optimal trajectory provides the minimum for the elastic energy functional

$$\mathcal{F}[\mathbf{u}(\tau)] = \frac{1}{2}\int_0^L d\tau[u_1^2(\tau) + u_2^2(\tau)]. \tag{B.7}$$

Here $\mathbf{u} = \{u_1, u_2\}$. By analogy with results of Appendix A, e.g. see Eqs. (A.8) and (A.9), it is possible to rewrite Eqs. (B.6) in the matrix form. Instead of Eq. (A.10), we obtain,

$$\frac{d}{dt}\begin{pmatrix} \cos\theta & -\sin\theta & x \\ \sin\theta & \cos\theta & y \\ 0 & 0 & 1 \end{pmatrix} = \begin{pmatrix} \cos\theta & -\sin\theta & x \\ \sin\theta & \cos\theta & y \\ 0 & 0 & 1 \end{pmatrix} \cdot \begin{pmatrix} 0 & -u_2 & u_1 \\ u_2 & 0 & 0 \\ 0 & 0 & 0 \end{pmatrix}. \tag{B.8}$$

Here, the matrices belonging to the Heisenberg–Lie group, are replaced by the matrices belonging to the Lie group $E(2)$. This group is a semi-direct product of translations and rotations of $\mathbf{R}^2$, that is $E(2) = \mathbf{R}^2 \ltimes S^1$. Interested readers can find another application of this group to polymer physics, [64]. In the meantime, following the same logic as given in Appendix A, the Lie algebra for $E(2)$ is easily obtainable. If E_{ij} represents 3×3 matrix which has 1 at ijth place and 0 in the rest of the places,

---

[b] Incidentally, the same problem has its utilization in robotics and computer vision [B.3]. In the case of robotics it can be formulated as follows. Consider a mobile robot moving on the plane $\mathbf{R}^2$. It can move backward and forward and rotate around itself. The state of the robot is described in terms of coordinates (x, y) of its center-of-mass and by the angle of orientation θ. Given the initial and terminal state of the robot, its path (the length) is measured in space $M = (x, y, \theta)$. The task lies in finding the shortest path from the initial to final state.

the Lie algebra $\mathfrak{e}(2)$ can be represented as $\mathfrak{e}(2) = \text{span}\{\mathbf{e}_1, \mathbf{e}_2, \mathbf{e}_3\}$ with $\mathbf{e}_2 = E_{21} - E_{12}, \mathbf{e}_1 = E_{13}, \mathbf{e}_3 = -E_{23}$. Therefore, instead of the Lie algebra, Eq. (A.18), now we obtain

$$[\mathbf{e}_1, \mathbf{e}_2] = \mathbf{e}_3, \quad [\mathbf{e}_2, \mathbf{e}_3] = \mathbf{e}_1, \quad [\mathbf{e}_1, \mathbf{e}_3] = 0. \tag{B.9}$$

Notice that if $L = \text{span}\{\mathbf{e}_1, \mathbf{e}_2, \mathbf{e}_3\}$, then $L \supset L^{(1)} = \text{span}\{\mathbf{e}_2, \mathbf{e}_3\} \supset L^{(2)} = 0$. Thus, the Lie algebra $\mathfrak{e}(2)$ is solvable but not nilpotent.

Let

$$\mathbf{X} = \begin{pmatrix} \cos\theta & -\sin\theta & x \\ \sin\theta & \cos\theta & y \\ 0 & 0 & 1 \end{pmatrix},$$

so that $\mathbf{X} \in E(2)$. Then, the analog of Eq. (A.21) can be written as

$$\mathbf{X}^{-1}\dot{\mathbf{X}} = u_1\mathbf{e}_1 + u_2\mathbf{e}_2. \tag{B.10}$$

This allows us to write (taking Eq. (6.3.10a) into account)

$$\dot{\mathbf{r}}(\tau) = u_1 X_1 + u_2 X_2, \tag{B.11}$$

where the operators X_i, $i = 1$–3 obey the same commutation relations as in Eq. (B.9) and can be realized as

$$X_1 = \cos\theta\frac{\partial}{\partial x} + \sin\theta\frac{\partial}{\partial y}, \quad X_2 = \frac{\partial}{\partial\theta}, \tag{B.12}$$

and since $[X_1, X_2] = X_3$, we also obtain

$$X_3 = \sin\theta\frac{\partial}{\partial x} - \cos\theta\frac{\partial}{\partial y}. \tag{B.13}$$

It is of interest to check the correctness of Eq. (B.11) following the same methodology as it was done in Eq. (6.3.9). This time we have

$$\dot{\mathbf{r}}(\tau) = \dot{x}\partial_x + \dot{y}\partial_y + \dot{\theta}\partial_\theta. \tag{B.14}$$

This should be compared against Eq. (B.11). To do so, we should do the following. First, we formally rewrite Eq. (B.11) as

$$\begin{aligned}\dot{\mathbf{r}}(\tau) &= u_1 X_1 + u_2 X_2 + u_3 X_3 \\ &= u_1\left(\cos\theta\frac{\partial}{\partial x} + \sin\theta\frac{\partial}{\partial y}\right) + u_2\left(\frac{\partial}{\partial\theta}\right) + u_3\left(\sin\theta\frac{\partial}{\partial x} - \cos\theta\frac{\partial}{\partial y}\right) \\ &= (u_1\cos\theta + u_3\sin\theta)\frac{\partial}{\partial x} + (u_1\sin\theta - u_3\cos\theta)\frac{\partial}{\partial y} + u_2\left(\frac{\partial}{\partial\theta}\right).\end{aligned} \tag{B.15}$$

By comparing Eqs. (B.14) and (B.15) it follows that

$$\begin{aligned}\dot{x} &= u_1 \cos\theta + u_3 \sin\theta,\\ \dot{y} &= u_1 \sin\theta - u_3 \cos\theta,\\ \dot{\theta} &= u_2.\end{aligned} \tag{B.16}$$

From here we obtain:

$$\begin{aligned}u_1 &= \dot{x}\cos\theta + \dot{y}\sin\theta,\\ u_2 &= \dot{\theta},\\ u_3 &= \dot{x}\sin\theta - \dot{y}\cos\theta.\end{aligned} \tag{B.17}$$

It remains to demonstrate that $u_3 = 0$. Using Eqs. (B.6) we multiply the first equation by $\sin\theta$, the second by $\cos\theta$. Then, we subtract one of these equations from the other providing us with the nonholonomic constraint

$$\dot{x}\sin\theta - \dot{y}\cos\theta = 0 \tag{B.18a}$$

leading to the horizontal 1-form ω

$$\omega = \sin\theta dx - \cos\theta dy. \tag{B.18b}$$

By direct computation we obtain the already familiar nonintegrability condition $\omega \wedge d\omega = -dx \wedge dy \wedge d\theta \neq 0$. Thus, the $\ker\omega \rightarrow \omega = 0$ provides us with the horizontality condition guaranteeing that $u_3 = 0$ thus returning back to Eq. (B.10). The vector fields X_1 and X_2 are fixed by this 1-form since by construction they obey the horizontality condition, Eq. (6.3.8): $\omega(X_1) = \omega(X_2) = 0$. This means that we can write $\Gamma(\mathcal{D}_p) = \text{span}\{\mathbf{e}_1, \mathbf{e}_2\} = \text{span}\{X_1, X_2\}$ while $\text{span}\{X_1, X_2, X_3\}$ generates the tangent space for semidirect product $\mathbf{R}^2 \ltimes S^1$. Vector X_3 is surely vertical. It is the Reeb vector since $\omega(X_3) = 1$, e.g. see Eq. (6.1.2a). Use of Eqs. (6.3.16) and (6.3.18) leads us to the kinetic energy, Eqs. (B.7), which is exact analog of earlier derived Eq. (6.3.21).

To understand these results better, we need to construct the Pontryagin-type analog of the Hamiltonian, Eq. (A.2.6). In the present case we obtain

$$\hat{H} = p_x u_1 \cos\theta + p_y u_1 \sin\theta + p_\theta u_2 - \frac{1}{2}(u_1^2 + u_2^2). \tag{B.19}$$

Using it, we get

$$\begin{aligned}\dot{x} &= \frac{\partial \hat{H}}{\partial p_x} = u_1 \cos\theta,\\ \dot{y} &= \frac{\partial \hat{H}}{\partial p_y} = u_1 \sin\theta,\\ \dot{\theta} &= \frac{\partial \hat{H}}{\partial p_\theta} = u_2.\end{aligned} \tag{B.20}$$

Also, by analogy with Eq. (A.27), we obtain,

$$\frac{\partial \hat{H}}{\partial u_1} = \frac{\partial \hat{H}}{\partial u_2} = 0.$$

That is

$$p_x \cos\theta + p_y \sin\theta = u_1 \quad \text{and} \quad p_\theta = u_2. \tag{B.21}$$

Using these results back in Eq. (B.19) we obtain the exact analog of the Hamiltionian Eq. (6.2.4) for Euler's elastica

$$\mathcal{H} = \frac{1}{2}(p_x \cos\theta + p_y \sin\theta)^2 + \frac{1}{2}p_\theta^2. \tag{B.22}$$

With such a Hamiltonian we can follow the protocol described in Chap. 6. Specifically, Eqs. (6.3.3) for normal geodesics now acquire the following form

$$\begin{aligned}
\dot{x} &= (p_x \cos\theta + p_y \sin\theta)\cos\theta, \\
\dot{y} &= (p_x \cos\theta + p_y \sin\theta)\sin\theta, \\
\dot{\theta} &= p_\theta, \\
\dot{p}_x &= \dot{p}_y = 0, \Rightarrow p_x = C_1,\ p_y = C_2, \\
\dot{p}_\theta &= -(p_x \cos\theta + p_y \sin\theta)(p_y \cos\theta - p_x \sin\theta).
\end{aligned} \tag{B.23}$$

By combining the third and the sixth equations we obtain the following Newton's equation of motion

$$\ddot{\theta} = -\frac{1}{2}\frac{\partial}{\partial\theta}(p_x \cos\theta + p_y \sin\theta)^2 \equiv -\frac{\partial V}{\partial\theta}. \tag{B.24}$$

Since both p_x and p_y are constants, the above equation can be further rearranged as follows. First, notice that

$$p_x \cos\theta + p_y \sin\theta = A\sin(\theta + \alpha), \tag{B.25}$$

where $A = \sqrt{p_x^2 + p_y^2}$ and $\alpha = \tan\frac{p_x}{py}$. Next, it is always possible to eliminate the phase α and the amplitude A while solving Eq. (B.24). Finally, we shall also use the identity, Eq. (B.25), in the sub-Riemannian Hamiltonian, Eq. (B.22). Using these results and the fact that $\sin^2\theta = \frac{1}{2}(-\cos 2\theta + 1)$, Newton's Eq. (B.24) becomes the equation for Euler's elastica

$$\ddot{\Theta} + \sin\Theta = 0 \tag{B.26}$$

which is just properly rescaled Eq. (8.3.3.9) of the main text in which we have to put $J = 0$.

The obtained results allow us to switch back and forth between classical and quantum mechanical treatments of the elastica problem. This can be achieved following the same methodology as was used in the main text for treatment of motion of the charged particle in the magnetic field as described by the Hamiltonian, Eq. (6.2.4). To go from classical Hamiltonian to quantum requires us to replace the generalized momenta in Eq. (6.2.4) by their operator forms given in Eq. (6.1.1). It is general rule that the sub-Riemannian Hamiltonian is quadratic form in the associated generalized momenta (classical or quantum). This is so because in sub-Riemannian spaces we have to use the Carnot–Carateodory distance, Eq. (6.3.24), so that the fundamental

Theorem B.2.1. (Rashevski–Chow theorem, [167]) *If D_p is bracket-generating distribution (at $p \in M$) on a connected manifold M, then any two points a and b of M can be joined by a horizontal piecewise curve*

can be applied. Notice, that the commutator, Eq. (6.1.2a), coincides with that given by Eq. (A.18) so that the Heisenberg group in mathematics is larger than the Weyl group (e.g. see Eq. (6.1.2.b)) used in physics. The same is true in the case of the commutator algebra, Eq. (B.9), realized by the operators Eqs. (B.12) and (B.13). Thus, if we would like to relate the classical Hamiltonian, Eq. (B.22), to its quantum counterpart, we have to require for the quantum counterpart to "live" on wave functions of the type $\psi(x, y, \theta) = \varphi(\theta) \exp(ip_x x + ip_y y)$. In such a case using Eq. (B.25) and the fact that $\sin^2\theta = \frac{1}{2}(-\cos 2\theta + 1)$, the quantum version of $\mathcal{H}$ is given (up to a constant additive tern and up to proper rescaling) by

$$\check{\mathcal{H}} = -\frac{1}{2}\left(\frac{d}{d\theta}\right)^2 - \kappa \cos 2\theta, \tag{B.27}$$

where κ is some (known) constant.

More on elastica, our readers could find in the recent review paper by Zelikin [B.4] where, among other things, they will find that Euler's elastica was a subject matter of Max Born's PhD. thesis of 1906. Not surprisingly, the knowledge acquired from this work had helped him later on in formulation (along with his younger research associates Pascual Jordan, Werner Heisenberg and Wolfgang Pauli) of the basic laws of quantum mechanics in 1925–1926.

Appendix C

Quantum Computers and Quantum Random Walks

C.1 Classical Computers (Basics)

All computers (classical and quantum) require some *algorithm* (a set of instructions) for their operation. By applying an algorithm to its *input*, eventually some *output* is obtained. Typically, both inputs and outputs are some *strings*. These are made of finite sequence of symbols taken from some *finite alphabet*. Normally, but not always, an algorithm is executed with the help of the device called *Turing* machine. This is some kind of *finite state automaton* (e.g. read http://en.wikipedia.org/wiki/Deterministic_finite_automaton).

A Turing machine (TM) is made of the following components:

(a) a finite alphabet S,
(b) a *blank* element $\emptyset \in S$,
(c) a subset $A \subset S$ called the *external alphabet*, the symbol $\emptyset \notin A$,
(d) a finite set Ω whose elements are called *states* of the TM,
(e) an initial state $\omega_0 \in \Omega$,
(f) a *transition function* f: $\Omega \times S \to \Omega \times S \times \{-1, 0, 1\}$.

To better understand (f) we shall assume that the TM has a tape divided into *cells* so that each cell carries one symbol from the alphabet S. Typically, the tape is infinite only to the right. The content of the tape is some infinite sequence $\Sigma = \{\sigma_0, \sigma_1, \ldots, \sigma_i, \ldots\}, \sigma_i \in S$. Also, the TM has a read–write head that moves along the tape and changes σ_i's. The head can read the symbol, say, $\sigma_i(k)$ (where k is position of the kth cell along the tape) and replace it by another symbol also from the set S. Since the TM is finite state automaton, at each step of computation, it is in some state $\omega_j \in \Omega$.

The action of the automaton depends upon (a) the state ω_j and (b) upon $\sigma_i(k)$. The value of the transition function $f(\omega_j, \sigma_i(k)) = (\omega_l, \sigma_m(k); \Delta k)$ contains a new state $\omega_l \in \Omega$, a new symbol $\sigma_m(k) \in S$, and the shift $\Delta k = \{-1, 0, 1\}$, where -1 denotes the shift to the left, $+1$ to the right and 0 no shift at all. The essential feature of such TM lies in the prescription: whenever $k + \Delta k < 0$, the machine stops. The work of finite automaton can be visualized by using some kind of finite directed graph [C.1] in which the vertices are labelled by the set of internal states $\{\omega_j\}$. With such a graph we can associate the universal covering space also a graph (but without cycles) called a *circuit.* Many examples of covering graphs can be found in our work, [C.2]. Already from this observation the connection between computational processes and random walks is evident.

C.2 Boolean Algebra and Number Theory

From popular scientific literature it is known that all digital computers use binary system of numbers for computation. It is also known that the Boolean algebra plays the major role in such computations. But the Boolean algebra is set-theoretic (or logic-theoretic) concept while representation of integer numbers in binary form is number-theoretic concept. Since even the best books on quantum computation, e.g. see [C.3, C.4], use the Boolean algebra for computation without explaining its connections with number theory, we would like to fill in this gap now. We begin with the Boolean algebra.

Let $\mathcal{M}$ be some set on which three operations are defined: $\cup, \cap$ and $\neg$ (that is union of two sets, intersection of two sets and complement of a set). Then, the set-theoretic analog to the set of integers $\mathbf{Z}$ with two binary operations $+$ and $\times$ is the set $\mathcal{P}(\mathcal{M})$ of all parts of $\mathcal{M}$ with just introduced three set-theoretic operations. The properties of addition and multiplication in $\mathbf{Z}$ makes $\mathbf{Z}$ a *commutative ring.* Analogously, the set-theoretic rules for just defined three operations applied to $\mathcal{P}(\mathcal{M})$ makes it a *Boolean algebra.* In most physics books only the simplest Boolean algebra made of just one element, say, $\mathcal{M} = \{a\}$ is considered. In such a case we can relabel a as 1, and $\neg a$ as 0. Then the simplest Boolean algebra is defined by the following multiplication tables

$\cap$	0	1
0	0	0
1	0	1

,

$\cup$	0	1
0	0	1
1	1	1

and

$\neg$	
0	1
1	0

Remark C.2.1. Although the Boolean algebra was just defined set-theoretically, when it is used in the logic and computer science, the notations change. For instance, operations $\cup, \cap$ and $\neg$ are denoted by $\vee$ (OR), $\wedge$ (AND) and $\neg$ (NOT) respectively. In logic it is common to associate 1 with "*true*" and 0 with "*false*" while in the TM 1 is "yes" and 0 is "no". In engineering the symbol "$\vee$" is replaced by "+" so that (using the tables (above)) $0 + 0 = 0$, $0 + 1 = 1 + 0 = 1 + 1 = 1$, [C.1], Chap. 6. However, in higher algebra the symbol "+" is used for mod2 addition, e.g. $0 + 0 = 0$, $1 + 1 = 0$, $0 + 1 = 1 + 0 = 1$. In engineering the same mod 2 addition is denoted by $\oplus$. The same symbol is used (below) for operation of "direct sum" for sets. Next, in engineering the operation $\wedge$ is replaced by $\cdot$ and, sometimes, the dot is also omitted. Our readers should be alert to such notational conventions.

From this example it is clear that one can design the Boolean algebra with two or more elements so that the above multiplication tables should be redone accordingly. The question arises: What makes the above 1-element Boolean algebra so special in physics other than the technological simplicity? Formulated differently: Can we use the Boolean algebras composed of more than one element for arithmetic computations done by computers? According to Kitaev *et al.* [C.3], p. 13, the TM with arbitrary number of states and with *alphabets of any size* can be encoded by strings over the fixed alphabet. Since this property is stated without proof in [C.3], we would like to provide some evidence that this is indeed taking place. In doing so we follow [C.1]. A quick introduction into the number theory and advanced algebra can be found in our work, [C.5].

Definition C.2.1. A ring is an algebraic system $R = [R, +, \times]$ with two binary operations + and $\times$ which is: (a) an Abelian group under +, (b) a monoid under $\times$, and (c) which satisfies the distribution laws $a(b + c) = ab + ac$ and $(a + b)c = ac + bc \; \forall \, a, b, c \in R$.

Definition C.2.2. A monoid $[S, \beta]$ is a set with binary operation $\beta(x, y) = xy$ satisfying the following conditions: (a) $x(yz) = (xy)z \forall \, x, y, z \in S$ (associativity), (b) $\exists 1 \in S$ and $\forall x \in S$ the following equalities are satisfied $1x = x1 = x$.

It is intuitively well known that the integers $\mathbf{Z}$ form a ring $\mathbf{Z} = [\mathbf{Z}, +, \times]$ in which division is not always possible. Among rings the simplest are commutative rings, that is such that $ab = ba \; \forall \, a, b \in R$. These include $\mathbf{Z}$ and $\mathbf{Z}_p$, where the last one is the set of integers mod p where p is some prime number. If ring can be extended to include division, it becomes a field. A field has no zero divisors. Familiar examples of fields are $\mathbf{Q}$: the field of rational numbers and, $\mathbf{C}$: the field of complex numbers. It can be proven that only if p is some prime number, $\mathbf{Z}_p$ becomes a field. Simple and natural generalization of rings and fields of numbers are the *polynomial* rings and fields [C.5]. This can be understood if we recall that, say, in mod 2 number system any number $N \in \mathbf{Z}$ can be represented as some kind of a polynomial, e.g. $N = a_0 + a_1 2^1 + a_2 2^2 + \cdots + a_n 2^n$, provided that $a_i \in [0,1] \forall i$. Thus, we can define a polynomial ring $F_2[x]$ if we replace 2's by x in the above expression, etc. Thus, in general, the set of all polynomials with coefficients in the commutative ring R is a polynomial ring $R[x]$. Polynomial rings are used in computer science for coding [C.1].

Exercise C.2.1. Investigate how to adopt the formalism of quantum computation for multiplication of polynomials.

Definition C.2.3. Let R and S be any two rings, then their *direct sum* $R \oplus S$ is the set of all ordered pairs (x, y), $x \in R, y \in S$. Addition, subtraction and multiplication are defined in $R \oplus S$ by

$$(r,s) \pm (r',s') = (r \pm r', s \pm s'),$$
$$(rs)(r's') = (rr', ss').$$

The unit element in $R \oplus S$ is $(1_R, 1_S)$.

Using these definitions let us design the direct sum of Boolean rings which are intimately connected with the Boolean algebras. For this purpose consider a set of h-component vectors $\mathbf{x} = (x_1, x_2, \ldots, x_h)$, where $x_i \in [0,1], i = 1 - h$. This set can be considered either as the Boolean algebra $\mathcal{B}^h = [2^h, \wedge, \vee, \neg, O, I]$, where O and I are the so-called *universal bounds*,[a] or as the Boolean ring $\mathbf{Z}_2^h$ with operations

$$\mathbf{x} \cdot \mathbf{y} = \mathbf{x} \wedge \mathbf{y} = (x_1 y_1, \ldots, x_n y_n) = (x_1 \wedge y_1, \ldots, x_h \wedge y_h) \quad \text{and}$$
$$\mathbf{x} + \mathbf{y} = (\mathbf{x} \wedge \neg \mathbf{y}) \vee (\neg \mathbf{x} \wedge \mathbf{y}).$$

It should be noted finally, that multiplication in the Boolean ring is idempotent, that is $\mathbf{x} \cdot \mathbf{x} = \mathbf{x}$. This information in our hands is sufficient for solving problems 3 and 4 from Exercises C, [C.1], p. 301. These are formulated as

Exercise C.2.2. (a) Show that any two-element Boolean ring is isomorphic with $[\mathbf{Z}_2, +, \times]$. (b) Show that if R is any finite Boolean ring, then $R \cong \mathbf{Z}_2$

[a] These can be be easily understood on a specific example. Let M be a set, $M = [1, 2, \ldots, n]$ then the Boolean algebra $B^n = [\mathcal{P}(M), \wedge, \vee, \neg, \emptyset, M]$.

$\cdots \mathbf{Z}_2 = \mathbf{Z}_{2^n}$ for some $n \in \mathbf{N}$. The product of $\mathbf{Z}_2 \cdots \mathbf{Z}_2$ contains r-factors and $r \neq n$. The factor r depends on the number of elements in finite Boolean ring.

Solution of these exercises supplies proofs to statements made on p. 13 of [C.3]. Thus, from now on we shall be dealing only with $[\mathbf{Z}_2, +, \times]$ rings or their direct sums.

C.3 Classical Computers: Turing Machines versus Classical Circuits

In C.1 we noticed that the TM can be replaced by circuits. Topologically, these are covering graphs which are acyclic. They always contain vertices of degree zero serving as inputs. All other vertices are logical gates (or just gates). The gates are selected from some fixed set $\mathcal{A}$ of Boolean functions. A circuit C over $\mathcal{A}$ is a sequence of assignments [C.3]. They involve n input variables $x_1, \ldots, x_n$ and several auxiliary variables $y_1, \ldots, y_m$. The jth assignment has the form $y_j := f_j(u_1, \ldots, u_r)$. Here $f_j \in \mathcal{A}$ and each of u_i's is either some x_i or an auxiliary variable that precedes y_j. Evidently the input variables are assigned at the very beginning and only then variables y_1, y_2, etc. are being assigned. Under such conditions the value(s) of the last auxiliary variable(s) is(are) the *result of computation.* As a basis set for $\mathcal{A}$ it is possible to choose only three Boolean functions: $NOT(x) = \neg x$, $OR(x_1, x_2) = x_1 \vee x_2$, $AND(x_1, x_2) = x_1 \wedge x_2$. It can be proven that such a basis is *complete* [C.3], p. 19. In fact, it is even overcomplete. Instead, it is possible to choose either $[\neg, \vee]$ or $[\neg, \wedge]$. Sometimes, it is more convenient to choose $[\oplus, \wedge]$, etc. It should be noted that these are theoretical results. If we want to implement them in real physical world, then the selection might be somewhat different but surely reducible to those just listed. This forced by-pass is caused exclusively by technological considerations [C.6].

C.4 Reversible Computation (Conservative Logic) and Billiard Ball Computers

Since ultimately we are interested in quantum computers (and quantum random walks), we have to look at classical reversible computation based on conservative logic (to be defined momentarily). It comes closest to quantum computers.

Since any circuit is a graph topologically, the question arises: Is it possible, based on the output data, to recover the input data? The answer is: Not always. For example, following Feynman [C.6], p. 153, consider the classical AND logical gate. It has two inputs and one output. If the output is in state 0, then one of the three possible states could lead to it since the "truth table" for such a gate is

A	B	$A \wedge B$
0	0	0
0	1	0
1	0	0
1	1	1

Such a gate is called "irreversible". The OR gate is irreversible too. In addition, the irreversibility might also come from the irreversible heat losses during computation. Careful analysis of computational steps demonstrated that the entropy loss (and, hence, the heat loss) comes from resetting the computer needed for its next round of computation. Quantum computers perform their logical operation reversibly (if decoherence is ignored) because quantum computation is always associated with rotation performed by a sequence of unitary transformations (gates) of some initial (input) vector to its final state, the output is in the Hilbert space. Unitarity insures reversibility. Since quantum computers are reversible, the question arises: Are there classical computers which are reversible? If the answer to this question is "yes", then the topology of classical circuit and quantum circuit must be the same because, by definition, quantum computer should be capable of making the same kind of computations which classical reversible computer can do. Thus, the task lies in designing classical reversible computer explicitly. It was completed by Fredkin and Toffoli in 1982 [C.7]. They come up with beautifully simple billiard ball model (BBM) to be described momentarily. The rules for this model are the same as for the Maxwellian ideal gas whose molecules/atoms are made of absolutely rigid hard balls of equal size so that the collisions between atoms or between atoms and walls are elastic. Since the collision between two balls depends upon their initial location and direction and may or may not be facilitated by the presence of walls (reflectors), it can be described in terms of the logical gates. When the BBM is initialized appropriately, *the sequence of states that appear at successive integer time-steps is equivalent to discrete digital dynamics.* Hard balls travel on the two-dimensional plane which is a lattice of exactly the same type as in Sec. 8.3.4. That is to say, the initial square lattice is counterclockwise rotated by 45° and the balls are moving along the diagonals of squares of the original lattice or, which is equivalent, along the edges of the rotated lattice: from one lattice site to the next, at each tick of the clock. If the length of edges is chosen to be 1, then the ball radius is $1/\sqrt{2}$. Such a choice excludes the head-on collisions so that each elastic collision changes direction of the ball by 90°. *Then the path of a ball becomes the same as for the massive particle in the discretized Dirac propagator as described in* Sec. 8.3.4. The presence or absence of a ball at a given point of a grid is encoded by the binary variable taking values 1 and 0 respectively at each integral value of time. The simplest logical gate (the "interaction gate") is depicted below, in Fig. 1. When there is no ball at A, but there is ball at B, it will continue along the straight path. The same happens if A is present but B is not. If balls are present at both A and B, then we shall get scattering as depicted. Incidentally, the idea to use scattering for logical operations was implemented at the quantum mechanical level only very recently in [C.8]. In the same reference this idea was further developed culminating in suggested design of quantum computer. Unfortunately, the gate shown in Fig. 1 is not sufficient for realistic computation even at the classical level. Accordingly, another gate (called the "switch gate") depicted in Fig. 2 was designed by Ressler in his MIT master's thesis [C.9]. The "truth" tables for both the interaction and switch gates are given below:

Interaction gate

A	B	1	2	3	4
0	0	0	0	0	0
0	1	0	1	0	0
1	0	0	0	1	0
1	1	1	0	0	1

Switch gate

A	B	1	2	3
0	0	0	0	0
0	1	1	0	0
1	0	0	0	1
1	1	0	1	1

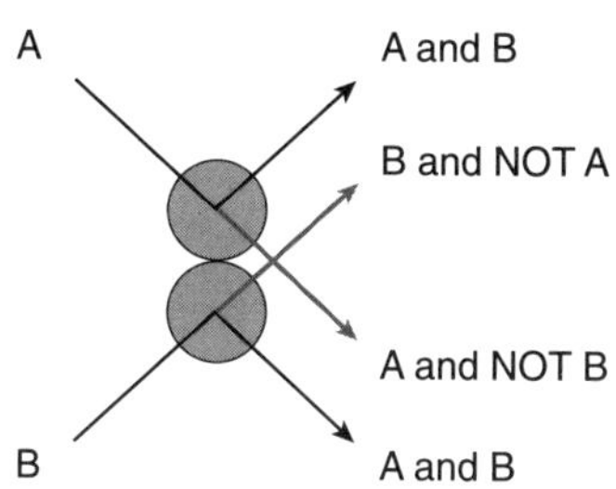

Fig. 1 Logical structure of the interaction gate.

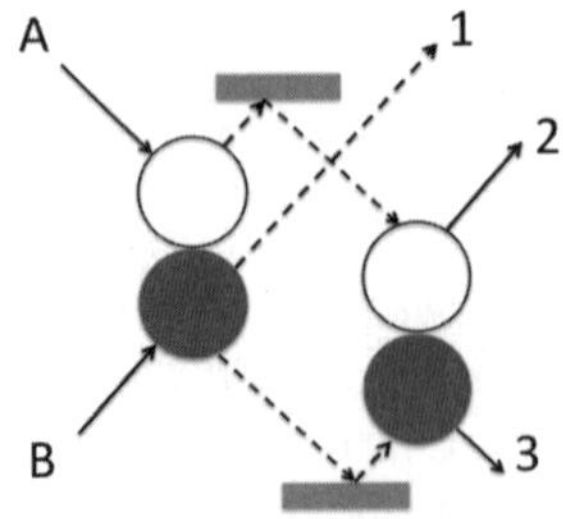

Fig. 2 The Ressler–Feynman switch gate.

Finally, using some combination of switch gates the most important for our purposes, the universal Fredkin gate for the BBM, was originally designed by Ressler and Feynman [C.7].[b] It shares with the switch gate the following important control property which can be formulated as follows. Irrespective of presence or absence of the ball B, the output 3 is 1 if the A ball is present and 0 if it is absent. The A ball is called the "control" ball.

The Fredkin gate can be visualized as box with 3 inputs, say, a, b and c and 3 outputs, say, a', b' and c'. One of the inputs, say, c is control whose value is not changed by the action of the gate, that is *always* $c = c'$. By design, if $c = 0$, then $a = a'$ and $b = b'$, while if $c = 1$, then a and b are swapped, that is $a' = b$ and $b' = a$. The truth table for the Fredkin gate is given below:

Fredkin gate

a	b	c	a'	b'	c'
0	0	0	0	0	0
0	0	1	0	0	1
0	1	0	0	1	0
0	1	1	1	0	1
1	0	0	1	0	0
1	0	1	0	1	1
1	1	0	1	1	0
1	1	1	1	1	1

From the above table it follows that topologically Fredkin's gate is acting as braiding operator for colored braids. Evidently, there are two colors in the present case: 0 and 1. Information about the colored braids can be

[b] Another universal gate, the Toffoli gate [C.4], can be constructed from the Fredkin's gate. It takes 3 Fredkin's gates to make one Toffoli's gate and vice versa [C.10].

found in [C.11]. Thus both *classical (reversible) and quantum computations are topological in nature.* The question arises: How does such described topological computation related to that advocated by Kitaev [C.12]? The answer to this question was recently discussed in some detail by Zhang [C.13] who explained the distinction between (a) the "topological quantum computing" [C.12] and (b) the "integrable quantum computing". We shall define (b) momentarily. Before doing so, we would like to mention a few related things. In particular, we would like to mention the recent paper by Garnerone *et al.* [C.14] where colored braid formalism was used for quantum computation of the colored Jones polynomial.

Exercise C.3. Use results of [C.14] in order to improve results by Zhang [C.13] whose work does not involve colored braids.

Next, we would like to finish our discussion of the Fredkin gate. For this purpose, we recall from Sec. C.3 that classical circuits involve uses of AND, OR and NOT gates and those which can be obtained using just mentioned in various combinations. Since the gates for such operations are all irreversible, the question arises: How Fredkin's manifestly reversible gate is related to the gates just mentioned? Another related question is: Is it possible to implement the above irreversible gates using BBM model? It happens that all the above-mentioned irreversible gates can be obtained *directly* from the reversible Fredkin's gate. In the case of BBM model the construction of the above gates was recently discussed in [C.15]. At more abstract level the same construction is outlined in [C.4]. Here we provide this construction, including some details missing in [C.4].

To create the AND gate, look first at the "truth table" for $\cap$. Then, for the Fredkin gate choose as input c and b for construction of AND gate while make a' as output. Furthermore, set once and for all $a = 0$ as auxiliary input. Such type of auxiliary bit (that is a whose value is fixed) is called *ancilla* (or "work") bit.

To create the NOT gate, look first at the truth table for NOT. Then, make c as input and a' as output. In addition, make $b = 0$ (this is another ancilla). This gate is also working as FANOUT gate [C.6]. By definition, an input for such a gate gets copied (duplicated) at the output. In the present case, for such a gate we have to choose c as input and b', c' as output while keeping $b=0$. According to Sec. C.3 the $\mathcal{A}$ basis is complete with just AND and NOT gates. This means that Fredkin's gate is universal gate in the sense that all other gates, including Toffoli's gate, can be built out of Fredkin's gate.

The moral of this story is the following. First, we have to have a sufficient supply of ancillas in the prescribed states: 0 or 1. Second, we have to be prepared to receive some of unwanted outputs from Fredkin's gates called *garbage* in literature. The irreversibility is caused by the preparation of ancillas in prescribed states and by the necessity to remove the garbage from the output. Dealing with these problems in full generality is at the forefront of current research [C.16].

The question emerges: Is it possible to replace the Fredkin gate by just the switch gate, e.g. that depicted in Fig. 2? In more abstract terms, by the gate with two inputs a and b and two outputs $a' = b$ and $b' = a$? If this would be possible, this would be equivalent of replacing colored braids by colorless. Clearly, this is not possible because such switch gate would lose its status of being the universal gate. Nevertheless, an attempt to build a computer (quantum computer) using such non-universal gate was announced in [C.17] and subsequently attracted a lot of attention. An attempt to realize this theoretical program using quantum dots was made in equally popular [C.18]. The latest progress in this direction is summarized in [C.19]. These results allow us to provide promised explanation of what should be understood as "integrable quantum computing" — term introduced by Zhang [C.13].

C.5 Boolean Algebra, Directed Random Walks, Yang–Baxter Equations and Integrable Quantum Computation

We begin with discussion of Exercises in Sec. C.2. Let n be some non-negative integer. Furthermore, let $n = \mathcal{M}$ and introduce $\mathcal{P}(\mathcal{M})$ as in Sec. C.2. Now we shall prove that $\mathcal{P}(\mathcal{M})$ contains exactly 2^n elements. This is the size of the n-fold Cartesian product $\{0,1\}^n = \{0,1\} \times \cdots \times \{0,1\}$. Said differently, there is a bijection between $\mathcal{P}(\mathcal{M})$ and all n-tuples of 0's and 1's. To prove this, following [C.20], let us consider a specific example. Let $n = 21$ and $A = \{1, 3, 4, 6, 7, 10, 12, 13, 16, 19\} \subset \mathcal{P}(21)$. We can depict A by a system of parenthesis which we form by the rule: if $i \in A$ the parenthesis associated with i is right, otherwise it is left. Specifically, in our case we have:

$$\begin{array}{ccccccccccccccccccccc}) & (&) &) & (&) &) & (& (&) & (&) &) & (& (&) & (& (&) & (& (\\ 1 & 2 & 3 & 4 & 5 & 6 & 7 & 8 & 9 & 10 & 11 & 12 & 13 & 14 & 15 & 16 & 17 & 18 & 19 & 20 & 21. \end{array}$$

We can represent the same by a sequence of 0's and 1's, e.g. 1011011001011 00100100. This allows us to make the additional bijection onto directed random walk of exactly the same nature as in the main text, in Sec. 8.3.4. Specifically, we rotate the two-dimensional square lattice counterclockwise by 45°. In such rotated lattice we choose the line perpendicular to the original x-axis as time and let the directed walk go along the diagonals of squares of the original lattice: if the step is taken to the left with respect to time axis, we assign 1 to it while if it is taken to the right, we assign 0. Rotate once again the axes so that the time direction is now along the positive x-axis. Then, 1 will correspond to one step "up" while 0 to one step "down". Evidently, such a walk is in bijective correspondence with $\mathcal{P}(\mathcal{M})$. Furthermore, we can order different walks by the principle (the Gray code) that encoding of neighboring walks differ by just one element. Such an order corresponds to some *Hamiltonian cycle* in the n-dimensional hypercube. The cycle is called "Hamiltonian" if it visits exactly once each node of a hypercube (or, more generally, the node of some graph). The task of finding the Hamiltonian cycle is directly related to solving the *traveling salesman problem.*[c] In 1994, Adleman suggested solving Hamiltonian cycle problem using DNA computer.[d] Since DNA computing belongs to the domain of classical computers discussed in C.1, we shall not cover this topic further in this Appendix. Instead, we shall connect just obtained results with classical and quantum Yang–Baxter equations. To do so several steps are required.

First, we begin with yet another bijection. To describe it, we need to define the Young diagram (tableaux). For this purpose, let n be some positive integer. A partition $\lambda = \{\lambda_1, \dots, \lambda_l\}$ of n shall denoted as $\lambda \vdash n$. More accurately, [173], a partition of n is defined by the set of all solutions in non-negative integers of the equation $\lambda_1 + \cdots + \lambda_l = n$. The number $N(n, l)$ of such solutions is $\binom{n+l}{l}$. The Young diagram is associated with such a partition.[e] To construct this diagram, on the square lattice, we choose $n \times l$ rectangle so that the Young diagram will occupy the left upper part of this rectangle with the upper left vertex of the rectangle being chosen as the origin of xy coordinate system whose y-axis is directed downwards while the x-axis eastwards. Then, the south–east boundary of the Young diagram can be interpreted as directed random walk which begins at $(0, -l)$ and ends at $(n, 0)$. *Such defined walk completely determines the Young diagram.* The walk can be described by the sequence of 0's and 1's (from here we again obtain a bijection!). Let, say, we denote by 0 the x-step move while by

[c] E.g. read http://en.wikipedia.org/wiki/Travelling_salesman_problem
[d] E.g. read http://en.wikipedia.org/wiki/DNA_computing
[e] See also Appendix D, Sec. D.2.10.

1 the y-step move. We shall use a notation $\omega(\lambda) := (\omega_1, \ldots, \omega_n)$. Evidently, $\omega_1 + \cdots + \omega_n = l$. The totality N of the Young diagrams can be placed into rectangle is in one-to-one correspondence with the number of arrangements of 0's and 1's. An easy calculation provides: $N = N(n, l)$.

Next, we associate with each partition λ the Schur polynomial $s_\lambda(\mathbf{x})$. To define $s_\lambda(\mathbf{x})$ explicitly, introduce $\mathbf{x}^{\mathbf{a}} := x_1^{a_1} \cdots x^{a_n}$. Then, we have to decide how to use the partition $\lambda \vdash n$ for filling of boxes of the Young diagram so that the shape of the diagram is determined by the partition λ. The Young tableaux T is *semistandard* if the filling of boxes is weakly increasing across each row and is strictly increasing down each column. Let us consider an example. The tableaux for partition $\lambda = \{3, 2, 2, 1\}$ can be made out of three boxes in the first row, two in the second and third, and of one in the fourth. At the same time, the filling μ of the same tableaux can also be encoded by $\mu = \{2, 3, 2\}$ meaning that we can choose 2 1's , 3 2's and 2 3's filling boxes of such diagram in accordance with the semistandard prescription. Let $x_i^{a_i}$ in $\mathbf{x}^{\mathbf{a}}$ be assigned to keep track of the number a_i of i's in T (e.g. 2 is the number of 1's, 3 is the number of 2's, etc.), then

$$s_\lambda(\mathbf{x}) = \sum_{\mathrm{T}} \mathbf{x}^{\mathrm{T}}. \tag{C.1}$$

For instance, let $\lambda = (2, 1)$ so that $n = 3$ and the Schur polynomial is given by

$$s_{(2,1)}(x_1, x_2, x_3) = x_1^2 x_2 + x_1^2 x_3 + x_1 x_1^{2`} + 2x_1 x_2 x_3 + x_1 x_3^{2`} + x_2 x_3^{2`} + x_2^2 x_3. \tag{C.2}$$

In addition, the Schur polynomials are characterized by their scalar product

$$\langle s_\lambda, s_\mu \rangle = \delta_{\lambda,\mu} \tag{C.3}$$

and by their fusion (composition) rule

$$s_\lambda \circ s_\lambda = \sum_\nu C^\nu_{\lambda\mu} s_\nu, \tag{C.4}$$

where the Littlewood–Richardson (L-R) coefficients $C^\nu_{\lambda\mu}$ can be formally defined by $C^\nu_{\lambda\mu} = \langle s_\lambda \circ s_\lambda, s_\nu \rangle$. There are many ways to calculate the L-R coefficients. For instance, in our work, [173], they had been calculated using the so-called Heisenberg honeycombs. These are combinatorial objects which were introduced originally by Heisenberg in his first seminal work on quantum mechanics. Clearly, Heisenberg was not using such a terminology. Instead, following Bohr, he begins with Bohr's postulate for energy. In Heisenberg's notations it can be written as

$$\omega(n, n-\alpha) = \frac{1}{\hbar}[E(n) - E(n-\alpha)], \tag{C.5}$$

where both n and $n-\alpha$ are some integers. Using Eq. (C.5) he obtained the following fundamental composition law

$$\omega(n-\beta, n-\alpha-\beta) + \omega(n, n-\beta) = \omega(n, n-\alpha-\beta) \tag{C.6a}$$

which he used *instead* of the commutator $[\hat{x}, \hat{p}_x] = i\hbar$ for development of quantum mechanics. To do so, Eq. (C.6a) should be rewritten in the manifestly symmetric form

$$\omega(n, m) + \omega(m, k) + \omega(k, n) = 0 \tag{C.6b}$$

in view of the fact that $\omega(m, k) = -\omega(k, m)$. Knutson and Tao, [C.22] without any reference to work of Heisenberg on quantum mechanics and without any reference to quantum mechanics used Eq. (C.6) for rederiving of the L-R fusion rule, Eq. (C.4). They named Eq. (C.6) as *the honeycomb condition*. It can be visualized graphically as 3-valent vertex graph. Collection of such vertices with properly assigned gluing rules http://www.math.ucla.edu/˜tao/java/Honeycomb.html indeed forms a honeycomb. Dirac [C.21] noticed that the above honeycomb condition can be interpreted as the condition on frequencies for the Fourier components of observables. As such, it leads to the multiplication rule for the Fourier amplitudes

$$a(nm)b(mk) = ab(nk). \tag{C.6c}$$

Since in general

$$ab(nk) \neq ba(nk), \tag{C.6d}$$

he concluded (in accord with Heisenberg who provided Dirac a copy of his manuscript prior to its publication) that the above multiplication rule is characteristic for matrices. From this observation to arrive at (now known as quantization postulate) $[\hat{x}, \hat{p}_x] = i\hbar$ is almost straightforward and, using it, at the rest of quantum mechanics. The quantization condition $[\hat{x}, \hat{p}_x] = i\hbar$ which is considered as postulate in all known textbooks on quantum mechanics was actually *derived* by Heisenberg. In our paper, [173], this derivation is provided in full detail.

Exercise C.5.1. Read [173] along with Heisenberg's original paper on quantum mechanics. Investigate the conditions of validity for the quantization rule $[\hat{x}, \hat{p}_x] = i\hbar$ obtained in Heisenberg's paper and find when these conditions break down. Think about what could happen to the quantization postulate once such conditions are violated. Can such an approach shed some light on the physical foundation of the deformation quantization?[f]

Reading of this book should allow our readers to by-pass entirely Heisenberg's arguments. Various quantization conditions are obtained nonperturbatively. They are topological in nature. The analogous results were achieved in our works, [173, 275], where combinatorics of honeycombs and coagulation-fragmentation processes were used for derivations of quantum mechanics, quantum field and string theories from the same mathematical standpoint.

The unexpected outcome of such a treatment lies in the recognition of the fact that combinatorics formalism makes distinction between classical and quantum realities for all practical purposes not essential since both are being treated with the same set of mathematical tools. For instance, in the case of classical reversible computers their "quantum" nature is hidden in the protocol of how measurements are made. From Sec. C.4. we recall that "When the BBM is initialized appropriately, *the sequence of states that appears at successive integer time-steps is equivalent to discrete digital dynamics*". This fact causes us to think about the analog of the honeycomb condition, Eq. (C.6b), for classical reversible computers. Such an analog is actually easy to find based on results of our already cited work, [173]. Indeed, the orthogonality condition for Schur polynomials, Eq. (C.3), supplies us with the (classical) wave functions, Schur polynomials, living is the (classical?) Hilbert-like space. The fusion rule, Eq. (C.4), defining the L-R coefficients is exactly of the same nature as quantum mechanical Clebsch–Gordan (C–G) fusion rule for addition of angular momenta (or spins, or both) and can be looked upon as the honeycomb condition. In notations of Landau–Lifshitz book [247] the C–G fusion rule can be written as

$$\psi_{jm} = \sum_{m_1,m_2} C^{jm}_{m_1,m_2} \psi_{j_1 m_1} \otimes \psi_{j_2 m_2}, \tag{C.7}$$

where j is the absolute value of the angular momentum, m is its projection onto z-axis and $C^{jm}_{m_1,m_2}$ are the C-G coefficients. Although this form of the C-G expansion is most known in literature on quantum mechanics, it cannot be immediately used for comparison with Eq. (C.4). Fortunately, the same authors notice that the matrix $C^{jm}_{m_1,m_2}$ of C-G coefficients is unitary. This

[f]Read more on this topic in Appendix A.

fact is implying that it is possible to rewrite Eq. (C.7) in equivalent form

$$\psi_{j_1m_1}\otimes\psi_{j_2m_2}=\sum_{j=|j_2-j_1|}^{j_1+j_2}\tilde{C}^{jm}_{m_1,m_2}\psi_{j,m_1+m_2}. \tag{C.8}$$

Here, in view of unitarity, $\tilde{C}^{jm}_{m_1,m_2}$ is the matrix whose coefficients are complex conjugate to $C^{jm}_{m_1,m_2}$. Clearly, Eqs. (C.4) and (C.8) are of the same mathematical nature. Furthermore, the symbol $\circ$ can (and should) be replaced by $\otimes$. The simplest application of Eq. (C.7) is for the spin-1/2 wave functions.

From quantum mechanics we know that $|0,0\rangle=\frac{1}{\sqrt{2}}(|\uparrow\rangle_1|\downarrow\rangle_2-|\downarrow\rangle_1|\uparrow\rangle_2)$ is a singlet while $|1,0\rangle=\frac{1}{\sqrt{2}}(|\uparrow\rangle_1|\downarrow\rangle_2+|\downarrow\rangle_1|\uparrow\rangle_2)$ is the part of the wave function for a triplet. In theory of quantum computers these wave functions correspond respectively to $|\beta_{01}\rangle$ and $|\beta_{11}\rangle$ Bell (or EPR) states/basis (EPR stands for Einstein, Rosen and Podolsky) [C.4], p. 25. Even though there are two additional Bell states: $|\beta_{00}\rangle=\frac{1}{\sqrt{2}}(|\uparrow\rangle_1|\uparrow\rangle_2+|\downarrow\rangle_1|\downarrow\rangle_2)$ and $|\beta_{10}\rangle=\frac{1}{\sqrt{2}}(|\downarrow\rangle_1|\downarrow\rangle_2-|\uparrow\rangle_1|\uparrow\rangle_2)$ which do not immediately fit the traditional protocol of theory of representation of groups, already the first two states $|\beta_{01}\rangle$ and $|\beta_{11}\rangle$ provide examples of the entangled states.

Definition C.5.1. A state of a composite system that cannot be written as a product of states of its component systems is called *entangled state.*

Remark C.5.1. The Bell basis is an orthonormal basis for the two-qubit state space as can be checked by direct computation.

Evidently, the property of states to be entangled should be looked upon as some corollary in the theory of group representations. This means that entanglements should be possible at the classical level since the permutation group works well already at the classical level. This conclusion will be further reinforced in Appendix D, Sec. D.2.18. Nevertheless, we would like to investigate if the entanglement is possible at the level of classical reversible computation based on information provided thus far. It is clear that each Schur polynomial is associated with some binary number, say, with the input. Because of the orthogonality property, Eq. (C.3), for Schur polynomials, such a polynomial is also a wave function! In quantum computations we use a sequence of unitary transformations to bring the initial (input) wave function into its final (output) state. More on this will be said in the next section on quantum random walks. But at this moment we are dealing with (seemingly) classical computation. If this is so, then what is the analog of unitary transformation for the "classical" reversible computation? To provide an answer, we need to introduce the following:

Definition C.5.2. The symmetric group S_n has the following presentation in terms of generators s_i and (Coxeter) relations

$$\begin{aligned} s_i^2 &= 1, && \forall\, i \text{ (reflection property)}, \\ s_i s_j &= s_j s_i && \text{for } |i-j| > 2, \\ s_i s_{i+1} s_i &= s_{i+1} s_i s_{i+1} && \text{(braiding property)}. \end{aligned} \tag{C.9}$$

If there is a set of n elements, the generator s_i interchanges an element i with $i+1$ so that $s_1, \ldots, s_{n-1}$ generates S_n.

Remark C.5.2. Without the requirement $s_i^2 = 1, \forall\, i$ the above presentation coincides with that for the braid group B_n introduced by Artin [C.23].

Based on such definition and the remark following it the question arises: Is it possible to embed B_n into S_n? [C.23] provides an affirmative answer. Specifically, there is a surjective homomorphism $B_n \to S_n$ that is $\pi(\sigma_i) = s_i \forall\, i, i = 1, 2, n-1$. Here σ_i is generator of the braid group.

Definition C.5.3. The homeomorphism π from the set A to the set B is surjective if each element of B is image of at least one element of A.

This homomorphism provides classical analog of unitary transformation and connects combinatorics with topology of braids. Evidently, the topological aspect can be realized macroscopically while the unitarity is present only implicitly since Schur's polynomials formally are the "classical" wave functions.

These results cause us to focus our attention on the permutation group S_n first. Clearly, there are $n!$ permutations in the set of n elements. If the initial (ordered) state is assigned, then any other state can be reached by the successful application of permutation generators to this state so that the "word" $w = s_{a_1} s_{a_2} \cdots s_{a_l}$ can be identified with such (final) state. Here the indices $a_1, a_2, \ldots, a_l$ represent a subset of a set of $n-1$ elements. Evidently, the final state can be reached in many ways so that it makes sense to introduce the *reduced word* w whose length $l(w)$ is minimal. Being armed with these results Lascoux and Scutzenberger [C.24] introduced the operator Δ_i via

$$\Delta_i := \frac{1 - s_i}{x_i - x_{i+1}}. \tag{C.10}$$

It is acting on monomials $\mathbf{x}^{\mathbf{a}}$ (e.g. those entering Schur's polynomials) so that the generator s_i acting on $x_i^{a_i} x_{i+1}^{a_i+1}$ converts it into $x_i^{a_i+1} x_{i+1}^{a_i}$. By design, an action of this operator on monomial is zero if $a_i = a_{i+1}$. Otherwise, it diminishes the degree of monomial by 1. In addition, these authors introduce the operators

$$\bar{\pi}_i := \frac{1 - s_i}{x_i - x_{i+1}} x_{i+1} \tag{C.11}$$

and $\pi_i = 1 + \bar{\pi}_i$. Using these operators it is convenient to introduce the operator $T_i(p, q, r) = p\Delta_i + q\bar{\pi}_i + rs_i$ with p, q and r being some numbers. Evidently, such an operator obeys the braid-type relations. If relation $s_i^2 = 1$ is included into consideration, it is being replaced now by

$$T_i^2 = qT_i + r(q + r). \tag{C.12a}$$

The above relation (along with the braid group relations) with constants p, q and r properly chosen defines the Iwahori–Hecke algebra H_n. Specifically, the standard form [C.23] of $H_n(q)$ is given by

$$T_i^2 = (q - 1)T_i + q\mathbf{1}, \tag{C.12b}$$

where $\mathbf{1}$ is just the identity element. From this definition it is clear that $H_n(q)$ is a deformation of S_n since for $q = 1$ we re-obtain S_n. Thus, we just obtained a sequence of homomorphisms $B_n \to S_n \to H_n$. It can be proven that $\forall\, w \in S_n \,\exists\, T_w$ such that $T_w = T_{i_1} \cdots T_{i_r}$ whenever $w = s_{i_1} \cdots s_{i_r}$ is the reduced word for w.

The question arises: If $H_n(q)$ is deformation of S_n, what will happen with the L-R fusion rule, Eq. (C.4)? The answer happens to depend upon the values the parameter q entering Eq. (C.12b) can take. If q is not, say, m-th root of unity, then the deformation will preserve the fusion rule. The Schur polynomials (they are characters of S_n) will be replaced by the characters of Hecke algebra (see details and references in our honeycomb paper, [173]) which will smoothly go back to s_λ when $q \to 1$.

The well known in knot and braid theory, the Temperley–Lieb algebra $A_n(a)$ defined by the set of relations [C.23]

$$\begin{aligned} e_i e_j &= e_j e_i \quad \text{for } |i - j| > 2, \\ e_i e_j e_i &= e_i \quad \forall\, i, j = 1, 2, \ldots, n-1, |i - j| > 1, \\ e_i^2 &= a e_i, \end{aligned} \tag{C.13}$$

where a is some number. It is related to $H_n(q)$ as can already be seen from comparison between the defining relations. Furthermore, it can be demonstrated [C.23] that $e_i = \frac{a}{q+1}(T_i + 1)$. The surjective homomorphism $H_n(q) \to A_n(a)$ has kernel defined by

$$1 + T_i + T_{i+1} + T_i T_{i+1} + T_{i+1} T_i + T_i T_{i+1} T_i = 0. \tag{C.14}$$

Remark C.5.3. Evidently, the operator $T_i(p, q, r)$ by design is ideally suited for the role of the switch gate *in the classical reversible computations.* Because such a computation is topological, there should be the quantum mechanical analog of such type of computations. Such quantum mechanical analog does indeed exist. We would like to describe it now.

First, following Dunkl, [C.25], it is convenient to replace the operator Δ_i by

$$b_{ij} := -\frac{1 - s_{ij}}{x_i - x_j}. \tag{C.15}$$

By construction $b_{ij} = -b_{ji}$ and, when acting on polynomials made of $x_1, \ldots, x_n$, it permutes indices for x_i and x_j. This means that for $s_{i,i+1}$ we can simply write s_i. Next, we define the *rational* Dunkl operators D_i by

$$D_i = \partial_i + k \sum_{j \neq i} b_{ij}, \tag{C.16}$$

where $\partial_i = \frac{\partial}{\partial x_i}$ and k is some auxiliary number. By design, such defined operator possesses obvious property: $w D_i w^{-1} = D_{w(i)} \forall\, w \in S_n$. Less trivial is to prove that $[D_i, D_j] = 0$. This condition is the well-known zero curvature condition known for all exactly integrable classical systems [173]. By direct computation it can be demonstrated that vanishing of the commutator is contingent upon the constraint

$$[b_{12}, b_{13}] + [b_{12}, b_{23}] + [b_{13}, b_{23}] = 0 \tag{C.17}$$

on operators b_{ij}, Eq. (C.15). The above equation is *known as the classical Yang–Baxter (Y–B) equation.* Solutions of this equations are called the *classical* b-matrices. These were obtained and classified by Belavin and Drinfel'd in [C.26]. Surprisingly, the derivation of all solutions to Eq. (C.17) is extremely sophisticated.

Second, Belavin and Drinfeld found three types of solutions for b_{ij}. The simplest is the "rational" case: $b_{ij}(z) = \frac{\Omega_{ij}}{z}$ where $[\Omega_{ij}, \Omega_{kl}] = 0$ for $i \neq j \neq k \neq l, \Omega_{ij} = \Omega_{ji}$ and $[\Omega_{ij}, \Omega_{ik} + \Omega_{jk}] = [\Omega_{ij} + \Omega_{ik}, \Omega_{jk}]$. These are the infinitesimal pure braid relations. By combining these results with the definition of the operator D_i, the Knizhnik–Zamolodchikov (K–Z) equations are obtained [173]. Other solutions of the classical Y–B equations, trigonometric and elliptic, produce their own K–Z equations which we are not going to discuss in this book.

Third, consider now b_{ij} for the case when $j = i + 1$. Then $b_{i,i+1} = -\Delta_i$. By analogy with $T_i(p, q, r)$ it is now possible to design an operator $\hat{s}_i = s_i + hb_{i,i+1}$ which can also be interpreted as the deformation of s_i. It should be noted that (at least initially) the defining Eq. (C.12) for the Hecke algebra was made for noncomplex parameter q while in the present case the deformation is caused by the complex parameter h. So, this is a different kind of deformation! This leads us to the following:

Definition C.5.4. The *degenerate affine Hecke algebra* for the root system A_{n-1} is the algebra $\tilde{H}_n$ over $\mathbf{C}$ spanned by its two subalgebras $\mathbf{C}[S_n]$ and $\mathbf{C}[x_1, \ldots, x_n] \equiv \mathbf{C}[x]$ of polynomials with complex coefficients defined by relations

$$\begin{aligned} x_{i+1}s_i - s_i x_i &= s_i x_{i+1} - x_i s_i = h, \\ x_i s_j &= s_j x_i \quad \text{if } \neq j, j+1, \end{aligned} \tag{C.18}$$

where $h \in \mathbf{C}$ is fixed constant. In literature its value is typically $i\hbar$ where $\hbar$ is Planck's constant. This is caused by the fact that the first line in Eq. (C.18) can be looked upon as discretized version of more familiar $[\hat{x}, \hat{p}_x] = i\hbar$. When $h = 0$, this algebra is reduced to the semidirect product $\mathbf{C}[S_n] \ltimes \mathbf{C}[x]$ so that for the nonzero h this algebra can be considered as deformation of this semidirect product. This means that when s_i in Eq. (C.18) is replaced by $\hat{s}_i$, with such a replacement Eqs. (C.18) must still hold. Also, the braiding relation

$$\hat{s}_1\hat{s}_2\hat{s}_1 = \hat{s}_2\hat{s}_1\hat{s}_2 \tag{C.19}$$

obeyed by the elements of the Hecke algebra must hold as well. Since $\hat{s}_i = s_i + hb_{i,i+1}$ cannot be brought into the form $T_i(p, q, r)$, it is clear that we are dealing now with the *degenerate affine Hecke algebra.*

Fourth, having this fact in mind, let $R_{12} = s_1\hat{s}_1, R_{23} = s_2\hat{s}_2$ and, finally, $R_{13} = s_1 R_{23} s_1 = s_2 R_{12} s_2$ then, the braid relation for $\hat{s}_i$ implies the (quantum) *Yang–Baxter equation* (YBE) for R_{ij}. That is

$$R_{12}R_{13}R_{23} = R_{23}R_{13}R_{12}. \tag{C.20}$$

Let $R_{ij} = 1 + hb_{ij}$ (or $R_{ij} \simeq \exp(hb_{ij})$ for $h \to 0$), then by substitution of this result into Eq. (C.20) we recover back the classical YBE, Eq. (C.17) as required.

With these results in our hands we still have to find a relationship between these quantum/topological results (seemingly) for quantum computers and those obtained for classical reversible computers. Most likely the attentive reader already noticed that the answer begins with the relationship between the classical and the quantum YBE. Furthermore, as demonstrated by Jimbo [C.27], the representations for both the Hecke algebra and the YBE are interrelated (and even coincide!). Furthermore, according to *Drinfeld–Kohno theorem* [C.28], Theorem 19.4.1, the monodromy representation for the K-Z equation is equivalent to the YBE representation. Thus, the solutions of the K-Z equation can replace the "classical" Schur polynomials (and vice versa) since, as we already mentioned, the characters of Hecke algebra (of the A_{n-1} type — the only type we are using in this and next appendix[g]) in the Coxeter–Dynkin-type classification scheme) for generic values of q will deform back to Schur's polynomials. This also means that the L-R fusion rule can still be used. The new problems emerge only when q is some root of 1. This case is truly quantum as explained by Goodman and Wenzl [C.29]. Under such condition the fusion rule coincides with that for the Verlinde algebra for su(k) Wess–Zumino–Witten (WZW) model.[h] But solutions of the WZW model are solutions of the K-Z equations [C.30] which are related to representations of Hecke algebra. Thus, we come to a conclusion that the borderline between quantum and classical lies exactly when the parameter q is a root of unity.

Remark C.5.4. Solutions of YBE satisfying Hecke algebra, Eq. (C.12b), in which q is not complex (e.g. is not the root of 1) were obtained by Jones [C.31] and Turaev [C.32]. For such solutions the classical L-R fusion rule, Eq. (C.4), holds. Wenzl [C.33] studied much more difficult case of q being the root of unity (see also [C.29]). As we just stated, the fusion rule, Eq. (C.4), should be modified in such a case. This modification is discussed in detail [173]. It results in the "quantum" L-R coefficients which are in fact the Gromov–Witten invariants. A sketch of computation of these invariants is given in the same paper. It can also be shown that in such quantum case the Schur polynomials are being replaced by the Schubert polynomials [D.31, D.36].

In classical reversible computers we dealt with the billiard-like system of hard balls. How can we translate the properties of such a classical system into quantum language? Fortunately, this can be done! We begin by

[g]The could be other types as well (e.g. read [173]).

[h]The factor k is related to the root of 1 in some known way [C.29, 173].

reproducing some excerpts from [C.34]. First, we are interested to know what are quantum analogs of the Toffoli and Fredkin gates. That is to say, what are the quantum analogs of the classical universal gates? Since in quantum computing every computational step is associated with some kind of unitary transformation, we have to look for the quantum analog of universal gates among unitary transformations. An arbitrary finite-dimensional unitary transformation can be built out of a relatively small set of the so-called *primitives* consisting of one or more (that is product of) two-dimensional unitary transformations and four-dimensional transformations swapping two qubits. Otherwise the unitary transformation is called *imprimitive* [C.35]. Brulinski and Brulinski [C.36] demonstrated that

Theorem C.5.1. *A two-qubit gate is universal if and only if it is entangling.*

According to definition C.5.1 this means that a gate G is entangling if for any vector $|\alpha,\beta\rangle = |\alpha\rangle\otimes|\beta\rangle \in V\otimes V$ we can get a vector $G|\alpha,\beta\rangle$ which is not decomposable as tensor product of two qubits. The question arises: How is G related to the solutions of the YBE? The answer depends upon physical assumptions. For instance, Kauffman and Lomonaco in [C.34] made an assumption that pattern of topological linking (that is linking in S^3) and quantum entanglement are in direct correspondence with each other. Such an assumption is plausible since the key concept in the construction of quantum link invariants is the association of the Yang–Baxter operator R to each elementary crossing in the link diagram. Since the operator R is a linear mapping $R : V\otimes V \to V\otimes V$ defined on twofold tensor product of a vector space V, (typically) it is realizing the swap gate (when V represents a vector space on which one qubit is acting). *In topological applications such transformations need not to be unitary.* This fact follows from the discussion after Definition C.5.3. Nevertheless, the authors of [C.34] assumed that such transformations *must* be unitary if and when they are being used for quantum computing. Since such an assumption seems to be convincing, it is worthwhile to investigate its consequences. The first thing which is required in such a case is to find all unitary solutions of the YBE in dimension 4. This task was brought to completion in a remarkable paper by Dye [C.37]. She was able to find and to classify all such unitary solutions. With such an achievement, we still have to ask the following questions: (a) How are these solutions related to those obtained from the unitary representation of the braid group? (b) How are these solutions related to representations of the Hecke and degenerate Hecke algebras?

The answer to (a) is given in [C.38]. The answer to (b) is given in Remark C.5.4. Following [C.38] let b_i and $\check{R}_i$ be respectively the braid and the YBE representation of operators both acting on the tensor product $V_i \otimes V_i$ of vector spaces. Both b_i and $\check{R}_i$ satisfy the braid-type operator equations

$$b_i b_{i+1} b_i = b_{i+1} b_i b_{i+1} \quad \text{and} \tag{C.21a}$$

$$\check{R}_i(x)\check{R}_{i+1}(xy)\check{R}_i(y) = \check{R}_{i+1}(y)\check{R}_i(xy)\check{R}_{i+1}(x). \tag{C.21b}$$

The spectral parameter x is chosen in such a way that

$$\check{R}_i(x = 0) = b_i. \tag{C.22}$$

Definition C.5.5. The YBE, Eq. (C.21b) is called *braiding* YBE.

Introduce the swap (permutation) operator S and consider the composition $S\check{R} \equiv R$. Then, the matrix R also satisfies the YBE.

Definition C.5.6. The YBE in which the matrix R is being used is called *algebraic* YBE.

With this information in our hands it is possible either: (a) to construct the braid group representation (BGR) from the $\check{R}$-matrix or, (b) to reduce $\check{R}(x)$ to the known BGR b-matrix.

Definition C.5.7. Process of construction of $\check{R}(x)$ from BGR is called *Yang–Baxterization.*

Now, suppose that the process of Yang–Baxterization is completed successfully so that $\check{\mathrm{R}}$ is some unitary matrix. The next step is then to figure out what physical model leads to the just obtained result. This can be figured out via the following chain of arguments. Since $\check{R}$ is unitary we can formally define $\psi(x) = \check{R}(x)\psi_0$. Differentiating this relation over x we obtain the following Schrödinger-like equation ($\hbar = 1$)

$$i\frac{\partial}{\partial x}\psi(x) = \check{H}(x)\psi(x) \quad \text{where } \check{H}(x) = i\frac{\partial \check{R}(x)}{\partial x}\check{R}(x)^{-1}. \tag{C.23}$$

Solutions, that is the Hamiltonians, for such Schrödinger-type equations for various $\check{R}(x)$ were obtained in [C.38]. These are the Hamiltonians for the six- and eight-vertex models. Unfortunately, such models can hardly be realized as quantum computers. This problem was realized and studied by Zhang in [C.13], one of the authors of [C.38]. Discussing his and related works will provide us with the definition and properties of integrable quantum computation. Furthermore, this discussion will bring into correspondence the classical BBM with its quantum analog. Indeed, let us begin with just this task first. In doing so we follow the very recent paper by Bose and Korepin, [C.39]. The paper begins with discussion of two spinless bosons on a line whose behavior is described by the following Lieb–Linger-type Hamiltonian

$$\hat{H} = -\frac{\partial}{\partial x_1^2} - \frac{\partial}{\partial x_2^2} + 2c\delta(x_1 - x_2). \tag{C.24}$$

The interaction constant c can be both positive or negative. Physical meaning for each case is discussed in detail in our paper [181]. Suppose that both particles (assumed to be indistinguishable like in BBM) are moving in the same direction with momenta $p_2 > p_1$, then the scattering matrix S (the S-matrix) extracted from the wave functions with these momenta is given by

$$S(p_2, p_1) = \frac{p_2 - p_1 - ic}{p_2 - p_1 + ic}. \tag{C.25}$$

Evidently, all other possibilities for the momenta can be treated analogously. As in the classical case of the Fredkin gate, for example, in quantum case the spinless bosons scattering from each other happen to be useful for logical computations. Because of this, such bosons are called "*flying qubits*". Thus, let us consider a line and let such a qubit with momentum towards *right* direction be called "qubit A" while the same but towards *left* direction be called "qubit B". The logical states of such a system of two flying qubits are convenient to characterize by the magnitudes of the momenta in the same direction. Specifically, for the qubit A, the momenta states $|\lambda_0\rangle$ and $|\lambda_1\rangle$ stand respectively for the qubit states $|0\rangle_A$ and $|1\rangle_A$, while for the qubit B, we have respectively $|-\lambda_0\rangle \to |0\rangle_B$ and $|-\lambda_1\rangle \to |1\rangle_B$. For definiteness, let $0 < \lambda_0 \ll c$ while $\lambda_1 \gg c$. Under such conditions, following philosophy of the classical paper by Zamolodchikov brothers [C.40], and taking indistinguishability of particles into account, we can write the following set of relations

$$\begin{aligned}
|0\rangle_A|0\rangle_B &= S(\lambda_0, -\lambda_0)|0\rangle_A|0\rangle_B = -|0\rangle_A|0\rangle_B, \\
|0\rangle_A|1\rangle_B &= S(\lambda_0, -\lambda_1)|0\rangle_A|1\rangle_B = |0\rangle_A|1\rangle_B, \\
|1\rangle_A|0\rangle_B &= S(\lambda_1, -\lambda_0)|1\rangle_A|0\rangle_B = |1\rangle_A|0\rangle_B, \\
|1\rangle_A|1\rangle_B &= S(\lambda_1, -\lambda_1)|1\rangle_A|1\rangle_B = |1\rangle_A|1\rangle_B.
\end{aligned} \tag{C.26a}$$

As shown by Zamolodchikov brothers (see also [C.41], p. 5), the S-matrix for exactly integrable systems must obey the braiding YBE. In the present case the S-matrix is diagonal. Therefore it is also a solution of the algebraic YBE. Explicitly,

$$Z = \begin{pmatrix} 1 & 0 & 0 & 0 \\ 0 & 1 & 0 & 0 \\ 0 & 0 & 1 & 0 \\ 0 & 0 & 0 & -1 \end{pmatrix}. \tag{C.26b}$$

The reason for using letter Z will be explained shortly below. This matrix-solution of the algebraic YBE was discussed as the first example in the paper by Kauffman and Lomonaco [C.34]. Theorem 0 of this reference states that CNOT (controlled NOT [C.6]) universal quantum gate $U_{\mathrm{CNOT}} = QSQ^{-1} \equiv QZQ^{-1}$. Here $Q = H \otimes I$ where the Hadamard matrix H (to be also discussed in the next subsection) is given by

$$H = \frac{1}{\sqrt{2}} \begin{pmatrix} 1 & 1 \\ 1 & -1 \end{pmatrix} \tag{C.27}$$

and I is just the unit matrix.

Exercise C.5.2. Demonstrate that $Q^{-1} = Q$.

It should be noted though that in classical computations CNOT is not the universal gate. The universal gate is CONTROLLED CNOT (or CCNOT). Fredkin's gate is almost the same gate as CCNOT [C.6]. According to [C.4], p. 189, any unitary operation can be approximated to arbitrary accuracy using H, CNOT, phase and $\pi/8$ gates. Phase and $\pi/8$ gates are single qubit gates explicitly defined in [C.4]. Thus, seemingly, we had just pinpointed a situation in which classical reversible and quantum unitary computations come closest to each other. Could these schemes be brought even closer to each other? Addressing this issue brings us to the notion of *integrable quantum computation.* This kind of quantum computation is facilitated by the properties of exactly integrable systems which are realized in real world. The Lieb–Linger model is based on the multiparticle generalization of the two-particle Hamiltonian, Eq. (C.24). It describes the one-dimensional interacting Bose gas which can be realized in laboratory. Our paper, [181], provides many references on this topic. Examples of other one-dimensional physically realizable systems include, for instance, the XXX and XXZ one-dimensional magnets and also semiflexible polymer chains. Details are provided in Chap. 8.

To bring our discussion to completion, we still have to explain several issues. For instance, (a) How do results of Eqs. (C.26) lead to the entangled final states? Also, (b) How is the degenerate affine Hecke algebra (e.g. see Eqs. (C.18)) related to the Lieb–Linger model?

To answer (a) some facts from the book by Nielsen and Chuang, [C.4], are helpful. We have already introduced the Bell states. These are the basic entangled states. Now we would like to discuss how such states can be obtained from the basic set of nonentangled states such as $|00\rangle \equiv |0\rangle|0\rangle$, $|01\rangle \equiv |0\rangle|1\rangle, |10\rangle \equiv |1\rangle|0\rangle$ and $|11\rangle \equiv |1\rangle|1\rangle$. [C.4] suggests that first we have to act on each of the basic states by the Hadamard gate. But such a gate is 1-qubit gate! Because of this, to do our calculations we have to follow some conventions. For instance, if 1 qubit is given by $\alpha|0\rangle + \beta|1\rangle$, we have to associate with it the 1 column matrix $[\alpha, \beta]^T$ so that when H-gate is acting on such a matrix (from the left), we obtain $\alpha\frac{|0\rangle+|1\rangle}{\sqrt{2}} + \beta\frac{|0\rangle-|1\rangle}{\sqrt{2}}$. When dealing with, say, the vector $|00\rangle$ in a similar fashion, we obtain, $\mathrm{H}|00\rangle = \frac{|0\rangle+|1\rangle}{\sqrt{2}}|0\rangle$. This is 2-qubit state on which we want to act by the CNOT gate U_{CN} (from the left). The gate is given by

$$U_{CN} = \begin{pmatrix} 1 & 0 & 0 & 0 \\ 0 & 1 & 0 & 0 \\ 0 & 0 & 0 & 1 \\ 0 & 0 & 1 & 0 \end{pmatrix} \tag{C.28}$$

so that the question arises how it is acting on $\frac{|0\rangle+|1\rangle}{\sqrt{2}}|0\rangle$? The authors of [C.4] provide us only with the final answer. The result is: $|\beta_{00}\rangle$. The answer, of course, can be obtained in a standard way by taking into account that, say, $|0\rangle|0\rangle = |0\rangle \otimes |0\rangle$, etc. We mention this only to alert our readers to the conventions used in different books. To avoid ambiguities, the standard way of doing calculations (that is by direct use of direct products) is recommended. The anticipated outcome of such described calculations — the entangled Bell states — is obtained from the nonentangled basis set. The procedure just outlined requires preparation of the basic set by passing them initially through the H-gate. In connection with Theorem C.5.1, the question arises: To what extent the entangling property of 2-qubit entangling gates depends upon the initial set of nonentangled states? To illustrate the nontriviality of this question, let us use the controlled Z-gate, Eq. (C.26), and apply it to the basis set made of $|0\rangle_A|0\rangle_B$, etc. vectors.

The reason to call Z-matrix, Eq. (C.26), as "controlled-Z", or CZ (for short) gate is explained in [C.4]. Basically, controlled gates are always at least 2-qubit gates. Let, for the sake of argument, this be a 2-qubit gate. Then, by definition, it is not acting on the first qubit which is "control qubit" and second qubit is acted upon by the matrix U, that is by

$$U = \begin{pmatrix} a & b \\ c & d \end{pmatrix}$$

so that controlled-U gate looks like this

$$CU = \begin{pmatrix} 1 & 0 & 0 & 0 \\ 0 & 1 & 0 & 0 \\ 0 & 0 & a & b \\ 0 & 0 & c & d \end{pmatrix}.$$

In our case, U is Pauli-Z gate and, therefore, the net result, is the CZ gate given by the Z-matrix.

Application of CZ gate to vectors from the basis set does not lead to entanglement as can be seen by direct computation. However, if following [C.39], we prepare the initial vector as

$$\frac{|0\rangle_A + |1\rangle_A}{\sqrt{2}} \frac{|0\rangle_B + |1\rangle_B}{\sqrt{2}}$$

and act upon it by the CZ gate from the left, the resulting state is indeed entangling as can be directly checked. Thus, Theorem C.5.1 should actually be read as follows:

A 2-qubit gate is universal if there is at least one nonentangled state such that when acted upon by 2-qubit gate the resulting state is entangling.

Finally, consider the equation $Z\Psi = \tilde{\Psi}$. Since $Q^2 = 1$, we might as well write $QZQ^{-1}Q\Psi = Q\tilde{\Psi}$ or $U_{CN}\Phi = \tilde{\Phi}$. Since application of the CZ gate to vectors from the basis set does not lead to entanglement, this means that if we identify Ψ with the basis set, then the vector $\tilde{\Phi}$ is not entangling. This observation puts the significance of entanglements in quantum computations into question in accord with conclusions of [223]. This fact is compatible with Theorem C.5.1 though.

Finally, to answer (b), we need to explain the place which the degenerate affine Hecke algebra occupies in the Lieb–Linger (L-L) model of non-ideal one-dimensional Bose gas formalism. Because this issue was exhaustively clarified at different levels of mathematical rigor, we shall only comment on the most relevant references leaving details outside the scope of this Appendix. At the physical level of rigor the place of the affine Hecke algebra in the L-L model formalism was explained in the paper by Hikami [C.42]. In turn, this paper is based on some results taken from the Annals of Mathematics paper, [C.43], by Heckman and Opdam. Interested readers might also want to consult [C.25, C.44] prior to reading [C.43].

Alternative approach to the same problematics can be found in [275]. Such an approach has some physical flavor because of its connection with theory of arrangements [D.21]. We used this theory in connection with the new mathematical description of Veneziano amplitudes [C.45]. Study of Veneziano amplitudes is linked to studies of dual resonance models of hadron physics which gave birth to all kinds of string-theories [C.46].

Following [C.47], we would like to explain the meaning of arrangements using the L–L model as example. This discussion is aimed at bringing classical reversible and quantum computations as close as possible to each other.

As in BBM consider a system of N particles in one dimension which interacts only at the moments of collisions. If treated quantum mechanically, the wave function Ψ defined in $\mathbf{R}^N = \{\mathbf{x}_1, \ldots, \mathbf{x}_N\}$ should be the same as the wave function of free particles outside the union of hyperplanes of collision $\{\mathbf{x}_i = \mathbf{x}_j, 1 \leq i < j \leq N\}$. These hyperplanes divide $\mathbf{R}^N$ into $N!$ disconnected chambers C_i, $i = 1{-}N$, labelled by elements s_i of permutation group S_N. Such system of hyperplanes is called *hyperplane arrangement.* In each chamber Ψ is just a linear combination of plane waves subject to the boundary conditions in nontrivial way known as Bethe ansatz. Details can be found in [C.47]. Connections between different chambers can be interpreted in terms of orbits of the Weyl reflection groups [C.43]. These will be discussed in Appendix D. Because of this connection with reflection groups, beautifully explained in [C.48], there is yet another way to interpret the obtained results. This time in terms of just one particle inside the billiard. Brief introduction to billiards can be found in [C.49]. But such an interpretation brings us back to the classical BBM model!

C.6 From Classical to Quantum Random Walks

In Secs. C.1 and C.5 we had already mentioned about intimate connection between random walks (RW) and computations done either by classical or by quantum computer. Evidently, every classical algorithm should somehow be embedded into quantum algorithm. The RW formalism is ideally suited for demonstration that this indeed should be the case.

We begin our discussion with excerpts from review paper by Kendon [C.50]. Both classical and quantum computers require hardware and software. The design of software for quantum computers is contingent upon natural requirement which can be explained as follows.

Select the most efficient classical algorithms and try to estimate if there are quantum algorithms which can do a better job. The advantage of quantum RW for design of quantum algorithms lies in their build in rigidity to be defined shortly below. In computer science researchers attribute the rigidity to the quantumness of RW. Another paper by Kendon [C.51] addresses the question: What exactly is quantum about the quantum walk? In Chap. 8 we provide arguments aimed at convincing our readers that such a vision of quantum RW can be broadened with help of some facts from polymer physics, more specifically, conformational statistics of semiflexible polymers. Let us now make some elementary comparisons between the one-dimensional RW's.

Classical random walk

1. Starts at $x = 0$
2. Toss a coin: result is head or tail
3. Move one unit, left or right according to the coin state:
   (tail): $x \to x - 1$,
   (head): $x \to x + 1$.
4. Repeat steps 2 and 3 t-times
5. Measure position $x = x(t)$
6. Repeat steps 1–5 many times to obtain the probability distribution $P(x, t)$
7. Calculate $\sqrt{\langle x^2 \rangle} \propto \sqrt{t}$

Quantum random walk

1. Starts at $x = 0$
2. Toss a qubit (quantum coin) using, for example, the Hadamard gate, Eq. (C.27).
   $\mathrm{H}|x, 0\rangle = (|x, 0\rangle + |x, 1\rangle)/\sqrt{2}$
   $\mathrm{H}|x, 1\rangle = (|x, 0\rangle - |x, 1\rangle)/\sqrt{2}$
3. Move one unit, left or right dependent on the qubit state:
   $\mathrm{S}|x, 0\rangle \to |x - 1, 0\rangle$,
   $\mathrm{S}|x, 1\rangle \to |x + 1, 1\rangle$.
4. Repeat steps 2 and 3 t-times
5. Measure position $x = x(t)$
6. Repeat steps 1–5 many times to obtain the probability distribution $P(x, t)$
7. Calculate $\sqrt{\langle x^2 \rangle} \propto t$

In polymer terminology step 7 for classical and quantum RW's is equivalent to comparing conformational properties of fully flexible polymers against those for semiflexible polymers. The property $\sqrt{\langle x^2 \rangle} \propto t$ for quantum RW is equivalent to the property of semiflexible polymers to become fully rigid. For this reason we can talk interchangeably either about quantum RW's or about classical semiflexible polymers. Thus, quantum RW are rigid in *this* sense.

The above steps 1–7 describe the so-called *discrete-time* classical and quantum RW's. Their *continuous-time* analogs do not use coins! Instead, the coin toss and the conditional shift (steps 2 and 3) are replaced by the hopping rate γ per unit time for the probability of moving into neighboring location. Step 4 is replaced by the continuous evolution in time.

Specifically, without loss of generality, consider first the continuous classical RW on some graph G whose vertex set is V and edge set is E, that is $G = (V, E)$. The adjacency matrix $A_{j,k}$ for G is defined by

$$A_{j,k} = \begin{cases} 1 & \text{if } (j,k) \in E \\ 0 & \text{if } (j,k) \notin E \end{cases}$$

$\forall\, j, k \in V$. To describe random walks the matrix $\mathbf{H}$ is used instead of $\mathbf{A}$. Specifically,

$$H_{j,k} = \begin{cases} d_j\gamma & \text{if } i = j, \\ -\gamma & \text{if } (j,k) \in E, \\ 0 & \text{if } (j,k) \notin E. \end{cases} \tag{C.29}$$

Here d_j is the degree of the jth vertex and γ is the probability of jumping between neighboring nodes per unit time. If $p_j(t)$ is the probability of locating fictitious particle at vertex j, then its evolution is described by the following equation:

$$\frac{d}{dt}p_j(t) = -\sum_{k\in V} H_{j,k}p_k(t). \tag{C.30a}$$

Since the probability is conserved, that is $\frac{d}{dt}\sum_j p_j(t) = 0$, using this Eq. (C.29) in Eq. (C.30), we obtain

$$0 = \sum_{j\in V}\sum_{k\in V} H_{j,k}p_k(t) = \gamma\left(\sum_{j=k\in V} d_j p_j(t) - \sum_{j\in V}\sum_{k\neq j\in V} p_k(t)\right).$$

To check the correctness of this result, it is sufficient to consider the one-dimensional regular lattice as an example. Then, Eq. (C.30a) acquires the form

$$\frac{d}{dt}p_j(t) = \gamma(p_{j-1}(t) + p_{j+1}(t) - 2p_j(t)) \tag{C.30b}$$

making previous identity to hold trivially. Using Taylor's series expansion for the first and the second terms on the R.H.S. of Eq. (C.30b) the usual continuous diffusion equation is obtained. Without such an expansion, the formal solution of Eq. (C.30a) is obtained as

$$\mathbf{p}(t) = \exp(-\mathbf{H}t)\mathbf{p}(0). \tag{C.31}$$

With this result in our hands, to obtain the connection between the classical and quantum RW's is straightforward. It is very much the same as between the Bloch and Schrödinger's equations, that is between the finite temperature and zero temperature quantum mechanics/quantum field theories. In the present case we have to replace Eq. (C.31) by

$$|\psi(t)\rangle = \exp(-i\mathbf{H}t)|\psi(0)\rangle \equiv U(t)|\psi(0)\rangle, \tag{C.32}$$

where evolution of the wave function $|\psi(t)\rangle$ is controlled by the Schrödinger equation

$$i\frac{d}{dt}|\psi(t)\rangle = \mathbf{H}|\psi(t)\rangle \tag{C.33}$$

compatible with Eq. (C.32). The connection with quantum RW's is established if, instead of the wave function $|\psi(t)\rangle$, we use the (quantum) amplitude $\langle j|\psi(t)\rangle = a_j$. Then, Eq. (C.30a) is being replaced by

$$\frac{d}{dt}a_j(t) = \sum_{k\in V} H_{j,k}a_k(t). \tag{C.34}$$

Clearly, this result is not in disagreement with Eq. (C.31). This fact is helpful for connecting the discrete-time and the continuous-time quantum RW's. Indeed, according to [C.52] both can be written as

$$|\psi(t)\rangle = U^t|\psi(0)\rangle, \tag{C.35}$$

where in the continuous case we have to use the unitary operator $U = \exp(-i\mathbf{H})$ while in the discrete case we have to use $U = SH$. Very recently Childs [C.53] presented detailed study of interrelationship between the continuous–discrete-time quantum RW's on arbitrary graphs. He demonstrated that the continuous-time quantum RW is obtainable as an appropriate limit of the discrete-time quantum RW. This limiting procedure is discussed in Chap. 8, Sec. 8.3.4, using the discretized Dirac propagator as an example. In the meantime, following [C.50] let us consider the simplest (Hadamard) quantum RW. As it was already stated, the Hadamard gate, Eq. (C.27), plays a role of quantum coin toss operator. So, we begin with, say, the state $|0,0\rangle$.[i]

[i]Surely, we could start with the state $|0,1\rangle$ instead or use both states simultaneously. The choice in our example is purely arbitrary and is made for illustrative purposes only.

$$
\begin{array}{lll}
t=0 & & |0,0\rangle^{t=0} \\
t=1 & \text{(a)} & H|0,0\rangle \rightarrow (|0,0\rangle + |0,1\rangle)/\sqrt{2} \equiv |0,0\rangle_H^{t=1} \\
 & \text{(b)} & S|0,0\rangle_H^{t=1} \rightarrow (|-1,0\rangle + |1,1\rangle)/\sqrt{2} \equiv \psi_H^{t=1} \\
t=2 & \text{(a)} & H\psi_H^{t=1} \rightarrow (|-1,0\rangle + |-1,1\rangle + |1,0\rangle - |1,1\rangle)/2 \equiv |0,0\rangle_H^{t=2} \\
 & \text{(b)} & S|0,0\rangle_H^{t=2} \rightarrow (|-2,0\rangle + |0,1\rangle + |0,0\rangle - |2,1\rangle)/2 = \psi_H^{t=2} \\
t=3 & \text{(a)} & H\psi_H^{t=2} \rightarrow (|-2,0\rangle + |-2,1\rangle + 2|0,0\rangle + |2,0\rangle - |2,1\rangle)/\sqrt{8} \equiv |0,0\rangle_H^{t=3} \\
 & \text{(b)} & S|0,0\rangle_H^{t=3} \rightarrow (|-3,0\rangle + |-1,1\rangle + 2|-1,0\rangle + |1,0\rangle - |3,1\rangle)/\sqrt{8}
\end{array}
$$

In step 3, the component $|0,1\rangle$ is totally eliminated so that the whole process appears as if only the state $|0,0\rangle$ gets enhancement. Such a pattern is attributed in literature to effects of quantum mechanics. A quantum walk becomes a classical walk when decohered. In practice, such a decoherence is implemented by measurements of the path of the walker. Interestingly enough, such measurements introduce non-unitarity (which is to some extent essential for semiflexible polymers discussed in Chap. 8). The degree of intrusion into quantumness is characterized by the parameter p: $0 \leq p \leq 1$, so that technically one can speak about interpolating between quantum and classical RW's. Such an interpretation made in [C.50] is misleading though. In reality, quantumness of the walk is manifestation of its rigidity (as explained above). The less quantum walk is rigid, the more it is flexible — a well-known feature in the theory of semiflexible polymers. So, the parameter p is responsible for (physical) rigidity. Not surprisingly, presence of the nonzero parameter p destroys unitarity. That is quantum walk/quantum algorithm becomes classical! Thus, we end up with a question: What makes a quantum walk algorithm more efficient than a classical algorithm? This question can be resolved in terms of energy cost required in each computational step (recall the reversible classical computation discussed above), and with the amount of memory required. This is so because any physical computer (quantum or classical) has finite size and can process only finite amount of calculation steps per unit time. Should we be able to simulate quantum algorithm efficiently on a classical computer, then there would be no need for quantum algorithms. Furthermore, quantum algorithms with no efficient classical simulation may still be equalled or bettered by a classical algorithm using a different method. To prove an algorithmic speed up for a quantum algorithm, it is required to prove that no classical algorithm of any type can do the job as good as its quantum counterpart. In Sec. 8.1, papers by Jozsa and Linden [223] and Valliant [224] are mentioned. Both papers make such proofs especially difficult. The interpolation between quantum and classical RW's was recently elevated to general principle of how quantum computations should be done by Jozsa [C.54]. The latest attempts in this direction are summarized in [C.55].

Going back to our example of quantum walk the question arises: what makes the Hadamard gate for such a walk so special? The answer is: nothing much, except computational convenience. This was recognized in the paper by Chandrashekar *et al.* [C.56], see also [C.57]. The H-matrix can be replaced by the SU(2) matrix which we already introduced in the main text, e.g. see Eqs. (6.5.25)–(6.5.28). To match the results of [C.56, C.57] with Eq. (6.5.27) we have to replace $\theta/2$ by $-\theta$, relabel the combination $(\psi+\varphi)/2$ as ξ and $(\psi-\varphi)/2=-\zeta$. After that, we obtain the coin operator $\mathbf{C}$ as

$$\mathbf{C}(\xi,\theta,\zeta)=\begin{pmatrix} e^{i\xi}\cos\theta & e^{i\zeta}\sin\theta \\ -e^{-i\zeta}\sin\theta & e^{-i\xi}\cos\theta \end{pmatrix}. \tag{C.36a}$$

The Hadamard operator H (up to a global phase) is obtained if we choose $\zeta=\xi=\pi/2, \theta=\pi/4$. Now, however, we are not restricted by such a choice and, having in mind results of Sec. 8.3.4, we would like to choose instead $\xi=0, \zeta=\pi/2$ so that we obtain the following final result

$$\mathbf{C}\left(0,\theta,\frac{\pi}{2}\right)=\begin{pmatrix} \cos\theta & i\sin\theta \\ i\sin\theta & \cos\theta \end{pmatrix}. \tag{C.36b}$$

Appendix D

The Measurement Protocol. Geometry and Topology of Entanglements

D.1 The Measurement Protocol

There is a profound difference between formulations of quantum mechanics by Schrödinger and Heisenberg. While in Schrödinger's formulation it is necessary to postulate the Schrödinger's equation and then to worry about the meaning of the wave function etc. Heisenberg's formulation is based on almost trivial observation which can be stated as follows. Suppose that physically observed spectrum is made out of an ordered sequence $\omega_1, \omega_2, \ldots, \omega_l, \ldots, \omega_{k,\ldots}$. Define the difference $\omega_{ij} = \omega_i - \omega_j$. Then, evidently,

$$\omega_{ij} + \omega_{jk} = \omega_{ik}. \tag{D.1}$$

The triviality of this equality is misleading though because *just this statement* allows us to reconstruct both Heisenberg's and Schrödinger's formulations of quantum mechanics as explained in our work on Heisenberg's honeycombs, [173].[a] Thus, from the very beginning, Heisenberg's formulation operates only with physically measurable/observable quantities. These are the spectra of operators. It is believed that for any observable O there is an operator $\hat{O}$ acting in the complex Hilbert space and having the spectral decomposition

$$\hat{O} = \sum_i \lambda_i \hat{P}_i, \ \hat{P}_i = |\lambda_i\rangle\langle\lambda_i|, \ \hat{P}_i^2 = \hat{P}_i, \sum_i \hat{P}_i = \mathbf{I}. \tag{D.2}$$

[a]Read also Appendix C, Sec. C.5.

Here the projection operator $\hat{P}_i$ projects an arbitrary vector $|v\rangle$ onto (eigen)vector $|\lambda_i\rangle$ in such a way that the difference $|v\rangle - \hat{P}_i|v\rangle$ is orthogonal to $|\lambda_i\rangle$. Typically measurements are made at finite temperatures causing us to introduce the density matrix $\hat{\rho}$. The idea of introducing such a matrix is the same as in standard probability theory, that is the average of the operator is given by $\langle\hat{O}\rangle = \sum_i p(i)\lambda_i$. Following von Neumann, this result should be understood as follows. Introduce by analogy with Eq. (D.2) an operator $\hat{\rho} = \sum_i \rho_i|\psi_i\rangle\langle\psi_i|$ then,

$$p(i) = \langle\lambda_i|\hat{\rho}|\lambda_i\rangle = \mathrm{tr}(\hat{P}_i\hat{\rho}) = \sum_i \rho_i\langle\lambda_i|\psi_i\rangle\langle\psi_i|\lambda_i\rangle \geq 0. \tag{D.3}$$

Let us ignore for the moment the effects of temperature and suppose that we made a measurement and its outcome is λ_i. What can be said about the state vector immediately after the measurement? To be specific, let us consider a qubit as an example. In this case we obtain: $|\psi\rangle = a|0\rangle + b|1\rangle$, where $|0\rangle = |\uparrow\rangle = \begin{bmatrix}1\\0\end{bmatrix}$ and $|1\rangle = |\downarrow\rangle = \begin{bmatrix}0\\1\end{bmatrix}$. Both $|0\rangle$ and $|1\rangle$ are eigenfunctions of the Pauli z-matrix operator σ_z. Thus, in the present case the representation $\hat{O} = \sum_i \lambda_i\hat{P}_i$ acquires the following form

$$\sigma_z = \begin{pmatrix}1 & 0\\0 & -1\end{pmatrix} = 1|0\rangle\langle 0| - 1|1\rangle\langle 1|, \tag{D.4}$$

where the projectors $|0\rangle\langle 0|$ and $|1\rangle\langle 1|$ are defined by

$$|0\rangle\langle 0| = \begin{bmatrix}1\\0\end{bmatrix}\cdot[1,0] = \begin{pmatrix}1 & 0\\0 & 0\end{pmatrix} \quad \text{and} \quad |1\rangle\langle 1| = \begin{bmatrix}0\\1\end{bmatrix}\cdot[0,1] = \begin{pmatrix}0 & 0\\0 & 1\end{pmatrix}.$$

Thus, the probability of finding $|\psi\rangle$ in the state $|0\rangle$ or $|1\rangle$ is determined by the absolute values of a and b. That is $|0\rangle$ can be found with probability $|a|^2$ while $|1\rangle$ with probability $|b|^2$. Since we are making just one measurement, clearly, we cannot determine both a and b. And, if this is so, we cannot locate the position of the Bloch vector on the Bloch sphere! However, once the measurement was made, we know that we had just measured, say, $|0\rangle$ since we were able to detect the eigenvalue 1. The wave function Ψ *immediately after measurement* is given by

$$\Psi_i = \frac{|\lambda_i\rangle\langle\lambda_i|\psi\rangle}{[\langle\psi|\lambda_i\rangle\langle\lambda_i|\psi\rangle]^{\frac{1}{2}}}. \tag{D.5}$$

It is normalized as required by Heisenberg's rules. In the present case if the outcome of the measurement was 1, the wave function is $\frac{a}{|a|}|0\rangle$. Since the wave function is always defined with accuracy up to a complex phase, the factor $\frac{a}{|a|}$ can be discarded. What we just described is called *projective measurement* [C.4]. There are other types of measurements. They are used in more complicated situations, e.g. when we are interested in measuring a qubit in the multiqubit wave function, or if we are working at finite temperatures, etc. This topic is discussed in some detail in [C.4].

Remark D.1. From the discussion above it follows that if the wave function $|\psi\rangle$ is initially unknown, then as a result of its measurement the information collapses to its classical values 0 and 1. It can be proven [330] that:

Given an unknown qubit state $|\psi\rangle$, on average only $[2\ln 2]^{-1}$ bits of information can be obtained by a single projective measurement.

When measurements are made at finite temperature and the density matrix is used, the question emerges: What happens to the results just described? That is to say, can we recognize the results of projective measurements at finite temperatures? To answer this question it is convenient to introduce a concept of *pure* state. Following [230], we find that the state is pure if $\hat{\rho}^2 = \hat{\rho}$, otherwise it is called *mixed*. Since $\hat{\rho} = \sum_i \rho_i |\psi_i\rangle\langle\psi_i|$ we obtain $\hat{\rho}^2 = \sum_i \rho_i^2 |\psi_i\rangle\langle\psi_i|$ implying that $\rho_i^2 = \rho_i$. That is to say ρ_i is either 0 or 1. Also, since by construction $\mathrm{tr}\,\hat{\rho} = 1$ this leaves us with the choice: $\hat{\rho} = |\psi_i\rangle\langle\psi_i|$. This result can be understood in simple physical terms. At finite temperature we have to replace the internal energy U by the free energy $\mathcal{F} = U - TS$. For $T \to 0$ we are left with $\mathcal{F} = U$. If the ground state $\mathcal{E}$ is not degenerate we indeed obtain $\hat{\rho} = |\mathcal{E}\rangle\langle\mathcal{E}|$. As an example let us consider the density matrix for a qubit. That is

$$\hat{\rho} = |\psi\rangle\langle\psi|, \quad \text{where } |\psi\rangle = a|0> +b|1\rangle. \tag{D.6a}$$

Evidently, it is given by

$$\hat{\rho} = \begin{pmatrix} a \\ b \end{pmatrix} \cdot (a^*, b^*) = \begin{pmatrix} |a|^2 & ab^* \\ ba^* & |b|^2 \end{pmatrix}. \tag{D.6b}$$

From here we obtain: $\mathrm{tr}\,\hat{\rho} = 1$. Because of this, we also obtain

$$\begin{aligned} \hat{\rho}^2 &= \begin{pmatrix} |a|^2 & ab^* \\ ba^* & |b|^2 \end{pmatrix} \cdot \begin{pmatrix} |a|^2 & ab^* \\ ba^* & |b|^2 \end{pmatrix} \\ &= \begin{pmatrix} |a|^4 + |a|^2|b|^2 & ab^*(|a|^2 + |b|^2) \\ ba^*(|a|^2 + |b|^2) & |b|^4 + |a|^2|b|^2 \end{pmatrix} = \hat{\rho}. \end{aligned} \tag{D.6c}$$

That is the density matrix for a qubit describes a pure state.

Next, let us go back to Sec. 6.5. and let us use Eqs. (6.5.25) and (6.5.28). Evidently, the column vector $\mathbf{z}$ in Eq. (6.5.25) can be used as a qubit and such a qubit can be represented with help of Eq. (6.5.28) in which the angle ψ should be suppressed. Let such a qubit be labeled as $|\psi(\theta,\varphi)\rangle$, that is

$$|\psi(\theta,\varphi)\rangle = \begin{pmatrix} e^{-i\frac{\varphi}{2}}\cos\frac{\theta}{2} \\ e^{i\frac{\varphi}{2}}\sin\frac{\theta}{2} \end{pmatrix}. \tag{D.7}$$

Because of the Hopf map $S^3 \hookleftarrow S^2$, the Bloch vector $\tilde{\mathbf{n}}$ defined in Eq. (8.4.3.10) can now be used for construction of the scalar product $\tilde{\mathbf{n}}\cdot\tilde{\sigma}$. Explicit calculation produces

$$\tilde{\mathbf{n}}\cdot\tilde{\boldsymbol{\sigma}} = \begin{pmatrix} \cos\theta & e^{-i\varphi}\sin\theta \\ e^{i\varphi}\sin\theta & -\cos\theta \end{pmatrix}. \tag{D.8}$$

Using this result consider an eigenvalue equation

$$\tilde{\mathbf{n}}\cdot\tilde{\boldsymbol{\sigma}}|\psi(\theta,\varphi)\rangle = \lambda|\psi(\theta,\varphi)\rangle. \tag{D.9}$$

In [230] by direct calculation it is demonstrated that $\lambda \equiv 1$ is the only eigenvalue for this equation. Furthermore, by direct computation, using Eq. (D.8) we obtain,

$$(\tilde{\mathbf{n}}\cdot\tilde{\boldsymbol{\sigma}})^2 = \mathbf{I}, \tag{D.10}$$

where $\mathbf{I}$ is the unit 2×2 matrix. Since the qubit describes pure state and since $|\psi(\theta,\varphi)\rangle$ is a qubit, we can construct the density matrix using these data. That is we obtain,

$$\begin{aligned}
\rho(\theta,\varphi) &= |\psi(\theta,\varphi)\rangle\langle\psi(\theta,\varphi)| \\
&= \begin{pmatrix} \cos^2\frac{\theta}{2} & e^{-i\varphi}\cos\frac{\theta}{2}\sin\frac{\theta}{2} \\ e^{i\varphi}\cos\frac{\theta}{2}\sin\frac{\theta}{2} & \sin^2\frac{\theta}{2} \end{pmatrix} \\
&= \frac{1}{2}\mathbf{I} + \frac{1}{2}\begin{pmatrix} \cos\theta & e^{-i\varphi}\sin\theta \\ e^{i\varphi}\sin\frac{\theta}{2} & -\cos\theta \end{pmatrix} \\
&= \frac{1}{2}\mathbf{I} + \frac{1}{2}\left(\sin\theta\cos\varphi\begin{pmatrix} 0 & 1 \\ 1 & 0 \end{pmatrix}\right. \\
&\quad \left. + \sin\theta\sin\varphi\begin{pmatrix} 0 & -i \\ i & 0 \end{pmatrix} + \cos\theta\begin{pmatrix} 1 & 0 \\ 0 & -1 \end{pmatrix}\right) \\
&= \frac{1}{2}(\mathbf{I} + \tilde{\mathbf{n}}\cdot\tilde{\boldsymbol{\sigma}}).
\end{aligned} \tag{D.11}$$

Because of Eq. (D.10), we can write as well

$$\rho^2(\theta,\varphi) = \left[\frac{1}{2}(\mathbf{I} + \tilde{\mathbf{n}}\cdot\tilde{\boldsymbol{\sigma}})\right]^2 = \frac{1}{4}\mathbf{I} + \frac{1}{4}(\tilde{\mathbf{n}}\cdot\tilde{\boldsymbol{\sigma}})^2 + \frac{1}{4}(\mathbf{I}\tilde{\mathbf{n}}\cdot\tilde{\boldsymbol{\sigma}} + \tilde{\mathbf{n}}\cdot\tilde{\sigma}\mathbf{I}) = \rho$$

so that, indeed, just constructed density matrix describes pure state. Consider generalization of this result to the mixed state. In such a case the density matrix should look as follows [230].

$$\hat{\rho}(\theta, \varphi) = \frac{1}{2}(\mathbf{I} + \tilde{\mathbf{u}} \cdot \tilde{\boldsymbol{\sigma}}), \tag{D.12}$$

where the vector $\tilde{\mathbf{u}}$ should coincide with $\tilde{\mathbf{n}}$ for pure states but otherwise it should still be determined. This can be done based on two conditions: (a) $\operatorname{tr} \hat{\rho} = 1$ and (b) $\det \hat{\rho} \geq 0$. The last requirement has the same meaning as in Eq. (D.3). With both such requirements we obtain:

$$\det \hat{\rho} = \frac{1}{4}(1 - \tilde{\mathbf{u}}^2) \geq 0. \tag{D.13}$$

From here, it follows that $\tilde{\mathbf{u}}^2 \leq 1$ and that when $\tilde{\mathbf{u}}^2 = 1$ the vectors $\tilde{\mathbf{n}}$ and $\mathbf{u}$ coincide. The inequality $\tilde{\mathbf{u}}^2 \leq 1$ defines the *Bloch ball.* Whenever $\tilde{\mathbf{u}}^2 < 1$, the density matrix describes mixed states. In such a case, the Hopf mapping breaks down and, with it, we are losing the connections with differential geometry of curves (elastic rods, semiflexible polymers). This connection is lost because the equation $\tilde{\mathbf{u}}^2 = 1$ has differential-geometric meaning as explained in the main text, in Chap. 8.

Remark D.2. Although the connections with differential geometry of curves are lost, the connections with sub-Riemannian geometry still hold as detailed calculations done in [261, 262] indicate.

The above discussion of measurements might leave our readers a bit puzzled since the projective measurements on a qubit (seemingly) provide even less information than the classical bit. Furthermore, an impression is created that since a single measurement cannot deliver information about a's and b's the notion of the Bloch sphere is almost useless. We would like now to provide some arguments that this is not the case. For this purpose we have to discuss an action of the unitary operator $\exp\{\pm i(-\sigma_z p + \sigma_x m)t\}$ defined by Eq. (8.3.5.16) on the qubit. To do so, we need to use the standard formulas, e.g. see [C.4], such as

$$R_x(\theta) = \exp\left(-i\frac{\theta}{2}\sigma_x\right) = \mathbf{I}\cos\frac{\theta}{2} - i\sigma_x \sin\frac{\theta}{2} = \begin{pmatrix} \cos\frac{\theta}{2} & -i\sin\frac{\theta}{2} \\ -i\sin\frac{\theta}{2} & \cos\frac{\theta}{2} \end{pmatrix}, \tag{D.14a}$$

$$R_z(\theta) = \exp\left(-i\frac{\theta}{2}\sigma_z\right) = \mathbf{I}\cos\frac{\theta}{2} - i\sigma_z \sin\frac{\theta}{2} = \begin{pmatrix} \exp(-i\frac{\theta}{2}) & 0 \\ 0 & \exp(i\frac{\theta}{2}) \end{pmatrix} \tag{D.14b}$$

and, for completeness, also

$$R_y(\theta) = \exp\left(-i\frac{\theta}{2}\sigma_y\right) = \mathbf{I}\cos\frac{\theta}{2} - i\sigma_y\sin\frac{\theta}{2} = \begin{pmatrix} \cos\frac{\theta}{2} & -\sin\frac{\theta}{2} \\ \sin\frac{\theta}{2} & \cos\frac{\theta}{2} \end{pmatrix}. \tag{D.14c}$$

For the sake of argument, we suppose that our qubit was prepared as $|0\rangle = [1,0]^T$ and consider the case when there are no external forces, that is $p = 0$. Then, we obtain

$$\exp\{\pm i\sigma_x mN\}|0\rangle = \begin{pmatrix} \cos 2mN & \pm i\sin 2mN \\ \pm i\sin 2mN & \cos 2mN \end{pmatrix}\begin{bmatrix} 1 \\ 0 \end{bmatrix}$$
$$= \begin{bmatrix} \cos 2mN \\ \pm i\sin 2mN \end{bmatrix} = |\psi(N)\rangle. \tag{D.15}$$

The above result is rewritten in polymer's language introduced in Sec. 8.4.3. Suppose now that we would like to measure the observable σ_z. Then $|\psi(N)\rangle$ should be rewritten as

$$|\psi(N)\rangle = \cos 2mN|0\rangle \pm i\sin 2mN|1\rangle. \tag{D.16}$$

If now $2mN = 2\pi n, n = \pm1, \pm2, \ldots,$ we shall end up with the initial state. In general, however, we shall have the probability of spin-up $\mathrm{P}_\uparrow(N) = \cos^2 2mN$ and spin-down $\mathrm{P}_\downarrow(N) = \sin^2 2mN$. Clearly, by using the Hopf mapping we can calculate the components of the Bloch vector. These are

$$x = 0, \quad y = \sin 4mN, \quad z = \cos 4mN.$$

The Boch vector in this case is lying (moving) on the circle $y^2+z^2 = 1$ of the Bloch sphere where it can be located experimentally. A circle S^1 is better than nothing but, still, it is not the Bloch sphere. Can we do something to extend just obtained result? The answer is "yes" but, as it is always the case, we have to pay some price for doing this. Let us investigate this issue in some detail.

Following [C.4], it is convenient to introduce the T-gate. It can be obtained from $R_z(\theta)$ if we write $\theta = \pi/2$. In such a case we obtain

$$\mathrm{T} = \exp\left(i\frac{\pi}{8}\right)\begin{pmatrix} \exp(-i\frac{\pi}{8}) & 0 \\ 0 & \exp(i\frac{\pi}{8}) \end{pmatrix} \tag{D.17}$$

so that the factor $\exp\left(i\frac{\pi}{8}\right)$ can be dropped. Using Eqs. (D.8) and (D.11) we define a rotation of the Bloch sphere around the axis defined by the vector $\tilde{\mathbf{n}}$:

$$R_{\mathbf{n}}(\theta) = \exp\left(-i\frac{\theta}{2}\tilde{\mathbf{n}}\cdot\tilde{\boldsymbol{\sigma}}\right) = \cos\left(\frac{\theta}{2}\right)\mathbf{I} - i\sin\left(\frac{\theta}{2}\right)[\tilde{\mathbf{n}}\cdot\tilde{\boldsymbol{\sigma}}]. \tag{D.18}$$

Surprisingly, from [C.4] we find that the combination

$$\exp\left(-i\frac{\pi}{8}\sigma_z\right)\exp\left(-i\frac{\pi}{8}\sigma_x\right)$$
$$= \cos^2\left(\frac{\pi}{8}\right)\mathbf{I} - i\left[\cos\left(\frac{\pi}{8}\right)(\sigma_z+\sigma_x) + \sin\left(\frac{\pi}{8}\right)\sigma_y\right] \tag{D.19}$$

corresponds to the rotation of the Bloch sphere about the axis $\tilde{\mathbf{n}} = \left(\cos\left(\frac{\pi}{8}\right), \sin\left(\frac{\pi}{8}\right), \cos\left(\frac{\pi}{8}\right)\right)$ through the angle determined from the equation $\cos\left(\frac{\theta}{2}\right) = \cos^2\left(\frac{\pi}{8}\right)$. That is up to a phase the factor $R_{\mathbf{n}}(\theta)$, defined by Eq. (D.18), coincides with Eq. (D.19) written for the specific direction $\tilde{\mathbf{n}}$ and specific angle θ. Even more surprising is the fact that with help of the Hadamard gate, Eq. (C.27), and the T-gate, Eq. (D.17), we can achieve the same result (up to a phase factor). That is the combination (HTH) T is equivalent to Eq. (D.19)! This fact allows us to re-obtain the Hadamard gates by just using $R_x(\theta)$ and $R_z(\theta)$ gates! Following [D.1] we write

$$R_z\left(\frac{\pi}{2}\right) = \exp\left(i\frac{\pi}{4}\right)\begin{pmatrix}1 & 0\\ 0 & e^{i\frac{\pi}{2}}\end{pmatrix} \sim \begin{pmatrix}1 & 0\\ 0 & i\end{pmatrix} \equiv \mathrm{S}.$$

The S-gate was introduced in [C.4]. Next, we have

$$R_x\left(\frac{\pi}{2}\right) = \begin{pmatrix}\cos\frac{\pi}{4} & -i\sin\frac{\pi}{4}\\ -i\sin\frac{\pi}{4} & \cos\frac{\pi}{4}\end{pmatrix} = \frac{1}{\sqrt{2}}\begin{pmatrix}1 & -i\\ -i & 1\end{pmatrix}.$$

Using these results we obtain

$$R_z\left(\frac{\pi}{2}\right)R_x\left(\frac{\pi}{2}\right)R_z\left(\frac{\pi}{2}\right) \simeq \frac{1}{\sqrt{2}}\begin{pmatrix}1 & 1\\ 1 & -1\end{pmatrix} = \mathrm{H}. \tag{D.20}$$

With such results in our hands, following [C.4] it can be proven that the repeated iteration of $R_{\mathbf{n}}(\theta)$ with θ defined through equation $\cos(\frac{\theta}{2}) = \cos^2(\frac{\pi}{8})$ and $\tilde{\mathbf{n}} = (\cos(\frac{\pi}{8}), \sin(\frac{\pi}{8}), \cos(\frac{\pi}{8}))$ can reproduce any rotation with arbitrary pre-assigned accuracy. Alternatively, it is sufficient to use the T and H gates only to reproduce with whatever accuracy any unitary gate and, thus, to reach any point on the Bloch sphere. These results leave open the design of the 2-qubit gates needed for universal computation. The design of the 2-qubit gates is a much more delicate process. We initiated its description in Appendix C and now we would like to add more details.

The design of 2-qubit gates and states is associated with uses of Kronecker's direct product. Mathematically, this is a completely harmless operation used routinely in the theory of group representations. From our Honeycomb paper, [173], it follows that such a product is deeply rooted into combinatorics. For instance, if we are dealing with an assembly of spins, the combinatorial effects will begin to show up at the level of just two spins. The first issue comes from the fact that, for instance, in atomic and molecular physics carriers of these spins (electrons) are indistinguishable. If this is so, let us recall that for two electrons with spins we can normally have four wave functions. The first two (in qubit notations) are

$$|L=0,L_z=0\rangle=\frac{1}{\sqrt{2}}\{|0\rangle_1|1\rangle_2-|1\rangle_1|0\rangle_2\}, \tag{D.21a}$$

$$|L=1,L_z=0\rangle=\frac{1}{\sqrt{2}}\{|0\rangle_1|1\rangle_2+|1\rangle_1|0\rangle_2\}. \tag{D.21b}$$

These two are easily recognizable as two of the four Bell states (Appendix C). But then, there are two other given by

$$|L=1,L_z=1\rangle=|0\rangle_1|0\rangle_2 \quad \text{and} \quad |L=1,S_z=-1\rangle=|1\rangle_1|1\rangle_2. \tag{D.21c,d}$$

Evidently, the last two are pure states while the first two are entangled states. Furthermore, we can construct two more (entangled) Bell states out of these pure states. They are:

$$|L=?,L_z=?\rangle\frac{1}{\sqrt{2}}\{|0\rangle_1|0\rangle_2\pm|1\rangle_1|1\rangle_2\}. \tag{D.21e, f}$$

The first question which comes to mind: What operator will measure these two states. According to Di Vincenzo criteria formulated in Sec. 8.2.1 each qubit should be accessible for loading and unloading of information. Can his criteria be applied to the states given by Eqs. (D.21.e, f)? Consider an action of the spin operator σ_{z_1} on such a combination. The outcome will be very much like that for a single qubit: that is we are going to get the spin values ± 1. What happens with the wave function as result of such a measurement? By looking at Eq. (D.5) we recognize that immediately after measurement the wave function should look like $|0\rangle_1|0\rangle_2$ so that by acting with σ_{z_2} on such a wave function the outcome should be 1. Thus, measurement of the first qubit (of such an entangled pair) has a side effect on the other qubit *even without its measurement*! This is fine, but how do we know that these qubits are entangled? The answer comes again from our experiences with the algebra of angular momentum operators. Indeed, we used the Kroneker's product to build the Hilbert space of two operators. Let us formalize this observation. For a composite system made of the Hilbert spaces H_1 and H_2, the total Hilbert space H^T should be of the form

$$H^T = H_1 \otimes H_2$$

while the normalized wave functions $|\psi\rangle$ living in such a space must be of the form

$$|\psi\rangle = \sum_{ij} c_{ij} |\varphi_{1i}\rangle \otimes |\varphi_{2j}\rangle. \tag{D.22}$$

Although the examples considered above are consistent with this result, for the sake of developing the concept of the Bloch sphere/ball it is instructive to look at this problem from the standpoint of qubits. Clearly, we can construct pure states by using Kronecker's product of qubits of the type $|\psi\rangle = a|0\rangle + b|1\rangle$. By design, these are pure states. Expansion given by Eq. (D.22) also allows the mixed/entangled states (even at zero temperature!). It should be clear that *any* state (pure or not) in H^T can be represented in the form

$$|\psi\rangle = c_0|00\rangle + c_1|01\rangle + c_2|10\rangle + c_3|11\rangle \tag{D.23}$$

where $|00\rangle = |0\rangle|0\rangle$, etc. Since the state vector $|\psi\rangle$ should be normalized, we obtain:

$$\sum_{i=0}^{3} |c_i|^2 = 1. \tag{D.24}$$

This is an equation for S^7! So, for the two-spin case S^3 should be replaced by S^7. Only a subset of S^7 will correspond to pure states. It is determined by the following equations: $a_1a_2 = c_0, a_1b_2 = c_1, b_1a_2 = c_2, b_1b_2 = c_3$. The rest of the states will be entangled. Under such circumstances what can be said about the analog of the Bloch sphere S^2 and the Bloch ball? An early attempt to provide an answer to this question for two qubits was made in [D.2] while in [D.3] these results were generalized to multiqubits. Alternative (but similar) treatment is given in [D.4]. A good summary of these and other efforts can be found in [257]. In short, the unitary group SU(2) used for one qubit should be replaced by the group SU(2^n), $n = 2, 3, \ldots$ in such a way that the notion of the Bloch vector, the Bloch sphere and the Bloch ball survives. As it follows from our discussion of the Bloch density matrix for one qubit, its derivation *does not* reveal the connection with the Hopf map whatsoever. Accordingly, such a connection becomes even less apparent when generalization to the multiqubit case is made in references just cited. We are concerned with the Hopf mapping in view of the following question:

Is there a way to extend the quatum-classical (differential-geometric) correspondence described in Sec. 8.2.2 *to multiqubit/multilevel case*?

An attempt to do so was made in [D.5] based on earlier work by Altafini [D.6]. Later on a very comprehensive analysis of the problem done in [D.7] revealed the intrinsic inability to describe the dynamics on the whole Bloch sphere starting already with general case of interacting two qubits. This conclusion is based on results of works [D.2–D.4] in which the Hopf mapping was by-passed altogether, just like in our derivation for $\rho(\theta, \varphi)$. *Is it possible to approach the whole problem from the point of view of Hopf mapping*? An attempt to do so was made by Mosseri and Dandoloff [D.8] who considered the case of two qubits. In such a case the group SU(2) is being replaced by SU(4), accordingly, the sphere S^3 is replaced by S^7 while the Hopf fiber S^1 by S^3 with the base S^4. Following these authors, Bernevig and Chen, [D.9], extended results of [D.8] still further to three qubits. In such a case S^7 is replaced by S^{15}, the fiber S^3 by S^7 and the base by S^8. To achieve these results the authors of [D.8] used quaternions while in [D.9] the octonions were used. In the same [D.9] the authors provided convincing arguments explaining why the Hopf fibration method cannot be extended beyond three qubits. Since S^3, S^7 and S^{15} are odd-dimensional spheres, methods of sub-Riemannian geometry, that is of optimal control, are fully applicable to these cases. Recall that such sub-Riemannian analysis for S^3 was outlined in Sec. 8.2.6. Much more detailed calculations were made in [261, 262] and also in [D.10] for two- and three-level quantum systems, including dissipation. At the moment of writing of this book the latest attempt to extend methods of sub-Riemannian geometry to multilevel systems employing Hopf fibration was made in [D.11]. In this work and in [D.8–D.10] as well as in Sec. 8.2.6, the control of entanglements, absolutely essential for designing of universal gates (e.g. see Appendix C), was left unexplained. Incidentally, such a control is ultimately linked with our ability/inability to extend methods of optimal control to multiqubits. Furthermore, as results of [D.5–D.7] indicate, the connections between spin dynamics and differential geometry of curves/rolling balls become very difficult to develop/implement already at the level of two rolling balls. Thus, it is necessary reconsider to the approaches discussed thus far. In doing so we would like to preserve methods of optimal control (e.g. in the form of Pontryagin optimal principle (Appendices A and B, etc.) unchanged. This seems like an impossible task. Fortunately, this is not true as explained below.

D.2 Geometry and Topology of Entanglements

D.2.1 *Classical Mechanics Inside Quantum Mechanics. Basic Facts*

In this section we are not going to talk about delicate inter-relationship between classical and quantum mechanics, e.g. the WKB method known to everybody who has physics background, e.g. read Chap 9. Instead, we would like to discuss those aspects of quantum mechanics which are known to mathematicians for sometime but, to our knowledge, have not yet found their rightful place in physics literature.

We begin with the normalization condition, Eq. (D.24). More generally, following Arnol'd, [136], we define the Hermitian quadratic form

$$\langle \xi, \eta \rangle = \sum_i \xi_i \bar{\eta}_i, \quad \xi_i, \bar{\eta}_i \in \mathbf{C}, \tag{D.25}$$

where the overbar denotes complex conjugation. The normalization condition, Eq. (D.24), can now be written as $\langle \xi, \xi \rangle = 1$ or, alternatively, as

$$\sum_{i=1}^{n+1} |\xi_i|^2 = 1. \tag{D.26a}$$

Since physically nothing changes if we replace ξ_i by $e^{i\varphi}\xi_i$, the above equation defining the sphere S^{2n+1} in $\mathbf{C}^{n+1}$ should be replaced by the quotient

$$\left\{ \sum_{i=1}^{n+1} |\xi_i|^2 = 1 \right\} \Big/ \ (\xi_i \to e^{i\varphi}\xi_i) \tag{D.26b}$$

defining the complex projective space $\mathbf{CP}^n$. This is the space used in quantum mechanics. It should be noted that there are several definitions of $\mathbf{CP}^n$ [D.12] so that Eq. (D.26) is just one of them. It is the most convenient for our tasks. In such a model each point of $\mathbf{CP}^n$ should be identified with a circle S^1 (one-dimensional torus) on S^{2n+1}. Following [D.13], it is convenient to describe $\mathbf{CP}^n$ in terms of dynamical systems of classical mechanics. Specifically, let us consider classical Hamiltonian H for $n+1$ uncoupled harmonic oscillators[b]

$$\mathrm{H} = \frac{1}{2} \sum_{j=1}^{n+1} (p_j^2 + q_j^2) \equiv \sum_{j=1}^{n+1} \mathrm{H}_j. \tag{D.26c}$$

[b]Evidently, all exactly integrable systems expressed in the appropriate coordinates and momenta will be described by such Hamiltonian.

By introducing complex variables $\xi_j = p_j + iq_j$ and by assuming that the energy E of such a system is conserved and equal, say, to 1/2, Eq. (D.26a) can be re-interpreted as the statement about conservation of energy. To move forward, we notice that introduction of complex variables $\xi_j = p_j + iq_j$ and using them in Eq. (D.26) defines a map $\Phi : \mathbf{C}^{n+1} \to \mathbf{R}^{n+1}$ given by

$$\{\xi_1, \ldots, \xi_{n+1}\} \to \frac{1}{2}\{|\xi_1|^2, \ldots, |\xi_{n+1}|^2\} \tag{D.27}$$

which is an example of the *moment map*. The dynamics is created by the Hamiltonian flow $\xi_j \to e^{i\varphi_j}\xi_j$ so that the quotient space (also called the *reduced space* or *orbit space*) is $\mathbf{CP}^n$. For the Hamiltonian H the symplectic 2-form $\omega = \sum_i dp_i \wedge dq_i$, when it is rewritten in terms of ξ_i's, acquires the following look [136]:

$$\omega = \frac{i}{2}\sum_j d\xi_j \wedge d\bar{\xi}_j \tag{D.28a}$$

making it manifestly invariant with respect to the Hamiltonian flow. The complex vector space $\mathbf{C}^{n+1}$ with this 2-form is known as Kähler. Since such a form can exist on manifolds other than $\mathbf{C}^{n+1}$, e.g. on $\mathbf{CP}^n$, in general case it can be written as

$$\omega = \frac{i}{2}\sum_{i,j} h_{ij}(\boldsymbol{\xi})d\xi_i \wedge d\bar{\xi}_j. \tag{D.28b}$$

It was proven [D.12] that, provided that this form is closed, that is $d\omega = 0$, it is always possible to bring the 2-form, Eq. (D.28b), to its canonical form given by Eq. (D.28a). This means that $h_{ij}(\boldsymbol{\xi}) = \delta_{ij} + O(|\boldsymbol{\xi}|)$ in the vicinity of any point of such complex manifold. Since this is essentially the complex version of the Darboux theorem in mechanics, we come to conclusion that every Kähler manifold possesses a symplectic structure. Furthermore, it can be demonstrated that $h_{ij}(\boldsymbol{\xi})$ can be obtained with the help of the Kähler potential $\rho(\boldsymbol{\xi})$ defined via

$$h_{ij}(\boldsymbol{\xi}) = \frac{\partial}{\partial \xi_j}\frac{\partial}{\partial \bar{\xi}_j}\rho(\boldsymbol{\xi}). \tag{D.29}$$

This potential is easily obtainable for $\mathbf{CP}^n$. Indeed, since the sphere S^2 is a complex projective line $\mathbf{CP}^1$ [D.12], we can rewrite the metric of the sphere accordingly. For the sphere of unit radius the metric is known to be

$$(ds)^2 = 4\frac{dx^2 + dy^2}{(1 + x^2 + y^2)^2}.$$

By rewriting this result in complex notations the 2-form is obtained as follows:

$$\omega_{S^2} = \frac{2idz \wedge d\bar{z}}{(1+|z|^2)^2}. \tag{D.30}$$

Up to a constant factor this result provides us with the 2-form living on $\mathbf{CP}^1$. It is immediately generalizable to $\mathbf{CP}^n$ allowing us to construct the Kähler potential $\rho(\boldsymbol{\xi})$ for the Fubini–Study (FS) 2-form. To do so we have to recall that S^2 can be covered by two coordinate patches. Such a patching is naturally extendable to $\mathbf{CP}^n$ so that if we use $n+1$ coordinate patches we get for the ith patch U_i

$$\rho_i(\boldsymbol{\xi}) = \ln \frac{|\xi|^2}{|\xi_i|^2}, \tag{D.31}$$

where $|\xi|^2 \equiv \sum_{i=1}^{n+1} |\xi_i|^2$. Since the difference $\rho_i(\boldsymbol{\xi}) - \rho_j(\boldsymbol{\xi})$ defined on $U_i \cap U_i$ is real part of analytical function, it will not contribute to $h_{ij}(\boldsymbol{\xi})$ so that such constructed metric is well-defined on $\mathbf{CP}^n$. The normalization constant is determined by the requirement [D.13]

$$\int_{\mathbf{CP}^n} (\omega_{FS})^n = (2\pi)^n. \tag{D.32}$$

Hamiltonians $H_j = \frac{1}{2}|\xi_j|^2$ defined in Eq. (D.26c) should be regarded now as "living" on $\mathbf{CP}^n$. Because of this it becomes possible to visualize $\mathbf{CP}^n$ with help of the moment map, Eq. (D.27). Indeed, by redefining ξ_j as $\xi_j = \sqrt{t_j}\exp(i\varphi_j)$, Eq. (D.26a) acquires the following form:

$$t_1 + \cdots + t_{n+1} = 1. \tag{D.33}$$

It defines an n-simplex spanned by the points $(1,0,0,\ldots), (0,1,0,\ldots), \ldots,$ $(0,\ldots,0,1)$. It is the deformation retract of $\mathbf{CP}^n$ [D.14]. From the same reference it is known that any complex projective algebraic variety can be restored from its deformation retract which in the most general case is some convex polytope living in $\mathbf{R}^n$. We shall explain the physical meaning of this polytope below, in the next subsection. In the meantime, we would like to return to closely related topic stemming from observation that every Kähler manifold possesses symplectic structure. Since we know already that $\mathbf{CP}^n$ is Kähler manifold, let $\mathcal{V}$ be some complex manifold other than $\mathbf{CP}^n$. To find if such a manifold is symplectic, we shall try to embed it into $\mathbf{CP}^n$ Following [D.15] it is appropriate at this point to provide the following:

Definition D.2.1.1. A *projective variety* $\mathcal{V}$ is a non-empty closed subset of $\mathbf{CP}^n$. Accordingly, an *affine variety* $\mathcal{V}$ is a non-empty closed set in $\mathbf{C}^n$.

Definitions of these varieties are rather formal. In what follows, we shall provide more details whenever it will become appropriate. In the meantime, we would like to recall the theorem by Kodaira [D.16].

Theorem D.2.1.1. *Compact complex manifold is a projective variety which is algebraic if and only if this manifold is of Hodge-type.*

Definition D.2.1.2. A variety $\mathcal{V}$ is *algebraic* if it is represented by the joint set of zeros of homogenous polynomials.

Remark D.2.1.2. The Fubini–Study 2-form is an example of the Hodge form [D.16]. Thus, the $\mathbf{CP}^n$ is manifold of the Hodge type. By adjusting normalization once and for all, e.g. as it is done in Eq. (D.32), the period integral (that is integral over any closed two-dimensional cycle) is always an integer for manifolds of Hodge-type. This result plays the central role in theory of *geometric quantization* [D.17].

Remark D.2.1.3. Because of one-to-one correspondence between Kähler and symplectic manifolds, the above requirement is in fact exactly equivalent to the Bohr–Sommerfeld quantization condition $\oint pdq = \iint dp \wedge dq = n$, $n = 1, 2, \ldots$ in the system of units where $h = 1$.

In [136], Appendix 3, Arnol'd proved the following:

Theorem D.2.1.2. *Let j: $\mathcal{V} \to \mathbf{CP}^n$ be the projective embedding of $\mathcal{V}$ into $\mathbf{CP}^n$ and let ω_{FS} be the 2-form on $\mathbf{CP}^n$, then the symplectic 2-form Ω on $\mathcal{V}$ is determined by the pullback $\Omega = j^*\omega_{FS}$.*

Remark D.2.1.3. In [D.17], p. 261, it is being argued that the ω_{FS} and even the Kähler 2-form defined by Eq. (D.28b) is SU(n)-invariant. Thus, the classical mechanics is "firmly entangled with quantum". Given this, we want to accomplish more — to demonstrate that such an "entanglement" reveals the "true" nature of quantum entaglement.

To reach this goal we need to develop further just described mechanical analogy. We begin with the design of the toric varieties. The protocol for doing so requires several steps.

D.2.2 *From Zonotopes to Fans and Arrangements*

We are going to be guided by the fact that all projective varieties can be built from toric varieties [D.14]. Our discussion will uncover a remarkable connection between these varieties and polytopes. Following [D.18], we start with design of a zonotope. To do so, we have to use a p-dimensional cube C_p defined by

$$C_p = \{\mathbf{x} \in \mathbf{R}^p, -1 \leq x_i \leq 1, i = 1 - p\}$$

and the surjective map $\pi : \mathbf{R}^p \to \mathbf{R}^d$ made with help of $d \times p$ matrix W represented in the vector form as $W = \{\mathbf{v}_1, \ldots, \mathbf{v}_p\}$. Thus, $\pi : \mathbf{x} \to W\mathbf{x} + \mathbf{z}$ or, explicitly, $\mathbf{x}' = \mathbf{z} + \sum_i \mathbf{v}_i x_i$ with $\mathbf{x}' \in \mathbf{R}^d$ and $-1 \leq x_i \leq 1$.

Definition D.2.2.1. A zonotope $Z(W)$ is the image of a p-cube under affine projection π, that is $Z(W) = WC_p + \mathbf{z}$.

It can be shown [D.18] that such created zonotope is a centrally symmetric polytope.

Remark D.2.2.1. Because of this observation, it is always possible to construct a mirror (dual) polytope [D.19]. This construction was implemented in string theory.

Exercise D.2.2.1. Read [D.19] and decide if such noticed mirror symmetry can be of any use in designing of quantum computers. In [D.15] the Calabi–Yau (C-Y) manifolds (also used in string theory along with mirror symmetry) were discussed. These are compact Kähler manifolds with vanishing first Chern class.

Remark D.2.2.2. Recently Heydari, [D.20], used the C-Y manifolds to describe quantum entanglements. Our treatment of entanglement problem is different. Moreover, we shall demonstrate that C-Y manifolds are not suitable for developing quantum control.

It is convenient to describe zonotopes in terms of fans. It is helpful to describe both the zonotope and its fan graphically. Figure 1 depicts the formation of the simplest two-dimensional zonotope Z from C_3 while Fig. 2 depicts the fan associated with such a zonotope.

As Fig. 2 suggests, such a fan is collection of cones. These cones can be constructed by drawing semi-infinite strips whose semi-infinite edges are perpendicular to the corresponding edges of Z. The cones are obtained by excluding Z and these strips from 2-plane and bringing what is left together to one point. Thus formed fan is *complete* if upon parallel translation of apexes of the cones to a point in $\mathbf{R}^2$ (or $\mathbf{R}^d$ in general case) the resulting picture spans $\mathbf{R}^2$ (or $\mathbf{R}^d$). For zonotopes, in view of their central symmetry, the fans are always complete. This allows us to connect subsequently the obtained results with the results from Lie algebras. We initiate this process with the following set of definitions.

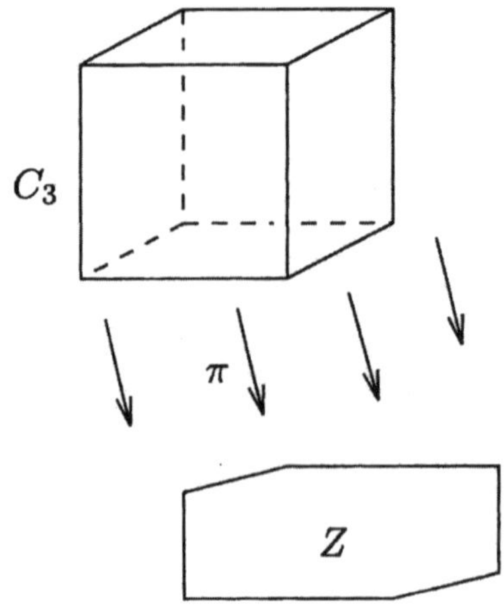

Fig. 1. Two-dimensional zonotope Z from three-dimensional cube C_3.

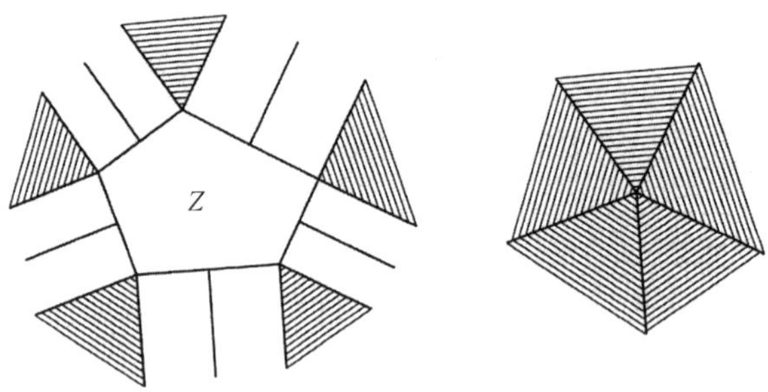

Fig. 2. Construction of the complete fan associated with Z. As it can be seen from the picture, since Z iz centrally symmetric, its fan can be made out of rays perpendicular to each face and emanating from the center of symmetry.

Definition D.2.2.2. Let $\mathcal{R} \subset \mathbf{R}^d$ be a subset of $\mathbf{R}^d$ made of finite set of vectors $\mathbf{y}_1, \ldots, \mathbf{y}_k$. Then a *convex polyhedral cone* σ is a set

$$\sigma = \sum_{i=1}^{k} r_i \mathbf{y}_i \in \mathcal{R}, \quad r_i \geq 0. \tag{D.34a}$$

If the scalar product $\langle , \rangle$ in $\mathbf{R}^d$ is defined as usual, that is as $\langle \mathbf{x}, \mathbf{y} \rangle = \sum_{i=1}^{d} x_i y_i$, the above definition can be used to define the *dual cone* $\sigma^\vee$ via

$$\sigma^\vee = \{\mathbf{u} \in \mathcal{R}^* : \langle \mathbf{u}, \mathbf{y} \rangle \geq 0 \; \forall \, \mathbf{y} \in \sigma\}. \tag{D.34b}$$

Finally, we define a face.

Definition D.2.2.3. A *face* τ of σ is determined by the condition:

$$\tau = \sigma \cap \mathbf{u}_\alpha = \{\mathbf{y} \in \sigma : \langle \mathbf{u}_\alpha, \mathbf{y} \rangle = 0\}. \tag{D.34c}$$

Equation $\langle \mathbf{u}_\alpha, \mathbf{y} \rangle = 0$ defines the *supporting hyperplane* $\mathbf{H}_\alpha$ for some $\mathbf{u}_\alpha \in \mathcal{R}^*$.

Let $\mathcal{P} \subset \mathbf{R}^d$ and $\mathbf{Z}^d$ is some lattice in $\mathbf{R}^d$, then we obtain the following:

Definition D.2.2.4. A convex hull of the intersection $\mathcal{P} \cap \mathbf{Z}^d$ is called *integral polytope.* Accordingly, if vectors $\mathbf{y}_1, \ldots, \mathbf{y}_k \in \mathbf{Z}^d$, the cone σ is called *rational polyhedral cone.*

Remark D.2.2.3. Since physically interesting complex projective varieties are of Hodge-type, we shall from now on work only with integral polytopes and rational cones.

There is a one-to-one correspondence between the faces and supporting hyperplanes. This observation allows us to introduce yet another:

Definition D.2.2.5. A finite set $\mathcal{A} := \{H_1, \ldots, H_k\}$ of supporting hyperplanes is called an *arrangement.* $\mathcal{A}$ decomposes $\mathbf{R}^d$ into complete fan.

Remark D.2.2.4. The above definitions allow us to identify zonotopes with complete fans and with arrangements. Such correspondence is absolutely essential for establishing links between these objects and those in the theory of reflection groups and Lie algebras. Although it is always possible to translate the issues from one language, say, of zonotopes, to another, say, of fans, etc. the theory of arrangements, for example, had become a separate branch of mathematics with its own terminology [D.21]. Further developments brought this theory under the umbrella of theory of *matroids* [D.22] and, in particular, of *oriented matroids* [D.23].

Remark D.2.2.5. The above results admit yet another interpretation originating from theory of linear programming. In particular a subset of $\mathbf{R}^d$ is called polyhedron $\mathcal{P}$ if there exist an $m \times d$ matrix $\mathbf{M}$ (with $m < d$) and a vector $\mathbf{b} \in \mathbf{R}^d$ such that

$$\mathcal{P} = \{\mathbf{x} \in \mathbf{R}^d | \mathbf{M}\mathbf{x} \leq \mathbf{b}\}.$$

Since each component of inequalities $\mathbf{M}\mathbf{x} \leq \mathbf{b}$ determines the half-space while each component of the equality $\mathbf{M}\mathbf{x} = \mathbf{b}$ determines the hyperplane, we are getting another:

Definition D.2.2.6. A polyhedron is an intersection of *finitely many* half-spaces.

Based on this information, the following problem can be formulated as *The major problem of linear programming*:

For the linear functional $\tilde{H}[\mathbf{x}] = \langle \mathbf{c}, \mathbf{x} \rangle$ and pre-assigned vector $\mathbf{c}$, find max $\tilde{H}[\mathbf{x}]$ on $\mathcal{P}$.

In symplectic and algebraic geometry this problem was solved with the help of Morse theory (see Sec. D.2.2.13) by Atiyah and, independently, by Guillemin and Sternberg (AGS). A condensed summary of their results can be found, for example, in the book by Audin [D.24]. Instead of discussing these remarkable results in full generality, we would like to explain the findings of these authors by discussing the simplest benchmark example in some detail.

D.2.3 *What Classical Dynamics on Two-Sphere Can Teach Us?*

[D.24] contains the following:

Exercise D.2.3.1. The symplectic 2-form ω defined on S^2 can be written in coordinates $(x, y, z) \in \mathbf{R}^3$ as

$$\omega = xdy \wedge dz + ydz \wedge dx + zdx \wedge dy. \tag{D.35a}$$

Endow S^2 with the cylindrical (Darboux) coordinates θ, z so that

$$(\theta, z) \to (\sqrt{1 - z^2}\cos\theta, \sqrt{1 - z^2}\sin\theta, z) \tag{D.35b}$$

and demonstrate that in these coordinates $\omega = d\theta \wedge dz$.

Solution. Notice that we already obtained the 2-form, Eq. (D.35a), when we discussed monopoles. Equation (5.6.3.6.) when adapted to sphere S^2 of unit radius produces ω (up to unimportant normalization constant). With this in mind, the result $\omega = d\theta \wedge dz$ is obtained by straightforward substitution of Eq. (D.35b) into (D.35a).

Using this 2-form we can write Hamilton's equation as

$$dH = i_{X_H}\omega. \tag{D.36}$$

Here the vector X_H is defined in a usual way: $X_H = \frac{\partial H}{\partial z}\frac{\partial}{\partial \theta} - \frac{\partial H}{\partial \theta}\frac{\partial}{\partial z} = \dot{\theta}\frac{\partial}{\partial \theta} + \dot{z}\frac{\partial}{\partial z}$ (e.g. compare with Eq. (6.3.10a)). With help of these equations we obtain Hamilton's equations in their traditional form:

$$\dot{\theta} = \frac{\partial H}{\partial z}, \quad \dot{z} = -\frac{\partial H}{\partial \theta}. \tag{D.37}$$

Typically, e.g. see [D.15], p. 86, one chooses $H = z$. In such a case $X_H = \frac{\partial}{\partial \theta}$ so that the Hamiltonian flow $\phi_H(\theta)$ rotates each circle (parametrized by z) with constant speed. This flow has two fixed points: the North and the South poles, whose significance we shall discuss repeatedly. At this moment, though, we would like to introduce the symplectic volume form $\Omega = \frac{\omega}{2\pi} = \frac{d\theta \wedge dz}{2\pi}$ normalized in such a way that

$$\int_{S^2} \Omega = \frac{1}{2\pi}(4\pi) = 2. \tag{D.38}$$

Remark D.2.3.1. Such a choice of normalization is physically meaningful since 2 is just the length (the "volume") of the interval $[-1, 1]$. This interval is our polytope! In fact, it is the simplest example of the so-called *cross-polytope.*[c] In two dimensions it is a diamond-shape square while in 3 it is an octahedron. Below, in Sec. D.2.15, we are going to demonstrate that the *entanglement space for two qubits is an octahedron.*

Remark D.2.3.2. Since $S^2 = \mathbf{CP}^1$ and it is a manifold of the Hodge-type [D.16] the normalization chosen in Eq. (D.38) is not only physically meaningful but also mathematically correct.

Next, we need the following:

Definition D.2.3.1. *A moment* map is a smooth map $\Phi : M \to \mathfrak{g}^*$ that is *equivariant* with respect to the G-action on M and the coadjoint action on $\mathfrak{g}^*$ and such that Φ^{ξ_M} will satisfy Hamilton's equation

$$d\Phi^{\xi_M} = \imath_{\xi_M}\omega. \tag{D.39}$$

Remark D.2.3.3. Based on this definition the moment map Φ should be identified with the Hamiltonian H which in our case it is z. The points ± 1 are the extrema of this Hamiltonian. Such, almost ovbious, observation was extended by AGS to any convex polytope. They showed that the moment map is also the Morse function. More exactly, the Morse–Bott function, e.g. see Sec. D.2.13. The convex hull of critical points of the moment map defines the *moment polytope.* At critical points (in our case ± 1 for the cross-polytope or the North and South poles for S^2) the 2-form (in our case Ω) is zero. Thus, the critical points of the Morse function are those at which the 2-form vanishes. This fact has very deep physical implications which we shall discuss later, in Secs. D.2.13 and D.2.17.

[c] http://en.wikipedia.org/wiki/Cross-polytope

Going back to Definition D.2.3.1, notice that we defined the term "coadjoint" in Sec. 8.2.3, while the term "equivariant" is definied now with the help of the following:

Definition D.2.3.2. If the group $\mathfrak{G}$ is acting on two manifolds V and W and if $\varphi : V \to W$ is a smooth map, such a map φ is *equivariant map* if $\forall x \in V$ and $\forall g \in \mathfrak{G}$, $\varphi(g \cdot x) = g \cdot \varphi(x)$.

This terminology is extension of a concept of "invariance". In the last case $\varphi(g \cdot x) = \varphi(x)$. In the quantum field theory equivariance is used routinely (without use of this terminology) for description of transformational properties of quantum fields, e.g. read [D.25].

Consider now the simplest example of the torus action on S^2 described by Eq. (D.36). E.g. see Fig. 3. In this case the torus T rotates S^2 about the z-axis by an angle $\theta(t) = 2\pi t$. Group-theoretically, S^2 is a homogeneous space with respect to the action of $\mathrm{SO}(3, \mathbf{R})$ group. The stabilizer (the isotropy subgroup) fixing the North pole, $\mathrm{SO}(2, \mathbf{R})$, is embedded into $\mathrm{SO}(3, \mathbf{R})$ as a subgroup of matrices of the form

$$T_N(t) = \begin{pmatrix} \cos t & -\sin t & 0 \\ \sin t & \cos t & 0 \\ 0 & 0 & 1 \end{pmatrix}. \tag{D.40a}$$

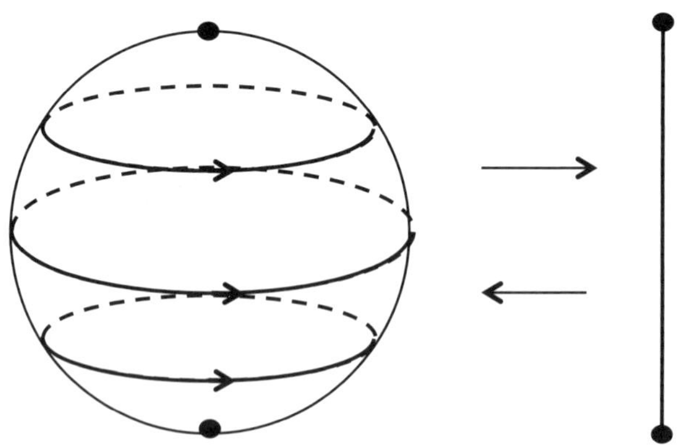

Fig. 3. Torus (S^1) dynamics on the sphere S^2. The quotient $S^2/S^1 = \mathbf{CP}^1$ is the simplest example of *symplectic reduction*. The arrow to the right points to the deformation retract of $\mathbf{CP}^1$: the segment $[-1, 1]$. It is the simplest moment polytope. The arrow to the left indicates that $\mathbf{CP}^1$ can be restored from its deformation retract. Details are explained in Exercise D.2.4.1. The fixed points of the torus action are represented by dots.

To treat the North and South poles on equal footing one has to introduce yet another matrix (the antipodal map)

$$T_S(t) = \begin{pmatrix} -\cos t & \sin t & 0 \\ \sin t & \cos t & 0 \\ 0 & 0 & -1 \end{pmatrix}. \tag{D.40b}$$

If we fix our attention only to the fixed points, then $T_N(t)$ trivially fixes the North pole while $T_S(t)$ exchanges the North and South poles. This simplest example contains the seed of the entire theory of reflection groups and, with it, the theory of Lie groups and Lie algebras. Why do we need such a theory at this moment of our exposition? It is because we want to reconstruct the symplectic manifolds ($\equiv$ complex projective varieties of Hodge-type) from the moment polytopes. The example just described can be interpreted group-theoretically as follows. The T_N can be identified with the identity map while T_S, with the reflection with respect to the plane perpendicular to the segment $[-1, 1]$ and crossing it through the point 0 located on this segment. Is this information sufficient for restoration of $T_N(t)$ and $T_S(t)$? To answer this question requires several sections which follow.

D.2.4 *Designing Toric Varieties from Fans*

Definition D.2.4.1. A *monoid* S is *finitely generated* if there exist a set $a_1, \ldots, a_k \in S$, called *generators* of S, and such that

$$S = Z_{\geq 0}a_1 + \cdots + Z_{\geq 0}a_k.$$

Here $Z_{\geq 0}$ are integers greater than or equal to zero. The monoid $S_\sigma = \sigma \cap \mathbf{Z}^d$, where σ is rational polyhedral cone is obviously finitely generated. Consider next a set of the Laurent monomials of the type $\mathbf{z}^\alpha = z_1^{\alpha_1} \cdots z_n^{\alpha_n}$ forming a closed polynomial subring with respect to addition and multiplication and make the crucial step: $\alpha \in S_\sigma$. This identification leads us to the following mapping:

$$u_i := z_i^{\alpha_i} \tag{D.41}$$

with α_i being one of generators of the monoid S_σ and $z_i \in \mathbf{C}_i$. Such a mapping provides an isomorphism between the additive group of exponentials α_i and the multiplicative group of Laurent monomials. Recall that

Definition D.2.4.2. The function ϕ is considered to be quasi-homogeneous of degree d with exponents $l_1, \ldots, l_n$ if

$$\phi(\lambda^{l_1}x_1, \ldots, \lambda^{l_n}x_n) = \lambda^d \phi(x_1, \ldots, x_n) \tag{D.42}$$

provided that $\lambda \in \mathbf{C}^* \equiv \mathbf{C}\backslash\mathbf{0}$. By applying this result to $\mathbf{z}^\alpha$ we re-obtain Eq. (D.34a) for the cone:

$$\sum_j (l_j)_i \alpha_j = d_i \tag{D.43}$$

or

$$\mathfrak{L} \cdot \boldsymbol{\alpha} = \mathbf{d}. \tag{D.44}$$

Here the index i is numbering different monomials. Equation (D.43) can be looked upon as scalar product with $(l_j)_i$ living in the space dual to α_j's. Thus, the set of $(l_j)_i$'s can be considered as the set of generators for the dual cone $\sigma^\vee$. Next, we consider a polynomial ring made of polynomials of the type

$$f(\mathbf{z}) = \sum_{\alpha \in S_\sigma} \lambda_{\boldsymbol{\alpha}} \mathbf{z}^{\boldsymbol{\alpha}} \equiv \sum_{\mathbf{l}} \lambda_{\mathbf{l}} \mathbf{u}^{\mathbf{l}}. \tag{D.45}$$

The ideal for such a ring is constructed based on the observation that for fixed d_i's and the assigned set of cone generators $\{\alpha_i\}$ there is always more than one set of generators for the dual cone [D.26]. This redundancy produces relations of the type

$$u_1^{l_1} \cdots u_k^{l_k} = u_1^{\tilde{l}_1} \cdots u_k^{\tilde{l}_k}. \tag{D.46}$$

Since $u_i^{l_i} \in \mathbf{C}_i$, the above equation belongs to the ideal $I(\mathbf{z})$. Collection of such relations is called the affine (or projective) hypersurfaces (varieties).[d] These are known as toric varieties and will be denoted as $X_{\sigma^\vee}$. The generators $\mathbf{u} = \{u_1 \cdots u_k\} \in \mathbf{C}^k$ are coordinates for $X_{\sigma^\vee}$. They represent the same point in $X_{\sigma^\vee}$ if and only if $\mathbf{u}^{\mathbf{l}} = \mathbf{u}^{\tilde{\mathbf{l}}}$. Thus formed toric variety corresponds to just one dual cone. In view of Fig. 2, if we need to build a complete toric variety we should be equipped with a set of gluing maps $\{\Psi_{\sigma^\vee \breve{\sigma}^\vee}\}$ as it is always done, e.g. when we assemble furniture, etc. If Σ represents a complete fan and $\amalg_{\sigma^\vee \in \Sigma} X_{\sigma^\vee}$ is the disjoint union of toric varieties associated with the dual cones, then the toric variety X_Σ is built out of this disjoint union with the help of gluing maps.

Exercise D.2.4.1. Design X_Σ for $\mathbf{CP}^n$. Notice that this procedure is opposite to making the deformation retract for $\mathbf{CP}^n$ described in Sec. D.2.1. Hint: Look for solution on pp. 232 and 279 of [D.26]. Consider a special case $\mathbf{CP}^1 = S^2$ of general construction of $\mathbf{CP}^n$ described on these pages

[d] An example of affine variety is an equation of the type $z_1^2 + z_2^2 = \epsilon, \epsilon \in \mathbf{C}$. An example of projective variety is an equation of the type $z_1^n + z_2^n = z_3^n$, etc.

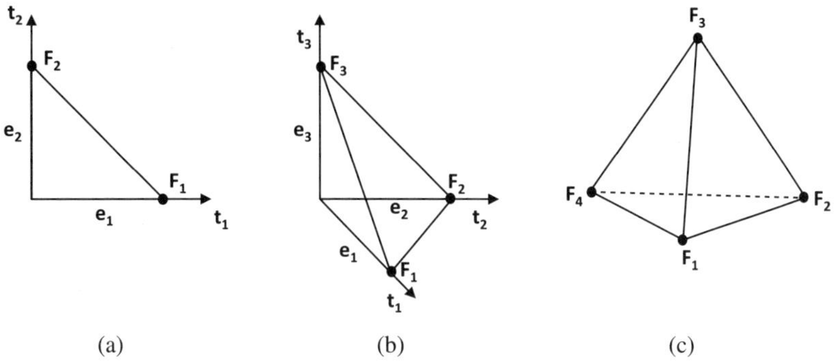

Fig. 4. Deformation retracts for (a) $S^2 = \mathbf{CP}^1(\mathbf{C})$: one-dimensional simplex, (b) $\mathbf{CP}^2$: two-dimensional simplex and (c) $\mathbf{CP}^3$: three-dimensional simplex pictured as subspace of four-dimensional vector space. The labels are explained in Sec. D.2.15.

and compare the design of S^2 given in [D.26] with that discussed in [D.15], pp. 149 and 150. After that, go ahead and design $\mathbf{CP}^2$ using either [D.26] or [D.15], pp. 150 and 151. Remember that the deformation retract (the moment polytope) for S^2 is a segment $[-1, 1]$ (e.g. see Figs. 3 and 4(a)) while that for $\mathbf{CP}^2$ is the equilateral triangle (Fig. 4(b)). The retract for $\mathbf{CP}^3$ is a tetrahedron (Fig. 4(c)). It represents the entire four-dimensional projective Hilbert space of two qubits. Incidentally, it also coincides with the projective twistor space PT. This fact will be used in Sec. D.2.16.

Remark D.2.4.1. The construction of toric varieties just described is valid for both integral and nonintegral polytopes. The question arises: Is there some differences in designing varieties from integral and nonintegral polytopes? That is to say, what makes integral polytopes so special? The answer to this question was given in the work by Delzant discussed in detail in [D.15, D.24]. A Delzant polytope Δ in $\mathbf{Z}^n$ is a polytope satisfying the following requirements: (a) there are n edges meeting at each vertex; (b) the edges meeting at each vertex p are rational in the sense that each edge is of the form $p + tu_i$ with $u_i \in \mathbf{Z}^n$, (c) for each vertex , the corresponding set $u_1, \ldots, u_n$ can be chosen as $\mathbf{Z}$-basis of $\mathbf{Z}^n$. From here it follows, for example, that tetrahedron is the right type of 3D Delzant polytope while the pyramid is not.

D.2.5 *Torus Action and Its Invariants*

Definition D.2.5.1. The set $T := (\mathbf{C}\backslash 0)^n \equiv (\mathbf{C}^*)^n$ is called *complex algebraic torus*. Each $z \in \mathbf{C}^*$ can be represented as $z = r\exp(i\theta)$ so that for each $r > 0$ the fiber $\{z \in \mathbf{C}^* || z| = r\}$ is a circle of radius r. We can represent T as a product $(R_{>0})^n \times (S^1)^n$. The product of n circles $(S^1)^n$ is indeed the *topological torus.* It is a deformation retract of T. Following [D.14] we shall call $(S^1)^n$ a *compact torus.* Thus, whatever we can prove for the deformation retract we can extend to full T.

Examples.

(a) Any unitary matrix $U(n)$ can be brought to the diagonal form

$$T_{U(n)} = \mathrm{diag}(e^{i\theta_1}, \ldots, e^{i\theta_n}) \tag{D.47a}$$

so that the *rank* of $U(n)$ is n. Such a representation is called *maximal torus* in $U(n)$.

(b) For SU(n) the maximal torus of rank $n - 1$ is given by

$$T_{\mathrm{SU}(n)} = \{\mathrm{diag}(e^{i\theta_1}, \ldots, e^{i\theta_n}) : \theta_1 + \cdots + \theta_n = 0\}. \tag{D.47b}$$

Clearly, SU(n) is a subgroup of $U(n)$. In fact $U(n) \simeq \mathrm{SU}(n) \times S^1$.

(c) The matrix $T_N(t)$, Eq. (D.40a), is the maximal torus for SO(3).

Definition D.2.5.2. The *torus action* is a continuous map

$$T \times X_{\sigma^\vee} \to X_{\sigma^\vee}, \quad (t, \mathbf{z}) \to t \cdot \mathbf{z} := (t^{a_1} z_1, \ldots, t^{a_k} z_k). \tag{D.48}$$

Let $\Psi : X_\Sigma \to X_{\tilde{\Sigma}}$ be a map and $\alpha : T \to T'$ a homomorphism, then for the torus action there could be two possibilities

$$\text{(a) } \Psi(T\mathbf{z}) = \alpha(T)\Psi(\mathbf{z}) \quad \text{or} \quad \text{(b) } \Psi(T\mathbf{z}) = \Psi(\mathbf{z}). \tag{D.49}$$

Here $\alpha(T)$ is the character of the algebraic torus group. With help of Definition D.2.3.2 we classify (a) as *α-invariant* (or equivariant) while (b) as invariant of the torus group action. To obtain invariants we need to study the orbits of the torus action. For this purpose we need to make a few definitions:

Definition D.2.5.3. *Orbits, stabilizers and normalizers.* Let M be some smooth manifold and G is some Lie group (also a manifold) The *orbit* of the point $m \in M$ is defined by $G \cdot m = (a \cdot m | a \in G\} \subseteq M$. The *stabilizer* (the *isotropy subgroup*) of the point $m \in M$ is a subgroup G_m of the group G such that $G_m = (a \cdot m = m | a \in G\}$. The *normalizer* $N_G(H)$ of the subgroup $H \subset G$ is defined as $N_G(H) = \{a \in G | aHa^{-1} = H\}$. The group action is said to be *free* if $G_m = \{e\}$: the unit element. The *evaluation map* $\mathrm{ev}_m : a \to a \cdot m$ induces a bijection from the quotient G/G_m to the orbit $G \cdot m$. Differently said, it can be proven [D.24], p. 11, that the quotient

G/G_m is the injective immersion into M. Therefore $\forall m \in M$ the orbit $G \cdot m \sim G \cdot (G/G_m)$. In such a case the group action is *transitive*, that is $\forall m, n \in M$ we have $a \cdot m = n$. Manifolds of the type G/G_m are called *homogeneous spaces*. If G is compact Lie group, its orbits are submanifolds of M. The dimension of such homogeneous submanifold is the same as the dimension of the associated with it orbit $\mathcal{O}$. That is $\dim \mathcal{O} = \dim(G/G_m) = \dim G - \dim G_m$. In addition, every quotient G/G_m can be considered as the fiber bundle with G being the total manifold and G_m being the fiber.

Examples.

(a) From Sec. D.2.3 we learned that when S^1 is acting on S^2 the quotient $S^2/S^1 = [-1, 1]$ is one-dimensional cross-polytope. The stabilizer is $T_N(t)$ and the centralizer is $T_S(t)$. Here $T_S(t) = G$ and $T_N(t) = G_m$. Thus, $\dim O = \dim S^2 - \dim S^1 = 1$ since S^2 is described in terms of two angles (parameters of the Lie group) while S^1 depends upon just one angle.

(b) When S^1 is acting on $S^3 \simeq \mathrm{SU}(2)$, the quotient $S^3/S^1 = S^2$. So that, indeed, S^3 is a fiber bundle with S^2 being a base as we already discussed in Sec. 6.5. Also, $\dim \mathcal{O} = \dim S^3 - \dim S^1 = 2$. Indeed, as we discussed in Sec. 6.5, the $\mathrm{SU}(2) \sim S^3$ is characterized by three parameters. From here we obtain $\dim \mathcal{O} = 2$.

(c) The stabilizer for the torus action is given by

$$t^a x = x. \tag{D.50}$$

Apart from the "trivial" solutions : $x = 0$ and $x = \infty$, there is a less trivial solution $x = 1$ implying: $t^a = 1$. For integer a's this is a cyclotomic equation whose nontrivial $a - 1$ solutions all lie on the circle S^1. This defines a one-dimensional lattice in the following sense. The circle $S^1 = \mathbf{R}/\mathbf{Z}$ can be identified with the exponential map: $\mathbf{R} \to \mathbf{C}$ that sends x to $\exp(i2\pi x)$ whose kernel is $\mathbf{Z}$. It converts $\mathbf{R}$ into a lattice $\mathbf{R}/\mathbf{Z}$ or, in general, $(S^1)^n = (\mathbf{R}/\mathbf{Z})^n$. If, instead of 2π we would have $2\pi/k$, where k is some integer, as required by the cyclotomic equation, we would still have a lattice. Under such circumstances every lattice Λ has its dual Λ^* so that if $\alpha \in \Lambda^*$ and $\xi \in \Lambda$, we would have $\langle a, \xi \rangle = 2\pi\mathbf{Z}$. In solid state physics the *dual lattice* is known as the *reciprocal lattice*. We shall return to these results in Sec. D.2.15.

Remark D.2.5.1. Notice that in the case of S^2 (Sec. D.2.3) the (cross)-polytope (the segment $[-1, 1]$) is integral and that the dimension of the torus group S^1 is half of the dimension of the symplectic manifold, that is of S^2. This fact was noticed and generalized by Delzant. He proved the following (e.g. read [D.24]).

Theorem D.2.5.1. *If the dimension d_T of the compact torus $(S^1)^{d_T}$ is 1/2 of the dimension d_M of the symplectic manifold M, then the moment polytope $\Phi(M)$ determines the Hamiltonian manifold (up to isomorphism).*

Remark D.2.5.2. The above theorem implies that *both* the Delzant and the complete fan constructions allow us to reconstruct the toric variety but if these varieties are *not* of Delzant type, then the AGS results cannot be used and the connection with mechanics becomes problematic. This issue is still an area of active research.

Exercise D.2.5.1. Check that Delzant's theorem is compatible with Kodaira's embedding Theorem D.2.1.1. Alternatively, check that Delzant's theorem is compatible with requirements of geometric quantization, [D.17].

Now, finally, we can return to (D.49) in order to construct the invariants of torus action. They can be qualified either as *equivariantly projective toric varieties* or just as *projective* toric varieties. The latter can be constructed from the former as the ratio of two equivariant mappings. Details of these constructions can be found in [D.26]. This reference treats these invariants only from the point of view of algebraic geometry. For better understanding of what follows next, we need in addition the group-theoretic understanding of these invariants. This is given in the next section.

D.2.6 *The Remarkable Duality (Beginning)*

For the sake of space, we shall provide only the most basic facts needed for uninterrupted reading. Those of our readers who are interested in more comprehensive knowledge of this problematics are encouraged to read our long review article, [D.27], and references therein.

Let V be the n-dimensional complex vector space $\mathbf{C}^n$ with the basis $\{e_1, \ldots, e_n\}$ (e.g. the finite Hilbert space in quantum mechanics). Introduce the *dual space* V^* which is a *polynomial ring* $K[\mathbf{z}]$, e.g. see Eq. (D.45), where indeterminates $\{z_1, \ldots, z_n\}$ are in one-to-one correspondence with the basis vectors $\{e_1, \ldots, e_n\}$. Let G be some compact Lie group. Without loss of generality (this will be explained below), we can replace the action of a group G by the maximal torus action for the same group. In such a case we shall be dealing with the equations of type given by Eq. (D.50). The eigenvalues in such a case are roots of unity. This is characteristic for *complex pseudo-reflection groups* [D.27, D.28]. The eigenvalues of G in V^* are just complex conjugates of those in V. Denote these eigenvalues as $c_1, \ldots, c_n$. Let t be some indeterminate and consider a product: $\det(1-tG) = (1-tc_1)\cdots(1-tc_n)$. This decomposition allows us to obtain the following generating function

$$\frac{1}{\det(1-tG)} = (1 + c_1 t + (c_1 t)^2 + \cdots)\cdots(1 + c_n t + (c_n t)^2 + \cdots)$$
$$= \sum_{k\geq 0}\left(\sum_{k_1+\cdots+k_n=k} c_1^{k_1}\cdots c_n^{k_n}\right) t^k. \tag{D.51}$$

By looking at Eq. (D.49) the following identification can be made: $\alpha(T) = c_1^{k_1}\cdots c_n^{k_n}$, provided that T is acting on the basis monomials of the type $z_1^{k_1}\cdots z_n^{k_n}$, and $k_1+\cdots+k_n = k$. This last result is useful to compare with Eq. (D.33) defining simplex. The results of Sec. D.2.2 indicate that information about the polyhedron/simplex is encoded in the complete fan, e.g see Fig. 2. Alternatively, the polyhedron can be made out of hyperplanes. The correspondence with hyperplanes can also be developed in the present case.

Indeed, since we are dealing with the complex pseudo-reflection groups the characters $c_i^{k_i}$ are of the form $\omega_i^{k_i} = \exp(\frac{2\pi i}{m_i}k_i)$, where m_i are some integers. Complex pseudo-reflection groups are isometries of $\mathbf{C}^n$ and, since locally $\mathbf{CP}^n$ and $\mathbf{C}^n$ are the same this means that they are also isometries of $\mathbf{CP}^n$. This fact makes such groups physically important. It is permissible [D.28] for the indeterminate t in Eq. (D.51) to be a complex number. Because of this circumstance and in view of the definition of $\alpha(T)$, Eq. (D.49), we can safely write $\sum_i \frac{k_i}{m_i} = k$ which is of exactly the same type as Eq. (D.43) defining a hyperplane. The word "reflection" should be understood in the present context as some kind of reflection symmetry connecting a given hyperplane with the next (either in polyhedron or in the fan (or in its dual) and so on). Although we leave details of the derivation to our readers, the simplest example of this phenomenon was already provided in Sec. D.2.3. In fact, because of this, we would like to do more in this regard to help our readers to develop needed skills. For this purpose we follow [D.29] where our readers will be able to find (among many other things) the following:

Theorem D.2.6.1. *Any* $g \in GL_n\mathbf{C}$ *can be uniquely factorized as* $g = UB$ *where* $U \in U(n)$ *and* B *is an upper triangular matrix with positive real diagonal entries.*

Definition D.2.6.1. The B matrices belong to the subgroup of $GL_n\mathbf{C}$ known as *Borel subgroup.*

This means that if B belongs to this subgroup, then $U(n) \cap B = T$, where $T = T_{U(n)}$, e.g. see Eq. (D.47a). Based on this information, it becomes possible also to prove the following

Theorem D.2.6.2. *The map $U(n)/T \to GL_n\mathbf{C}/B$ is homeomorphism.*

The above theorem is of profound importance for the rest of our exposition but its content is not immediately illuminating at this stage. To reveal its significance we begin with the following:

Definition D.2.6.2. The homogeneous space $U(n)/T \simeq GL_n\mathbf{C}/B$ is the space of *flags* in $\mathbf{C}^n$. A flag $\mathcal{F}$ is defined by

$$\mathcal{F} : 0 = V_0 \subset V_1 \subset \cdots \subset V_n = \mathbf{C}^n, \tag{D.52a}$$

where $\dim(V_i) = i$.

In Appendix B (Sec. B.1) we introduced the nilpotent and solvable Lie groups. It happens that Borel's B matrices belong to solvable groups. For such type of groups there is the following remarkable.

Theorem D.2.6.3. *Let G be solvable group acting on a projective variety X. Then G has a fixed point in X.*

Simply stated, this theorem implies that flags are fixed (points) by (for) matrices of B-type.

D.2.7 *The Remarkable Duality (Continuation). Delicate Inter-relationship between Flags and Grassmannians*

Equation (D.48) still holds if some of t-factors are replaced by 1's. This means that we should account for situations when one, two, etc. t-factors are replaced by 1's. This leads to the consideration of the torus action on toric subvarieties. By design, X_Σ is the disjoint union of finite number of torus orbits which can be identified with subvarieties of X_Σ. To make sense out of this, let us rewrite Eq. (D.48) in more familiar form, that is in the form of an eigenvalue problem

$$\begin{pmatrix} t^{a_1} & & & \\ & \cdot & & \\ & & \cdot & \\ & & & \cdot \\ & & & & t^{a_k} \end{pmatrix} \begin{pmatrix} z_1 \\ \cdot \\ \cdot \\ \cdot \\ z_k \end{pmatrix} = \begin{pmatrix} t^{a_1} z_1 \\ \cdot \\ \cdot \\ \cdot \\ t^{a_k} z_k \end{pmatrix}. \tag{D.53}$$

Evidently, the vector $(z_1, \ldots, z_k)$ forms a basis of k-dimensional complex vector space V while the vector $(z_1, \ldots, z_i), i \leq k$, forms a basis of a subspace V_i. This observation brings us back to the notion of a flag introduced by Eq. (D.52). Following (in part) [D.24], p. 50, we would like to discuss now an

Example. (Motivated by quantum mechanics) Consider matrices from the unitary group $\mathrm{U}(n)$ acting by conjugation on the space $\mathcal{H}$ of Hermitian matrices $M_n(\mathbf{C})$

$$\mathcal{H} = \{X \in M_n(\mathbf{C}) | \bar{X}^T = X\}. \tag{D.54a}$$

Since the Hermitian matrices represent the observables in quantum mechanics, their eigenvalues are real by design. They are diagonalizable with help of $\mathrm{U}(n)$ so that

$$\mathcal{H}_\lambda = \{A \in \mathcal{H} \mid \text{spectrum } \lambda_A = (\lambda_1, \ldots, \lambda_n)\}. \tag{D.54b}$$

Evidently, the spectrum λ_A could be made out of at most n distinct values. To complicate matters, following Gelfand and Serganova, [D.30], we shall allow for the eigenvalue multiplicities.[e] For this purpose we need to introduce some notations. Let $I_n = \{1, \ldots, n\}$, then $B(I_n)$ is the set made of all subsets of I_n. From now on we shall assume that the eigenfunctions/eigenvectors of A are mutually orthogonal so that they can be looked upon either as the basis vectors in Hilbert space of size n or, which is the same, as the basis vectors in $\mathbf{C}^n$. Thus, let $\{\mathbf{e}_1, \ldots, \mathbf{e}_n\}$ be some vector basis in $\mathbf{C}^n$. Select some $J_p = \{i_1, \ldots, i_p\} \in B(I_n)$ and denote as $\mathbf{C}^J$ the subspace of $\mathbf{C}^n$ made of vectors $\{\mathbf{e}_{i_1}, \ldots, \mathbf{e}_{i_p}\}$, $0 \leq p \leq n$.

Definition D.2.7.1. Let $G_k(\mathbf{C}^n)$ be the manifold made of *all* k-dimensional subspaces of $\mathbf{C}^n$, then $\mathrm{Gr}_k(\mathbf{C}^n)$ is called *Grassmannian* (associated with $\mathbf{C}^n$).

Remark D.2.7.1. Thus, in quantum mechanics Grassmannians should be considered whenever there is an eigenvalue multiplicity (not to be confused with the standard degeneracy!). Notice that such a situation requires reconsideration of Heisenberg's formulation of quantum mechanics (e.g. as summarized in [173]) since it causes us to find examples when such a multiplicity is realized in the Nature. *But this situation takes place exactly when*

[e] Exactly this fact will be used essentially below for solution of the entanglement problem.

the quantum entanglement is present! The purpose of this Appendix is to explain this fact.

Consider an action of the Cartan subgroup (it is the same as the torus group) $\mathbf{t}$ on $G_k(\mathbf{C}^n)$. Evidently, we must look at the orbit $\mathbf{t}{\cdot}X$ of the point $X \in G_k(\mathbf{C}^n)$. It can be shown [D.30] that the closure of $\mathbf{t}{\cdot}X$ made of *finite* number of such orbits is also a toric variety.

Definition D.2.7.2. Two subspaces X and Y of $\mathbf{C}^n$ are *equivalent* if $\dim(X \cap \mathbf{C}^J) = \dim(Y \cap \mathbf{C}^J)$. Based on this definition, the subspaces of Grassmannian (the equivalence classes) will be called *strata*. Each of these strata Γ can be described in terms of its dimension. That is let $\Gamma = \{X \in G_k(\mathbf{C}^n) | \dim(X \cap \mathbf{C}^J) = s(J)\}$.

This definition of strata is equivalent to more traditional definition of the *Schubert cell* [D.31], p. 105. These are the precursors of the so-called *Schubert calculus*. Now, instead of the flag defined by Eq. (D.52a), we can write $0 \subset \mathbf{C}^{J_1} \subset \mathbf{C}^{J_2} \subset \cdots \subset \mathbf{C}^{J_n} = \mathbf{C}^n$, where $|J_i| = i$. The totality of these flags is $n!$ How is this number related to the number of strata in the Grassmannian? How Grassmannians and flags are related to each other? It is essential to explain these topics before we enable them with physical content. We begin with explanation of the number $n!$ of flags.

D.2.8 *The Bruhat Decomposition and Bruhat Order*

For this purpose let us introduce a sequence $\boldsymbol{\pi} = (\pi(1), \ldots, \pi(n))$ which is a permutation of $(1, 2, \ldots, n)$. If $\pi(i)$'s can be looked upon as column vectors then permutations can be realized by the $n \times n$ matrices such that each of them is made out of zeros, except for that in row i where the entry for $\pi(i) = 1$. With help of this matrix it is possible to construct (to be described) the $n \times n$ matrix w in the *reduced echelon form*. It is the matrix in which: (a) each column w_i ends in a 1, say, in the π_ith row and, (b) the entries to the right of each 1 are all zero, i.e. $w_{\pi(i),j} = 0$ if $j > i$. For example, for 4×4 matrix w we get,

$$w = \begin{pmatrix} * & 1 & 0 & 0 \\ * & 0 & * & 1 \\ 1 & 0 & 0 & 0 \\ 0 & 0 & 1 & 0 \end{pmatrix}. \tag{D.55}$$

Such a matrix is obtained by permuting the columns of an upper triangular matrix n given by (in the 4×4 case)

$$\mathbf{n} = \begin{pmatrix} 1 & 0 & * & 0 \\ 0 & 1 & * & * \\ 0 & 0 & 1 & 0 \\ 0 & 0 & 0 & 1 \end{pmatrix}. \tag{D.56}$$

Thus, $\mathbf{w} = \mathbf{n}\boldsymbol{\pi}$. The following theorem (to be compared with Theorem D.2.6.1) of central importance follows from just obtained result:

Theorem D.2.8.1. *Any $g \in GL_n C$ can be decomposed as $g = n\pi B$, where the matrix n belongs to the subgroup N of upper-triangular matrices with* 1*'s on the diagonal, π is a permutation matrix and B belongs to the subgroup of upper-triangular matrices described in Theorem* D.2.6.1.

Let $\tilde{N}$ be subgroup of lower-triangular matrices with 1's on the diagonal. Consider an intersection $N \cap \pi \tilde{N} \pi^{-1} \equiv N_\pi$. Then, the matrix n in Theorem D.2.8.1 is taken from those which belong to N_π. In view of Definition D.2.5.3, we can talk about an orbit $\mathcal{O}_\pi$ of the type $N \cdot (GL_n\mathbf{C}/B)$. It can be demonstrated that $\mathcal{O}_\pi = N_\pi$. Since N is the subgroup of upper triangular matrices, w also belongs to this subgroup. The *number of variable entries in w* is determined by the number of pairs $i < j$ such that $\pi_j > \pi_i$. What does the statement $\pi_j > \pi_i$ actually mean? This requires us to introduce the concept of *Bruhat order*. Incidentally, the decomposition $g = n\pi B$ is known as the *Bruhat decomposition.*

To determine the Bruhat order we need to use the results of Appendix C, more specifically, we need to use Definition C.5.2 of the symmetric group S_n. From the same appendix we know that any permutation $\mathfrak{w}$ can be written as product $\mathfrak{w} = s_{a_1} s_{a_2} \cdots s_{a_l}$ of simple transpositions s_i. Evidently, the final state can be reached in many ways so that the *reduced word* $\mathfrak{w}$ whose length $l(\mathfrak{w})$ is minimal should be introduced. Using this concept the *weak Bruhat order* is defined as follows:

Definition D.2.8.1. (Weak Bruhat order) It is assumed that $\mathfrak{v}$ precedes $\mathfrak{w}$ (or $\mathfrak{v} \preceq \mathfrak{w}$ for short) if $l(\mathfrak{w}) = l(\mathfrak{v}) + 1$ and, in addition, $\mathfrak{w} = \mathfrak{v} s_i$.

Weak Bruhat order can (and should) be readily visualized. Indeed, let us place in $\mathbf{R}^n$ a set of vectors $\boldsymbol{\pi} = (\pi(1), \dots, \pi(n)), \pi(i) \in S_n$, this time describing the location of vertices of polyhedron. Connect by a segment the points (the vertices) corresponding to two permutations, one preceding the other. The end result of such a process-formation of the polyhedron Π_{n-1} of dimension $n - 1$ known in literature as *permutohedron.* We shall discuss below how these permutohedra are related to the entanglement problem. To get some feeling of these objects, following [D.32], let us consider two the most typical examples.

Examples of permutohedrons.

(a) In $\mathbf{R}^3$ according to just described rules we are allowed to permute just three points $(1, 2, 3)$. Based on this information, we create the vertex set $V_3 = \{(123), (132), (213), (231), (312), (321)\}$. Viewed as points in $\mathbf{R}^3$ such a set will have six vertices. Each vertex is described in terms of coordinates x, y and z. Notice that *for each vertex* $x + y + z = 6$. This is an equation for the plane! It can be shown [D.32] that the vertices are actually located at the intersection of this plane with the sphere $x^2+y^2+z^2 = 14$. The resulting permutahedron Π_2 is a regular hexagon inscribed in a circle as depicted in Fig. 5. In view of Fig. 2, Π_2 is also the simplest zonotope.

(b) In $\mathbf{R}^4$ the situation is slightly more complicated. The resulting permutohedron Π_3 is depicted in Fig. 6.

This is a three-dimensional polyhedron. It is obtained as result of intersection of the hyperplane $x + y + z + t = 10$ with S^3 whose equation is: $x^2 + y^2 + z^2 + t^2 = 30$. Π_3 is also a zonotope, [D.14], p. 200.

In general, the permutohedron Π_{n-1} can be obtained either as solution of two coupled equations

$$x_1 + \cdots + x_n = \frac{(n+1)n}{2} \quad \text{and} \quad x_1^2 + \cdots + x_n^2 = \frac{n(n+1)(2n+1)}{6}$$

or as an affine projection of the hypercube of dimension $p = \binom{n}{2}$ living in $\mathbf{R}^{d\times p}$ onto $\mathbf{R}^d$, $d = n - 1$. In the last case it is also known as zonotope Z_d. Recall that we discussed zonotopes in Sec. D.2.2.

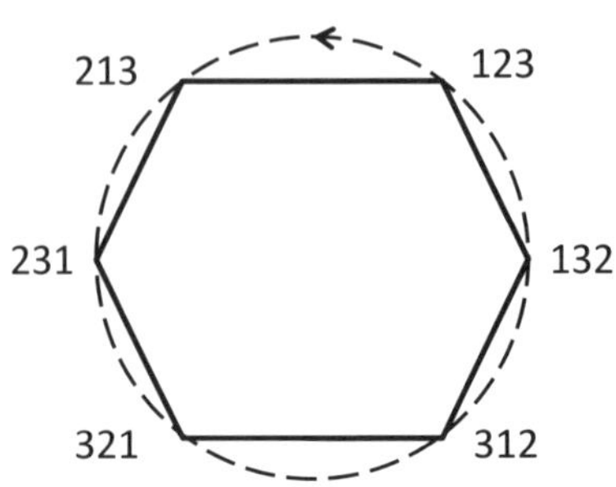

Fig. 5. Two-dimensional permutohedron Π_2.

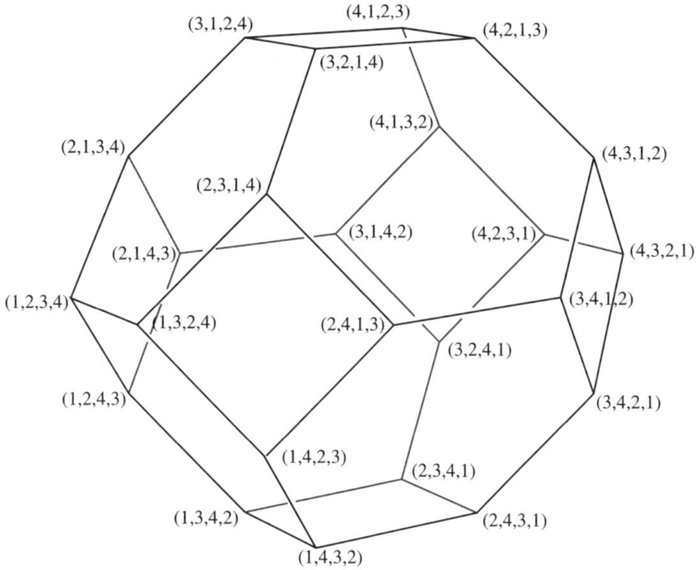

Fig. 6. Three-dimensional permutohedron Π_3.

Remark D.2.8.1. Although we followed [D.32] to obtain the three-dimensional permutohedron Π_3, such a derivation is formal. It is detached from physics. Fortunately, in the present case there is a chance to connect the obtained results with physics by noticing that what is known in mathematics as permutohedron is known in physics as the first Brillouin zone[f] for the face centered cubic (fcc) lattice (http://lamp.tu-graz.ac.at/~hadley/ss1/bzones/fcc.ph). Some crystals with fcc lattices are Al, Cu, C (diamond), Si, Ge, Ni, Ag, Pt, Au, Pb, NaCl. Because of this, it is possible to tessellate three-dimensional Euclidean space with such type of structures. The results of such an activity provide an input into the outstanding (both physically and mathematically) problem about the densest possible packing [D.33] in Euclidean (hyperbolic, etc.) space(s). Incidentally, such a problem is also of interest in the theory of data storage and transmission. Our work on the packing problem, [D.34], allowed us to find physically motivated way (in [D.35]) for reproving some results of Grisha Perelman

[f]The construction of Brillouin zones (associated with the respective Bloch vectors) is discussed at the beginning of any course in solid state physics.

(http://en.wikipedia.org/wiki/Grigori_Perelman) which he used for proving the Poincaré and geometrization conjectures.

Remark D.2.8.2. In addition to the weak Bruhat order, there is also *strong Bruhat order*. It is defined similarly to the weak case. Indeed, it is assumed that $\mathfrak{v}$ precedes $\mathfrak{w}$ (or $\mathfrak{v} \preceq \mathfrak{w}$ for short) if $l(\mathfrak{w}) = l(\mathfrak{v}) + 1$ and, in addition, there exist integers $i < j$ so that $\mathfrak{w} = \mathfrak{v}\tau_{ij}$. The transposition τ_{ij} is exchanging ith place with jth.

Going back to the subgroup N_π and the matrix w, the number of variable entries l_π in it is subject to the strong Bruhat order. l_π can be evaluated graphically as depicted in Fig. 7.

Thus, the Lie group N_π is homeomorphic to the complex space $\mathbf{C}^{l_\pi}$. The flag manifold $GL_n\mathbf{C}/B$ is made of orbits $\mathcal{O}_\pi \simeq N_\pi$ and, since these are permutations, there are $n!$ of them. Each orbit is a cell $C_\pi \simeq N_\pi \simeq \mathbf{C}^{l_\pi}$. This result provides us with an answer regarding the meaning of $n!$ number. We still have not explained how this result is related to the number of strata in the Grassmannian.

D.2.9 *The Remarkable Duality (End). Weyl Groups, Flags and Grassmannians*

Let us begin with the notion of the Weyl group. It can be developed based on Definition D.2.5.3 and Theorem D.2.6.3. To be specific, we fix our attention at the unitary group $U(n)$ for which we obtained $U(n) \cap B = T$. Using Definition D.2.5.3, let $N_U(T)$ be the normalizer of $U(n)$. Evidently, it is just a permutation which shuffles the diagonal entries of $U(n)$. Next, consider

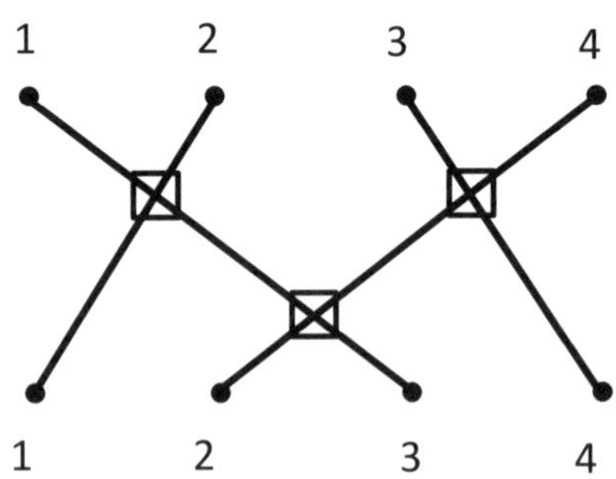

Fig. 7. Graphical calculation of l_π. In the upper row we have the standard order $\{1, 2, 3, 4\}$ while in the lower row we have a permutation $\pi = \{3, 1, 4, 2\}$. The number of crossings is l_π. In the present case, $l_\pi = 3$.

a quotient $W = N_U(T)/T$. It can be visualized as polytope made of points related to each other by permutation, e.g. just like that in Figs. 5 and 6. More on this will be presented in Sec. D.2.17.

Definition D.2.9.1. The quotient $W = N_U(T)/T$ is called the *Weyl group.*

Since we already know that the Bruhat decomposition is given by $g = n\pi B$, it can be shown [D.17], p. 155, that, in fact

$$GL_n\mathbf{C} = \bigsqcup_{w\in W} BwB. \tag{D.57a}$$

Since B fixes flag $\mathcal{F}$, let $Bw = \mathcal{F}_w$ so that, in view of the above equality and the fact that $\mathcal{F} = U(n)/T \simeq GL_n\mathbf{C}/B$, we obtain,

$$\mathcal{F} = \bigsqcup_{w\in W} \mathcal{F}_w = \bigsqcup_{w\in W} N_{\pi(w)} \simeq \bigsqcup_{\mathfrak{w}\in W} \mathbf{C}^{l(\mathfrak{w})}. \tag{D.57b}$$

Now we are left with only the relationship between Grassmannians and flags. Let us focus our attention on the subspace $\dim V_p = p$ of the complex vector space $\mathbf{C}^n$. The first question which comes to mind is this: Given the total dimension n of $\mathbf{C}^n$ and the number p, how many subspaces of dimension p can one create out of $\mathbf{C}^n$? Evidently, the number of subspaces (stratas) is $n!/p!(n-p)!$. This number can be understood as follows. Let $\{\mathbf{e}_1, \ldots, \mathbf{e}_n\}$ be some vector basis in $\mathbf{C}^n$. Out of this basis we can construct the exterior products of the type $\mathbf{e}_1 \wedge \mathbf{e}_2, \ldots, \mathbf{e}_1 \wedge \cdots \wedge \mathbf{e}_n$. As before, consider a subset $J_p = \{i_1, \ldots, i_p\}$ and associate with it the set of basis vectors $\{\mathbf{e}_{i_1}, \ldots, \mathbf{e}_{i_p}\}$. We can construct all kinds of exterior products of the type $\mathbf{e}_{i_1} \wedge \cdots \wedge \mathbf{e}_{i_p}$ so that there are exactly $n!/p!(n-p)!$ ways to select the basis vectors $\{\mathbf{e}_{i_1}, \ldots, \mathbf{e}_{i_p}\}$ from the set $\{\mathbf{e}_1, \ldots, \mathbf{e}_n\}$ and to make the exterior products $\mathbf{e}_{i_1} \wedge \cdots \wedge \mathbf{e}_{i_p}$ out of selected vectors. According to Definition D.2.7.1 the Grassmannian $Gr_p(\mathbf{C}^n)$ is the manifold made of all p-dimensional subspaces of $\mathbf{C}^n$. Moreover, from Definition D.2.7.2, we know that the strata are the equivalence classes of subspaces of $Gr_p(\mathbf{C}^n)$. This means that for fixed p we are having $n!/p!(n-p)!$ stratas. We defined the flag $\mathcal{F}$ in $\mathbf{C}^n$ while all previous discussions involved $\mathbf{CP}^n$. Based on the information just presented, it should be clear that $Gr_p(\mathbf{C}^n)$ is associated with only one subdivision: $\dim V_p = p$ of $\mathcal{F}$. Therefore, the entire flag $\mathcal{F}$ should be made out of product of Grassmannians, that is (still in $\mathbf{C}^n$) we should have the following correspondence

$$\mathbf{C}^{J_p} \rightleftarrows Gr_{J_p}(\mathbf{C}^n), \quad 0 \le p \le n \tag{D.58a}$$

causing us to write [D.36]

$$\mathcal{F} \subset Gr_{J_1}(\mathbf{C}^n) \times \cdots \times Gr_{J_n}(\mathbf{C}^n). \tag{D.58b}$$

To transfer this information to $\mathbf{CP}^N$ of some dimensionality N is very natural at this point. To do so requires us to make just a few additional steps.

For instance, we already know that S^2 is the same object as $\mathbf{CP}^1$ (e.g. see Fig. 4(a)). This result should be understood as follows. Consider $\mathbf{CP}^1$ made out of equivalence classes $[z_0 : z_1] = c[z_0 : z_1]$. This means that each equivalence class is characterized by a nonzero complex number c which we shall consider as a point (kind of a coordinate) in $\mathbf{CP}^1$. Thus, there is a one-to-one correspondence between points in $\mathbf{CP}^1$ and lines in $\mathbf{C}^2$. For $z_0 \neq 0$, we construct a ratio z_1/z_0, (alternatively, we can write z_0/z_1 if $z_1 \neq 0$). Clearly, such a ratio is scale-invariant and, therefore, it represents a point in $\mathbf{CP}^1$. The same is true for any other ratio. So, S^2 can be covered by two overlapping patches made of complex two-planes, that is of two copies of $\mathbf{C}$'s. The overlap between patches yields the transition function. Thus, S^2 can now be identified with the $\mathbf{CP}^1$. Consider now $Gr_1(\mathbf{C}^n)$. From Definition D.2.7.1, it follows that $Gr_1(\mathbf{C}^n)$ represents an assembly of lines, that is the set of one-dimensional subspaces of $\mathbf{C}^n$ and, therefore, an assembly of points in $\mathbf{CP}^n$. Thus, $Gr_1(\mathbf{C}^n)$ can be identified with $\mathbf{CP}^n$. A point in $\mathbf{CP}^n$ is represented now by $\{c_1, \ldots, c_n\}$. How we can extend this result to all k-dimensional subspaces of $\mathbf{C}^n$ representing $Gr_k(\mathbf{C}^n)$?

Consider a vector $\mathbf{e}_{i_1} \wedge \cdots \wedge \mathbf{e}_{i_p}$ representing a member of a stratum $\{p\}$. If in the complex space $\mathbf{C}^n$ we are interested in changing the basis from, say, $\{\mathbf{e}_1, \ldots, \mathbf{e}_n\}$ to $\{\mathbf{e}'_1, \ldots, \mathbf{e}'_n\}$, this can be achieved with the help of $n \times n$ matrix A_{ij} in a usual way. When dealing with the vector $\mathbf{e}_{i_1} \wedge \cdots \wedge \mathbf{e}_{i_p}$, we obtain instead:

$$\mathbf{e}'_{i_1} \wedge \cdots \wedge \mathbf{e}'_{i_p} = A_{i_1 j_1}\mathbf{e}_{j_1} \wedge \cdots \wedge A_{i_p j_p}\mathbf{e}_{j_p} = \det(A_{ij})\mathbf{e}_{i_1} \wedge \cdots \wedge \mathbf{e}_{i_p}. \tag{D.59}$$

Remark D.2.9.1. (a) Since the equivalence relation, Eq. (D.59), is of the same nature as that used for deriving $\mathbf{CP}^1$ from $\mathbf{C}^2$, it provides us with the coordinate component of a point in the projective space $\mathbf{CP}^N$. The dimensionality of such a space is determined by $N = (n!/p!(n-p)!) - 1$. When $p = 1$, we obtain $N = n - 1$. In particular, for $\mathbf{C}^2$ we re-obtain $N = 1$ as required.

Just described (*Plücker*) *embedding* of the Grassmannian into complex projective space of dimension N is naturally extendable to the product of

Grassmannians, Eq. (D.58b), resulting in the product of projective spaces of respective dimensions. In view of Eq. (D.53), the torus action will result in the formation of a determinant made out of indeterminates. It can be identified with the character of symmetric group S_n. Such a character, known as the Schur polynomial $s_\lambda(\mathbf{x})$, was already discussed in Appendix C. It is well known that $s_\lambda(\mathbf{x})$ is represented by the determinant [D.31] (actually the ratio of determinants but the denominator cancels out). The Littlewood–Richardson fusion rule, Eq. (C.4), plays the central role in the *Schubert calculus*, e.g. read [D.31, D.36]. It was discussed in detail from various angles in our papers, [173] and [D.37]. The information in this subsection explains the origins of correspondence between the spaces V and V^* introduced in Sec. D.2.6, thus explaining the "remarkable duality" existing between these spaces.

D.2.10 *Connections between the Grassmannians and Exactly Integrable Systems. Many Uses of Plücker Embedding: From Robotics and Computer Vision to Twistor Theory*

The connection between Grassmannians and exactly integrable systems is discussed in detail in remarkable book by Miwa, Jimbo and Date, [D.38]. In [D.39] Flaschka explains how the torus action formalism matches that developed in [D.38]. In [D.38], the Schur polynomials (Plücker coordinates) are interpreted in terms of the tau functions of the Sato Grassmannian designed for description of the Kadomtsev–Petviashvili (KP) hierarchy of exactly integrable systems while the Plücker relations (to be described) are in one-to-one correspondence with the Hirota bilinear equations which, in turn, are in one-to-one correspondence with the KP-type equations.

In our work, [D.37], we discussed the same type of problems while connecting them with the properties of directed random walks (Appendix C). Specifically, we reinterpreted the combinatorial properties of Schur functions in terms of combinatorial properties of an assembly of directed ("vicious"/fermionic) random walkers and, by doing so, we established a connection between the walkers and the exactly integrable systems, e.g. those related to the Witten–Kontsevich model. Since the Plücker relations are responsible for the embedding of Grassmannians into projective spaces and since such coordinates in these spaces are essentially the same as the Schur functions, we would like now to describe the idea behind the Plücker embedding using terminology and examples familiar to every physicist.

We begin with an observation that for the Grassmannian $Gr_p(\mathbf{C}^n)$ the duality-type relation

$$Gr_p(\mathbf{C}^n) = Gr_{n-p}(\mathbf{C}^n) \tag{D.60}$$

is quite natural. It is originating from the fact that $n!/p!(n-p)! = n!/(n-p)!p!$ causing the exterior products of the type $\mathbf{e}_{i_1} \wedge \cdots \wedge \mathbf{e}_{i_p}$ and those of the type $\mathbf{e}_{i_{p+1}} \wedge \cdots \wedge \mathbf{e}_{i_{n-p}}$ to be related to each other in the way known from the Hodge theory of differential forms [D.16, D.40]. Following [D.36] we shall write $\mathbf{e}_{i_1} \wedge \cdots \wedge \mathbf{e}_{i_p} \in \overset{p}{\wedge}\mathcal{E}$ and $\mathbf{e}_{i_1} \wedge \cdots \wedge \mathbf{e}_{i_{n-p}} \in \overset{n-p}{\wedge}\mathcal{E}$ meaning that in both cases we use the basis $\{\mathbf{e}_1, \ldots, \mathbf{e}_n\}$ so that $\mathbf{e}_i \in \mathcal{E}\,\forall\, i$. If this is so, we can construct the products of the type $\overset{p}{\wedge}\mathcal{E} \wedge \overset{n-p}{\wedge}\mathcal{E}$ and then require that

$$\bigwedge^{p}\mathcal{E} \wedge \bigwedge^{n-p}\mathcal{E} \sim \bigwedge^{p+1}\mathcal{E} \wedge \bigwedge^{n-p-1}\mathcal{E}. \tag{D.61a}$$

The sign $\sim$ means "equivalence". Evidently, we can write as well

$$\bigwedge^{p}\mathcal{E} \wedge \bigwedge^{n-p}\mathcal{E} \sim \bigwedge^{p+2}\mathcal{E} \wedge \bigwedge^{n-p-2}\mathcal{E} \tag{D.61b}$$

and so on. The set of equivalence relations of such type is in fact the same as Plücker's conditions for embedding into projective space. This is formally clear and, should this be a book just for mathematicians, we would stop exactly at this point. Having in mind physicists as well, we would like to illustrate now how these abstract results find their uses in real life.

Example. Consider the most well-studied Grassmannian $Gr_2(\mathbf{C}^4)$. For it, we have $\{\mathbf{e}_0, \ldots, \mathbf{e}_3\}$ as our basis for $\mathcal{E}$. Using it, we construct $\overset{2}{\wedge}\mathcal{E}$

$$\bigwedge^{2}\mathcal{E} \sim \{\mathbf{e}_0 \wedge \mathbf{e}_1, \mathbf{e}_0 \wedge \mathbf{e}_2, \mathbf{e}_0 \wedge \mathbf{e}_3, \mathbf{e}_1 \wedge \mathbf{e}_2, \mathbf{e}_1 \wedge \mathbf{e}_3, \mathbf{e}_2 \wedge \mathbf{e}_3\}. \tag{D.62a}$$

If $\mathbf{x}^p = \sum_{i=0}^{3} x_i \mathbf{e}_i$ and $\mathbf{y}^p = \sum_{i=0}^{3} y_i \mathbf{e}_i$ we obtain[g]

$$\mathbf{x}^p \wedge \mathbf{y}^p = \sum_{(i,j)\in J} (x_i y_j - y_i x_j)\mathbf{e}_i \wedge \mathbf{e}_j, \tag{D.62b}$$

where $J = \{(0,1), (0,2), (0,3), (1,2), (1,3), (2,3)\}$. The combinations $l_{ij} = x_i y_j - y_i x_j$ are known as *Plücker coordinates*. These are the determinants $l_{ij} = \begin{vmatrix} x_i & x_j \\ y_i & y_j \end{vmatrix}$. The subset J is made out of six elements. The number six

[g]The meaning of the subscript p will be explained shortly below.

is just $\frac{4!}{2!2!}$. It corresponds to six 2-planes in $\mathbf{C}^4$. Thus, the dimensionality N of the embedding projective space $\mathbf{CP}^N$ is 5. This result is standard but formal and detached from life. It can be understood in practical terms nevertheless as follows.

Consider two points $\mathbf{x}$ and $\mathbf{y}$ on a line $\mathcal{L}$ in 3D space. Since the coordinates $\mathbf{x} = \{x_1, x_2, x_3\}$ and $\mathbf{y} = \{y_1, y_2, y_3\}$ of these points on $\mathcal{L}$ are taken with respect to some fixed origin, these are vectors. We want to demonstrate that Plücker's coordinates for the line $\mathcal{L}$ are in fact given by $\mathbf{Z} = \mathbf{y} - \mathbf{x}$ and $\mathbf{L} = \mathbf{x} \times \mathbf{y}$. These can be constructed as follows: 1. Enlarge the embedding Euclidean space, e.g. consider $\mathbf{x}^p = \{x_0, \mathbf{x}\}, \mathbf{y}^p = \{y_0, \mathbf{y}\}$. With such an enlargement the vectors $\mathbf{x}^p$ and $\mathbf{y}^p$ can be looked upon either as vectors in $\mathbf{R}^4$ or as points in $\mathbf{P}^3$ (recall our discussion of S^2 as projective space). 2. Consider the exterior product $(x_0, x_1, \ldots, x_3) \wedge (y_0, y_1, \ldots, y_3) \equiv (l_{01}, l_{02}, l_{03}, l_{23}, l_{31}, l_{12})$ with l_{ij} defined as before. 3. Choose $x_0 = y_0 = 1$, then $l_{01} = y_1 - x_1 = Z_1, \ldots, l_{23} = x_2y_3 - y_2x_3 = L_1, \ldots, l_{12} = x_1y_2 - y_1x_2 = L_3$.

Remark D.2.10.1. (a) Plücker's coordinates are describing (directed) lines in 3-space. (b) Since the vector $\mathbf{L}$ can be interpreted as the angular momentum, Plücker's coordinates are useful for description of kinematics of rigid bodies (e.g. as discussed in Chap. 8). Because of this, they find their extensive uses in robotics (e.g. read http://en.wikipedia.org/wiki/Plücker_coordinates), and in computer vision (http://en.wikipedia.org/wiki/Grassmannian). (c) In such physical language the Plücker embedding acquires familiar form

$$\mathbf{Z} \cdot \mathbf{L} = 0. \tag{D.63}$$

Explicitly,

$$l_{01}l_{23} + l_{02}l_{31} + l_{03}l_{12} = 0. \tag{D.64}$$

Written in terms of $\mathbf{x}$ and $\mathbf{y}$ variables, this relation is known in literature as Klein's quadric [D.40]. Felix Klein obtained this result in 1868.

Remark D.2.10.2. Exactly this result for Klein's quadrics happen to play central role in Roger Penrose twistor program, e.g. read [D.41]. All known exactly integrable systems can be re-interpreted and can be solved in terms of twistor formalism [D.42]. All of them can be obtained as some reductions of the self-duality equations for pure Yang–Mills fields. Since these equations are conformally invariant, they are massless. Therefore, the twistor formalism is also applicable (rigorously speaking) to massless particles only. In terms of this formalism both the self-dual Yang–Mills and pure

Einsteinnian gravity equations can be re-obtained and solved. It happens that some of the solutions for both are identical. The full advantage of this coincidence was exploited in our work, [181], where solution of the mass gap problem[h] for pure Yang–Mills fields was obtained. This result leading to the topological mass generation is complementary to that discussed in Chap. 4.

Exercise D.2.10.1. Write the Plücker embedding condition for the "multiparticle case" using the "single particle" result, Eq. (D.63). Think about physical devices (e.g. some robots, linkages, etc.) in which such a condition can be effectively utilized physically. Hint: Read [D.43].

While the significance of Klein's discovery for twistors will be discussed in some detail below, in Secs. D.2.15 and D.2.16, to finish this section, we need to keep focus on the following problem. Plücker coordinates can be obtained as determinants of columns i and j of the 2×4 matrix

$$\mathbf{M} = \begin{pmatrix} x_0 & x_1 & x_2 & x_3 \\ y_0 & y_1 & y_2 & y_3 \end{pmatrix} \tag{D.65a}$$

obtainable by replacing $\mathbf{x}^p = \sum_{i=0}^{3} x_i \mathbf{e}_i$ and $\mathbf{y}^p = \sum_{i=1}^{3} y_i \mathbf{e}_i$ expansions by a single equation

$$x_i^p = \sum_{j=0}^{3} \mathrm{M}_{ij} \mathbf{e}_j, \quad i = 1, 2. \tag{D.65b}$$

In such a form this result can be obviously generalized to any dimension. By doing so, the matrix $\mathbf{M}$ acquires the following form:

$$\mathbf{M} = \begin{pmatrix} \mathbf{A}_{d\times d} & \mathbf{B}_{(n-d)\times d} \\ \mathbf{O}_{d\times(n-d)} & \mathbf{C}_{(n-d)\times(n-d)} \end{pmatrix}. \tag{D.65c}$$

The submatrices $\mathbf{A}$, $\mathbf{B}$, $\mathbf{C}$ and $\mathbf{O}$ respectively are having dimensions $d \times d$, etc. The matrix $\mathbf{O}$ is made of zeros. Clearly, the matrix $\mathbf{M}$ is $n \times n$. It can be made of complex entries and can be identified with the matrix $g \in GL_n\mathbf{C}$ so that Theorem D.2.8.1 can be applied resulting in the decomposition $g = n\pi B$. Since the matrix $\mathbf{O}_{d\times(n-d)}$ is made of zeros, this is helpful. Equation (D.65b) is working either in the complex projective space or in the complex space. The projective space can be covered by patches (as it was done for S^2) made of complex spaces whose dimension should still be calculated. This calculation is based on the fact that $\mathbf{M}$ describes transformations appropriate for the Grassmannians. This means that the part of $\mathbf{M}$, the

[h] E.g. see http://www.claymath.org/millennium/Yang-Mills_Theory/

matrix **A**, cannot be changed since **A** fixes subspaces of dimensionality d. This leaves us only with the matrix **B** of dimension $d(n-d)$ describing transformations in subspaces of dimension complementary to d. Since **A** fixes subspaces of dimensionality d it can always be chosen in the form of the unit $d \times d$ matrix. Because of the duality, Eq. (D.60), the same must be true for **C** which should also be diagonal. In such a case the matrix **M** is upper triangular and is of the same type as **n** in Eq. (D.56). Therefore, it can be brought to the reduced echelon form w, Eq. (D.55), by acting on it by permutation matrices. In the case of $Gr_2(\mathbf{C}^4)$ we obtain the following reduced echelon matrices

$$\begin{aligned}
&(a)\begin{pmatrix}0&1&0&0\\1&0&0&0\end{pmatrix}, \quad (b)\begin{pmatrix}0&*&1&0\\1&0&0&0\end{pmatrix}, \quad (c)\begin{pmatrix}0&*&*&1\\1&0&0&0\end{pmatrix},\\
&(d)\begin{pmatrix}*&0&1&0\\ *&1&0&0\end{pmatrix}, \quad (e)\begin{pmatrix}*&0&*&1\\ *&1&0&0\end{pmatrix}, \quad (f)\begin{pmatrix}*&*&0&1\\ *&*&1&0\end{pmatrix}.
\end{aligned} \tag{D.66}$$

What is the significance of this result? We would like to provide an answer to this question following our work, [D.44], and that by Richard Stanley, [D.45]. Consider an equation

$$n_0 + \cdots + n_k = N. \tag{D.67}$$

Since it is of the same type as Eq. (D.43), everything to be mentioned below is applicable to toric varieties as well. The Grassmannian is just one of such varieties [D.30].

Without loss of generality we can order entries in Eq. (D.67) as follows $n_0 \geq n_1 \geq \cdots \geq n_k \geq 1$.[i] We shall call $n \equiv \{n_0, \ldots, n_k\}$ a partition. If n is a partition of N, this is written as $n \vdash N$ or as $|n| = N$. With every partition it is common to associate the Young tableaux [D.31, D.36]. It is made of square boxes of unit size which fill in (partially or completely), say, the $n \times m$ rectangle. For instance, let, say, the rectangle be of size 9 by 5 and suppose that we are dealing with a partition $\{7, 5, 3, 1, 1\}$, then we place seven boxes (without gaps) into the first row so that the first box is placed into the upper left corner. Next, we place five boxes into the second row and so on.

Let now $p(N; k, m)$ be the number of partitions of N into $\leq k$ non-negative parts, each not larger than m. Let $\mathcal{F}(k, m|q)$ be the generating

[i]The last inequality: $n_k \geq 1$, is chosen only for the sake of comparison with the existing literature conventions.

function for $p(N;k,m)$, that is

$$\mathcal{F}(k,m|q)=\sum_{N=0}^{S}p(N;k;m)q^{N}. \tag{D.68a}$$

The upper limit S was determined in [D.44] as mk. This is just the area of $k\times m$ rectangle. Also, it was argued in [D.44] that $m+k=n$. In view of this, we write $\mathcal{F}(n,n-k|q)=\begin{bmatrix} n \\ k \end{bmatrix}_q$. The last result is known in combinatorics literature as the Gaussian coefficient

$$\begin{bmatrix} k \\ m \end{bmatrix}_q=\frac{(q^k-1)(q^{k-1}-1)\cdots(q^{k-m+1}-1)}{(q^m-1)(q^{m-1}-1)\cdots(q-1)}\rightarrow\begin{pmatrix} k \\ m \end{pmatrix}_{q=1}. \tag{D.68b}$$

It possesses an obvious property: $\begin{bmatrix} k+m \\ m \end{bmatrix}_q=\begin{bmatrix} k+m \\ k \end{bmatrix}_q$ which, for instance, is an origin of the duality relation Eq. (D.58) for the Grassmannian, etc.

The Gaussian coefficient happens also to be the Poincaré polynomial (see Sec. D.2.13) for the Grassmannian $Gr_m(\mathbf{C}^k)$. It is obtained on p. 293 of the book by Bott and Tu, [D.46].[j] In both [D.44] and [D.45], it was noticed that this result can also be interpreted number-theoretically. Those of our readers who are interested in the number-theoretic interpretation of the Gaussian coefficient are encouraged to read our long review paper, [D.47]. It provides enough number-theoretic background for reading and understanding of the original ground-breaking work by Andre Weil [D.48] where Grassmannian-related problems are treated number-theoretically.

Thus, since the generating function $\mathcal{F}(k,m|q)$ is related to combinatorial properties of Grassmannians, we want to demonstrate now how the echelon form matrices, Eq. (D.66), can be re-obtained combinatorially. Since the results to be obtained are not limited to $Gr_2(\mathbf{C}^4)$, we hope that our readers may be able to find other uses for them.

Following [D.45], suppose we are given a sequence of non-negative integers $1\leq a_1<a_2<\cdots<a_k\leq 4$. Having in mind applications to Eq. (D.66), we have to choose $k=2$. In such a case we obtain: $1\leq a_1<a_2\leq 4$. By looking at the matrices in Eq. (D.66) we shall let the first nonzero entry in the ith row to be in a_ith position. For instance, for the matrix (a) we obtain: $(a_1,a_2)=(1,2)$. Given this, let λ_i be the number of $*$'s in the ith row. It is given by $2-a_i+i$. Here $2=n-k, n=4, k=2$. Evidently, the numbers

[j] To make a comparison we have to replace respectively t^2 by q, k by m and n by m.

λ_i are also making a partition: $\lambda = (\lambda_1, \ldots, \lambda_k)$, so that $N = \sum_{i=1}^{k} \lambda_i$. λ is made out of partitions with $\leq k$ parts each of which is not larger than $n-k$. The partitions λ enumerated by $p(N; k, m)$ may be conveniently described by the Young tableaux which will fit into $(n-k)k$ rectangle. In our case: 2×2 quadrangle. Now we are having all the data needed for restoration of matrices given by Eq. (D.66). We start with the following table:

#	a_1	a_2
1	1	2
2	1	3
3	1	4
4	2	3
5	2	4
6	3	4

Using it, we construct yet another table:

#	λ_1	λ_2
1	2	2
2	1	2
3	0	2
4	1	1
5	0	1
6	0	0

This result allows us to write all Young tableaux. These are organized in accordance with the value of N. The results are summarized as follows:

N	λ_1	λ_2	$\lambda_1 + \lambda_2$	letter
0	0	0	0	(a)
1	0	1	1	(b)
2	0	2	2	(c)
2	1	1	2	(d)
3	1	2	3	(e)
4	2	2	4	(f)

allowing us to write the sequence of Young tableaux depicted in Fig. 8.

Thus, we just demonstrated that the Grassmannian is a purely combinatorial object. But it is also a toric variety [D.30] as well as a homogeneous

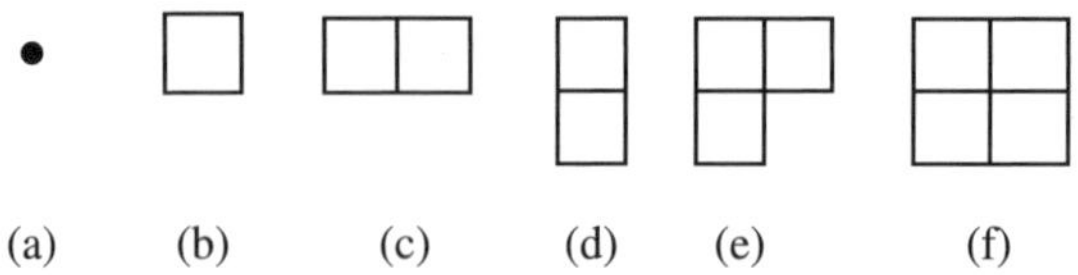

Fig. 8. Young tableaux associated with the reduced echelon matrices, Eq. (D.64).

space! With this information in our hands and that of preceding sections we are now in the position to formulate and solve the entanglement problem.

D.2.11 *The Problem of Quantum Entanglement (Statement and Goals)*

In [D.49, D.50] it was demonstrated that the problem of entanglement for the case of two qubits involves use of the Lie group SU(4) and the quotient SU(4)/SU(2) × SU(2). This quotient is nothing but the $Gr_2(\mathbf{C}^4)$, e.g. see [D.51], which we discussed in the previous section! More generally, for n-qubits study of entanglement should involve the quotient $\mathrm{SU}(2^n)/(\mathrm{SU}(2) \times \cdots \times \mathrm{SU}(2))$. It is a homogeneous flag manifold [D.51]. In this Appendix we formulate and solve

The entanglement problem. *To understand the mathematical and physical content of the quotient* SU(4)/SU(2) × SU(2), *provided that results of such a study are compatible with those discussed in Appendix C.*

The goals are: (a) to provide an answer to the problem just formulated in terms of mathematics developed for the torus actions on symplectic matifolds and (b) to find a place for methods of contact geometry developed in the main text in solving the entanglement problem. In the rest of this Appendix, we provide the solutions to these problems.

D.2.12 *Classical Mechanics Inside Quantum Mechanics (Revisited)*

In Sec. D.2.1 we used the familiar to every physics major formalism of finite-dimensional Hilbert space quantum mechanics and were able to reinterpret it in terms of concepts of classical mechanics. This formalism requires uses of complex vector spaces, Hermitian scalar products, linear operators (matrices) acting in such spaces, etc. In this subsection we want to reformulate the same results in terms of less familiar formalism. This will then allow us to extend the notion of the moment map initially introduced in Sec. D.2.3 for "truly mechanical systems" to dynamical systems described by matrices. We begin with some

Examples.

(a) The complex plane $\mathbf{C}$

Since any complex number $z = x + iy$ can be identified with a point (x, y) in $\mathbf{R}^2$ there is a way to identify z's with 2×2 matrices, e.g.

$$z = x + iy \in \mathbf{C} = \mathbf{R}^2 \rightleftarrows A = \begin{pmatrix} x & -y \\ y & x \end{pmatrix}.$$

The ordinary scalar product in $\mathbf{R}^2$, that is $\mathbf{x} \cdot \mathbf{y} = x_1 x_2 + y_1 y_2$ can now be rewritten either as $\mathbf{x}\cdot\mathbf{y} = \frac{1}{2}(z_1\bar{z}_2 + \bar{z}_1 z_2)$ or as $\mathbf{x}\cdot\mathbf{y} = \langle A_1, A_2\rangle = \frac{1}{2}\mathrm{tr}(A_1 A_2^T)$. In such a case, the squared length (the squared norm) $|z|^2 = \det A = |A|^2$.

Exercise D.2.12.1. Extend these two-dimensional results to $2n$ dimensions. This will lead you to the complex and almost complex structures, J-holomorphic (or pseudo-holomorphic) curves, etc. [D.52].

(b) The complex double $\mathbf{C}^2$

In this case we have to use the quaternions, e.g. $q = t + ix + jy + zk$. q can be identified with the point $(t, x, y, z) \in \mathbf{R}^4$. This observation allows us to think about re-interpretation of space–time concepts in terms of twistors [D.41, D.42]. So we obtain (Chap. 6, Sec. 6.5)

$$q = t + ix + jy + zk \in \mathbf{H} = \mathbf{R}^4 \rightleftarrows A = \begin{pmatrix} z_1 & -\bar{z}_2 \\ z_2 & \bar{z}_1 \end{pmatrix},$$

$$z_1 = t + ix, \quad z_2 = y + iz.$$

The scalar product $\mathbf{x} \cdot \mathbf{y}$ in $\mathbf{R}^4$ becomes $\langle A_1, A_2\rangle = \frac{1}{2}\mathrm{tr}(A_1 \bar{A}_2^T)$ and the squared norm is $\det A = |A|^2$. If we equate $|A|^2$ to 1, we obtain, as we already know the group isomorphism $\mathrm{SU}(2) \rightleftarrows S^3$.

Exercise D.2.12.2. Extend these results to $2n$ dimensions. This will lead you to the formalism of symplectic groups [D.53], etc.

(c) The $\mathrm{SU}(n)$

Our treatment of $\mathrm{SU}(n)$ mainly follows [D.54, D.55]. We begin with defining scalar product for $n \times n$ matrices $M_n(\mathbf{C})$ with complex entries. Thus, for $A \in M_n(\mathbf{C})$ the adjoint A^* is defined as in (b), that is as $A^*_{kl} = \bar{A}_{lk}$, with the overbar denoting the complex conjugation. An inner (scalar) product is defined now by $\langle A, B\rangle = \mathrm{tr}(A\bar{B}^T)$. For the purposes of this Appendix we need to discuss both the Lie group $\mathrm{SU}(n)$ and the Lie algebra $\mathfrak{su}(n)$. We already introduced the maximal torus T for $\mathrm{SU}(n)$

in Eq. (D.47b). Now we can identify its Lie algebra $\mathbf{t}$ with the space of traceless diagonal matrices

$$\mathbf{t} = \left\{ \left(\begin{matrix} \lambda_1 & & & \\ & \cdot & & \\ & & \cdot & \\ & & & \cdot \\ & & & & \lambda_n \end{matrix} \right) \middle| \sum_i \lambda_i = 0 \right\}. \tag{D.69}$$

Development of the formalism for $\mathrm{SU}(n)$ is based on nontrivial extension of the well-known formalism in physics for $\mathrm{SU}(2)$. To some extent this task was recently accomplished in lecture notes by Zuber, [D.56], which we would like to recommend to our readers as supplementary reading. Since the formalism for $\mathrm{SU}(n)$ will be used essentially in the rest of this Appendix, we need to provide some basics facts in this section.

We begin by comparing the $\mathfrak{su}(2)$ Lie algebra written in physics notations

$$[J_z, J_\pm] = \pm 2J_\pm, \quad [J_+ J_-] = J_z \tag{D.70a}$$

with the same algebra written in mathematics literature as

$$[H, X] = 2X, \quad [H, Y] = -2Y, \quad [X, Y] = H. \tag{D.70b}$$

Building a representation begins with the familiar eigenvalue equation

$$H\psi_0^{(\lambda)} = \lambda\psi_0^{(\lambda)} \tag{D.71a}$$

defining the highest (the *dominant*) *weight*. Here $H \rightleftarrows J_z$. Next, there is the vacuum state assumption/constraint:

$$X\psi_0^{(\lambda)} = 0. \tag{D.71b}$$

The excited states are obtained with help of raising/lowering operators

$$(Y)^k\psi_0^{(\lambda)} = \psi_k^{(\lambda)}.$$

The excited states are obtained by induction by using commutators in Eqs. (D.70b). This results in the following recursion relation:

$$X\psi_{k+1}^{(\lambda)} = [k\lambda - k(k-1) + (\lambda - 2k)]\psi_k^{(\lambda)}. \tag{D.72}$$

Next, the assumption that H can have only finitely many eigenvalues must be invoked. This is facilitated by the requirements $(Y)^k\psi_0^{(\lambda)} = \psi_k^{(\lambda)}$, $k \leq m$ but $\psi_{m+1}^{(\lambda)} = (Y)^{m+1}\psi_0^{(\lambda)} = 0$. From here it follows that $X\psi_{m+1}^{(\lambda)} = 0$. Using Eq. (D.70) produces eventually the following result: $(m+1)(\lambda - m) = 0$

implying $\lambda = m$. These $\mathfrak{su}(2)$ results can be formulated as theorem stating that every irreducible representation of $\mathfrak{sl}(2,\mathbf{C})$ (of which $\mathfrak{su}(2)$ is its compact part) is given by a nonzero linearly independent set of $m+1$ vectors $\psi_0^{(m)}, \ldots, \psi_m^{(m)}$ obeying the following set of equations:

$$\begin{aligned} H\psi_k^{(m)} &= (m-2k)\psi_k^{(m)}, \\ Y\psi_k^{(m)} &= \psi_{k+1}^{(m)}, \ k < m, \\ Y\psi_m^{(m)} &= 0, \\ X\psi_k^{(m)} &= k[m-k+1]u_{k-1}, \ k > 0, \\ X\psi_0^{(m)} &= 0. \end{aligned} \tag{D.73}$$

By comparing such a set with that known from physics literature, we notice that equation $Y\psi_k^{(m)} = \psi_{k+1}^{(m)}$, $k < m$ is written in physics as $Y\psi_k^{(m)} = a_k\psi_{k+1}^{(m)}$ while $X\psi_k^{(m)} = k[m-k+1]u_{k-1}$, $k > 0$ as $X\psi_k^{(m)} = b_k\psi_{k-1}^{(m)}$ with constants a_k and b_k to be determined. This discrepancy causes some differences in final results which can be brought into correspondence with each other by the appropriate normalization. Specifically, to translate the obtained results into physics language we let $m = 2j$ and absorb the factor of 2 into H. Then, we write $\psi_k^{(m)} \to \psi_{jm}$ meaning that we also had replaced k by $j - m$ implying $-j \leq m \leq j$. With these changes we are left with the familiar result: $J_z\psi_{jm} = m\psi_{jm}$. However now $J_-\psi_{jm} = (j+m)(j-m+1)\psi_{j,m-1}$ while in physics we would have instead $J_-\psi_{jm} = \sqrt{(j+m)(j-m+1)}\psi_{j,m-1}$. We leave with our readers the task of figuring out how to correct such a discrepancy.

The extension of obtained mathematical results to $\mathfrak{su}(n)$ does not require us to be complacent with physics notations since in the present case we are having different goals in mind. We need to keep in mind the following two observations instead.

(1) There is a mirror (reflection) symmetry in the "string" $j, j-1, ., 0, \ldots, -j$ of eigenvalues with respect to zero. This means that we can think about such a string geometrically. That is we can introduce a kind of one-dimensional finite (*root*) lattice having $j+1$ unit steps. The size of the lattice is regulated by j and, since j is pre-assigned, we shall remove the restriction for the lattice to be finite but, for further convenience, we shall fix the origin.

(2) As we know from quantum mechanics, j can be integer or half-integer. In the last case we shall have another (*weight*) lattice so that the root lattice can be obtained from the weight lattice. The weight lattice Λ_W

is *dual* to the co-root lattice $\Lambda_R^\vee$ generated by simple co-roots.[k] The situation here is exactly the same as in solid state physics where in crystals one encounters the direct and the reciprocal (dual) lattices [317].

Remark D.2.12.1. (a) The reflection symmetry for $\mathfrak{su}(2)$ just described is *exactly the same* as discussed in Sec. D.2.3. It is linked with the Weyl group symmetry as explained in detail in [D.54], pp. 176 and 177. From this observation it is possible to go back to classical mechanics interpretation of the obtained results as discussed in Sec. D.2.3. This will be explained in greater detail below. (b) Designing any kind of representation for semisimple Lie algebras other than $\mathfrak{su}(2)$ involves usage of multiple copies of commutation relations Eq. (D.70b). Because of this, just discussed results for a single copy of $\mathfrak{su}(2)$ survive. The whole theory can be built around clever reassessment of results for su(3) [D.54]. Incidentally, in physics both SU(3) and SU(4) symmetry (the last one is of major interest in this Appendix) was (is) studied in connection with designing of the *quark models of hadrons* [D.56].

The Weyl group for SU(n) is the permutation group S_n. Although such an outcome is in accord with results of Secs. D.2.8 and D.2.9, we shall explain the rationale behind this in yet greater detail below, in Sec. D.2.14. Incidentally, the same S_n group is discussed in Appendix C.

To use these results efficiently in mechanical setting, we have to make one major step forward — to introduce the moment map for matrices. From [D.24] we quote the following:

Definition D.2.12.1. The map $\mathcal{H} \to \mathbf{R}^n$ associating with any Hermitian matrix its diagonal entries is a *moment map for the torus action.*

To use with confidence this definition our readers should be equipped with the background which we still have to provide.

First, we need to generalize the results for $\mathfrak{su}(2)$ to $\mathfrak{su}(n)$. This is done by attaching a label α to the set of Eqs. (D.70b). That is we write

$$[H_\alpha, X_\alpha] = 2X_\alpha, \quad [H_\alpha, Y_\alpha] = -2Y_\alpha, \quad [X_\alpha, Y_\alpha] = H_\alpha, \\ \alpha = 1, 2, \ldots, n-1. \tag{D.74}$$

Instead of Eq. (D.71a), now we write (in the abbreviated form)

$$H_i\psi = \lambda(H_i)\psi, \quad i = 1, \ldots, n-1. \tag{D.75a}$$

[k]The co-roots (to be defined below) differ from roots only by a scale factor.

Accordingly, the vacuum condition, Eq. (D.71b), should now be changed to

$$X_1\psi = 0, \ldots, X_{n-1}\psi = 0. \tag{D.75b}$$

Remark D.2.12.2. It is, of course, possible to design trivial representations which are made of just direct sum or direct product of copies of $\mathfrak{su}(2)$. But, surely, there are other less trivial options [D.54, D.55]. These, less trivial options, presumably, could be classified/qualified as *entangled.* Thus, *designing irreducible representations for semisimple Lie groups is firmly rooted into entanglement problem for quantum computing.* We shall stick, however, only to the goals formulated in the previous section.

Remark D.2.12.3. In [D.44, D.47] we demonstrated how one can develop many-particle supersymmetric quantum mechanics using multiple copies of $\mathfrak{su}(2)$. The Hilbert space of such a mechanics is made out of Hodge-type differential forms on which the Dolbeault operators are acting. Due to the Lefshetz isomorphism theorem, [D.16], the algebra of Dolbeault operators acting on differential forms is isomorphic with $\mathfrak{su}(2)$. Because of this, the theory of representations of semisimple Lie algebras can be rewritten with the help of the formalism of supersymmetric quantum mechanics.

With every copy of $\mathfrak{su}(2)$ Lie algebra there is an associated one-dimensional weight/root lattice. Simultaneous uses of several copies of $\mathfrak{su}(2)$ in design of representations replaces these lattices by the multidimensional ones. While the weight lattice uses $\lambda(H_i)$ as components of the weight vector on such a lattice, equations

$$ad(H_\alpha)Z_\alpha \equiv [H_\alpha, Z_\alpha] = \alpha(H_\alpha)Z_\alpha, \quad H \in \mathbf{t}, \quad \alpha = 1, \ldots, n-1. \tag{D.76}$$

introduce the roots $\alpha(H_\alpha)$. If $\mathbf{t}$ is the Cartan subalgebra made of mutually commuting operators H_i (e.g. see Eq. (D.69)), then the semisimple Lie algebra, e.g. $\mathfrak{su}(n) \equiv \mathfrak{g}$, is presented symbolically as

$$\mathfrak{g} = \mathbf{t} \oplus \bigoplus_{\alpha \in R} Z_\alpha. \tag{D.77}$$

It is essential to define explicitly Z_α's at this point. Following [D.54, D.55] we introduce the $n \times n$ matrix E_{kl} in which 1 stands in (k, l) entry and 0 stands in the rest of entries. If we replace for a moment H_i by $H \in \mathbf{t}$, e.g. see Eq. (D.69), then (by direct computation) we obtain:

$$[H, E_{kl}] = (\lambda_k - \lambda_l)E_{kl}. \tag{D.78a}$$

Here $\lambda_i, i = 1, \ldots, n$ are defined either by Eqs. (D.75a) or by Eq. (D.69). If $k = l$, then E_{kl} does not have trace zero and, therefore, it does not belong to $\mathfrak{sl}(n; \mathbf{C})$. However, the combination $[X_\alpha, Y_\alpha] = E_{kk} - E_{ll} \equiv H_\alpha \equiv \alpha_{kl}$ does belong. This fact allows us to give new interpretation to the provisional roots $\lambda_k - \lambda_l \equiv \tilde{\alpha}_{kl}$ in Eq. (D.76a). Such an interpretation is needed for the following reason. By comparing Eqs. (D.74) with (D.76) we have to "reconcile irreconcilable". That is we have to require

$$\alpha(H_\alpha) = 2 \ \forall \alpha. \tag{D.79}$$

Next, we compare Eqs. (D.78a) and (D.79) and, again run into severe contradiction. The way out of it is nontrivial. Recall the scalar product for matrices $\langle A, B\rangle = \mathrm{tr}(A\bar{B}^T)$. Moment of thought allows us to write

$$\lambda_k - \lambda_l = \langle \alpha_{kl}, H\rangle, \quad H \in \mathbf{t} \tag{D.80}$$

so that for the very special case $H_\alpha = H$ we obtain $\langle \alpha_{kl}, \alpha_{kl}\rangle = 2$ as required. And what to do when we are not having the special case? We follow the ideology of [D.55–D.58]. For this purpose we introduce the Euclidean orthogonal basis e_i, $e_i \cdot e_j = \delta_{ij}$ with an extra constraint $\sum_i e_i = 0$ and notice that for the root $\tilde{\alpha}_{kl} = e_k - e_l$ we obtain $\tilde{\alpha}_{kl} \cdot \tilde{\alpha}_{kl} = (e_k - e_l) \cdot (e_k - e_l) = 2$. In view of this, following Kirillov, [D.57], we introduce e-operator $e_i(H) = \lambda_i, H \in \mathbf{t}$, provided that $\sum_i e_i = 0$. With such a trick we can rewrite Eq. (D.78a) as

$$[H, E_{kl}] = (e_k - e_l)(H)E_{kl}. \tag{D.78b}$$

This is going to work if by using $E_{kk} - E_{ll} \equiv H_\alpha$ we replace E_{kk} by e_k and E_{ll} by e_l and understand $(e_k - e_l)(H)$ as $(e_k - e_l) \cdot (H)$. Then a matrix H becomes a vector living in Λ_R, e.g. see [D.58], p. 103,

$$\Lambda_R = \left\{ \sum_{i=1}^{n} \lambda_i e_i, \sum_{i=1}^{n} \lambda_i = 0 \right\}. \tag{D.81}$$

Definition D.2.12.2. The A_{n-1} root system $\Delta \subset \Lambda_R$ for A_{n-1} (to which Lie algebra $\mathfrak{su}(n)$ belongs) is determined by

$$\Delta = \{e_i - e_j | i \neq j, 1 \leq i, j \leq n\}. \tag{D.82a}$$

The subscript $n - 1$ is just the dimension of the Cartan subalgebra $\mathbf{t}$ for $\mathfrak{su}(n)$. It is associated with a system of *fundamental roots* Σ determined by

$$\Sigma = \{e_1 - e_2, e_2 - e_3, \ldots, e_{n-1} - e_n\}. \tag{D.82b}$$

So that

$$e_i - e_j = (e_i - e_{i+1}) + (e_{i+1} - e_{i+2}) + \cdots + (e_{j-1} - e_j) \quad \text{and}$$
$$e_j - e_i = -(e_i - e_{i+1}) - (e_{i+1} - e_{i+2}) - \cdots - (e_{j-1} - e_j)$$

causing us to split Δ as $\Delta = \Delta^+ \cup \Delta^-$ with $\Delta^+ = \{e_i - e_j | i < j\}$ and $\Delta^- = \{e_i - e_j | i > j\}$.

Next, since $\alpha(H_\alpha) = \langle \alpha_{kl}, H_\alpha \rangle$ and $\alpha(H_\alpha) = 2$, we have no choice but to write

$$H_\alpha = 2\frac{\alpha}{\langle \alpha, \alpha \rangle} \quad \text{and} \quad \alpha = 2\frac{H_\alpha}{\langle H_\alpha, H_\alpha \rangle}. \tag{D.83}$$

From Eq. (D.83) it also follows that $\langle \alpha, \alpha \rangle \langle H_\alpha, H_\alpha \rangle = 4$.

Definition D.2.12.3. Such defined H_α is a *co-root*. In accord with the existing literature, we shall write $H_\alpha = \alpha^\vee$.

For A_{n-1} it is known, [D.58], p. 104, that $\Sigma = \Sigma^\vee$. Introduction of co-roots allows us to introduce the system of *fundamental weights* $\{\omega_1, \ldots, \omega_{n-1}\}$ which are just components of the weight vector living on the weight lattice. The weight lattice Λ_W is dual to the co-root lattice:

$$\alpha_i^\vee \cdot \omega_j = \delta_{ij}. \tag{D.84a}$$

In the case of A_{n-1} the fundamental weights ω_j (with respect to Σ) are given by

$$\omega_k = e_1 + \cdots + e_k, \tag{D.85}$$

provided that $e_1 + \cdots + e_n = 0$. The co-roots $\{e_i - e_{i+1}\}$ and the elements $\{e_1 + \cdots + e_k\}$ are dual to each other, that is

$$(e_i - e_{i+1}) \cdot (e_1 + \cdots + e_k) = \delta_{ik}. \tag{D.84b}$$

Example. The A_2 system

In this case we have fundamental weights $\omega_1 = e_1$ and $\omega_2 = e_1 + e_2$ and, since $e_1 + e_2 + e_3 = 0$ we obtain for fundamental roots $\alpha_1 = e_1 - e_2$ and $\alpha_2 = e_2 + (e_1 + e_2)$. This allows us to connect to co-roots with weights, that is $\alpha_1 = 2\omega_1 - \omega_2$ and $\alpha_2 = -\omega_1 + 2\omega_2$. These results can be represented succinctly with the help of the Cartan matrix $C_{ij} = \langle \alpha_i \cdot \alpha_j \rangle$, that is

$$\alpha_i = \sum_j C_{ij} \omega_j. \tag{D.85a}$$

In the present case we obtain,

$$\mathbf{C} = \begin{pmatrix} 2 & -1 \\ -1 & 2 \end{pmatrix}.$$

From here, we obtain its inverse

$$\mathbf{C}^{-1} = \frac{1}{3}\begin{pmatrix} 2 & 1 \\ 1 & 2 \end{pmatrix}$$

providing us with the relationship

$$\omega_i = \sum_j [C_{ij}]^{-1}\alpha_j. \tag{D.85b}$$

Explicitly we have $\omega_1 = \frac{1}{3}(2\alpha_1 + \alpha_2)$, $\omega_2 = \frac{1}{3}(\alpha_1 + 2\alpha_2)$. It is very helpful to depict the obtained results graphically. For this purpose we need to use the normalization condition, Eq. (D.79), and the orthogonality condition, Eq. (D.84). So, we obtain:

$\langle \alpha_1, \alpha_1 \rangle = \langle \alpha_2, \alpha_2 \rangle = 2$ but $\langle \alpha_1, \alpha_2 \rangle = |\alpha_1||\alpha_2| \cos\theta_{12} = -1$ (see Cartan's matrix). From here we obtain: $\cos\theta_{12} = -\frac{1}{2}$ so that $\theta_{12} = 120° = 2\pi/3$. Analogously, we calculate the square of the norm for the fundamental weights, that is $\langle \omega_1, \omega_1 \rangle = \langle \omega_2, \omega_2 \rangle = \frac{2}{3}$, and also $\langle \omega_1, \omega_2 \rangle = |\omega_1||\omega_2| \cos\tilde{\theta}_{12} = \frac{1}{3}$. From here, we obtain: $\cos\tilde{\theta}_{12} = \frac{1}{2}$, that is $\tilde{\theta}_{12} = \pi/3 = 60°$.

Remark D.2.12.4. At this point our readers probably have noticed that in both Eqs. (D.85) ω_i's and α_i's are being treated as scalars while the same quantities were treated as vectors in calculations we just made. This observation plays decisive role in visualization of just obtained results. For instance, if we treat equations $\omega_1 = \frac{1}{3}(2\alpha_1 + \alpha_2)$ and $\omega_2 = \frac{1}{3}(\alpha_1 + 2\alpha_2)$ as *vector* equations, then by adding them up we obtain $\boldsymbol{\alpha}_1 + \boldsymbol{\alpha}_2 = \boldsymbol{\alpha}_1 + \boldsymbol{\alpha}_2$. Finally, notice that the same equations produce $\langle \omega_2, \alpha_1 \rangle = \langle \omega_1, \alpha_2 \rangle = 0$. These results are summarized in Fig. 9. From this figure we can see that: (a) roots and weights are forming triangular lattices, (b) both lattices can be generated by reflections in the sides of the fundamental (elementary) triangle for the weight lattice, (c) the area of the fundamental triangle for the root lattice is three times larger than that for the weight lattice. Incidentally, the determinant of the Cartan matrix is 3. This result follows from the theory of lattices [D.27, D.34] which has number theory in its foundation.

Next, we formally introduce

$$[H_\alpha, X_\beta] = \beta(H_\alpha)X_\beta. \tag{D.86}$$

Can such an equation make sense? Well, if we write

$$\beta(H_\alpha) = \langle \beta, H_\alpha \rangle = 2\frac{\langle \beta, \alpha \rangle}{\langle \alpha, \alpha \rangle} = 2\frac{\langle H_\beta, H_\alpha \rangle}{\langle H_\beta, H_\beta \rangle} \tag{D.87}$$

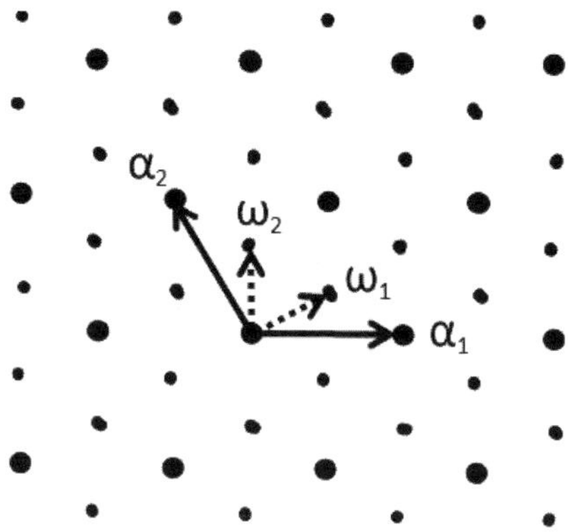

Fig. 9. Root and weight lattices for $\mathfrak{su}(3)$ designed with the help of calculations discussed in the text.

this would cause us to require that both combinations $2\frac{\langle\beta,\alpha\rangle}{\langle\alpha,\alpha\rangle}$ and $2\frac{\langle H_\beta,H_\alpha\rangle}{\langle H_\beta,H_\beta\rangle}$ be integers. This is indeed the case as demonstrated in [D.54], p. 170. This property leads us to introduction of the Weyl group. This introduction can be done purely formally based on the combined use of Eqs. (D.78a) and (D.80). But, we have to keep in mind that the notion of the Weyl group was already formulated in Definition D.2.9.1. Now it can be formulated as the adjoint action of the normalizer N_U on the Lie algebra $\mathfrak{g}$ restricted to $\mathbf{t}$. Following [D.54], pp. 174 and 175, we provide yet another

Definition D.2.12.4. The action of the Weyl group $W = N_U(T)/T$ on $\mathbf{t}$ denoted as $w \cdot H$ with $w \in W$ and $H \in \mathbf{t}$ is given by

$$w \cdot H = Ad_{N_U}(H). \tag{D.88}$$

Take

$$[H, X] = \langle\alpha, H\rangle X \tag{D.89}$$

and use the above definition, Eq. (D.88), in order to obtain (after few steps)

$$[H, Ad_{N_U}(X)\} = \langle Ad_{N_U}\alpha, H\rangle Ad_{N_U}(X) \equiv \langle w\cdot\alpha, H\rangle Ad_{N_U}(X).$$

Since $H_\alpha = 2\frac{\alpha}{\langle\alpha,\alpha\rangle}$, we obtain

$$w \cdot H_\alpha = 2\frac{w\cdot\alpha}{\langle\alpha,\alpha\rangle} = 2\frac{w\cdot\alpha}{\langle w\cdot\alpha, w\cdot\alpha\rangle} = H_{w\cdot\alpha}. \tag{D.90}$$

In deriving this result we used the fact that reflection caused by the Weyl group is isometry to the Euclidean space, e.g. see [D.58]. Obtained result allows us to write the action of W on roots. It is given by the following fundamental result:

$$w_\alpha \cdot \beta = \beta - 2\frac{\langle \alpha, \beta \rangle}{\langle \alpha, \alpha \rangle}\alpha = \begin{cases} -\alpha & \text{if } \beta = \alpha, \\ \beta & \text{if } \beta \neq \alpha. \end{cases} \tag{D.91}$$

The equation $\langle \alpha, \beta \rangle = 0$ is an equation for the reflecting hyperplane. It defines the boundary of the *Weyl chamber*. We introduced the notion of the rational polyhedral cone and the supporting hyperplane H_α in Definition D.2.2.3. Evidently, the equation $\langle \alpha, \beta \rangle = 0$ is exactly the same as for the supporting hyperplane. Analogously, the collection of cones makes a complete fan as depicted in Fig. 2. Now we simply rename the rational polyhedral cone by the Weyl chamber so that the supporting hyperplane is the boundary of the Weyl chamber. In view of Eq. (D.91) it is the "reflecting hyperplane". An assembly of Weyl chambers makes a complete fan. Further connections with previous subsections can be easily worked out now. For instance, Fig. 4 depicts deformation retracts of complex projective spaces. These are all simplexes. Now, following [D.55], p. 214, and [D.58], p. 18, we can demonstrate using Eqs. (D.81) and (D.82) that the Weyl group is symmetry group for these simplexes. We begin with a few facts from the theory of convex polytopes.

Definition D.2.12.5. ([D.26]) A set $S \subset \mathbf{R}^{n+1}$ is *convex* if $\forall \mathbf{x}, \mathbf{y} \in S$ and, $x \neq y$, the line segment $[x, y] := \{\varrho\mathbf{x} + (1-\varrho)\mathbf{y} | 0 \leq \varrho \leq 1\}$ is contained in S. From here it follows that $\mathbf{x}$ is *convex combination* of $\mathbf{x}_0, \ldots, \mathbf{x}_n \in \mathbf{R}^{n+1}$ if there exist $t_0, \ldots, t_n \in \mathbf{R}$ such that

$$\mathbf{x} = \sum_{i=0}^{n} \mathbf{x}_i t_i, \quad \sum_{i=0}^{n} t_i = 1, \quad t_i \geq 0 \ \forall i. \tag{D.92}$$

Going back to Eq. (D.81) (and increasing dimensionality of space by 1) we notice that the condition $\sum_{i=0}^{n} \lambda_i = 0$ can always be presented as $\sum_{i=1}^{n} \lambda_i = -\lambda_{n+1}$. Although the choice of λ_{n+1} is quite arbitrary, we can always order entries and relabel them so that this becomes permissible. Furthermore, we can expect λ_{n+1} to be negative. Next, we can simultaneously add to both sides of this equation some numbers so that we end up again with the equation for the $n-1$ simplex (e.g. see Eq. (D.33)). Thus, we just obtained the following result. If the location of vertices of the $n-1$ simplex Δ_{n-1} is given by the vectors $\{e_0, \ldots, e_n\}$ forming an orthogonal basis set, e.g. see Fig. 4, then Δ_{n-1} is defined by Eq. (D.92). The Weyl group W is acting as symmetry group for Δ_{n-1} in $\mathbf{R}^{n-1}$ but *not* in $\mathbf{R}^n$.

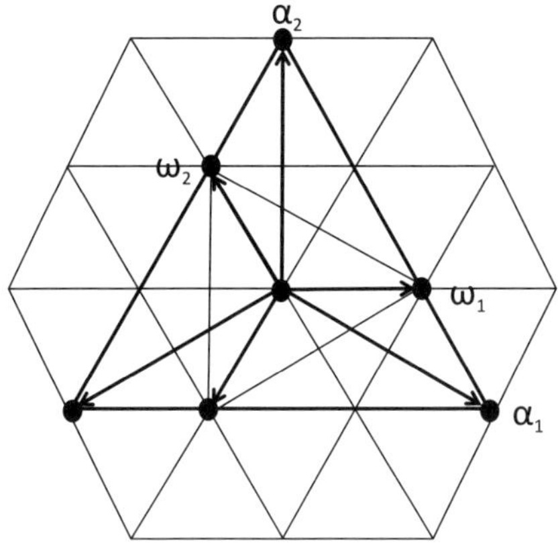

Fig. 10. Weight and root elementary cells for $\mathfrak{su}(3)$.

The symmetries of Δ_{n-1} are obtained by permuting the vertices of the simplex. Following [D.55] it is convenient to move the coordinate system into the barycenter (that is into the center of mass) of the simplex as it is always done in the mechanics of rotating bodies [235]. Figure 10 is drawn accounting for such simplicial symmetry. It only remains to demonstrate that it is not a coincidence that such noticed simplicial symmetry is symmetry of $\mathbf{CP}^n$.

D.2.13 *Some Lessons from Morse Theory*

We begin by working out all implications which follow from the Definition D.2.12.1. As good as this definition is, it is not immediately evident that the moment map obtained in such a way can serve as the Morse function (in accord with Remark D.2.3.3). In view of this, we have to bring some relevant results from the Morse theory. In doing so, we shall avoid extensive quotations from this theory because of the multitude of excellent literature on this subject.

Remark D.2.13.1. At the level accessible to any physics major the basics facts of Morse theory are beautifully explained in the book by Matsumoto, [D.59]. Matsumoto's book is sufficient for reading and understanding of the classical book by Milnor [D.60]. The thoughtful book by Nicolaescu [D.61] provides further details.

Thus, let $\mathcal{M}$ be a smooth manifold and let $f : \mathcal{M} \to \mathrm{R}$ be a smooth function on $\mathcal{M}$. A point $m \in \mathcal{M}$ is critical if $df|_m = 0$. As f is smooth, we can Taylor series expand $f(x)$ in the usual way around m (choosing m as origin)

$$f(x) = f(0) + \sum_i a_i x_i + \frac{1}{2}\sum_{i,j} a_{ij} x_i x_i + rest. \tag{D.93a}$$

If m is a critical point, then $\sum_i a_i x_i = 0$ while the matrix a_{ij} defines a symmetric bilinear form (the Hessian of f) on $T_m\mathcal{M}$.

Definition D.2.13.1. If m is a *critical point*, the *index* of m is defined as the number of negative eigenvalues of a_{ij}.

Definition D.2.13.2. The critical point is *nondegenerate* if the Hessian does not have zero eigenvalues.

Definition D.2.13.3. The function f is *Morse function* if every critical point is nondegenerate. It can then be proven (Morse lemma) that *without any approximation* there is a system of coordinates in which such a function can be presented as

$$f(x) = f(0) - \sum_{i=1}^{\lambda} x_i^2 + \sum_{i=\lambda+1}^{n} x_i^2. \tag{D.93b}$$

Remark D.2.13.2. The results of Morse theory for real spaces are transferable to complex spaces .

Example. Following Milnor, [D.60], p. 26 and also [D.62], let us discuss $\mathcal{M} = \mathbf{CP}^n$. Introduce $n+1$ real numbers $c_0, \ldots, c_n$ and construct a Morse function $f : \mathbf{C}\,\mathbf{P}^n \to \mathbf{R}$ given by

$$f(z_0, \ldots, z_n) = \frac{\sum_{i=0}^n c_i |z_i|^2}{\sum_{i=0}^n |z_i|^2}. \tag{D.94a}$$

In doing so, notice the following.

(a) The actual values of constants c_i are not important. Indeed, we can always add (or subtract) in the numerator of Eq. (D.94a) the factor of $c\sum_{i=0}^n |z_i|^2$. The new Morse function

$$f(z_0, \ldots, z_n) = \frac{\sum_{i=0}^n (c_i + c)|z_i|^2}{\sum_{i=0}^n |z_i|^2} \tag{D.94b}$$

will differ by constant c from the old one and will exhibit exactly the same critical behavior.

(b) If we choose the model of projective space as described by Eq. (D.26b), then we can get rid of the denominator in Eqs. (D.94).

(c) If we do keep the denominator, we can rewrite Eq. (D.94a) in the alternative form

$$f(t_0,\dots,t_n)=\sum_{i=0}^{n}c_it_i\equiv\langle\mathbf{c},\mathbf{t}\rangle;\quad \sum_{i=0}^{n}t_i=1,t_i\geq 0\ \forall i \tag{D.94c}$$

causing the functional $f(\mathbf{t})$ to look the same as previously introduced (Sec. D.2.2) functional $\tilde{H}[\mathbf{x}]=\langle\mathbf{c},\mathbf{x}\rangle$ used in the theory of linear programming. We also had almost the same result in Eq. (D.92) of previous section. To take the full advantage of this observation, we notice that taking into account Definition D.2.12.5, it is possible to prove that a function $f(\mathbf{x}):\mathcal{M}\rightarrow\mathbf{R}$ is convex if and only if for every convex combination $\mathbf{x}=\sum_{i=0}^{n}\mathbf{x}_it_i$, $\sum_{i=0}^{n}t_i=1, t_i\geq 0\ \forall i$ of points $\mathbf{x}_0,\dots,\mathbf{x}_n\in\mathbf{R}^{n+1}$

$$f(\mathbf{x})\leq t_0f(\mathbf{x}_0)+\cdots+t_nf(\mathbf{x}_n). \tag{D.95}$$

Without loss of generality we can always assume that $f(\mathbf{x}_0)\leq f(\mathbf{x}_1)\leq\cdots\leq f(\mathbf{x}_n)\equiv N$ where $N\in\mathbf{R}$. This then leads us to the bound: $f(\mathbf{x})\leq N$, $\mathbf{x}\in\mathbf{R}^{n+1}$. By using this result in Eq. (D.94c), in principle, we can reinterpret the AGS results either in terms of the theory of linear programming or in terms of the theory of Lie groups (algebras). Below, we shall take advantage of the latter opportunity.

To do so, let U_0 be the subset of $(z_0:z_1:\cdots:z_n)\in\mathbf{C}^{n+1}$ such that $z_0\neq 0$. U_0 can be looked upon as the affine chart. In it, set $\frac{z_j}{z_0}|z_0|=x_j+iy_j$. Using this result we obtain: $|z_j|^2=x_j^2+y_j^2$, and $|z_0|^2=1-\sum_j(x_j^2+y_j^2)$, since $\sum_{i=0}^{n}|z_i|^2=1$. Substitution of these results back into Eq. (D.94) we obtain

$$f(z_0,\dots,z_n)=c_0+\sum_{j=1}^{n}(c_j-c_0)(x_j^2+y_j^2), \tag{D.94d}$$

implying that in U_0 the only critical point of f is $p_0=(1:0:\cdots:0)$. This point is nondegenerate and has index equal to twice the number of j's for which $c_j<c_0$. Evidently, this requirement may or may not be fulfilled. It is definitely fulfilled in the case when $c_0,\dots,c_n$ are ordered as $c_0<\cdots<c_n$ since then, when the same analysis is applied to another affine patch, e.g. $U_i, 0\leq i\leq n$ covering $\mathbf{CP}^n$, the outcome will be the same. Thus, the critical points are: $p_0=(1:0:\cdots:0), p_1=(0:1:\cdots:0),\dots,p_n=(0:0:\cdots:1)$. With the ordering of c's just described the index of f at p_k's will be twice the number of j's with $c_j<c_k$ causing every possible index between 0 and $2n$ to appear exacly once. This information can be encoded into the *Morse polynomial* $M(t)$ designed as follows. Let m_i is the number of critical

points with index i, then $M(t) = \sum_i m_i t^i$. This result is normally being compared against the Poincaré polynomial $P(t)$. If b_i is ith Betti number: $b_i = \dim H_i$ where H_i is the ith homology group of the manifold M, then $P(t) = \sum_i b_i t^i$.

Definition D.2.13.4. A Morse function f for which $M(t) = P(t)$ is called *perfect.*

Remark D.2.13.3. It can be demonstrated [D.61, D.63] that if on compact manifold $\mathcal{M}$ the Morse polynomial contains only even powers of t, then f is perfect. In the example just discussed $M(t)$ is given by

$$M(t) = P(t) = \sum_{i=0}^{n} t^{2i} \tag{D.96}$$

so that the function $f(z_0, \ldots, z_n)$ is perfect. $M(t)$ is the Poincaré polynomial for $\mathbf{CP}^n$. It was obtained in the book by Bott and Tu, [D.46], p. 269, who used entirely different methods.

Remark D.2.13.4. The result for $\mathbf{CP}^n$ was obtained based on the assumption that parameters $c_0, \ldots, c_n$ are ordered. It leaves open the situation when some of them may coincide. Furthermore, because of Definition D.2.12.1 of the moment map these parameters can always be chosen as entries of diagonal matrices (e.g. those from $\mathbf{t}$ subgroup). In such a case, it appears that when all entries are different, the results just obtained make it impossible to recover any useful information (different polytopes and different symplectic manifolds for different matrices) as it appears, for example, in [D.62, D.64].

Is there a way out of this apparent contradiction? The answer is "definitely yes" as results presented at the end of previous section indicate. There are several ways to rectify the situation. We begin with contributions of Raoul Bott to Morse theory. In standard Morse theory critical points are manifolds of dimension zero. In the case if some of critical points merge they could form manifolds of dimensionality higher than zero.

Definition D.2.13.5. A function $f : \mathcal{M} \to \mathbf{R}$ is *a Morse–Bott function* if every critical point of $\mathcal{M}$ belongs to a nondegenerate critical manifold.

We shall skip technical explanation of nondegeneracy since anyway we are not going to derive conditions needed for existence of the Morse–Bott functions. In any case, we dealt with them already! For example, in Eq. (D.39) the Hamiltonian (moment map) function Φ^{ξ_M} is the Morse–Bott function. In fact, the AGS theory discussed in the book by Audin, [D.24], is aimed exactly at demonstrating the convexity of moment maps by methods of Morse–Bott theory.

Remark D.2.13.5. Should we choose this path, we would inevitably run into the subject matters of *equivariant homology/cohomology* (a subset of *K-theory*) which would require us to write at least another chapter. Fortunately, such a task had been accomplished by Michele Vergne, [D.65].

In this Appendix we choose a different path to our final destination. To provide some credibility to this alternative path, it is helpful to recalculate the results obtained in Sec. D.2.3. Thus, we discuss again the SU(2). Evidently, it is sufficient to consider its Lie algebra. In such a case we are left with σ_z. It is convenient to analyze $a\sigma_z$ instead with a being some parameter which could be put equal to 1 at the end of calculations. Thus, Eq. (D.94) acquires now the following form:

$$f(z_0, z_1) = \frac{a|z_0|^2 - a|z_1|^2}{|z_0|^2 + |z_1|^2}. \tag{D.97a}$$

By adding a combination $-a(|z_0|^2 + |z_1|^2)$ to the numerator and by discarding (for the moment an overall constant) we end up with

$$f(z_0, z_1) = \frac{-2a|z_1|^2}{|z_0|^2 + |z_1|^2}. \tag{D.97b}$$

If we choose $a = 1/4$ our result will coincide with that given in Example 1, [D.15], p. 99. Fortunately, this example was left undeveloped leaving us with a chance to provide missing details. This can be accomplished in more than one way. It is instructive to discuss all of them.

Thus, we begin with the first method. In it, we let $|z_0|^2 + |z_1|^2 = 1$ from the beginning. Then, we replace $|z_1|^2$ by $x^2 + y^2$, next we use the polar coordinates (r, φ). Our Hamiltonian (the moment map) now becomes $H = -2ar^2$. Following [D.62] it is convenient to rewrite H in terms of the fictitious mass (the group-theoretic weight, in fact) m, that is $H = -2ar^2 \to \frac{1}{2}mr^2$.[1] Mathematically, the existence of a mass is caused by the torus action, Eq. (D.48), written in a special form $t \cdot \mathbf{z} := (t^{a_1}z_1, \ldots, t^{a_k}z_k) \to t \cdot \mathbf{z} := (e^{-im_1}z_1, \ldots, e^{-im_k}z_k), z_j = r_j e^{i\theta j}, j = 1, \ldots, k$. In such a case $dH = mrdr$ and the Hamiltonian, Eq. (D.36), causes us to select the 2-form ω as $\omega = rdr \wedge d\theta$ and the vector field X_H as $-m\frac{\partial}{\partial\theta}$. The generalization

[1]Compare with Eq. (D.26c).

of this result to the multicomponent case is straightforward. That is we should deal with the Hamiltonian

$$\mathrm{H} = \frac{1}{2}\sum_i m_i(x_i^2 + y_i^2) + \text{const.}, \tag{D.98}$$

where, unlike as it happens in real physical world, the masses could be both positive and negative. Furthermore, the masses cannot be all the same since in such a case the Hamiltonian is no longer going to be the Morse function. Even this requirement happens to be not sufficient! The masses should be incommensurate (that is they cannot be multiples of each other) as we shall explain momentarily. The fact that the masses can be both positive and negative is admissible at the classical level but becomes problematic at the quantum level. That is to say, if we would not deal with quantum mechanics in the first place but, instead, we would consider the classical Hamiltonian, e.g. of the type given by Eq. (D.98), and make an attempt at quantization with negative masses in this Hamiltonian, such an attempt will surely fail as explained in detail in [D.62]. The way out of this difficulty is discussed in the same reference. The authors suggest to introduce the polarization vector $\boldsymbol{\epsilon}$ whose components are ± 1. This observation brings us to the second method.

The $+1$ components of ϵ always multiply the positive masses while -1 components multiply the negative masses. In such a case the 2-form $\omega = \sum_i dx_i \wedge dy_i$ should be replaced by $\omega = \sum_i \epsilon_i dx_i \wedge dy_i$. If we apply such introduced polarization prescription to $f(z_0, z_1)$, Eq. (D.97a), the result will be: $f(z_0, z_1) = a$. Surely, this is not a meaningful Morse function at all and this fact already suggests that equal (positive) masses cannot be used in H. Our readers may argue at this moment: If this is so, why then Eq. (D.26c) was used as an example? An answer is easily obtained based on Eq. (D.97a). Indeed, the moment mapping is always defined with accuracy up to a constant. Thus, we can rewrite Eq. (D.97a) with positive masses and with added combination $-a(|z_0|^2 + |z_1|^2)$ in the numerator. This results in the following

$$f(z_0, z_1) = \frac{a|z_0|^2 - a|z_1|^2}{|z_0|^2 + |z_1|^2} \to a(t_1 + t_2) - a \to t_1 + t_2 = 1.$$

The last result coincides with Eq. (D.33) as required. That is to say the Hamiltonian, Eq. (D.26), is both living in $\mathbf{CP}^n$ and defining $\mathbf{CP}^n$.

At this moment we came to the point when we are in the position to explain why masses must be incommensurate. Indeed, suppose that they are commensurate. Then, by applying the torus map action $t \cdot \mathbf{z} := (e^{-im_1} z_1, \ldots, e^{-im_k} z_k)$ repeatedly and taking into account that the vector field X_H originating from such an action is $-\sum_i m_i \frac{\partial}{\partial \theta_i}$, after several iterations we can always make all masses to have the same value. This will bring us back to $\mathbf{CP}^n$. At the same time, if the masses are incommensurate we obtain, using Eq. (D.94c),

$$\sum_i m_i t_i = \delta. \tag{D.99a}$$

Notice that this is exactly the same type of equation as we already obtained! E.g. see Eq. (D.43). In arriving at this result we also used the inequality Eq. (D.95). The actual value of δ is not too important since we can always make a shift : $m_i \to m_i - \delta$ to make Eq. (D.99a) to look like that for the hyperplane

$$\sum_i m_i t_i = 0. \tag{D.99b}$$

But from Definition D.2.2.5. it follows that the set of (reflecting) hyperplanes will define a complete fan (that is a complete set of Weyl chambers) and, therefore, a convex polytope as we shall see in the next section. The Weyl group is permuting m_i's in Eq. (D.99) so that, indeed, Eq. (D.99) is exactly the same as Eq. (D.43).

Let us also discuss the second method. In this case we write

$$H = f(z_0, z_1) = \frac{1}{2} \frac{|z_1|^2}{|z_0|^2 + |z_1|^2} = \frac{1}{2} \frac{1}{1 + |z|^2} \tag{D.100}$$

causing us to consider two fixed points: 0 and ∞ (e.g. see Eq. (D.50)), providing us with two critical values $H(0) = 1/2$ and $H(\infty) = 0$ for the moment map. They are replacing previously used 1 and -1, e.g read Remark D.2.3.1. Now we can introduce again polar coordinates so that

$$dH = d\frac{1}{2}\left(\frac{1}{1 + r^2}\right) = -\frac{rdr}{(1 + r^2)^2}. \tag{D.101}$$

In Sec. D.2.3, we obtained the 2-form $\omega_{\text{standard}} = d\theta \wedge dz$ while now it should be replaced by the Fubini–Study 2-form

$$\omega_{FS} = \frac{d\theta \wedge rdr}{(1 + r^2)^2}. \tag{D.102}$$

As before, e.g. see Eq. (D.38), the 2-form is proportional to the symplectic volume Ω. Without using the normalization needed for its calculation we obtain: $\int_{S^2} d\theta \wedge dz = 4\pi$. Now, we are having instead $\int_{S^2} \frac{d\theta \wedge r dr}{(1+r^2)^2} = 2\pi \int_0^\infty \frac{r dr}{(1+r^2)^2} = \pi$. Thus, we just obtained $\frac{1}{4}\omega_{\text{standard}} = \omega_{FS}$ in accord with [D.15], p. 105, where this result was given without proof.

D.2.14 *From Analysis to Synthesis: c-Phase Gate and Grassmannians*

At this point we have accumulated enough knowledge to begin our "final preparations for landing" at our destination-solution of the entanglement problem. As a warm up, we would like to complete our studies of $\mathfrak{su}(3)$ initiated in Sec. D.2.12. In this case the hyperplane, Eq. (D.69), acquires the following form:

$$\lambda_1 + \lambda_2 + \lambda_3 = 0. \tag{D.103}$$

From previous section we know that λ_i's can be chosen as (incommensurate) integers. There could be only two cases: (a) all λ_i's are different and, (b) two of λ_i's are the same. In case (a) there are $3! = 6$ combinations of λ_i's producing the same Eq. (D.103). Looking at Fig. 5 we realize that we are dealing with two-dimensional permutohedron. In view of Theorem D.2.6.2 and Definition D.2.6.2, we also realize that we are dealing with the homogeneous (flag) manifold [D.51, D.53, D.66] of the type

$$\frac{\mathrm{SU}(3)}{S(U(1)\times U(1)\times U(1))} \equiv \frac{\mathrm{SU}(3)}{T} \tag{D.104a}$$

where the symbol $S(U(1)\times U(1)\times U(1))$ stands for the 3×3 block-diagonal matrix whose diagonal entries are $U(1)''s$ while $\det \prod_i U(i) = 1$ [D.53].

The first issue which we have to resolve now is this: How can we reconstruct Fig. 5 as the image of the moment map? This issue was resolved in [D.64] with help of the method of *symplectic fibrations*. To understand what is involved we need to recall the results of Chap. 6, Sec. 6.5., and those of Chap. 8, Sec. 8.2.6, regarding the relationship between the group quotients and fiber bundles. From [D.53] we know that $U(3) \simeq \mathrm{SU}(3)\times U(1)$. Therefore, the orbit $\mathcal{O}^{\mathrm{SU}(3)}$ is a homogeneous space of the type

$$\frac{U(3)}{U(1)\times U(1)\times U(1)} = \frac{\mathrm{SU}(3)}{U(1)\times U(1)} \sim \mathcal{O}^{\mathrm{SU}(3)}. \tag{D.104b}$$

Evidently, this orbit is of the same type as in Eq. (D.98a). From Definition D.2.5.3, we know that $\dim \mathcal{O}^{\mathrm{SU}(3)} = \dim \mathrm{SU}(3) - 2\dim U(1) = 6$. But $6 = 3!$ so that the dimension of the orbit coincides with that for the flag manifold in accord with results of Sec. D.2.8. This number is also the number of vertices of two-dimensional permutohedron depicted in Fig. 5. Thus our theoretical predictions are correct. The $\mathcal{O}^{\mathrm{SU}(3)}$ orbit can be interpreted as well as the product

$$\mathcal{O}^{\mathrm{SU}(3)} \simeq \mathcal{O}_d^{\mathrm{SU}(3)} \times \mathcal{O}^{\mathrm{SU}(2)},$$

where

$$\mathcal{O}_d^{\mathrm{SU}(3)} = \frac{\mathrm{SU}(3)}{\mathrm{SU}(2) \times U(1)} \simeq \mathbf{CP}^2, \quad \text{and} \quad \mathcal{O}^{\mathrm{SU}(2)} = \frac{\mathrm{SU}(2)}{U(1)} \simeq \mathbf{CP}^1.$$

In accord with [D.64], the orbit $\mathcal{O}^{\mathrm{SU}(3)}$ is a fiber bundle with $\mathbf{CP}^2$ as the base and $\mathbf{CP}^1$ as the fiber. Since our readers might find the mathematics behind construction of such a fiber bundle challenging (even though its graphical implementation is in fact completely straightforward), we by-pass the justification of the graphical method described in [D.64]. Instead, we shall describe the obtained results using information provided in two previous sections and in Fig. 4. Specifically, we shall use the $\mathfrak{su}(3)$ diagonal matrix given by Eq. (D.69) in the moment map, Eq. (D.94a). In the most primitive situation (that is without uses of the Morse–Bott formalism) we will end up with the equilateral triangle, depicted in Fig. 4(b), representing the deformation retract of $\mathbf{CP}^2$. The triangle actually represents the degenerate $\mathcal{O}_d^{\mathrm{SU}(3)}$ orbit (the subscript d means "degenerate") originating from the fact that two entries in Eq. (D.103) are identical. The degeneracy can be lifted if all three entries into Eq. (D.103) are different. We would like to explain this phenomenon graphically following [D.64]. Since the degeneracy is concentrated at the vertices (superficially, this is so because there are three ways to select two objects out of three) it is necessary to "blow up" each vertex slightly. This is done as follows. To each vertex we attach a segment $[-1, 1]$ representing the retract of $\mathbf{CP}^1$ (depicted in Fig. 4(a)). It should be placed in parallel to the side of the triangle opposite to this vertex. Then, we push a little this segment inside the triangle parallel to itself till its end points (-1) and (1) will lie at the edges of a triangle (there are no actual sizes involved, just the topology!). Doing this with all three vertices brings us back to permutohedron of Fig. 5. The above procedure looks simple as long as we do not touch its mathematical justification. To explain things better, we need to use the results of Sec. D.2.12. Before doing so, we have to recognize that the purpose of this Appendix is to clarify the meaning of quantum entanglement. Thus, to avoid delays even further, we are going to focus on the main purpose of our exposition: the problem of entanglement.

From Chap. 8, Sec. 8.3.1, we know that any quantum computer will have the Hamiltonian, Eq. (8.3.1.1), realized by some physical device. Following [D.67] we select the form of the Hamiltonian which is typically used in the NMR-type quantum computers ($\hbar = 1$)

$$\hat{H} = \sum_i \omega_0^i \sigma_z^i + \sum_{i<j} 2\pi J_{ij} \sigma_z^i \sigma_z^j. \tag{D.105}$$

In the appropriate basis, e.g. read [D.67], the entangling part of the time evolution operator $U_J(t) = \exp(i2\pi J \sigma_z^1 \sigma_z^2 t)$ for two qubits is given by

$$U_J(t) = \begin{pmatrix} e^{-i\pi Jt/2} & 0 & 0 & 0 \\ 0 & e^{+i\pi Jt/2} & 0 & 0 \\ 0 & 0 & e^{+i\pi Jt/2} & 0 \\ 0 & 0 & 0 & e^{-i\pi Jt/2} \end{pmatrix}. \tag{D.106a}$$

From this result we instantly recognize the maximal torus for SU(4) given by Eq. (D.47b). Accordingly, we can therefore apply the whole machinery of previous sections. Before doing so, several technical comments are the most appropriate at this point. First, unlike in mathematics, in the "real world" one has to use the gate $U_J(t)$ taking into account time-dependence. Second, this fact then makes time $t = 1/2J$ the most physically appropriate. Third, for such defined time the evolution matrix acquires the form

$$U_J(t = 1/2J) = \begin{pmatrix} e^{-i\pi/4} & 0 & 0 & 0 \\ 0 & e^{+i\pi/4} & 0 & 0 \\ 0 & 0 & e^{+i\pi/4} & 0 \\ 0 & 0 & 0 & e^{-i\pi/4} \end{pmatrix}. \tag{D.106b}$$

At the Lie algebra level of Eq. (D.69) the hyperplane equation

$$\lambda_1 + \lambda_2 + \lambda_3 + \lambda_4 = 0 \tag{D.107}$$

will have two entries the same signalling to us that we are dealing with Grassmannians. Indeed, there are $4!/2!2! = 6$ ways to select two identical entries. This number is the same as the choice of two hyperplanes exhibited by Eq. (D.66) and in Fig. 8. To connect these results with the AGS results is nontrivial [D.63] since it requires uses of the Morse–Bott theory.

Exercise D.2.14.1. Combine the already obtained results, Eqs. (D.66), for the Grassmannian with results of [D.63] to obtain the Morse–Bott moment map resulting in an octahedron — the polytope synonymous with the two qubits entanglement. Hint: Read why this is so below, in this, and in the following subsections.

Following [D.66, D.68] we construct now a composite 2-qubit matrix $R_z^1(-\frac{\pi}{2}) \otimes R_z^2(-\frac{\pi}{2})$ with $R_z(\theta)$ matrices taken from Eq. (D.14b). Then, the *conditional phase* (also known in computer literature as the *c-phase*) *gate* is represented as follows:

$$U_{c\text{-phase}} = e^{-\frac{\pi}{4}}\left[R_z^1\left(-\frac{\pi}{2}\right)\otimes R_z^2\left(-\frac{\pi}{2}\right)\right]U_J(t=1/2J) = \begin{pmatrix} 1 & & & \\ & 1 & & \\ & & 1 & \\ & & & -1 \end{pmatrix}. \tag{D.108}$$

But this is the same (entangling) gate as discussed already in connection with scattering processes in the non-ideal Bose gas in Appendic C, Eq. (C.26b)! Thus, the c-phase entangling gate is (can be) used at least in the NMR-type, [D.67], the Josephson-junction-type, [D.68], and in the non-ideal Bose-gas-type, [C.39], quantum computer devices. We shall discuss implications of this fact in the rest of this Appendix and in the reminder of Chap. 8.

To proceed, we would like at this point to focus attention of our readers at the geometrical meaning of the just obtained results. This is initially achieved with help of Figs. 9 and 10. They allow us to construct the *weight diagrams* used in the theory of representations of compact Lie groups. Specifically, from Fig. 9, we notice that Eq. (D.81) defining the root lattice Λ_R can also be rewritten in terms of the weight lattice Λ_W (clearly, this is not essential and is done only for the sake of compliance with the group-theoretic literature). For some weight

$$w = \sum_{i=1}^{n} \lambda_i e_i, \quad \sum_{i=1}^{n} \lambda_i = 0$$

(associated with the matrix, Eq. (D.69)) in view of Definition D.2.12.5, there are three possibilities. They are depicted respectively in Figs. 11–13.

In these figures the dashed lines represent the reflecting boundaries of the respective Weyl chambers. These are coinciding with the directions of the respective weights as depicted in Fig. 9. All three weight diagrams are diagrams for the highest weight. This is achieved as a result of the convexity property summarized in Definition D.2.12.5. The highest weight in each case is depicted by a circle surrounding the dot. It represents the point the most distant from the origin (in the positive Weyl chamber).

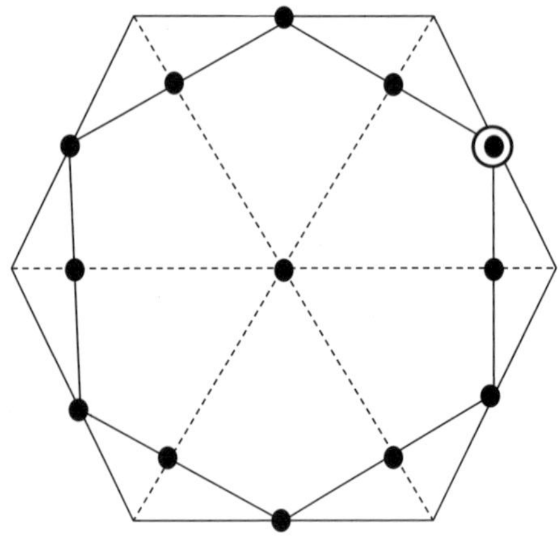

Fig. 11. Exceptional two-dimensional permutohedron representing the orbit $\mathcal{O}^{SU(3)}$. It is exceptional because its formation requires an unusually high symmetry.

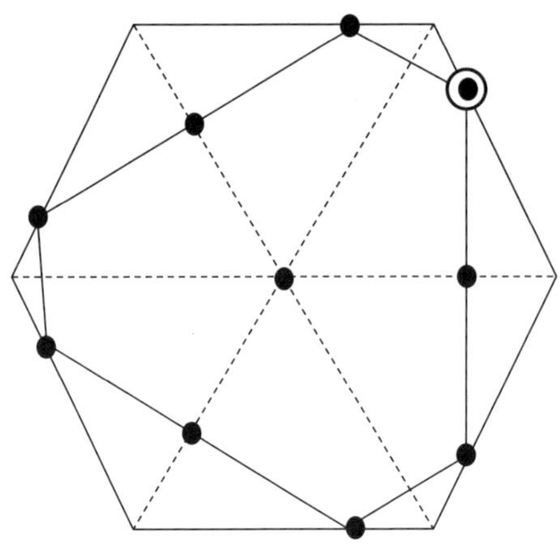

Fig. 12. Typical two-dimensional permutohedron.

The convexity exhibited in these diagrams links them with the dynamical systems in view of the AGS theorem. The degeneracy exhibited in Fig. 13. is coming from the fact that the highest weight is located at the mirror (at the boundary of the Weyl chamber). Alternatively, the mirror itself can be defined by the equation[m] $\lambda_i = \lambda_j$ as explained in [D.55]. The typical $\mathcal{O}^{SU(3)}$ orbit is depicted in Fig. 12. Thus, the symmetry of the orbit is

[m]In the case of $\mathfrak{su}(3)$ there are three ways to select two λ_i's out of three.

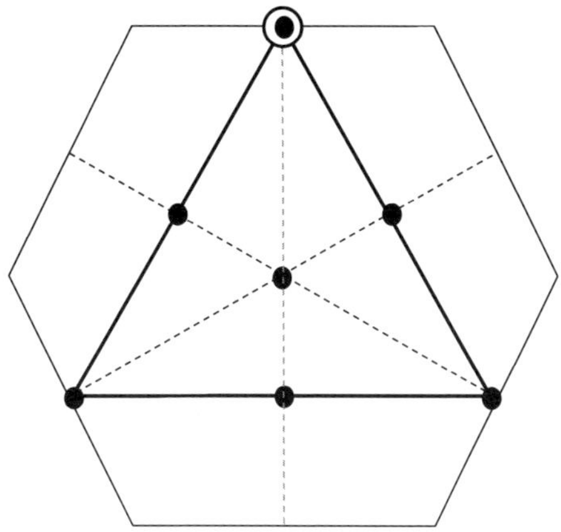

Fig. 13. Degenerate two-dimensional permutohedron. It represents the degenerate orbit $(\lambda_i = \lambda_j)$, $\mathcal{O}_d^{\mathrm{SU}(3)} \simeq \mathbf{CP}^2$.

determined by the entries in the matrix, Eq. (D.69), which, in the case of quantum computers, are experimentally adjusted. This represents the major experimental challenge since the number of non-entangling configurations is always much larger as compared with entangling.

All the three figures are designed with help of the same principles (in fact, they were already outlined in Sec. D.2.12). These are:

(a) each $\mathfrak{su}(n)$ representation is made with help of multiple copies of $\mathfrak{su}(2)$;
(b) each copy of $\mathfrak{su}(2)$ is depicted as the side of a polygon with evident symmetry with respect to the mirror, boundary of the Weyl chamber;
(c) every side of a polygon represents a string of eigenvalues (e.g. projections of either angular momenta or spin onto z-axis) depicted as black dots located symmetrically with respect to the mirror.

In the light of these observations, apparently, we can repeat the same protocol for SU(4) — the main object of our study. If, instead, we choose to use the method of symplectic fibrations and follow the prescriptions of [D.64], we have to begin with the tetrahedron, Fig. 4(c), representing $\mathbf{CP}^3$, and to use the $\mathbf{CP}^2$ triangles in the same way as we had used the $[-1, 1]$ segments (representing $\mathbf{CP}^1$). This will result in the truncated tetrahedron which is not yet exactly the same thing as the permutohedron depicted in Fig. 6. In fact, for SU(4) the number of possibilities is noticeably larger than that for $\mathfrak{su}(3)$. To describe all of them, in addition to [D.64], we follow [D.69–D.71] where these possibilities are clearly listed. Thus, we begin with the analog of Eq. (D.104a),

$$\mathcal{O}^{\mathrm{SU}(4)} \simeq \frac{\mathrm{SU}(4)}{S(U(1)\times U(1)\times U(1)\times U(1))} \equiv \frac{\mathrm{SU}(4)}{T} \tag{D.109a}$$

that is with the standard permutohedron Π_3 (Fig. 6). By analogy with Eq. (D.104b) the same result can be obtained as a quotient

$$\mathcal{O}^{\mathrm{SU}(4)} \simeq \frac{\mathrm{SU}(4)}{U(1)\times U(1)\times U(1)}. \tag{D.109b}$$

Indeed the dimension of the orbit $\mathcal{O}^{\mathrm{SU}(4)}$ in both cases is $\dim \mathcal{O}^{\mathrm{SU}(4)} = \dim \mathrm{SU}(4) - 3\dim U(1) = 12$. This number is also obtained in [D.64], p. 152. Notice that for Π_2 we had $3! = 6$ so that the generic orbit is going through all vertices of Π_2. In the case of Π_3 we have $4! = 4\cdot 3\cdot 2 = 24$ so that the generic 12-dimensional orbit is not visiting all the vertices! Furthermore, in this case there are four different degenerate orbits! These are (up to a permutation):

(a) $\mathcal{O}^{\mathrm{SU}(4)}_{d_{1_a}} \simeq \dfrac{\mathrm{SU}(4)}{\mathrm{SU}(2)\times U(1)\times U(1)}$,
(truncated tetrahedron: $\lambda_1 = \lambda_2 \neq \lambda_3 \neq \lambda_4$) (D.110a)

(b) $\mathcal{O}^{\mathrm{SU}(4)}_{d_2} \simeq \dfrac{\mathrm{SU}(4)}{\mathrm{SU}(3)\times U(1)} \simeq \mathbf{CP}^3$,
(tetrahedron: $\lambda_1 = \lambda_2 = \lambda_3 \neq \lambda_4$) (D.110b)

(c) $\mathcal{O}^{\mathrm{SU}(4)}_{d_3} \simeq \dfrac{\mathrm{SU}(4)}{S(U(2)\times U(2))} \simeq Gr_2(\mathbf{C}^4)$,
(octahedron: $\lambda_1 = \lambda_2 \neq \lambda_3 = \lambda_4$) (D.110c)

(d) $\mathcal{O}^{\mathrm{SU}(4)}_{d_{1_b}} \simeq \dfrac{\mathrm{SU}(4)}{S(U(1)\times U(2)\times U(1))}$.
(skew-cuboctahedron) (D.110d)

To obtain the orbit having 24 elements is nontrivial but physically very illuminating. Since this result is *not* contained in just cited literature, we shall briefly discuss it in Sec. D.2.17. For interested readers we would also like to mention that, unlike the $\mathcal{O}^{\mathrm{SU}(3)}$ case, it is now possible to design different fiber bundles using just listed degenerate manifolds as bases. For instance,

$$\begin{aligned}\mathcal{O}^{\mathrm{SU}(4)} &\simeq \mathcal{O}^{\mathrm{SU}(4)}_{d_1} \times \mathcal{O}^{\mathrm{SU}(3)} \\ &\simeq \frac{\mathrm{SU}(4)}{\mathrm{SU}(3)\times U(1)} \times \frac{\mathrm{SU}(3)}{\mathrm{SU}(2)\times U(1)} \times \frac{\mathrm{SU}(2)}{U(1)} = \mathbf{CP}^3 \times \mathbf{CP}^2 \times \mathbf{CP}^1\end{aligned}$$

and so on.

In this Appendix our goals are much more modest. We would only like to describe from several perspectives our main object of study: the octahedron. Not only the octahedron depicts graphically the essence of entanglement in 2-qubit quantum systems but, in addition, its description involves several features of independent physical interest worthy of mentioning. This will be done in the next section. In the meantime, we would like to develop the machinery needed for reproduction of figures analogous to Figs. 11–13 for $\mathfrak{su}(3)$. Formally, this can be achieved along the same lines as those culminating in Fig. 9. Details of such calculations are only briefly mentioned in [D.54] since their actual implementation is, in fact, cumbersome. Instead, we follow less detailed but, nevertheless, reliable method described in the book by Fulton and Harris, [D.55]. Its implementation is based on the set of principles already outlined in Sec. D.2.12.

First, the weight lattice for $\mathfrak{sl}(n)$ should be realized as lattice generated by the vertices of the regular simplex Δ_{n-1} centered at the origin of barycentric system of coordinates, e.g. see Fig. 10. Second, the root lattice can then be generated by the vector sums (or differences) between these vertices taken in pairs. Third, Eq. (D.81) for Λ_R now acquires the following look

$$\Lambda_R = \left\{ \sum_{i=1}^{n} \lambda_i e_i ; \lambda_i \in \mathbf{Z}, \sum_{i=1}^{n} \lambda_i = 0 \right\} \Bigg/ \left\{ \sum_{i=1}^{n} e_i = 0 \right\} \tag{D.111a}$$

while that for Λ_W is given by

$$\Lambda_W = \mathbf{Z}\{e_1, \dots, e_n\} \Bigg/ \left\{ \sum_{i=1}^{n} e_i = 0 \right\}. \tag{D.111b}$$

Here, $\mathbf{Z}\{e_1, \dots, e_n\}$ means "n-dimensional lattice".

Now we would like to illustrate these general principles using the $\mathfrak{sl}(4)$ (or its compact form $\mathfrak{su}(4)$) Lie algebra. Figure 14(a) depicts the elementary weight cell while Fig. 14(b) helps the eye to see the simplex symmetry. Based on these results, we come up with Fig. 15 depicting the elementary root cell, Figs. 16(a) and (b) are depicting the Weyl chamber and possible polyhedra listed in Eqs. (D.110) generated by mirror reflections.

As it can be seen from Fig. 15, the elementary root cell has the hexagonal-type symmetry which we encountered for $\mathfrak{su}(3)$. In fact, from [D.54], especially from the Plate 1 placed after p. 162, it can be seen that the A_3 root system can indeed be made out of a copy of A_2-type root system located in the plane and, another root system made of six roots

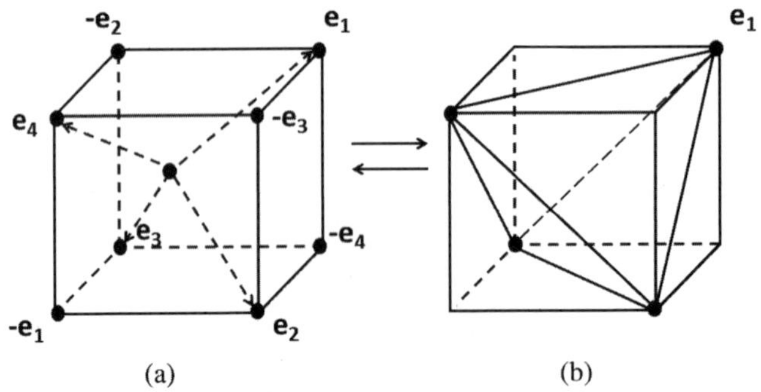

Fig. 14. Elementary weight cell (a) for $\mathfrak{su}(4)$ and the associated with it tetrahedron (b).

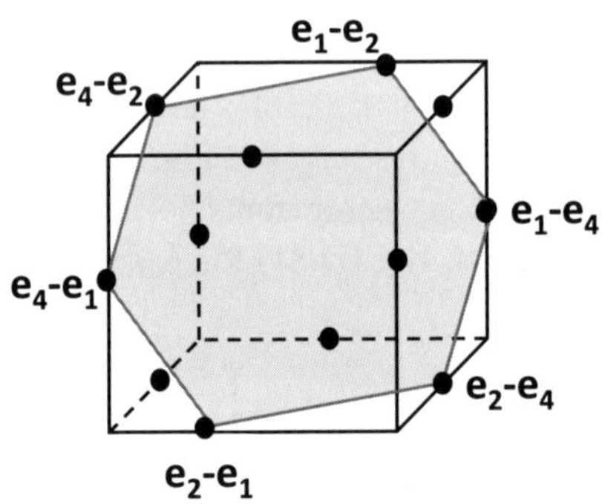

Fig. 15. Elementary root cell for $\mathfrak{su}(4)$ associated with the weight cell of Fig. 14.

centered at the same origin but emanating out of this plane so that the projection of this additional root system onto A_2-plane is made as well out of six elements. They are looking just like those for the A_2-system. Three of these elements are located above the A_2-plane and three below. To help the eye, look at Fig. 9, think about the roots α_1 and α_2 as lying in the plane while the weight ω_1 (which is yet another root in the A_3 case) is sticking out of the plane in "up" direction and that for ω_2 is sticking out of plane but in "down" direction. Use symmetry to obtain all six elements for both systems of roots.

Exercise D.2.14.2. Use the computational protocol we developed for the $\mathfrak{su}(3)$ case (summarized in Fig. 9) to re-obtain the results for roots of A_3-system. Notice that these results cannot be found in [D.55]. Why is this so? Hint: Consult [D.54], especially p. 262. From the same source you

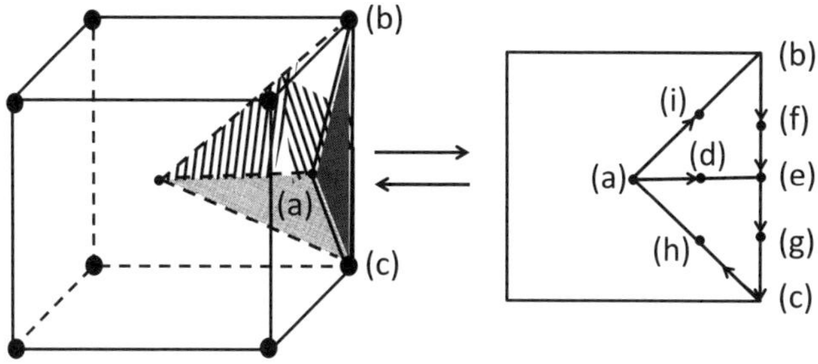

(a) (b)

Fig. 16. Elementary Weyl chamber for $\mathfrak{su}(4)$ (a). The side view (b) of the same chamber. Depending upon the location of the initial point: inside the chamber or on its boundary, the orbits generating the following polyhedra are obtained: (a) the octahedron ($\mathrm{Gr}_2(\mathbf{C}^4)$) — the entangling configuration for c-phase gate; (b) and (c) the tetrahedron ($\mathbf{CP}^3$) — the full entangling space for two qubits; (d) the permutohedron; (e) the cuboctahedron — the state describing all mesons (read below); (f) and (g) the skew-cuboctahedron; (i) and (h) the truncated tetrahedron. Arrows indicate an order in which these figures might be obtainable from each other.

will find that the Lie algebras $\mathfrak{su}(4)$ and $\mathfrak{so}(6)$ are isomorphic, that is they are having the same root systems and the same Weyl chambers.[n] It then follows that the root system for A_3 is equivalent to that for D_3 representing the dihedral group [D.58]. Can this isomorphism survive at the level of Lie groups? Hint: Look up the lecture notes, [D.72], pp. 32 and 33. These are also useful in the next subsection.

Remark D.2.14.1. Although Fig. 16(b) nicely illustrates/supplements Eqs. (D.110), it still does not exhaust all possibilities listed, for example, in http://en.wikipedia.org/wiki/Truncated_octahedron.

From this source we find that the permutohedron depicted in Fig. 6 is nothing but the truncated octahedron. This is just one out of the family of ten uniform octahedral polyhedra. That is to say, the octahedron, representing the entangled state, can break into ten different octahedral polyhedra! Some of them coincide in shape with the members of the family of eight uniform tetrahedral polyhedra. The most physically interesting among these (other than in quantum computer science) is cuboctahedron http://en.wikipedia.org/wiki/Cuboctahedron. It nicely keeps track of all mesons existing in Nature, e.g. see [D.56], p. 115.

[n]This isomorphism was used essentially in the already cited [D.69–D.71].

D.2.15 *Entangling Grassmannians with Twistors*

In Exercise D.2.14.1 we noticed that the entangling polytope, the octahedron, can be obtained from the Grassmannian with the theory of Morse–Bott functions. The role of moment map in the AGS formalism is being played by the Hamiltonian. Such a Hamiltonian is the Morse–Bott function at the same time [D.24]. To obtain such a function explicitly requires some extensions of results of Sec. D.2.12. Since these extensions are quite technical and since they are explained (to mathematicians by mathematicians) in [D.61, D.63], we decided to choose another, more physically illuminating, path to achieve the same goal. It is based on lecture notes by MacPerson, [D.73]. Following [D.73] we shall further extend our encounter with twistors initiated in Sec. D.2.10. In preparation for this task we recapitulate the main points of polytope design once again.

An action of n-torus $T \simeq (S^1)^n \equiv \mathbf{T}/\Lambda$ (here $\mathbf{T}$ is n-dimensional vector space and Λ is a lattice, e.g. see examples in Sec. D.2.5) on symplectic manifold X decomposes into orbits. Each orbit is k-torus where $k \leq n$.

Definition D.2.15.1. The k-skeleton of T acting on $X(T \circlearrowleft X)$ is the union of all orbits that are of dimension at most k. For example, the 0-skeleton is the union of fixed points while the n-skeleton is X itself. The k-skeleton is preserved as a set by the action of T on X so that k-skeleton itself is a space with T action.

Figure 3 depicts the action of $T = S^1$ on $X = S^2$ (that is $T \circlearrowleft S^2$). The orbits are the North and South poles and the circles of constant latitude $z = c$ where c is a constant between -1 and 1. MacPerson calls such a sphere with a torus action on it *a balloon.*

Definition D.2.15.2. A *balloon sculpture* is a space with a torus action such that it is a finite union of balloons (B_i's) assembled in such a way that any two of them are either disjoint or intersect at the fixed point of the torus action.

Definition D.2.15.3. If $T \circlearrowleft X$ is a balloon sculpture, then the quotient X/T is a graph whose vertices correspond to fixed points of T and whose edges correspond to the deformation retracts (that is to line segments) of the corresponding balloons. The 1-skeleton of $T \circlearrowleft X$ is a balloon sculpture.

Remark D.2.15.1. Based on results of previous sections the graph X/T represents an action of the Weyl reflection group. The graph is built on the assumption that each Lie algebra $\mathfrak{su}(n)$ is made of a set of copies of $\mathfrak{su}(2)$

and with each copy of $\mathfrak{su}(2)$ is associated a symmetric string of weights so that the edges of X/T are strings. The vertices of the graph are extrema of the convex domain.

Thus designed graph can be improved by assigning to each edge a direction. To do this, it is easy if we take a look at Fig. 3 and recognize that the circles of latitude carry with them an angular momentum whose direction coincides with the direction of z-axis: from South-to-North.

Definition D.2.15.4. The *moment graph* of $T \circlearrowright X$ is the linear graph obtained from the skeleton graph by associating the *direction vector* with each balloon.

In such a case this graph can be embedded into Euclidean vector space $\mathbf{V} = \mathrm{Hom}(T, \mathbf{R})$. That is the action of the torus T on the balloon is synonymous with the assigning of the direction to the balloon (better, its deformation retract) in the Euclidean space. This allows us to identify $\mathbf{T}$ with $\mathbf{R}^n$, Λ with $\mathbf{Z}^n$, and $\mathbf{V}$ with the *dual space* $\mathbf{V} = \mathbf{T}^* = (\mathbf{R}^n)^*$. The Euclidean scalar product $\langle \cdots \rangle$ can now be defined in a usual way as

$$\mathbf{T} \times \mathbf{V} \longrightarrow \mathbf{R} \qquad t \times v \rightarrow \langle t, v \rangle, \quad t = (t_1, \ldots, t_n), \quad v = (v_1, \ldots, v_n),$$
$$\langle t, v \rangle = t_1 v_1 + \cdots + t_n v_n.$$

If the standard basis for $\mathbf{V}$ is $\mathbf{e}_0, \ldots, \mathbf{e}_n$, then $\mathbf{e}_i$ represents a point where $v_i = 1$ and the rest of v_i's are zero. A crucial step now is based on the superposition of results of Sec. D.2.3 with Example in Sec. D.2.13. Equation (D.33) is an equation for n-simplex. It is the deformation retract of $\mathbf{CP}^n$. Representative examples are depicted in Fig. 4. In particular, we are interested in Fig. 4(a). It is clear now that vectors $\mathbf{e}_1$ and $\mathbf{e}_2$ are "living" in $\mathbf{V}$. In the projective space $\mathbf{CP}^1 = S^2/S^1$ there are two fixed points $F_1 = (1:0)$ on t_1-axis and, $F_2 = (0:1)$ on t_2-axis. They correspond to the points $p_1 = v_1 = 1$ and $p_2 = v_2 = 1$ in $\mathbf{R}^2$. Clearly, by the law of addition of vectors the segment $[F_1, F_2]$ should be assigned the direction, in our case $\mathbf{e}_1 - \mathbf{e}_2$. Now the standard action of the n-torus on the space $\mathbf{CP}^{n-1}$ is $z(x_1 : \cdots : x_n) = (z_1 x_1, \ldots, z_n x_n)$, where $z = (z_1, \ldots, z_n), z_j \in S^1 \subset \mathbf{C}$. The fixed points F_i are the n points where all homogeneous coordinates, except the ith, are zero. While it is 1 for the ith coordinate.

Remark D.2.15.2. The balloon B_{ij} in such a setting connects the fixed points F_i and F_j. Each balloon has three orbits: the two fixed points and one associated with the torus action as depicted in Fig. 3. *The direction vector* for B_{ij} is $\mathbf{e}_i - \mathbf{e}_j$. If we associate F_i to $\mathbf{e}_i$, then the straight line connecting F_i with F_j is parallel to $\mathbf{e}_i - \mathbf{e}_j$. Thus, the *moment graph* of $\mathbf{CP}^{n-1}$ is the

1-skeleton which is the $(n-1)$ simplex Δ_{n-1}. Δ_{n-1} is the convex hull of the basis vectors $\mathbf{e}_1, \ldots, \mathbf{e}_n$. Alternatively, $\Delta_{n-1} = \{(v_1, \ldots, v_n) | v_1 + \cdots + v_n = 1, v_i \geq 0\}$. Obviously, there is a complete duality between $\mathbf{T}$ and $\mathbf{V}$ spaces. A linear map $\mathbf{T} \rightarrow \mathbf{R}$ is given by $\langle (t_0, \ldots, t_j, \ldots, t_n), \mathbf{e}_j \rangle = t_j$.

Definition D.2.15.5. The $(2n-2)$-dimensional *quadric hypersurface* $\mathbf{Q}_{2n-2}$ is subset of $\mathbf{CP}^{2n-1}$ with homogeneous coordinates $(x_1 : \cdots : x_n : y_1 : \cdots : y_n)$ cut out by the equation

$$x_1 y_1 + \cdots + x_n y_n = 0.$$

The n-torus acts on $\mathbf{Q}_{2n-2}$ according to the rule:

$$z(x_1 : \cdots : x_n : y_1 : \cdots : y_n) = (z_1 x_1 : \cdots : z_n x_n : z_1^{-1} y_1 : \cdots : z_n^{-1} y_n).$$

This formula is compatible with the equivalence relation

$$(x_1 : \cdots : x_n : y_1 : \cdots : y_n) \simeq c(x_1 : \cdots : x_n : y_1 : \cdots : y_n)$$

for $\mathbf{CP}^{2n-1}$. It obviously preserves the equation for hypersurface $\mathbf{Q}_{2n-2}$.

Exercise D.2.15.1. Demonstrate that this is indeed the case.

Since the fixed points are points where exactly one homogeneous coordinate is nonzero, we can select F_i as such a point, where x_i is nonzero, and F_i', where y_i is nonzero. For every pair of homogeneous coordinates, except the n pairs $\{x_i, y_i\}$, the points in X where only that pair of homogeneous coordinates is nonzero is a *projective line*. Obviously, these lines are represented by the balloons. Thus, there is a balloon connecting any pair of fixed points, except the n pairs with the same index, that is F_i and F_i'. The total number of balloons in such moment graph is $\binom{2n}{2} - n = 2n(n-1)$. To get a feeling of what is going on, let us consider

Examples.

(a) $n = 2$

Without loss of generality [D.73], it is possible to replace complex projective space $\mathbf{CP}^3$ by the real one, that is by $\mathbf{RP}^3$ and, accordingly, to replace the complex quadric by the real one

$$x_1 y_1 + x_2 y_2 = 0$$

in $\mathbf{RP}^3$. Recall that $\mathbf{RP}^3$ can be covered by the affine charts of the type $\mathbf{R}^4 - \{0\} \rightarrow \mathbf{RP}^3$. Therefore, each of them contains $\mathbf{R}^3$ as dense subspace. This observation helps to visualize the quadric Q_2 as one sheet hyperboloid. It is depicted in Fig. 17(a) while the moment graph associated with it

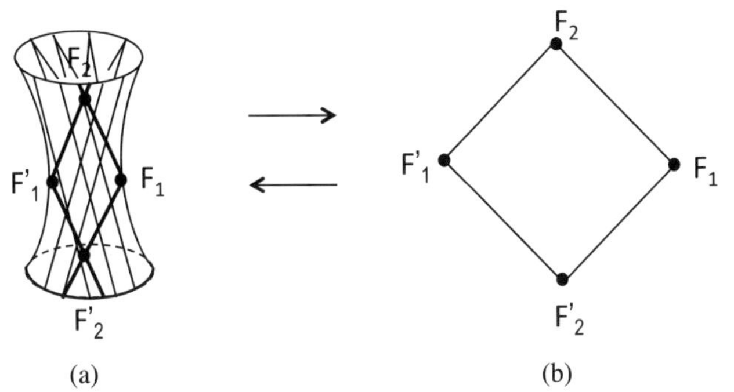

Fig. 17. (a) Quadric Q_2 with fixed points on it and (b) the associated moment graph.

is depicted in Fig. 17(b). The four fixed points F_1, F_2, F_1' and F_2' lie in $Q_2^{\mathbf{R}} \subset Q_2$. The four balloons are $\mathbf{CP}^1$'s intersecting $\mathbf{RP}^3$ in $\mathbf{RP}^1$(a circle). When confined to $Q_2^{\mathbf{R}} \subset \mathbf{R}^3$, the intersection is becoming a straight line. For a balloon joining F_i and F_j the direction vector is $\mathbf{e}_i - \mathbf{e}_j$, for that joining F_i' and F$'_j$ the direction vector is $-\mathbf{e}_i + \mathbf{e}_j$ while for the balloon joining F_i with F_j' it is $\mathbf{e}_i + \mathbf{e}_j$. In Sec. D.2.3, Remark D.2.3.1, we mentioned about the cross polytope. Now we would like to give a definition consistent with Fig. 17.

Definition D.2.15.6. The n-dimensional cross-polytope $C_rP(n)$ is a polyhedron in $\mathbf{R}^n$ defined by

$$C_rP(n) = \{(v_1, \dots, v_n) \in \mathbf{R}^n | |v_1| + \dots + |v_n| \leq 1\}.$$

The vertices of $C_rP(n)$ are located at the $2n$ points $\{\mathbf{e}_1, \dots, \mathbf{e}_n; -\mathbf{e}_1, \dots, -\mathbf{e}_n\}$ where $\mathbf{e}_i$ are standard basis vectors in $\mathbf{R}^n$.

From this definition it follows that there is an edge between any pair of vertices except for the n pairs $\{\mathbf{e}_i, -\mathbf{e}_i\}$. Thus the number of connected edges again is $2n(n-1)$ as expected.

(b) $n = 3$

The octahedron, the next in line cross polytope, can be built based on the rules formulated in (a). In this case we have the quadric Q_4

$$x_1y_1 + x_2y_2 + x_3y_3 = 0$$

which looks exactly the same as Klein's quadric, Eq. (D.64). And, surely, it is the same quadric. The general formula $2n(n-1)$ for the number of balloons in the moment graph is now reduced to 12. The moment graph — the octahedron $(Gr_2(\mathbf{C}^4))$ — is depicted in Fig. 18. It is the

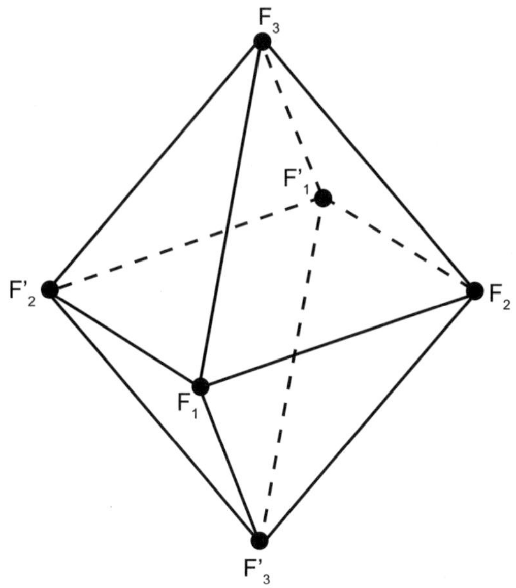

Fig. 18. The octahedron associated with the Klein quadric, Eq. (D.62). It is a deformation retract of the Grassmannian variety $Gr_2(\mathbf{C}^4)$.

entanglement configuration for two qubits. The same result can be obtained group-theoretically with help of Fig. 16. The results are depicted in Fig. 19. We would like to remind our readers that the same result, connecting the octahedron with Grassmannian, is obtainable with the help of the Morse–Bott methods developed in [D.63], with the help of methods of symplectic fibrations, [D.64] and, finally, with the help of theory of representations of Lie groups and Lie algebras [D.55].

From Fig. 16 it follows that the entanglement configuration, the octahedron, can easily decay into nearby configurations. Incidentally, not all of them can be obtained with help of methods we were using thus far. For instance, we were unable to obtain the 24-dimensional orbit for the permutohedron.

For the purposes of quantum computation it is essential to develop theory of optimal control which allows to reach the octahedron configuration from any other configuration in $\mathbf{CP}^3$. This task can be achieved without using the twistors. Since by using twistors we will be able to learn about many other physically interesting problems, it is worth mentioning such opportunities in the context of quantum entanglements.

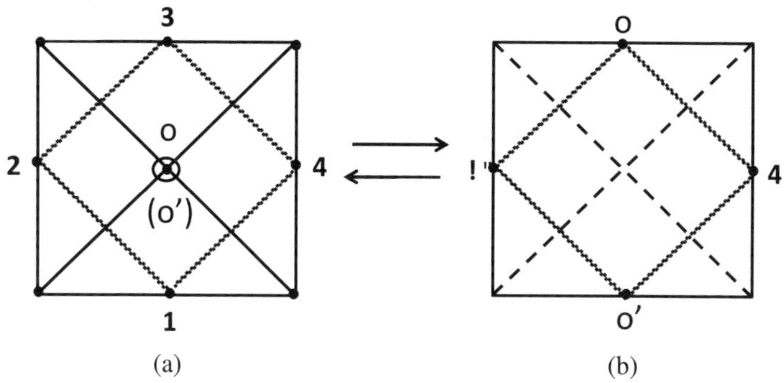

Fig. 19. (a) The top view of the cube displayed in Fig. 16 with Weyl chambers explicitly seen. The point O corresponds to point (a) in Fig. 16. Points 1, 2, 3, 4 are actually located at the centers of the respective sides of a cube. The point O' is at the center of cube's bottom. This is seen more clearly in (b) where the cross-section of the cube is displayed. Not surprisingly, the picture looks exactly the same as that in Fig. 17.

The development of twistor story takes place very much along the same lines as in Examples (a) and (b) in Sec. D.2.12. That is one goes back and forth between the matrices and vectors representing the same physical/mathematical reality whenever it is convenient. Our readers are encouraged at this point to read the original paper by Roger Penrose, [D.74], on twistors. The first section of this paper begins with the very basic physically motivated concepts of twistors which, incidentally, we had already encountered in this Appendix. Specifically, in Sec. D.2.10 we introduced the Plücker coordinates. They are describing directed lines. Twistors are also describing directed lines. Indeed, consider the null line L in the Minkowski space–time $\mathfrak{M}$ with coordinates x^i with respect to some origin O. Let l^i be the position (with respect to O) vector of some point on L and n^i be future-pointing tangent vector to L. Now we can introduce the Plücker coordinates m^{ij} as it was done in Sec. D.2.10.

$$m^{ij} = l^i n^j - n^j l^i.$$

Penrose considers instead the set (n^j, m^{ij}) as Plücker–Grassmann coordinates. The line L is *null* in view of the Minkowski constraint

$$n^j n_j = 0.$$

An analog of Eq. (D.63) now acquires the form

$$\varepsilon_{ijkl} m^{ij} n^k = 0$$

implying that

$$\varepsilon_{ijkl} m^{ij} m^{kl} = 0.$$

Clearly, the last equation is just Plücker's Eq. (D.64) defining Klein's quadric, the Grassmannian $Gr_2(\mathbf{C}^4)$ and, ultimately the octahedron, Fig. 18. The full proofs that, indeed, we are dealing with exactly the same Klein's quadric and, because of this, with $Gr_2(\mathbf{C}^4)$, are presented with great clarity and care in [D.40, D.41]. Seemingly, after this, there is nothing much to discuss. Fortunately, this is not at all the case! We explain why this is so in the next section.

D.2.16 *Complex Contact Geometry of Twistor Spaces and the Problem of Optimal Control of Entanglements*

Some of our readers who had enough patience to read through this Appendix till this point might have begun to wonder: If this is the book about contact geometry and topology, why then there is such a long Appendix which thus far had no mention of contact geometry whatsoever? This section is aimed at correcting this impression. In fact, it is for the sake of this section that most of previous sections were written! Thus, here we shall gradually bring our readers back to the realm of contact geometry. This time our readers will be able to understand why only now it is becoming possible to do so.

We begin with Fig. 16. It is telling us that all possible polytopes are living in $\mathbf{CP}^3$ so that if we are thinking about problems of optimal control of entanglements, these problems should be concerned with dynamics in $\mathbf{CP}^3$. In general, should we have n qubits, $n \geq 2$, we would replace $\mathbf{CP}^3$ by $\mathbf{CP}^{2n-1}$. From which it follows that, when dealing with qubits, we shall *always* be dealing with the odd-dimensional complex projective spaces. It happens that these are *complex contact* manifolds, [D.75]! Thus far in this book we studied only *real* contact manifolds. Naively, one might think that it is possible to transfer all results from real contact manifolds to complex unchanged. This happens *not* to be the case, e.g. read p. 234 of [D.75]. Because of this, the results of this section are going to be sketchy. Nevertheless, we hope that just mentioning certain topics might stimulate further research in this underdeveloped area.

Study of complex contact manifolds was initiated by Kobayashi in 1959. He proceeded in a traditional way, e.g. along the lines of Chap. 6, Sec. 6.5.

Specifically, let $\mathcal{M}$ be the complex contact manifold of $(2n+1)$ dimensions together with an open covering $\{\mathcal{O}_\alpha\}$ by coordinate neighborhoods so that

1. On each $\mathcal{O}_\alpha$ there is a holomorphic 1-form ω_α such that

$$\omega_\alpha \wedge (d\omega_\alpha)^n \neq 0. \tag{D.112}$$

2. On $\mathcal{O}_\alpha \cap \mathcal{O}_\beta \neq 0$ there is a nonvanishing holomorphic function $f_{\alpha\beta}$ such that $\omega_\alpha = f_{\alpha\beta}\omega_\alpha$.

   The subspaces $\{X \in T_m\mathcal{O}_\alpha : \omega_\alpha(X) = 0\}$ define the *horizontal subbundle* $\mathcal{H}$ of complex dimension $2n$. So, as in Chap. 6, we write

$$T\mathcal{M} = \mathcal{H} \oplus \mathcal{V},$$

   where $\mathcal{V}$ is the vertical subbundle. The quotient $T\mathcal{M}/\mathcal{H}$ is a complex line bundle L ($\sim \mathcal{V}$) defined by $\{f_{\alpha\beta}\}$. Kobayashi proved that the first Chern class (e.g. see Sec. 5.6.3) $c_1(\mathcal{M}) = (n+1)c(L)$. From this result it follows that for a compact complex manifold, a contact structure is determined by the *globally defined* 1-form ω if and only if $c_1(\mathcal{M}) = 0$. In this case all $f_{\alpha\beta} = 1$.

Remark D.2.16.1. Le Brun in 1995 [D.76] proved that "*A complex contact manifold of positive first Chern class is a twistor space if and only if it admits a Kahler–Einstein metric*". Projective spaces $\mathbf{CP}^{2n-1}$ surely fit this condition [D.77]. However, the nonvanishing of $c_1(\mathcal{M})$ is an obstruction for the 1-form ω to be globally defined. This fact makes complex contact manifolds noticeably different from the real ones and much harder to study. Developments in this field are still in their infancy [D.75].

Remark D.2.16.2. To illustrate this fact we notice the following. From Sec. D.2.1, we already learned that phase spaces of mechanical systems can be rewritten in terms of complex manifolds which are Kähler. In 1994 Ye, [D.78], p. 312, proved that a compact Kähler manifold with $c_1(\mathcal{M}) = 0$ has no holomorphic contact structure. This result is in apparent contradiction with that by Kobayashi illustrating that already at the very basic level one should be prepared for surprises in the field of complex contact manifolds.

Remark D.2.16.3. According to the existing definition of the Calabi–Yau-type manifolds, e.g. see http://en.wikipedia.org/wiki/Calabi-Yau_manifold, they are just those for which Ye obtained the negative result. Thus, in contradiction with results of [D.20], the Calabi–Yau manifolds cannot be used for development of optimal control of entanglements.

Instead, as we already mentioned, we have to focus our attention on the complex projective spaces of odd dimensionality. A contact 1-form for $\mathbf{CP}^{2n+1}$ (whose homogeneous coordinates are $(z_1 : \cdots : z_{n+1} : w_1 : \cdots : w_{n+1})$) is given by

$$\omega = \sum_{k=1}^{n+1}(z_k dw_k - w_k dz_k). \tag{D.113}$$

Comparison with the 1-form α, Eq. (5.5.1), for S^3 (but easily generalized to any S^{2n+1}) allows us to make a provisional identification: $x_k \rightleftarrows z_k, y_k \rightleftarrows w_k$. Thus, not totally surprisingly, the 1-form for $\mathbf{CP}^{2n+1}$ is obtainable from the 1-form for S^{2n+1} via identification we just made. The question arises: If this is so, can we find some *physical reasons* justifying our use of the 1-form for S^{2n+1} instead of that given by Eq. (D.113)? Notice that for $n = 1$ we are confronted with S^3 studied in detail in Chap. 8. The question arises: Can S^3 be used for two qubits case? That is to say: Is there a map between the $\mathbf{CP}^3$ and S^3? Looking at Eq. (D.113), surely, such a map exist. But what is the physics behind such a map? In [D.11] the standard path of Hopf fibration $S^3 \hookrightarrow S^7 \to S^4 = \mathbf{HP}^1$ (where **HP** stands for "quaternionic projective") was used for description of the 2-qubit entanglements. In this case we are dealing with the contact manifold S^7, which is the Hopf fiber bundle with S^4 being a base and S^3 being a fiber. The control is then done along the same lines as in the $S^3 \to S^2$ case discussed in Chap. 8. Remark D.2.16.1 allows us to replace this construction by another one coming from the theory of twistors. Although the book, [D.41], is ideal source for everybody seriously interested in twistors, we shall also use other references allowing us to arrive at the following chain of mappings

$$\mathbf{CP}^3 \to S^4 \to S^3$$

in the most economical way.

In Example (b) of Sec. D.2.12 we discussed the correspondence between quaternions $\mathbf{H}$ and $\mathbf{R}^4$. It can be extended to the Minkowski space. Following Baird, [D.79, D.80], let us begin with writing things in $\mathbf{R}^4$ first for the sake of the correspondence with his notations. Thus, we obtain:

$$(x^0, x^1, x^2, x^3) \rightleftarrows (x^{AB}) = \begin{pmatrix} x^0 + ix^1 & x^2 + ix^3 \\ -x^2 + ix^3 & x^0 - ix^1 \end{pmatrix}, \quad A, B = 0, 1, \tag{D.114a}$$

so that $\det(x^{AB}) = |z_1|^2 + |z_2|^2$ is the Euclidean norm since $z_1 = x^0 + ix^1$ and $z_2 = x^2 + ix^3$. Thus, we also have $(x^0, x^1, x^2, x^3) \rightleftarrows (z_1, z_2) \in \mathbf{C}^2$.

Using Example (b) of Sec. D.2.12, we have to identify $\mathbf{R}^4$ with $\mathbf{H}$. In Sec. D.2.9, we explained how the sphere S^2 can be identified with the complex projective space $\mathbf{CP}^1$. In complete analogy, now we are in the position to demonstrate that S^4 can be identified with the projective quaternionic space $\mathbf{HP}^1$. Why do we need this result? Well, take into account that the $\mathbf{R}^4$ and the Minkowski space $\mathfrak{M}$ can be obtained from $\mathbf{C}^4$. Recall the familiar from physics literature euclideanization (e.g. $t \to it$). In the non-relativistic path integrals one replaces time by the imaginary time to make integrals well behave etc.

Definition D.2.16.1. The *twistor space* $\mathbf{T}$ is $\mathbf{C}^4$ with coordinates $\mathbf{Z} = \{Z^0, Z^1, Z^2, Z^3\}$. The *projective twistor space* $\mathbf{PT}$ is $\mathbf{CP}^3$ with homogeneous coordinates $\{Z^0 : Z^1 : Z^2 : Z^3\}$.

Now, the field of quaternions $\mathbf{H}$ is isomorphic to $\mathbf{C}^2$. Therefore, $\mathbf{C}^4 \rightleftarrows \mathbf{H}^2$. If this is so, we can repeat almost word-for-word arguments of Sec. D.2.9. Indeed, we have to design a map from $\mathbf{PT} = \mathbf{CP}^3$, that is from $\{Z^0 : Z^1 : Z^2 : Z^3\}$, to $\mathbf{HP}^1$, that is to $\{q_1 : q_2\}$, remembering that the quaternions can be written as $q_1 = Z^0 + Z^1 j$ and $q_2 = Z^2 + Z^3 j$. Thinking by analogy, we have to remember that if $q_1 \rightleftarrows z_1$ and $q_2 \rightleftarrows z_2$, we can design a ratio $q_1/q_2 \rightleftarrows z_1/z_2$ and, since $z_1/z_2 \to S^2 = \mathbf{CP}^1$, we obtain now : $q_1/q_2 \to S^4 = \mathbf{HP}^1$ (because $\mathbf{R}^4 \rightleftarrows \mathbf{H}$). Thus, we just constructed a map $\mathbf{CP}^3 \to S^4$. In fact, it is the Hopf map for exactly the same reason for which the map $z_1/z_2 \to S^2 = \mathbf{CP}^1$ is the Hopf map for $S^3 \to S^2$, e.g. see [111]. What is the physical meaning of $\mathbf{CP}^3$ in the context of twistors? What is the physical meaning of just constructed map? It happens that these two questions are interrelated. To provide an answer, let us again take a closer look at the Hopf map: $z_1/z_2 \to S^2$. We begin with $\{z_1, z_2\} \in \mathbf{C}^2$, then we replace $\mathbf{C}^2$ with S^3 since for S^3 we have $|z_1|^2 + |z_2|^2 = 1$. Next, we construct the ratio z_1/z_2 and take into account that if we rescale simultaneously both z_1 and z_2 by the same scale factor the ratio will stay the same allowing us to replace $\{z_1, z_2\} \in \mathbf{C}^2 \simeq S^3$ by $\{z_1 : z_2\} \in \mathbf{CP}^1$. Finally, since both z_1 and z_2 can be written as $z_i = r_i \exp(i\varphi_i), i = 1, 2$, we obtain $z = z_1/z_2$. By adding one point at ∞ to z-plane we obtain (via stereographic projection): $z_1/z_2 \rightleftarrows S^2 \simeq \mathbf{CP}^1$.

Now we go to the quaternionic case. We begin with $\{Z^0, Z^1, Z^2, Z^3\} \in \mathbf{T}$, then we construct a quaternion $q = \frac{Z^0 + Z^1 j}{Z^2 + Z^3 j} \equiv \frac{f + gj}{h + kj}$. We used notations of [D.79] in the last equality. This fact allows us also to replace $\mathbf{T}$ by $\mathbf{PT}$.

Using this, we write

$$S^4 = \mathbf{H} \cup \{\infty\} \ni q = \frac{f + gj}{h + kj} \equiv z_1 + z_2 j \tag{D.115a}$$

or, which is equivalent,

$$f + gj - (h + kj)(z_1 + z_2 j) = 0. \tag{D.115b}$$

The last equation can be rewritten in the matrix form if we take into account the multiplication table for quaternions [D.53], p. 39. Thus, we obtain,

$$(h, k)\begin{pmatrix} z_1 & z_2 \\ -\bar{z}_2 & \bar{z}_1 \end{pmatrix} = (f, g). \tag{D.116a}$$

By looking at Eq. (D.114a) we can identify z_1 with $x^0 + ix^1$ and z_2 with $x^2 + ix^3$. h and k can then be identified with Z^2 and Z^3 while f and g with Z^0 and Z^1. In literature on twistors the spinor language is used causing us to identify Z^0 and Z^1 with the two-component spinor so that $Z^0 = \omega^0$, $Z^1 = \omega^1$, accordingly, $Z^2 = \pi_0$ and $Z^3 = \pi_1$. The fact that π_0 and π_1 are having lower indices while ω^0 and ω^1 having upper is the result of conventions used in theory of spinors. In any case, written in such notations Eq. (D.116a) acquires the following form:

$$\omega^A = x^{AB}\pi_B, \quad A, B = 0, 1 \tag{D.116b}$$

commonly used in mathematical physics literature, e.g. see [D.81]. In twistor literature, [D.41], this equation acquires special status because of the following.

Definition D.2.16.2. In twistor literature [148, D.38, D.74, D.81] this equation is known as the *incidence relation*. It connects points in $\mathbf{R}^4$ (or in Minkowski, or in compactified Minkowski spaces (see below)), with points in **PT**. Thus, the twistor formalism can be looked upon as *an* analog of the Fourier transform making some calculations much easier.

To obtain the same incidence relation in $\mathfrak{M}$ we only have to rewrite Eq. (D.114a) as follows

$$(x^0, x^1, x^2, x^3) \rightleftarrows (x^{AB}) = \begin{pmatrix} -i(x^0 - x^1) & x^2 + ix^3 \\ -x^2 + ix^3 & -i(x^0 + x^1) \end{pmatrix} \tag{D.114b}$$

causing the determinant $\det(x^{AB})$ to become $(x^0)^2 + (x^1)^2 + (x^2)^2 + (x^3)^2$. The light cone is obtained from the equation $\det(x^{AB}) = 0$, as usual. This result is useful to compare with that obtained in Eq. (D.13). The light cone condition $(x^0)^2 + (x^1)^2 + (x^2)^2 + (x^3)^2 = 0$ written in the affine form (that is for $x^0 \neq 0$) is the same as the condition determining the Bloch sphere ("celestial" sphere in twistors language [D.74]), i.e. $\boldsymbol{n}^2 = 1$ with $\boldsymbol{n}$ being the Bloch vector. The condition $\det(x^{AB}) = 0$ determining the light cone (or the Bloch vector) is also a condition for the density matrix to describe a pure state. Equation (D.6) is telling us that the density matrix for pure state can be represented in the factorized form. Exactly the same phenomenon of factorizability exist in twistor formalism. In complete analogy with Eq. (D.6b), we introduce a spinor $\xi = \{\xi^0 : \xi^1\} \in \mathbf{CP}^1 = S^2$, so that $x^{AB} = i\xi^A\bar{\xi}^B$, $A, B = 0, 1$. The i-factor located in front is written just to be in accord with the notations of [D.81]. The incidence relation will now look exactly the same as Eq. (D.116b) but, this time, we have to use the matrix given by Eq. (D.114b). It connects a point x^{AB} in $\mathfrak{M}$ (that is in Minkowski space) with the point $\{\omega : \pi\}$ in $\mathbf{PT}$. In the language of quantum computers, we can think of $\{\omega^0 : \omega^1\} \in \mathbf{CP}^1$ as representing the state of one qubit while $\{\pi_0 : \pi_1\} \in \mathbf{CP}^1$ is representing the state of another qubit. Then, the incidence matrix x^{AB} can be thought of as the "*entanglement tensor*", in terminology of [D.5]. Such an identification should be understood as follows. The state vector of two qubits $\{\omega^0 : \omega^1 : \pi_0 : \pi_1\} \in \mathbf{CP}^3$ may exist without any incidence relation. But, when such a relation does exist, these qubits become entangled! Our task is to provide a convincing evidence of correctness of this conjecture. This is accomplished below.

Let us begin with the observation that the matrix x^{AB} in the incidence relation (both the Euclidean and Minkowski forms) Eq. (D.116b) is not uniquely defined. If we replace x^{AB} by $y^{AB} + \lambda\pi^A\bar{\pi}^B$, $\lambda \in \mathbf{C}$, then nothing changes if $x^{AB} = y^{AB}$ since $\pi_A\pi^A = \varepsilon_{AB}\pi^B\pi^A = 0$. Here ε_{AB} is antisymmetric spinor [D.81]. Incidentally, spinors ω^A and π_A are also defined with accuracy up to similar additive terms and for the same reason. We shall not go into these details since we are not going to use them. Instead, following [D.42] we rewrite the incidence relation, Eq. (D.116b) as follows:

$$\begin{aligned} \tilde{z}Z^2 + wZ^3 &= Z^0, \\ \tilde{w}Z^2 + zZ^3 &= Z^1. \end{aligned} \tag{D.117}$$

In this expression we replaced the matrix x^{AB} by $\begin{pmatrix} \tilde{z} & w \\ \tilde{w} & z \end{pmatrix}$ and the variables $\tilde{z}$ and $\tilde{w}$ should *not* be considered as complex conjugates of z and w. As in [D.42], they will be specified differently for different situations. Rewritten in such a way the incidence relation acquires new meaning in view of the results of Sec. D.2.10. Indeed, using Eq. (D.117) the following matrix can now be read off these equations

$$\mathbf{M} = \begin{pmatrix} \tilde{z} & w & 0 & 1 \\ \tilde{w} & z & 1 & 0 \end{pmatrix}$$

so that Plücker's coordinates are $l_{01} = \tilde{z}z - \tilde{w}w$, $l_{02} = \tilde{z}, l_{03} = -\tilde{w}$, $l_{12} = w, l_{13} = -z$, $l_{23} = -1$. From here, the Plücker embedding equation is satisfied identically

$$(-1)(\tilde{z}z - \tilde{w}w) + z\tilde{z} - \tilde{w}w = 0.$$

This result is compatible with the requirement $z\tilde{z} - \tilde{w}w = 0$. Using matrix entries in Eqs. (D.107) we recognize the light cone equation $(x^0)^2 + (x^1)^2 + (x^2)^2 + (x^3)^2 = 0$ for the Minkowski case and $|z_1|^2 + |z_2|^2 = 0$ for the Euclidean case. Thus, we had begun with $\mathbf{CP}^3$ and ended up with $\mathbf{CP}^5$ for the Plücker embedded Grassmannian $Gr_2(\mathbf{C}^4)$. But $\mathbf{CP}^3$ is the space of twistors! Thus the Plücker embedding into $\mathbf{CP}^5$ is in fact an embedding into $\mathfrak{M}$!

The incidence relation, Eq. (D.116b), can be further looked upon from the point of view of projective geometry. Indeed each line in Eq. (D.117) is in fact an equation for a hyperplane

$$Z^\alpha A_\alpha = 0. \tag{D.118}$$

From projective geometry, [D.81–D.83], it is known that a hyperplane in n-dimensional projective space is an $(n-1)$-dimensional subspace of $\mathbf{P}^n$ (or $\mathbf{CP}^n$). So, if $\mathbf{PT} = \mathbf{CP}^3$, the above equation is for the hyperplane "living" in $\mathbf{CP}^2$.[o] Since according to Eq. (D.117) we are having two hyperplanes, the question arises: Under what conditions will these two hyperplanes intersect? Evidently, we have to study joint solutions (that is intersections) for the set of two hyperplanes

$$Z^\alpha A_\alpha = 0 \quad \text{and} \quad Z^\alpha B_\alpha = 0. \tag{D.119}$$

By analogy with the Euclidean geometry we expect that two planes are either parallel or will intersect each other in a line. To show this in the projective geometry is not difficult as well. Indeed, let $\mathcal{D}$ be a subspace of $\mathbf{P}^n$ (or $\mathbf{CP}^n$) of dimension k. Let points $p_0, p_1, \ldots, p_k$ span $\mathcal{D}$. This means that any point p of $\mathcal{D}$ can be represented as a linear combination $p = a_0p_0 + \cdots + a_kp_l$. For the line in $\mathbf{P}^n$ we should have $a_ip_i + a_jp_j + a_lp_l = 0$ where $i \neq j \neq l$. For the system of two equations, Eq. (D.119), standard

[o]This fact will be discussed from another point of view in the next section.

manipulations, e.g. [D.83], p. 83, demonstrate that joint solution is assured by the Plücker embedding identity, Eq. (D.64), and that the resulting four solutions are indeed the projective lines, that is they are $\mathbf{CP}^1 = S^2$. These lines are being parametrized in terms of Plücker coordinates so that all projective lines in $\mathbf{CP}^3$ are being parametrized by the Plücker coordinates.

It is of value to further analyze the Plücker embedding Eq. (D.64). Following [D.83] we can rewrite it in terms of new variables. These are:

$$\begin{aligned} l_{03} &= u_0 + u_3, l_{13} = u_1 + u_4, l_{23} = u_2 + u_5, \\ l_{12} &= u_0 - u_3, l_{20} = u_1 - u_4, l_{01} = u_2 - u_5. \end{aligned} \tag{D.120}$$

In new variables Eq. (D.64) acquires the following look

$$u_0^2 + u_1^2 + u_2^2 - u_3^2 - u_4^2 - u_5^2 = 0. \tag{D.121}$$

If for some reason $u_2^2 = u_5^2 = 0$, then the resulting equation is that known for the hyperboloid of one sheet in 3D space, e.g. see http://en.wikipedia.org/wiki/Quadric. It is already depicted in Fig. 17(a)! But now we can do more following ideas of Penrose and Rindler, [D.81] and Gindikin, [D.84]. Equation (D.121) is the equation for the *complex* quadric Q_4 in $\mathbf{CP}^5$. It is of physical interest to look at subspaces of Q_4 which are real. Clearly, we can begin with all u_i's being real. But this is trivial. Less trivial is to consider a situation when, say, u_0, u_1, u_2 and u_5, are real while $u_3 = iv_3, u_4 = iv_4$ are purely imaginary. Surely, the labelling of variables can be adjusted to physical problems at hand. Another situation is when only, say, $u_3 = iv_3$ is purely imaginary while the rest of variables are real. For these cases we obtain

$$u_0^2 + u_1^2 + u_2^2 + v_3^2 + v_4^2 - u_5^2 = 0, \quad \text{(S)} \tag{D.122a}$$

$$u_0^2 + u_1^2 + u_2^2 + v_3^2 - u_4^2 - u_5^2 = 0. \quad \text{(H)} \tag{D.122b}$$

The (S)-equation is an equation for a sphere S^4 written in projective form. The (H)-equation describes a hyperboloid of one sheet. It is associated with the way the Minkowski space is being compactified [D.81]. Both (S) and (H) quadrics have common part of Q_4 made of, say, $u_0^2 + u_1^2 + u_2^2 + v_3^2$. Evidently, this is S^3.

Exercise D.2.16.1. Let us consider the S-option first. (a) Using Eqs. (D.113) we obtain:

$$l_{01} = u_0 - iv_3, \quad l_{23} = u_0 + iu_3, \quad l_{20} = u_1 - iv_4, \quad l_{13} = u_1 + iv_4,$$
$$l_{03} = u_2 - u_5, \quad l_{12} = u_2 + u_5$$

implying

$$l_{23} = \bar{l}_{01}, \quad l_{13} = -\bar{l}_{20}, \quad \operatorname{Im} l_{03} = \operatorname{Im} l_{12} = 0. \tag{D.123}$$

Use of the already introduced (p. 408) Plücker coordinates in just obtained relations (with the appropriate relabeling of indices) demonstrates that the matrix $(x^{AB}) = \begin{pmatrix} \tilde{z} & w \\ \tilde{w} & z \end{pmatrix}$ can be identified with (x^{AB}) given by Eq. (D.114), whenever the conditions given by Eq. (D.123) can be written as restrictions $\bar{z} = \tilde{z}$ and $\bar{w} = -\tilde{w}$ on the matrix (x^{AB}) entries. This restriction leads to the Euclidean norm result $\det(x^{AB}) = |z|^2 + |w|^2$ as required. (b) Repeat the same calculations for the Minkowski case and demonstrate that, in view of Eq. (D.107b), the Minkowski analog of Eqs. (D.123) is given by the relations $\bar{w} = -\tilde{w}$, $\tilde{z} = -(\tilde{z})^*$, $z = -(z)^*$. Here, to avoid confusion, * denotes complex conjugation.

The results of this exercise indicate that the intersection of the Euclidean conditions $\bar{z} = \tilde{z}$ and $\bar{w} = -\tilde{w}$ with the Minkowski conditions $\bar{w} = -\tilde{w}$, $\tilde{z} = -(\tilde{z})^*$, $z = -(z)^*$, when translated into Eqs. (D.122), results in equation defining S^3. These conditions provide the mapping $S^4 \to S^3$ on the Grassmannian side. But what about the twistorial side? To provide an answer, recall that the incidence relation, Eqs. (D.109), has geometrical meaning given by Eq. (D.115). This means that twistors "live" in S^4 which can be described by *the same type of* S-*equation*. Each component of the twistor is some function of z and w. The explicit form of z, w-dependence is of no concern to us. However, it is obvious from Eq. (D.115) that locally, it should be possible to parametrize z and w by q.

To proceed, we need to inject some physical arguments. The sphere S^4 via stereographic projection is the compactified $\mathbf{R}^4$. This is the Euclidean space used in instanton calculations, e.g. read again Chap. 5, Sec. 5.6. Using instantons we can arrive at the monopoles (these are basically, Euclidean time-independent instantons) "living" in $\mathbf{R}^3$ or, upon compactification, in S^3, again via stereographic projection. The Minkowski space–time is essentially $S^3 \times S^1$ while the Euclidean space–"time" is essentially S^4. The twistor formalism happens to be the most effective when it comes to instantons, monopoles and other exactly integrable systems [D.42, D.148]. Thus, S^3 is the overlap space between the instantons and monopoles.

Remark D.2.16.4. The monopole case, actually the Dirac monopole (discussed in Chap. 3, Sec. 3.5), is known in the twistor theory under the name *Robinson's congruence* [D.74, D.81].

In view of Eqs. (D.115a), (D.116a) we notice the following. Equations (D.108a) is basically the map π: $\mathbf{CP}^3 \to S^4$. At the monoplole level we can map S^4 into S^3. Thus, we shall also consider π as a map from $\mathbf{CP}^3$ to S^3. This map allows us to replace the complex contact 1-form ω defined in Eq. (D.113) by much more familiar 1-form α, Eq. (5.2.2), which we also discussed in Chap. 8, Sec. 8.2.6. Consider now the inverse mapping $\pi^{-1} : S^3 \to \mathbf{CP}^3$. What is the condition for such a mapping to exist? [D.79, D.80] provides us an answer. It comes from imposition of the constraint on the quaternion q defined by Eq. (D.115a):

$$S^3 = \{\mathbf{H} \cup \{\infty\};\ q = -\bar{q}\}. \tag{D.124}$$

The logic behind this constraint is as follows. From example (b) of Sec. D.2.12 we know that $\mathbf{H}$-field of quaternions is isomorphic to $\mathbf{R}^4$. Since $q = t + ix + jy + zk$ $\mathbf{H} = \mathbf{R}^1 \oplus \mathbf{R}^3$, where $\mathbf{R}^1$ corresponds to real quaternions and $\mathbf{R}^3$ to purely imaginary ones. The condition $q = -\bar{q}$ singles out purely imaginary quaternions and, since the compactified $\mathbf{R}^3$ is S^3 (via stereographic projection), Eq. (D.124) follows. To write down this equation explicitly with the help of Eq. (D.115b) we have to remember only two rules for quaternions [D.53]. These are: (a) for any two quaternions q_1 and q_2, we have

$$\overline{[q_1 q_2]} = \bar{q}_2 \bar{q}_1$$

and (b) since $q = z_1 + j\bar{z}_2$, where $z_1 = r_0 + r_1 i$, $z_2 = r_2 + r_3 i$, we now also have $jq = \bar{q}j$.

With help of these two properties of quaternions by using both, Eqs. (D.115b) and (D.124), we arrive after some computation at the following condition:

$$\bar{h}f + k\bar{g} + h\bar{f} + \bar{k}g = 0. \tag{D.125a}$$

Immediately after Eq. (D.116a) we identified Z^0, Z^1, Z^2, Z^3 with h, f, g and k. Now we obtain:

$$\begin{array}{cccc} f \rightleftarrows Z^0 & g \rightleftarrows Z^1 & h \rightleftarrows Z^2 & k \rightleftarrows Z^3 \\ \bar{h} \rightleftarrows \bar{Z}^2 & \bar{k} \rightleftarrows \bar{Z}^3 & \bar{f} \rightleftarrows \bar{Z}^0 & \bar{g} \rightleftarrows \bar{Z}^1 \end{array} \tag{D.126}$$

allowing us to rewrite Eq. (D.125a) in the form used in twistor theory [D.74, D.81]

$$\Sigma(Z, \bar{Z}) = Z^0\bar{Z}^2 + Z^1\bar{Z}^3 + Z^2\bar{Z}^0 + Z^3\bar{Z}^1 = 0. \tag{D.125b}$$

In twistor theory to arrive at this result requires much longer explanations. It is remarkable that using Eq. (D.125a) allows us to by-pass these arguments. The result, Eq. (D.125b), is obtained at once when variables $\bar{h}, \bar{f}, \bar{g}$ and $\bar{k}$ are identified with the complex conjugates of h, f, g and k.

Since the significance of Eq. (D.125b) for twistor formalism is well documented, beginning with Penrose's seminal paper, [D.74], we shall concentrate attention of our readers on its possible uses for quantum computers. For this purpose it is instructive to rewrite the quadratic form $\Sigma(Z, \bar{Z})$ as follows:

$$\Sigma(Z, \bar{Z}) = \Sigma_{\alpha\beta} Z^{\alpha} \bar{Z}^{\beta}, \tag{D.125c}$$

where the 4×4 matrix $\Sigma_{\alpha\beta}$ is given by

$$\mathbf{\Sigma} = \begin{pmatrix} 0 & I_2 \\ I_2 & 0 \end{pmatrix}. \tag{D.125d}$$

Diagonalization of this matrix leads to

$$\Sigma(Z, \bar{Z}) = |Z^0|^2 + |Z^1|^2 - |Z^2|^2 - |Z^3|^2. \tag{D.125e}$$

Now Eq. (D.125b) can be rewritten as

$$|Z^0|^2 + |Z^1|^2 = |Z^2|^2 + |Z^3|^2 = \text{const.} \tag{D.127}$$

The constant can be selected with help of the normalization condition Eq. (D.24). So, it is 1/2. Evidently, both spinors are living on S^3 and, in fact, using the Hopf fibration we can make them to live in the same Bloch sphere S^2 in accord with results of [D.5]. The incidence relation, Eq. (D.109b), can now be rewritten as

$$\begin{pmatrix} Z^0 \\ Z^1 \end{pmatrix} = \begin{pmatrix} \tilde{z} & w \\ \tilde{w} & z \end{pmatrix} \begin{pmatrix} Z^2 \\ Z^3 \end{pmatrix}$$

so that, in fact, we are having a map $\{Z^0, Z^1, Z^2, Z^3\} \rightarrow \{\tilde{z}Z^2 + wZ^3, \tilde{w}Z^2 + zZ^3, Z^2, Z^3\}$. This result is in accord with general theory of Grassmannians discussed in Sec. D.2.7. Evidently, the map just described can be rewritten in terms of the eigenvalue Eq. (D.53) in which the diagonal 4×4 matrix has two identical entries.

D.2.17 *Nilpotent and Coadjoint Orbits, Lagrangian Fibrations and Lagrangian Grassmannians*

In Sec. D.2.14 (after Eq. (D.103d)) we noticed that methods of this section cannot be used to design an orbit made out of 24 elements needed for description of the permutohedron Π_3, Fig. 6. Because of its educational value, we begin this section by discussing this topic.

The story begins with the definition and Theorem D.2.6.2. Next, we should apply these general results to the specific Eqs. (D.104a), (D.104b) for SU(3) and, their SU(4) analogs, Eqs. (D.102a), (D.102b). Next, recall the Bruhat decomposition, Sec. D.2.8, and the subgroup N_π which is an orbit $\mathcal{O}_\pi = N \cdot (GL_n\mathbf{C}/B) \simeq N_\pi$. Each N_π is a cell $C_\pi \simeq N_\pi \simeq \mathbf{C}^{l_\pi}$. An example of graphical calculation of l_π is given in Fig. 7. We stated in Sec. D.2.8 that there are $n!$ cells of the type N_π. In [D.29], p. 67, it is shown that $|G/B| = \sum_{\pi \in S_n} |N_\pi|$ and, since S_n is the permutation group, we obtain: $|G/B| = n!$ This is the number we are looking for. In Sec. D.2.9, we linked this number with the size of the orbit of the Weyl group W. So, what remains to do is to demonstrate that the orbit of the Weyl group W goes over all vertices of Π_3, Fig. 6.

For this purpose, we use Definition D.2.12.4 and the associated with it Eqs. (D.88)–(D.90). To make progress, following [D.85], we notice that the $\mathfrak{su}(2)$ Lie algebra, Eq. (D.70b), can be realized by the matrices

$$H = \begin{pmatrix} 1 & 0 \\ 0 & -1 \end{pmatrix}, \quad X = \begin{pmatrix} 0 & 1 \\ 0 & 0 \end{pmatrix}, \quad Y = \begin{pmatrix} 0 & 0 \\ 1 & 0 \end{pmatrix}$$

called "elements" in [D.85]. In particular, H is called "neutral" while X and Y are respectively called "nilpositive" and "nilnegative". From here the following observation can be made.

Let $\mathfrak{g}$ be a semisimple Lie algebra. If $\mathfrak{b}$ is the Borel subalgebra (e.g. see Definition D.2.6.1), then $\mathfrak{b} = \mathfrak{h} \oplus \mathfrak{n}$ where $\mathfrak{h}$ is the Cartan subalgebra of $\mathfrak{g}$ and $\mathfrak{n} = [\mathfrak{b}, \mathfrak{b}]$ is the nilradical of $\mathfrak{g}$. It consists of nilpotent elements in $\mathfrak{b}$.

Since $\mathfrak{su}(2)$ Lie algebra can be used for designing of $\mathfrak{su}(n)$ (and $\mathfrak{sl}_n(\mathbf{C})$), the Borel subalgebra $\mathfrak{b}$ of $\mathfrak{sl}_n(\mathbf{C})$ should be made of the upper triangular matrices and $\mathfrak{n} \subset \mathfrak{b}$ of strictly upper triangular matrices. It can be proven that $\mathfrak{b}$ is solvable and $\mathfrak{n}$ is nilpotent subalgebras in accord with definitions of Appendix B, Sec. B.1. Being armed with these results we can go back to Eqs. (D.88)–(D.90) in order to state that the orbit of the Weyl group is nilpotent. For $\mathfrak{sl}_n(\mathbf{C})$ the following theorem was proven.

Theorem D.2.17.1. *Nilpotent orbits in $sl_n(\mathbf{C})$ are in one-to-one correspondence with the set $p(n)$ of partitions of n.*

Remark D.2.17.1. Our readers should be aware of the fact that the theory of nilpotent orbits is extremely involved. To those who are interested we recommend [D.85] and [D.86] as the point of departure. In this book for obvious reasons we shall not go into intricacies of this branch of science.

Remark D.2.17.2. In [D.87] Swann proved the following:

Let G be a compact semi-simple Lie group. A nilpotent orbit $\mathcal{O}$ in the complexified Lie algebra $g^{\mathbf{C}}$ is a quaternionic Kähler manifold whose twistor space is a homogeneous contact *manifold* $\mathbf{P}(\mathcal{O})$.

In the previous section we found S^4 as quaternionic Kähler manifold while $\mathbf{CP}^3$ as homogeneous twistor space. It is a Hopf-type fiber bundle with S^4 being a base and S^2 being a fiber. In [D.87] and in subsequent works Swann developed connections between the twistor and Kähler (and hyper-Kähler) spaces. For an up to date list of references on this topic, consult [D.75] and references therein. Connections of these results with instantons and monopoles can be found (in addition to those in [D.87]) in papers by Kronheimer [D.88] and Kovalyev [D.89]. Our readers might also want to consult the citation lists for these references.

To continue our efforts at demonstrating that the orbit of the Weyl group W goes over all vertices of Π_3 we need to connect the nilpotent orbits with the coadjoint ones. In such a case we are going to return to the mechanical analogies and, in particular, to the convex polytopes.[p] As in the simplest case discussed in Sec. D.2.3 and its generalization (e.g. the balloon sculptures of Sec. D.2.15) the vertices of respective polytopes are the fixed points of the torus action. The vertices are connected via action of the Weyl group W.

To make all this slightly more mathematical, we shall follow initially [D.24]. We begin with the Example of Sec. D.2.7. We would like to slightly modify the result expressed in Eq. (D.54a). Instead of using $U(n)$ acting by conjugation on space $\mathcal{H}$ of Hermitian matrices we shall also look at the action of $U(n)$ on its own Lie algebra $\mathfrak{u}(n)$ made of skew-Hermitian matrices, that is of matrices X obeying $X + \bar{X}^T = 0$, implying that $\mathfrak{u}(n) = i\mathcal{H}$. As in Example (b) of Sec. D.2.12, we introduce a symmetric bilinear form defined on $\mathcal{H}$

$$(X, Y) \to \mathrm{tr}(^T\bar{X}Y) = \mathrm{tr}(XY) \quad (\text{since}^T\bar{X} = X \text{ in } \mathcal{H}).$$

It is playing the same role here as the ordinary scalar product for vectors. Let now $ih \in i\mathcal{H}$ be some matrix from $\mathfrak{u}(n)$. Then, the dual space $\mathfrak{u}(n)^* \in \mathcal{H}$ can be identified with $i\mathfrak{u}(n)$. Introduce now a skew-symmetric bilinear form $\omega_h(X, Y)$, $X, Y \in \mathfrak{u}(n)$, via

$$\omega_h(X, Y) = \mathrm{tr}(i[X, Y]h). \tag{D.128a}$$

[p] Although the coadjoint orbits were introduced and somewhat discussed in Chap. 8, in this section we shall provide much more details.

Consider the orbit map $f_h : U(n) \to \mathcal{H}$ and its derivative (that is the tangent space) $\mathbf{T}f_h : \mathfrak{u}(n) \to \mathbf{T}_h\mathcal{H} \subset \mathcal{H}$. $\mathbf{T}_h\mathcal{H}$ is generated either by $[X, h]$ or by $[Y, h]$. In such a case the form $\omega_h(X, Y)$ can also be defined on the tangent space $\mathbf{T}_h\mathcal{H}$ via

$$\omega_h(X,Y) \to \tilde{\omega}_h([X,h],[Y,h]) \equiv \mathrm{tr}(i[X,Y]h), \quad X, Y \in \mathfrak{u}(n). \quad \text{(D.128b)}$$

Thus, the skew-symmetric bilinear form $\tilde{\omega}_h$ defined on any orbit of $\mathcal{H}$ is a candidate for the 2-form used in symplectic mechanics, provided that such a form is closed (read Sec. D.2.1). It indeed happens when

$$d\tilde{\omega}_h([X,h],[Y,h],[Z,h]) = 0 \quad \forall\, X, Y, Z \in \mathfrak{u}(n). \quad \text{(D.128c)}$$

To prove this result and to uncover its physical content requires us to make some detour into symplectic geometry/mechanics. We recommend our readers to consult Chap. 5 and Sec. D.2.1 for notations and sample calculations.

Following [D.17], we begin by introducing

Definition D.2.17.1. A vector field v is called *symplectic* if $\mathcal{L}_v\omega = 0$, where ω is the 2-form defined prior to Eq. (D.28a).

Example. (a) The skew-gradient of an infinitely differentiable function[q] $f^{(v)}$ is a symplectic vector field associated with the 1-form $-df^{(v)} = i_v\omega$ (e.g. look at Eq. (D.36)). Explicitly,[r]

$$v = (s - grad)f^{(v)} = \frac{\partial f^{(v)}}{\partial p_i}\frac{\partial}{\partial q_i} - \frac{\partial f^{(v)}}{\partial q_i}\frac{\partial}{\partial p_i}. \quad \text{(D.129a)}$$

By direct computation, we indeed obtain: $i_v\omega = -df^{(v)}$. Next, let us recall (see Eq. (5.5.9)) that $\mathcal{L}_v = d \circ i_v + i_v \circ d$. From here $\mathcal{L}_v\omega = -ddf^{(v)} = 0$ as required.

(b) Consider now another vector field w given by

$$w = (s - grad)f^{(w)} = \frac{\partial f^{(w)}}{\partial p_i}\frac{\partial}{\partial q_i} - \frac{\partial f^{(w)}}{\partial q_i}\frac{\partial}{\partial p_i}. \quad \text{(D.129b)}$$

Using it, we calculate $i_w \circ i_v\omega = i_w(-df^{(v)}) = i_w(-\frac{\partial f^{(v)}}{\partial p_i}dp_i - \frac{\partial f^{(v)}}{\partial q_i}dq_i) = \frac{\partial f^{(v)}}{\partial p_i}\frac{\partial f^{(w)}}{\partial q_i} - \frac{\partial f^{(v)}}{\partial q_i}\frac{\partial f^{(w)}}{\partial p_i} = \{f^{(v)}, f^{(w)}\}$ in which we instantly recognize the Poisson bracket! Thus,

$$\{f^{(v)}, f^{(w)}\} = i_w \circ i_v\omega \equiv \omega(v,w) \equiv vf^{(w)} \equiv -wf^{(v)}. \quad \text{(D.130a)}$$

[q] We write $f^{(v)}$ just to keep track of the fact that function f is associated with the field v.
[r] As usual, the summation over the repeated indices is assumed.

Consider now yet another symplectic vector field g associated with the function $f^{(g)}$. Using Eq. (D.130a) we obtain,

$$g\{f^{(v)}, f^{(w)}\} = g\omega(v,w) = \{f^{(g)}, \{f^{(v)}, f^{(w)}\}\}. \tag{D.130b}$$

At the same time, we can also look at

$$\begin{aligned}\omega(v,w) \to \omega([v,g],w) &= [v,g]f^{(w)} = vgf^{(w)} - gvf^{(w)} \\ &= v\{f^{(g)}, f^{(w)}\} - g\{f^{(v)}, f^{(w)}\} \\ &= \{f^{(v)}, \{f^{(g)}, f^{(w)}\}\} - \{f^{(g)}, \{f^{(v)}, f^{(w)}\}\}.\end{aligned} \tag{D.131}$$

Now we are ready to go back to Eq. (D.128c). By introducing notations $[X,h] = \tilde{X}, [Y,h] = \tilde{Y}$ and $[Z,h] = \tilde{Z}$ we obtain,

$$\begin{aligned}3d\tilde{\omega}_h([X,h],[Y,h],[Z,h]) = \tilde{X}\tilde{\omega}_h(\tilde{Y},\tilde{Z}) - \tilde{Y}\tilde{\omega}_h(\tilde{X},\tilde{Z}) + \tilde{Z}\tilde{\omega}_h(\tilde{X},\tilde{Y}) \\ + \tilde{\omega}_h([\tilde{X},\tilde{Y}],\tilde{Z}) + \tilde{\omega}_h([\tilde{Y},\tilde{Z}],\tilde{X}) + \tilde{\omega}_h([\tilde{Z},\tilde{X}],\tilde{Y})\end{aligned}$$

to which we have to apply Eqs. (D.130) and (D.131). After some calculation we obtain,

$$d\tilde{\omega}_h([X,h],[Y,h],[Z,h]) = 0,$$

which is the result of the use of Jacobi identity valid for Poisson brackets.

Exercise D.2.17.1. Suppose you are given two symplectic vector fields $v \equiv v(f^{(v)})$ and $w = w(f^{(w)})$. Using the computational tools developed in Examples (a) and (b) demonstrate that

$$\text{(a) } [v,w] = g(\{f^{(v)}, f^{(w)}\}), \quad \text{(b) } \{f^{(v)}, f^{(w)}\} = f^{([v,w])}.$$

Remark D.2.17.3. This exercise displays a remarkable connection between the classical and quantum mechanics. This connection is typically exploited with help of the *Kirillov, Kostant, Souriau* (KKS) *symplectic* 2-*form* which can be written in a number of equivalent ways. One of them is given in Eq. (D.128b). Examples of other formulations are given in [D.17].

Remark D.2.17.4. Part (b) of the above exercise can also be interpreted as the KKS form. Indeed, with the help of Eq. (D.130a), we obtain

$$\{f^{(v)}, f^{(w)}\} = \omega_f(v,w) = f^{([v,w])} \equiv \langle f, [v,w]\rangle \equiv f([v,w]). \tag{D.132}$$

This result is essentially the same as that in Eq. (D.128b). As the content of Sec. 8.2.3 of Chap. 8 indicates, Eq. (D.132) can be rearranged further. Following [D.53] we reintroduce the adjoint action Ad_g of the Lie group G on its Lie algebra $\mathfrak{g}$

$$Ad_g X = gXg^{-1}, g \in G,\ X \in \mathfrak{g}$$

and, accordingly,

$$ad_X Y = [X, Y].$$

The space G^* dual to G is always defined via properly defined scalar product. Specifically, let $\xi \in \mathfrak{g}^*$, $X \in \mathfrak{g}$, then the coadjoint analog of $Ad_g X$ is defined with the help of the scalar product as follows.

$$\langle (Ad_g^* \xi), X \rangle = \langle \xi, (Ad_{g^{-1}} X) \rangle. \quad (i)$$

Accordingly,

$$\langle (ad_X^* \xi), Y \rangle = \langle \xi, [X, Y] \rangle = \langle \xi, ad_X Y \rangle. \quad (ii).$$

By looking at Eq. (D.132) we recognize the correspondence $\xi([X, Y]) \rightleftarrows f([v, w])$ implying that this result can (and should) be further rewritten as $f([v, w]) = \langle (ad_v^* f), w \rangle$. But what is the physics behind such an identification?

If $\xi \in \mathfrak{g}^*$, then the coadjoint orbit $\mathcal{O}(\xi)$ is defined by

$$\mathcal{O}(\xi) = \{(Ad_g^* \xi), g \in G\}.$$

Using (i) and the definition of $\mathcal{O}(\xi)$ we obtain,

$$\langle (Ad^*_{\exp(-tX)} \xi), Y \rangle = \langle \xi, (Ad_{\exp(tX)} Y) \rangle.$$

Since the velocity $\mathfrak{v}(\xi^{(X)})$ vector can be defined by

$$\mathfrak{v}(\xi^{(X)}) = \left.\frac{d}{dt}\right|_{t=0} (Ad^*_{\exp tX} \xi), \quad X \in \mathfrak{g}, \tag{D.133a}$$

we can write

$$\left.\frac{d}{dt}\right|_{t=0} \langle (Ad^*_{\exp(-tX)} \xi), Y \rangle = \left.\frac{d}{dt}\right|_{t=0} \langle \xi, (Ad_{\exp(tX)} Y) \rangle$$
$$= \langle \xi, [X, Y] \rangle \overset{\text{(ii)}}{=} \langle (ad_X^* \xi), Y \rangle$$

implying

$$-\mathfrak{v}(\xi^{(X)}) = \left.\frac{d}{dt}\right|_{t=0} (Ad^*_{\exp tX} \xi) = (ad_X^* \xi). \tag{D.133b}$$

The question arises: is there any relationship between $\mathfrak{v}$ and v defined in Eq. (D.129a)? It does indeed exists! First, we take into account that, in view of previous notations, we should have the following identifications: $\langle \xi, [X, Y] \rangle = \xi([X, Y]) = \omega_\xi(X, Y) = \langle -\mathfrak{v}(\xi^{(X)}), Y \rangle$. Consider now a special case: $f^{(v)} = H$, $f^{(w)} = x(x = \{\mathbf{q}, \mathbf{p}\})$. Then, the equations of motion are: $\dot{x} = \{H, x\}$. Thus, we can make identifications: $X \rightleftarrows X_H$, $Y \rightleftarrows X_x, \xi([X,Y]) \rightleftarrows X_H(x) = \dot{x}$. Furthermore, if $-\mathfrak{v}(\xi^{(X)}) \rightleftarrows X_H$, then the equations of motion on a coadjoint orbit acquire the following form:

$$\dot{x} = -(ad^*_{X_H})x, \quad x \in \mathfrak{g}^*. \tag{D.134a}$$

Accordingly, the adjoint orbit $\mathcal{O}(X)$ is given by $Ad_g X = gXg^{-1}, g \in G, X \in \mathfrak{g}$ so that Heisenberg's equations of motion are

$$\dot{x} = -ad_{X_H}x, \quad x \in \mathfrak{g}. \tag{D.134b}$$

It is helpful to look at these results from the standpoint of the Maurer–Cartan equations [D.17, D.90].

We begin with the standard result for the exterior derivative of 1-form

$$d\theta(\xi, \eta) = \xi\theta(\eta) - \eta\theta(\xi) - \theta([\xi, \eta]). \tag{D.135a}$$

Suppose that the vector fields ξ, η are horizontal. That is such a form lives in the fiber bundle (e.g. read Sec. 6.5) to be described. Then, we obtain

$$d\theta(\xi, \eta) = -\theta([\xi, \eta]) = -i_\xi(-df^{(\eta)}). \tag{D.135b}$$

This result will be rederived now with help of the Maurer–Cartan formalism. This will provide an independent evidence that ξ, η are indeed horizontal vector fields. For this purpose we introduce the left-invariant 1-form Ω.

Remark D.2.17.5. To understand the left invariance is easy when dealing with matrix groups. Indeed, let

$$\Omega = B^{-1}\frac{\partial B}{\partial y^i}dy^i,$$

where B is an $n \times n$ matrix. If A is some fixed $n \times n$ matrix and $y^i = y^i$ $(x^1, \ldots, x^n)$ so that the coordinate transformation $y^i \to x^i$ is induced by the *left* multiplication $B(y^1, \ldots, y^n) = AB(x^1, \ldots, x^n)$, then we obtain

$$\begin{aligned} B^{-1}\frac{\partial B}{\partial y^i}dy^i &= B^{-1}(x)A^{-1}\frac{\partial}{\partial y^i}AB(x)dy^i \\ &= B^{-1}(x)\frac{\partial B}{\partial x^k}\frac{\partial x^k}{dy^j}dy^j = B^{-1}\frac{\partial B}{\partial x^i}dx^i. \end{aligned}$$

Now we can rewrite Ω as follows: $dB = B\Omega$. From this result we obtain zero curvature equation (analogous to Eq. (5.6.4))

$$ddB = 0 = B(d\Omega + \Omega \wedge \Omega) \to d\Omega + \Omega \wedge \Omega = 0. \tag{D.136a}$$

Remark D.2.17.6. These are the *Maurer–Cartan equations* for the matrix group. One obtained equations admit an interpretation in terms of gauge fields.

Suppose that we are dealing with the non-normalized matrix (or vector fields) Lie algebra $\mathfrak{g}$ so that by definition

$$[X_i, X_j] = c_{ij}^k X_k, \quad X_i \in \mathfrak{g}. \tag{D.136b}$$

On the other hand, suppose that the 1-form Ω is given by

$$\Omega = X_i \omega^i$$

where ω^i are left-invariant 1-forms. Then

$$d\Omega = X_i d\omega^i$$

while

$$\begin{aligned}\Omega \wedge \Omega &= X_i \omega^i \wedge X_j \omega^j = X_i X_j (\omega^i \wedge \omega^j) \\ &= \frac{1}{2}[X_i, X_j](\omega^i \wedge \omega^j) = \frac{1}{2} c_{ij}^k X_k (\omega^i \wedge \omega^j).\end{aligned}$$

Using these results in Eq. (D.136a) we arrive at the Maurer–Cartan equations

$$d\omega^l = -\frac{1}{2} c_{ij}^l \omega^i \wedge \omega^j = -c_{ij}^l \omega^i \wedge \omega^j, \quad i < j, \tag{D.136c}$$

relative to the basis $X_i \in \mathfrak{g}$. Comparison between Eqs. (D.135a), (D.136a) and (D.136c) provides confirmation of our fiber-bundle assumption and in addition the explicit form for $\theta([\xi, \eta])$. Furthermore, in obtaining this result we used essentially the defining Eqs. (D.136b) for the Lie algebra $\mathfrak{g}$. The classical mechanics analog of these equations is

$$\{x_i, x_j\} = c_{ij}^k x_k, \quad x \in \mathfrak{g}^*. \tag{D.136d}$$

With help of the results of Exercise D.2.17.1, this result can be rewritten as follows:

$$\{x_i, x_j\} = x([X_i, X_j]) = x(c_{ij}^k X_k) = c_{ij}^k x(X_k) = c_{ij}^k x_k.$$

This result was obtained based on the following assumptions:

(a) $\forall x \in \mathfrak{g}^*$ we can write an expansion $x = x_i X^i$;
(b) There is a scalar product $\langle , \rangle$ such that $\langle X^i, X_j \rangle = \delta_j^i$.

For instance X^i can be a differential 1-form in $\mathfrak{g}^*$ and X_j a vector field in $\mathfrak{g}$.

Remark D.2.17.7. It is clear that upon such an identification it becomes possible to replace the 1-form $\Omega = X_i\omega^i$ by $\Omega = x_i\omega^i$. These arguments provide independent justification of Eq. (D.134a).

Remark D.2.17.8. The obtained results belong to the basic toolkit used in the book, [D.91], for development of systematic Lie group treatment of optimal control problems. In fact, Appendices A and B provide practical physical examples of general formalism developed in [D.91].

To finish our detour into symplectic geometry we have to describe the fiber bundle associated with transition from Eq. (D.135a) to Eq. (D.135b). This description will bring us much closer to the completion of our main task — the description of Weyl group actions on coadjoint orbits.

We begin with Eq. (D.132) (or Eq. (D.128b)) in which we put $[v, w] = 0$ causing $\omega_f(v, w) = 0$. What does this result mean physically? To answer this question we go back to our benchmark example discussed in Sec. D.2.3 (e.g. see Fig. 3). The points where $\omega_f(v, w) = 0$ are the critical points in the sense of the Morse theory (Sec. D.2.13). They correspond to the vertices of corresponding polyhedra as discussed in Sec. D.2.15. Let us now investigate what such points mean mechanically. We start by considering the symplectic 2-form ω, i.e.

$$\omega = \sum_i dp_i \wedge dq_i,$$

originally defined on symplectic manifold M^{2n} but restricted to "live" in the submanifold M^n defined by the set of equations (e.g. read Sec. 5.2).

$$p_i = \frac{\partial S(\{q\})}{\partial q_i}, \quad 1 \leq i \leq n. \tag{D.137}$$

Only under such restriction can one talk, for instance, about the Bohr–Sommerfel'd-type quantization. At the same time, we also obtain

$$\omega = \sum_i \left(d\frac{\partial S}{\partial q_i}\right) \wedge dq_i = \sum_{i,j} \frac{\partial^2 S}{\partial q_i \partial q_j} dq_i \wedge dq_j = 0.$$

Definition D.2.17.2. For any smooth function $S(q)$ the graph $p = p(q)$ defined by Eq. (D.137) is called *Lagrangian manifold* L in the symplectic space M^{2n}. From here, it follows that the dimension of such a manifold is always half of that for symplectic manifold (compare this statement with that in Theorem D.2.5.1).

Based on this definition, it is possible now to design a fiber bundle known in mathematics literature as the *cotangent bundle* [D.92] T^*M while in physics literature [235] as the *phase space*. The base space is the *configuration space* $\{q_1, \ldots, q_n\} \in Q(q)$, the fibres are the Lagrangian manifolds — the momenta $\{p_1, \ldots, p_n\} \in P(p)$. The projection π is $\pi(p) = q$. Since the coadjoint orbit was already defined as $\mathcal{O}(\xi) = \{(Ad_g^* \xi), g \in G\}$ we can redefine it now (with the help of Definition D.2.5.3) as $G/\ker \omega_f$ where $\ker \omega_f = \omega_f(v, w) = 0$.

Definition D.2.17.3. ([D.92]) The set of all Lagrangian subspaces of a symplectic space of dimension $2n$ is a smooth manifold called *Lagrangian Grassmann manifold* $LG(n, 2n)$ whose dimension is $\dim LG(n, 2n) = n(n+1)/2$. There could be both real and complex Lagrangian Grassmann manifolds.

The above results can be understood in simple physical terms. Indeed, with every point x of the Lagrangian manifold L we can associate a tangent plane T_xL. By translating this point to the origin of the respective coordinate system we are also translating the plane T_xL. Thus, for each point $x \in L$ we are creating a Grassmannian made of these tangent planes. Therefore, by design there is one-to-one correspondence between the Lagrangian manifolds and the Lagrangian Grassmannians, [D.53], p. 60.

Definition D.2.17.4. A *Lagrangian fibration* is a smooth locally trivial fibration of a symplectic manifold all fibers of which are Lagrangian manifolds L. In particular, a cotangent bundle T^*M is a Lagrangian fibration.

Remark D.2.17.9. These concepts are being used in the theory of Maslov indices, theory of Lagrangian and Legendrian singularities, Lagrangian and Legendrian cobordisms, etc. all discussed in [D.92]. Since a very nice discussion of almost all of these topics done at the level of presentation given in this book can be found in [D.92–D.94], we advise our readers to read these references. Some of these concepts are also discussed in Chap. 9.

Remark D.2.17.10. The Lagrangian fibrations (fiber bundles) are only locally trivial (as most of other fiber bundles). Globally, they could be nontrivial. The degree of nontriviality is determined by the curvature 2-form which, up to a constant, coincides with that given by Eq. (D.136c), e.g. read [D.17], p. 13.

Remark D.2.17.11. In view of Eq. (D.132) the existence of Lagrangian manifolds is linked with commuting integrals of motion, i.e. $\{f^{(v)}, f^{(w)}\} = \omega_f(v, w) = 0$. From the definition of a Lagrangian manifold it follows that there could be no more than n integrals of motion if the dimensionality of symplectic manifold is $2n$. If there is exactly n integrals of motion the dynamical system becomes completely integrable. In which case the motion becomes restricted so that every component of the Lagrangian fibration becomes a torus $T \simeq \mathbf{R}^n/\Lambda$ (e.g. read Secs. D.2.5 and D.2.15) with Λ being some lattice. In the case of incomplete integrability, it is possible to transfer these ideas to lower dimensions, [D.92], p. 38.

This result can be easily understood at the level of the Landau–Lifshitz course, [235]. In the case of complete integrability the 2-form $dp \wedge dq$ is transformed into the action-angle 2-form $dI \wedge d\varphi$ so that the phase space is the algebraic torus (Definition D.2.5.1). Whenever the action variables are pre-assigned (e.g. on surfaces of constant energy which are the Lagrangian fibers), the motion indeed takes place on the torus T. Incidentally, on surfaces of constant energy one can use the Maupertuis principle (Chap. 6, Sec. 6.2) so that the dynamics of exactly integrable systems can be rewritten in terms of a geodesic motion on Riemannian manifolds [178]. We shall further discuss this topic both in the next subsection and in Chap. 8.

Being armed with all these results we are ready for our final move. The key Eq. (D.132) provides a crucial link between the exact integrability of a classical dynamical system — the set of Poisson commuting integrals of motion — with the Cartan subalgebra $\mathbf{t}$ defined in Eq. (D.77). Clearly, $\mathbf{t}$ is related to the maximal torus defined in Sec. D.2.5. Alternatively, we can look at Eq. (D.53). Taking into account results of Secs. D.2.8, D.2.9, D.2.12 and D.2.13, we are ready now to apply general rules to specific situations, e.g. to the Lie algebras $\mathfrak{su}(3)$ and $\mathfrak{su}(4)$. In both cases we can use Eq. (D.69). For $\mathfrak{su}(3)$, the Cartan subalgebra is made of matrices

$$\mathrm{M}_1(\mathfrak{su}(3)) = \begin{pmatrix} \lambda_1 & & \\ & \lambda_2 & \\ & & \lambda_3 \end{pmatrix}.$$

We shall assume that $\lambda_1 \neq \lambda_2 \neq \lambda_3$. $\mathrm{M}_1(\mathfrak{su}(3))$ is just one (reference) matrix of the Cartan subalgebra. All others are obtained by permutations. In the present case we are having $3! = 6$ permutations. These permutations coincide with those provided by the Weyl group for $\mathfrak{su}(3)$, [D.17], p. 303. Thus, the associated balloon sculpture has exactly six fixed points and its deformation retract is depicted in both Figs. 5 and 11. To transfer this formalism to $\mathfrak{su}(4)$ is straightforward. Now we are having the reference matrix $\mathrm{M}_1(\mathfrak{su}(4))$ with four entries $\lambda_1 \neq \lambda_2 \neq \lambda_3 \neq \lambda_4$ and, accordingly, $4! = 24$ Weyl's group permutations, leading to permutohedron Π_3 depicted in Fig. 6.

These results are seemingly irrelevant to the problem of quantum entanglements of two qubits. This happens to be not the case as we would like to explain now. According to Eq. (D.110c) there are $4!/2!2! = 6$ ways to select two eigenvalues out of four. Thus, in this case the Cartan subalgebra is made of six elements. The octahedron, Fig. 18, has six vertices. These are fixed points representing Lagrangian fibrations. How one should look at them? Since $Gr_i(\mathbf{C}^n)$ is the variety whose points are i-dimensional subspaces of $\mathbf{C}^n$, the n-torus is acting on $Gr_i(\mathbf{C}^n)$ through its action on $\mathbf{C}^n$, that is $z(x_1, \ldots, x_n) = (zx_1, \ldots, zx_n)$. As in Sec. D.2.7, we use the notations $I_n = \{1, 2, \ldots, n\}$ and $B(I_n)$. Then let $J_p = \{i_1, \ldots, i_p\} \in B(I_n)$ and denote as $\mathbf{C}^J \subset \mathbf{C}^n$ as subset made of vectors $\{\mathbf{e}_{i_1}, \ldots, \mathbf{e}_{i_p}\}, 0 \leq p \leq n$. Following Sec. D.2.9 with $J_p = \{i_1, \ldots, i_p\}$ we can associate all kinds of exterior products of the type $\mathbf{e}_{i_1} \wedge \cdots \wedge \mathbf{e}_{i_p}$ so that there are exactly $n!/p!(n-p)!$ of them. In our case, for $Gr_2(\mathbf{C}^4)$ there are six of them. There are six products of the type $\mathbf{e}_i \wedge \mathbf{e}_j$ representing six 2-planes (Sec. D.2.10). So that six vertices of the octahedron are representing six 2-planes [D.73]. They are fixed points since they remain invariant under the torus action. Equation (D.118) describes a hyperplane living in $\mathbf{CP}^2$. From Fig. 4(b) we know that the deformation retract of $\mathbf{CP}^2$ is a triangle. Each point inside of this triangle is a two-torus [D.14, D.73]. This result is easy to understand by generalizing the results of Exercise D.2.3.1. If $\mathbf{CP}^n$ is being realized as in Eq. (D.26b), then the Darboux coordinates are: $(z_1, \ldots, z_{n+1}) \rightarrow (\sqrt{t_1}e^{i\varphi_1}, \ldots, \sqrt{t_n}e^{i\varphi_n}, \sqrt{1 - \sum_j t_j})$. From here it follows that for $n = 2$ we are indeed getting a two-torus. Figure 20 depicts such a (Lagrangian) torus along with the limiting situations (e.g. see Fig. 3). Can we make physical

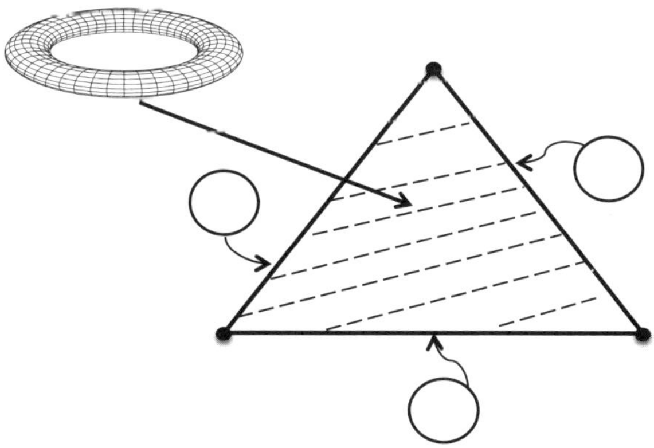

Fig. 20. Lagrangian tori living inside of $\mathbf{CP}^2$.

sense out of such tori? Fortunately, we can! This is discussed in the next subsection.

D.2.18 *Coupled Josephson Junctions and Linkages, Torus Knots and Entanglements, Legendrian Manifolds and Optimal Control: From Snakes and Airplanes to Quantum Hall Effect*

In Chap. 8 we obtained Eq. (8.3.3.9) describing the dynamics of a single Josephson's junction while in Appendix B we re-obtained this result, e.g. see Eq. (B.26), in the limit $J = 0$ using methods of sub-Riemannian geometry (and/or optimal control). Equation (B.26) describes the dynamics of both Euler's elastica (hence, the rolling ball and, therefore, the 2-level quantum system describing evolution of the qubit) and of mathematical pendulum. In Chap. 8 we discussed how Josephson's junctions can be used as qubits. Thus, Eq. (B.26) could be also understood in terms of dynamics of a qubit. This means that when two qubits are uncoupled, the dynamics of each of them can be described in terms of Eq. (B.26). How to couple these penduli so that their description in terms of qubit dynamics will have features characteristic of entanglement which we just have discussed? That is to say, can such a mechanical device describe the holly grail of quantum mechanics — the entanglement? We would like to provide the evidence that this is indeed the case.

For this purpose, following Arnol'd, [136], p. 248, we shall discuss the dynamics of a double pendulum. Here we find that the torus T^2 is the configurational space (that is the Lagrangian manifold) of planar double pendulum. Suppose we endow T^2 with some Riemannian metric. Then, among all closed curves on T^2 making m rotations around the parallel and n around the meridian, there exists a curve of shortest length (a geodesic). Without proof, Arnol'd provides the following:

Theorem D.2.18.1. *For any integers m and n there is a periodic motion of the double pendulum under which one segment (say, the first link*[s]*) makes m rotations while the other segment (the second link) makes n rotations. Furthermore, such periodic motions exist for any sufficiently large values of the total energy E (E must be larger than the potential energy at the highest position).*

[s]It is our clarification, since the translation from Russian of this part was done poorly.

Remark D.2.18.1. Closed geodesics on T^2 are torus knots. E.g. recall Theorem 2.5.1 of Chap. 2! At the same time T^2 could live inside of $\mathbf{CP}^2$ if we can construct a map $\mathcal{F}$: $T^2 \to \mathbf{CP}^2$. Such a map does exist. This can already be seen from the results of Sec. D.2.1 since any symplectic dynamical system when rewritten in terms of complex variables describes evolution on some Kähler manifold. Complex projective space *is* an example of Kähler manifold. A few additional details will be added below.

Remark D.2.18.2. Although in Arnold's book we find that *such periodic motions exist for any sufficiently large values of the total energy* E (E *must be larger than the potential energy at the highest position*), this happens to be a bit intricate as explained in the paper by Kozlov, [D.95], in which periodic solutions for planar multipendulum problem were discussed.

Following this reference, it is instructive to consider some aspects of multipendulum dynamics. We begin with the problem familiar to every student majoring in physics about the Lagrangian dynamics of a planar double pendulum, e.g. as desctibed both in [D.95] and in "Mechanics" by Landau–Lifshitz, [235]. Without loss of generality we shall be treating the case when both links are of the same length l. For such a case the Lagrangian reads:

$$\begin{aligned}\mathcal{L} = \frac{l^2}{2}[(m_1+m_2)\dot{\varphi}_1^2 + m_2\dot{\varphi}_2^2 + 2m_2\dot{\varphi}_1\dot{\varphi}_1\cos(\varphi_1-\varphi_2)] \\ + (m_1+m_2)gl\cos\varphi_1 + m_2 gl\cos\varphi_2. \end{aligned} \tag{D.138}$$

In this as well as in the multilinkage case the kinetic energy $T(\varphi,\dot{\varphi})$ can be rewritten in the following universal form

$$T(\varphi,\dot{\varphi}) = \sum a_{ij}(\varphi)\dot{\varphi}_i\dot{\varphi}_i$$

in which $a_{ij}(\boldsymbol{\varphi}) = a_{ji}(\boldsymbol{\varphi}) \geq 0$ and $a_{ij}(\boldsymbol{\varphi})$ is periodic function in variables $\varphi_1,\ldots,\varphi_n$ with the period 2π. The potential energy $V = V(\varphi_1,\ldots,\varphi_n)$ is also periodic function in each of its variables and also with the period 2π. Thus, the configuration space (the Lagrangian manifold) is a n-torus T (Sec. D.2.15). We can call it the "maximal torus" since it has exactly the same meaning from the point of view of dynamical systems as discussed in the previous subsection.

Notice that geometry of the problem is such that it is convenient to write $\mathcal{L} = T + V$ while for the total energy $E = T - V$. Since $T \geq 0$, the classical motion is allowed only in the region $\mathcal{D}(\boldsymbol{\varphi}) = \{E + V(\varphi) \geq 0\}$. The boundary $\partial\mathcal{D}(\varphi)$ is determined by the condition $\partial\mathcal{D}(\boldsymbol{\varphi}) = \{E + V(\varphi) = 0\}$. It is $(n-1)$-dimensional manifold which is smooth. The potential V has critical points (where it becomes zero) $\varphi = \{m_1\pi, \ldots, m_n\pi\}$, with $m_1, \ldots, m_n$ being integers. *Critical points correspond to the equilibrium positions of the n-link planar multipendulum.* This system can exibit two types of periodic solutions: rotations and librations.[t] Surely, we are interested in rotations only. Mathematically, librations are distinguished from rotations by the fact that in the case of rotations the trajectories intersect $\partial\mathcal{D}(\varphi)$ while in the case of librations they are having exactly two points of contact with the boundary (since, in fact, these are vibrations). The trajectories generated by solutions of the Lagrangian equations are *geodesics inside* the domain $\mathcal{D}(\varphi)$ in the Maupertuis–Jacobi-type metric. In Chap. 6 we discussed in detail geodesics in sub-Riemannian geometry using Maupertuis' principle. Since neither Landau–Lifshitz course, [235], nor even the book by Arnol'd, [138], provide sufficient details on geodesics in classical mechanics, we would like to include the full derivation into this section.

Thus, if $T(\mathbf{q}, \dot{\mathbf{q}}) = \frac{1}{2}\sum a_{ij}(\mathbf{q})\dot{q}_i\dot{q}_i$ is the kinetic energy and $V = V(\mathbf{q})$ is the potential energy of some dynamical system for which $\mathcal{L} = T - V$ and $E = T + V$, the Lagrangian equations of motion are given by

$$\frac{d}{dt}\frac{\partial \mathcal{L}}{\partial \dot{q}^i} - \frac{\partial \mathcal{L}}{\partial q^i} = 0, \quad i = 1, \ldots, n.$$

Explicitly,

$$\frac{d}{dt}(a_{ij}(\mathbf{q})q^j) - \frac{\partial a_{hk}}{\partial q^i}\dot{q}^h\dot{q}^k + \frac{\partial V}{\partial q^i} = 0. \tag{D.139a}$$

Since

$$\sqrt{\frac{2(E-V)}{a_{hk}(\mathbf{q})\dot{q}^h\dot{q}^k}} = 1,$$

the above equation can be rewritten as

$$\begin{aligned}
&\frac{d}{dt}\left(\sqrt{\frac{2(E-V)}{a_{hk}(\mathbf{q})\dot{q}^h\dot{q}^k}}a_{ij}(\mathbf{q})q^j\right) + \sqrt{\frac{2(E-V)}{a_{hk}(\mathbf{q})\dot{q}^h\dot{q}^k}}\frac{\partial V}{\partial q^i} - \sqrt{\frac{2(E-V)}{a_{hk}(\mathbf{q})\dot{q}^h\dot{q}^k}}\frac{\partial a_{hk}}{\partial q^i}\dot{q}^h\dot{q}^k \\
&\quad = \sqrt{\frac{1}{2a_{hk}(\mathbf{q})\dot{q}^h\dot{q}^k}}\frac{d}{dt}\left(\frac{2(E-V)a_{ij}(\mathbf{q})q^j}{\sqrt{2a_{hk}(\mathbf{q})\dot{q}^h\dot{q}^k}}\right) \\
&\qquad - \frac{1}{2}\frac{\partial}{\partial q^i}(2(E-V)a_{hk}(\mathbf{q}))\frac{\dot{q}^h\dot{q}^k}{2a_{hk}(\mathbf{q})\dot{q}^h\dot{q}^k} = 0.
\end{aligned} \tag{D.139b}$$

[t]For example of librations see http://en.wikipedia.org/wiki/Libration_(molecule).

By introducing the Maupertuis–Jacobi-type metric $ds^2 = 2(E-V)a_{ij}(\mathbf{q})dq^i\, dq^j \equiv g_{ij}dq^i dq^j$, the above Lagrangian equations can be rewritten as follows:

$$\frac{d}{ds}\left(g_{ij}\frac{dq^j}{ds}\right) - \frac{1}{2}\frac{\partial g_{hk}}{\partial q^i}\frac{dq^h}{ds}\frac{dq^k}{ds}$$
$$= g_{ij}\frac{\partial^2 q^j}{\partial s^2} + \frac{1}{2}\left(\frac{\partial g_{ih}}{\partial q^k} + \frac{\partial g_{ik}}{\partial q^h} - \frac{\partial g_{hk}}{\partial q^i}\right)\frac{dq^h}{ds}\frac{dq^k}{ds} = 0.$$

By multiplying this expression by g^{ij} we finally obtain the equation for geodesics

$$\frac{\partial^2 q^k}{\partial s^2} + \Gamma^k_{ij}\frac{dq^i}{ds}\frac{dq^j}{ds} = 0, \quad k = 1, \ldots, n \tag{D.139c}$$

on Riemannian manifold whose metric is g_{ij}. Here Γ^k_{ij} is the Levi–Civita (or Christoffel) symbol for connection compatible with metric g_{ij}. Equation (D.139c) coincides with Eq. (6.3.5) of Chap. 6 as required. It is helpful also to present these equations in the form compatible with Eqs. (6.3.6) describing normal geodesics in the sub-Riemannian case. Following [139], we write

$$\begin{aligned} \dot{q}_k &= v_k, \\ \dot{v}_k &= -\Gamma^k_{ij}v_i v_j, \quad k = 1, \ldots, n. \end{aligned} \tag{D.139d}$$

The vector field associated with this system of equations is given by

$$v = \sum_{i=1}^{n} v_i \frac{\partial}{\partial q_i}. \tag{D.139e}$$

Going back to our multipendulum problem, now let $\mathbf{a}, \mathbf{b} \in \mathcal{D}(\varphi)$. Then, in complete analogy with results of Chap. 6, Sec. 6.3, we introduce the distance $l(p, q)$ (*this time, not Carnot Caratheodory, but Riemannian*). Using it, we can determine the distance of a given point to the boundary $\partial\mathcal{D}(\varphi)$ as follows. Let $\mathbf{a} \in \mathcal{D}(\varphi)$ while $\mathbf{b} \in \partial\mathcal{D}(\varphi)$, then $\partial\mathbf{a} = \inf l(\mathbf{a}, \mathbf{b})$ defines the distance of $\mathbf{a}$ from $\partial\mathcal{D}(\varphi)$. Kozlov proves the following:

Theorem D.2.18.2. *If* $l(a, b) < \partial a + \partial b$, *then Eq.* (D.139c) *has a solution* $\varphi(t), t \in [t_1, t_2]$, *such that* $\varphi(t_1) = \mathbf{a}$ *and* $\varphi(t_2) = \mathbf{b}$. *In addition, the distance between* a *and* $\mathbf{b}$ *is exactly* $l(\mathbf{a}, \mathbf{b})$.

For the trajectories defined by $\mathcal{D}(\varphi) = \{E + V(\varphi) \geq 0\} \subset T$ it is convenient to define

$$E^- = \min_T V \quad \text{and} \quad E^+ = \max_T V.$$

This then allows us to completely separate librations (in which we are not interested) from rotations — an object of our primary interest. For this purpose, we now look at the energies $E > E^+$. Under such conditions $\partial\mathcal{D}(\varphi) = \emptyset$ so that *by design* the periodic motions could only be rotations. So, if the kth link makes N_k complete turns/rotations per unit of time, then we can assign the set of integers $N_1, \ldots, N_n$ describing rotations of n-links per unit time. How many rotations can exist in such a system? At first sight the problem looks very similar to that discussed in Sec. D.2.10 in connection with Grassmannians. In reality, though, the counting is done differently. Kozlov proves the following

Theorem D.2.18.3. *For any set of fixed integers $N_1, \ldots, N_n$ and any $E > E^+$ there are 2^{n-1} different rotations with total energy E, whose trajectories on n-torus T pass through pairs of critical points of the potential V.*

Independent of whether or not the integers $N_1, \ldots, N_n$ are coprime, the idea of the proof is as follows. Using Theorem D.2.18.2, we are going to assume that **a** is one of the critical points of V and **b** is another. Then, these two points can be connected by a geodesic. Thus, the potential V having n critical points will have n pairs of them which should be joined by geodesics. However, in view of the fact that the total energy is conserved, the number of available critical points is actually $n - 1$. The rest of arguments depends upon whether or not the numbers $N_1, \ldots, N_n$ are coprime resulting finally in the theorem above.

Remark D.2.18.3. From the proof of this theorem it follows that all geodesics will correspond to the motion with the same energy E. This result differs from that obtained in Sec. 6.5 of Chap. 6 where the fiber bundle construction was used.[u] Thus, in the present case, *all torus-type knots for two-link planar pendulum will have the same energy thus representing the same 2-qubit entanglement.* This result is plausible since it is not mixing the topology with mechanics. Thus, in accord with results of Appendix C, we are coming again to the conclusion: *The entanglements are of topological origin.*

The generalization of the obtained results to multiqubits at this point should not cause any additional difficulties.

[u]Notice that in Chap. 6 we discussed the sub-Riemannian problem for which it is typical to have the number of geodesics joining two points to be greater than one. In the Riemannian case the problem is treated differemtly.

The problem of designing maps from complex (Riemann) surfaces into $\mathbf{CP}^n$ was discussed in a comprehensive mathematical review by Eells and Wood [D.96]. Exactly the same results in terms of familiar to every physicist language of *nonlinear sigma models* [41] were discussed in [D.97]. At the classical (saddle point) level in both cases one is obtaining the same set of equations known in mathematics as *harmonic maps.* These are having their origin in the theory of minimal surfaces [134], that is surfaces whose extrinsic curvature H is zero. So, if one uses harmonic maps, one should be interested in embedding of a torus "living", say, in $\mathbf{R}^2$ (e.g. see Eq. (D.138)) into $\mathbf{CP}^2$ as *minimal surface.* This may be permissible mathematically but could violate physics, e.g. mechanical constraints imposed on the Lagrangian surfaces. Fortunately, this is not the case. In the literature one can find both *minimal Lagrangian* submanifolds (surfaces), say, in $\mathbf{CP}^2$, as well as the Hamiltonian-stationary Lagrangian (HSL) submanifolds in $\mathbf{CP}^2$. The HSL manifolds in the Kähler space are the Hamilton-stationary (*Hamiltonian-mininal*) if variations of the volume (area) of such a submanifold along the Hamiltonian flow fields are equal to zero. Examples of HSL tori in $\mathbf{CP}^2$ are given in [D.98]. In another paper (cited in [D.98]) the construction of the Lagrangian minimal tori was also discussed. By design, the Lagrangian minimal submanifolds are HSL submanifolds as well. Thus, both situations are allowed physically. Theorem D.2.1.2 of Sec. D.2.1 provides an independent support of the results just cited.

To conclude this section (and this Appendix) we would like to discuss several related topics. First, we would like to discuss many uses of linkages.

In polymer physics (e.g. read Chap. 8) linkages of the type discussed in this section sometimes used to model conformational behavior of fully flexible polymer chains in the so-called athermal solvents (when the effects of excluded volume can be neglected). Models of this type are known as *freely-jointed polymer chains* http://www.statmech.org/rsh/freejoint.pdf. They can be planar, as in the case of multipendulum we had just described, or they can live in three dimensions. They can be circular and knotted or they can be anchored by one end, as it is done above-all depends upon the physical problem in question. Incidentally, all these problems make physical sense. In mathematics the same models are being used, for instance, for development of *topological robotics* [D.99]. In this discipline the linkages are studied by means of symplectic and toric geometry (among other methods). Other representative papers include those by Kapovich and Millson in which the symplectic geometry of closed linkages is developed both in the plane [D.100] and in 3D Euclidean space [D.101]. In addition, in the pedagogically written paper by Knutson and Hausmann [D.102], these results are connected with Grassmannians.

Second, we would like to discuss the Legendrian manifolds and optimal control in the context of what was just presented. The Legendrian manifolds *in contact geometry* are the exact analogs of the Lagrangian manifolds *in symplectic geometry.* In fact, we had encountered such manifolds when we reformulated thermodynamics in terms of contact geometry in Chap. 5, Sec. 5.3. Since problems of optimal control employ contact geometry as their tool, it makes sense to re-introduce Legendrian manifolds here if we can find a way to formulate the optimal control problems for multilinkages, e.g. planar double pendulum, etc. That such types of problems can be formulated for multilinkages, can be seen from the number of recent papers which is worth mentioning. The optimal control type problems for multilinkages range from the design of optimal air traffic control systems (problem of optimal collision avoidance) to design of unmanned aerial vehicles [D.103], from the problem of swimming of micro-organisms to problem of movement of snakes [D.104], etc. Mechanics of snake-like motion without uses of optimal control methods was studied with help of planar linkages model some time ago in [D.105]. In this reference the authors noticed the following.

"The system of compound pendula that was studied extensively [136, 201, D.95][v] is somewhat similar to ours in that it is just a chain (of linkages[w]) suspended by one end in a gravitational field. The crucial difference between this system and the free chain considered here is the lack of a preferred inertial frame for the free chain. It is therefore desirable to reduce the dimension of the problem by eliminating the reference to an inertial frame; the reduced equations will be in terms of angles θ_i between the neighboring links and will not contain any variables characterizing the orientation or position with respect to an inertial frame."

The optimal control of snake's motion discussed in [D.104] is relying on methods of sub-Riemannian geometry. Since it involves practically the same mechanical set up as in [D.105] the question arises: Is it possible to adopt the optimal control scheme of [D.104] for the motion of snakes to that of multipendulum describing quantum entanglements? The answer is: "No". That is, not immediately! This is so because of physical differences summarized in [D.105] which we just cited. If this is the case, what else

[v] We replace the numeration in this paper by that in our book.
[w] Our clarification.

can we do? Very fortunately-a lot! Let us begin with just one link *planar* pendulum and consider the situation suggested by Arnol'd in Theorem D.2.18.1. That is we exclude librations (that is oscillations) and leave only rotations. But then, we are left with the situation which is *exactly the same* as for the charged particle in constant magnetic field we had discussed in detail in Chap. 6! This is the sub-Riemannian problem. If we add yet another link, formally, we shall obtain exactly the same problem, this time for the second link. There is something missing, still, since if the motion of links becomes completely decupled, then, accordingly, there could be no entanglement which we can put under control. Fortunately, to correct this situation is not too difficult if we use some analogies from physics. This time, from physics of quantum Hall effect (QHE), e.g. see [64, D.106], for a quick introduction. In the absence of electron–electron interactions (this is permissible as explained by Laughlin [D.107]) the classical Hamiltonian $\mathcal{H}$ for an assembly of n electrons in constant external magnetic field is given by n copies of the Hamiltonian, Eq. (6.2.4). Following [A.5] and [D.107] and using our Eq. (6.2.4), the Hamiltonian for n particles in the constant magnetic field can be written as

$$\mathcal{H} = \frac{1}{2}\sum_{j=1}^{n}\{(\xi_{2j-1} + 2a_j x_{2j}\theta)^2 + (\xi_{2j} - 2a_j x_{2j-1}\theta)^2\}. \tag{D.140}$$

Here $0 < a_1 \leq a_2 \leq \cdots \leq a_n$ are some positive constants (absolute values of charges). In the context of planar multipendulum it is clear that these constants can be related to the set $N_1, \ldots, N_n$. As in the case of one particle in the magnetic field, e.g. see Eq. (6.1.2a), we introduce now the vector fields $X_1, \ldots, X_n, Y_1, \ldots, Y_n$ so that this time we obtain

$$[X_j, Y_j] = -4a_j\frac{\partial}{\partial t}, \quad j = 1, \ldots, n. \tag{D.141}$$

Definition D.2.18.1. The set of commutators defined by Eqs. (D.141) defines the Lie algebra of the n-dimensional Hesenberg group $\mathbf{H}_n$.

The contact nature of $\mathbf{H}_n$ is revealed as follows. As for $\mathbf{H}_1$ (e.g. see Eqs. (6.1.1)) we introduce the vector fields

$$X_j = \frac{\partial}{\partial x_j} + 2a_j y_j\frac{\partial}{\partial t}, \quad Y_j = \frac{\partial}{\partial y_j} - 2a_j x_j\frac{\partial}{\partial t}.$$

This time, however, in agreement with Eq. (D.140) it is useful to relabel these fields as

$$X_{2j-1} = \frac{\partial}{\partial x_{2j-1}} + 2a_j x_{2j}\frac{\partial}{\partial t}, \quad X_{2j} = \frac{\partial}{\partial x_{2j}} - 2a_j x_{2j-1}\frac{\partial}{\partial t}$$

so that 2×2 matrix Λ_j given by

$$\Lambda_j = \begin{pmatrix} 0 & 2a_j \\ -2a_j & 0 \end{pmatrix}, \quad \Lambda = \mathrm{diag}[\Lambda_1, \ldots, \Lambda_n], \quad \Lambda + \Lambda^T = \mathbf{0}$$

can be used in order to rewrite the above set of vector fields in terms of just one matrix equation

$$\mathbf{X} = \mathbf{I}\frac{\partial}{\partial \mathbf{x}} + \Lambda \cdot \mathbf{x}\frac{\partial}{\partial t}$$

meaning

$$X_j = \partial_{x_j} + 2\sum_{k=1}^{n} a_{jk} x_k \partial_t, \quad j = 1, \ldots, 2n. \tag{D.142}$$

Definition D.2.18.2. The horizontal distribution is defined in exact analogy with Eqs. (6.3.9), (6.3.10), $\mathrm{H}_x\mathcal{M} = \mathrm{span}\{X_1, \ldots, X_{2n}\}$, and by the horizontal curve $c(t)$ such that its velocity $\dot{c}$ is given by

$$\begin{aligned} \dot{c} &= \sum_{j=1}^{2n} \dot{x}_j \partial_{x_j} + \dot{t}\partial_t \\ &\equiv \langle \dot{x}, \partial_x \rangle + \dot{t}\partial_t \\ &= \langle \dot{x}, \partial_x + \Lambda x \partial_t - \Lambda x \partial_t \rangle + \dot{t}\partial_t \\ &= \langle \dot{x}, X \rangle - \langle \dot{x}, \Lambda x \rangle + \dot{t}\partial_t \\ &= \langle \dot{x}, X \rangle + (\dot{t} - \langle \dot{x}, \Lambda x \rangle)\partial_t \\ &= \langle \dot{x}, X \rangle; \quad \dot{t} - \langle \dot{x}, \Lambda x \rangle = 0. \end{aligned} \tag{D.143}$$

From here we obtain the4 horizontal 1-form ω

$$\omega = dt - 2\sum_{j=1}^{n} a_j (x_j dy_j - y_j dx_j). \tag{D.144a}$$

The $2a_j$ factors can be absorbed into x_j (and/or y_j or both). After that, the obtained contact 1-form acquires the standard (textbook) look [139]

$$\omega = dt + \sum_{j=1}^{n} (x_j dy_j - y_j dx_j) \tag{D.144b}$$

known for the contact 1-forms on $\mathbf{R}^{2n+1}$. To connect this result with the standard Legendrian manifolds (e.g. read again Sec. 5.3 of Chap. 5) requires us to perform a contactomorphic transformation, [139], p. 52, of the type

$$f(\mathbf{x}, \mathbf{y}, t) = ((\mathbf{x} + \mathbf{y})/2, \quad (\mathbf{y} - \mathbf{x})/2, t + \mathbf{x} \cdot \mathbf{y}/2)$$

bringing the 1-form Eq. (D.144b) to its Legendrian form

$$\omega = dt - \sum_j y_j dx_j. \tag{D.144c}$$

For $dt = 0$ this 1-form coincides with the symplectic 1-form discussed in Sec. D.2.17. So that if the Lagrangian manifold is defined by the conditions

$$y_i = \frac{\partial t}{\partial x_i}(x_1, \ldots, x_n) \quad \text{that is by the projection}$$
$$\pi(x_1, \ldots, x_n; y_1, \ldots, y_n) \to (x_1, \ldots, x_n), \tag{D.145a}$$

the Legendrian manifold (fibration) is defined by the projection

$$\pi(x_1, \ldots, x_n; y_1, \ldots, y_n; t) \to (x_1, \ldots, x_n; t). \tag{D.145b}$$

Other than in Chap. 5, Sec. 5.3, examples can be found on p. 367 of [136]. It is useful to understand these results, say, in the thermodynamic language. For instance, let us re-interpret the theorem (on p. 367) which reads as follows:

Theorem D.2.18.4. *For any partition* $\mathbf{I} + \mathbf{J}$ *of the set* $\{1, \ldots, n\}$ *in two disjoint subsets and for any function* $S(x_{\mathbf{I}}, y_{\mathbf{J}})$ *of n variables* $x_i, i \in \mathbf{I}, j \in \mathbf{J}$, *the formulas*

$$y_{\mathbf{I}} = \frac{\partial S}{\partial x_{\mathbf{I}}}, \quad x_{\mathbf{J}} = -\frac{\partial S}{\partial y_{\mathbf{J}}}, \quad z = S - x_{\mathbf{I}}\frac{\partial S}{\partial x_{\mathbf{I}}}$$

define a Legendre submanifold of $\boldsymbol{R}^{2n+1}$. *Conversely, every Legendre submanifold of* $\boldsymbol{R}^{2n+1}$ *is defined in a neighborhood of every point by these formulas for at least one of the* 2^n *possible choices of the subset* $\mathbf{I}$.

From here, it follows that the Legendre manifold is determined by the condition $dz + xdy = 0$. Indeed, $d(z + x_{\mathbf{I}}y_{\mathbf{I}}) = y_{\mathbf{I}}dx_{\mathbf{I}} - x_{\mathbf{J}}dy_{\mathbf{J}}$.

Example. Identify the free energy $\mathcal{F} - U - TS$ with $S(x_{\mathbf{I}}, y_{\mathbf{J}})$. Since $d\mathcal{F} = -SdT - PdV$ we have $\mathcal{F} = \mathcal{F}(T, V)$ so that $T = x_{\mathbf{I}}$ and $V = y_{\mathbf{J}}$. Also, $\frac{\partial \mathcal{F}}{\partial T} = \frac{\partial S}{\partial x_{\mathbf{I}}} = y_{\mathbf{I}} = -S$ and $-\frac{\partial \mathcal{F}}{\partial V} = -\frac{\partial S}{\partial y_{\mathbf{J}}} = P = x_{\mathbf{J}}$. From these results we obtain: $\mathcal{F} + TS = U \to \mathcal{F}(T, V) - x_{\mathbf{I}}y_{\mathbf{I}} = \mathcal{F} + ST = U$. Thus, the Legendre manifold is determined by the condition $dU - TdS + PdV = 0$ in agreement with Eq. (5.3.1).

Evidently, replacement of the Lagrangian manifold by the Legendrian makes it possible to develop the theory of optimal control. In connection with this observation, going back to our initial multipendulum problem let us consider two cases: (a) $a_1 = a_2 = \cdots = a_n$; (b) anything else. Such a subdivision is motivated by the results obtained in [D.107]. The case (a) is physically relevant for the QHE problem. This is so because the numbers a_i (or the numbers N_i) can be associated with the effective charges of the particles moving on the plane in the constant magnetic field. Since all electrons are having the same charges, it makes situation described by (a) appropriate for studying of QHE by methods of sub-Riemannian geometry.

Exercise D.2.18.1. Taking into account that for the case (a) there are finitely many geodesics joining points $(0,0)$ and (x,t), [D.108], p. 657, among which there is the shortest (that is of Carnot–Caratheodory-type), think about what this result could mean physically for the QHE. Hint: For the case (a) each family of geodesics — from $(0,0)$ to $(0,t)$ — is parametrized by the parameters describing the $2n-1$ sphere, [D.108], p. 656.

Let us adopt just discussed result to the case of planar two-linkage pendulum. It allows us to replace the five-dimensional contact manifold for this problem by the three-dimensional contact manifold! But we studied sub-Riemannian dynamics on S^3 already in Chap. 6, Sec. 6.5 and in Chap. 8, Sec. 8.2.6! Thus, at least in the case (a) we had brought the entanglement problem under control. This result is also in accord with that obtained in Sec. D.2.16 by entirely different methods.

The case (b) is also discussed using general protocol of sub-Riemannian calculations in [A.5] and [D.108]. In this case calculations become much lengthier but, still under control. It remains to squeeze some physics out of these formal results.

Exercise D.2.18.2. Interpret results of Exercise D.2.18.1 in terms of the Cauchy–Riemann and complex hyperbolic geometries [214].

Exercise D.2.18.3. Read Appendix 10 of the book by Arnol'd, [136], entitled "Multiplicities of characteristic frequencies, and ellipsoids depending on parameters" in which Arnol'd discusses the role of symmetry in creating/destroying multiplicities of eigenvalues in spectra of some mechanical systems. Think about the usefulness of this information for building mechanical analogs for studying the entanglement problems occurring in quantum computation.

Remark D.2.18.4. The mechanical planar double pendulum, used for description of entanglements in this subsection, was recently adapted for description of the Laudau–Zener-type tunneling in [D.109]. Alternative mechanical treatment of the L-Z tunneling is discussed in Chap. 8. Thus, for all quantum processes we were able to find their analogs in classical mechanics. It remains to utilize this observation in designing classical "quantum" computers.

References

[1] Leggett A., *Quantum Liquids. Bose Condensation and Cooper Pairing in Condensed-Matter Systems* (Oxford Univ. Press, Oxford, 2006).

[2] Andreussi T., Morrison P. and Pegoraro F., Hamiltonian magnetohydrodynamics: Helically symmetric formulation, Casimir invariants, and equilibrium variational principles, *Phys. Plamas* **19**, 052102 (2012).

[3] Dean D. and Hjorth-Jensen M., Pairing in nuclear sytems: From neutron stars to finite nuclei, *Rev. Mod. Phys.* **75**, 607–658 (2003).

[4] Braun-Minziger P. and Wambach J., Phase diagram of strongly interacting matter, *Rev. Mod. Phys.* **81**, 1031–1050 (2009).

[5] Maeda K., Baym G. and Hatsuda T., Simulating dense QCD matter with ultracold atomic boson-fermion mixtures, *Phys. Rev. Lett.* **103**, 085304 (2009).

[6] Kozlov V., *General Theory of Vortices* (Springer-Verlag, Berlin, 2003).

[7] Martins A. and Pinhero M., Fluidic electrodynamics: Approach to electromagnetic propulsion, *Phys. Fluids* **21**, 097103 (2009).

[8] Maddocks J. and Pego R., An unconstrained Hamiltonian formulation for incompressible fluid flow, *Comm. Math. Phys.* **170**, 207–217 (1995).

[9] Ballard E. and Kholodenko A., Topological character of hydrodynamic screening in suspensions of hard spheres: An example of universal phenomenon, *Physica A* **388**, 3024–3062 (2009).

[10] Landau L. and Lifshitz E., *Fluid Mechanics* (Pergamon Press, London, 1959).

[11] Frohlich H., Macroscopic wave functions in superconductors, *Proc. Phys. Soc.* **87**, 330–332 (1966).

[12] Chandrasekhar S. and Kendall P., On force-free magnetic fields, *Astrophys. J.* **126**, 457–460 (1957).

[13] Lifshitz, E. and Pitaevski L., in *Statistical Physics*. Part 2, eds. Landau L. and Lifshitz E., Course in Theoretical Physics, Vol. 9 (Pergamon Press, London, 1980).

[14] London H. and London F., The electromagnetic equations for supraconductor, *Proc. Roy. Soc. Lond. A* **149**, 71–88 (1935).

[15] London F., *Superfluids*, Vol. 1. *Macroscopic Theory of Superconductivity*, Vol. 2. *Macroscopic Theory of Superfluid Helium* (Wiley, New York, 1950–1954).

[16] Abrikosov A., Gor'kov L. and Dzyaloshinskii I., *Methods of Quantum Field Theory in Statistical Physics* (Dover, New York, 1963).
[17] Jost J., *Riemannian Geometry and Geometric Analysis* (Springer-Verlag, Berlin, 2005).
[18] Nicolaescu, L., *Notes on Seiberg–Witten Theory* (Amer. Math. Soc., Providence, 2000).
[19] Kronheimer P. and Mrowka T., *Monopoles and Three-Manifolds* (Cambridge Univ. Press, Cambridge, 2007).
[20] Marsh G., *Force-Free Magnetic Fields: Solutions, Topology and Applications* (World Scientific, Singapore, 1996).
[21] Taylor J., Relaxation of toroidal plasma and generation of reverse magnetic fields, *Phys. Rev. Lett.* **33**, 1139–1141 (1974).
[22] Taylor J., Relaxation and magnetic reconnection in plasmas, *Rev. Mod. Phys.* **58**, 741–763 (1986).
[23] Landau L. and Lifshitz E., *Electrodynamics of Continuous Media*, Course in Theoretical Physics, Vol. 8 (Pergamon Press, London, 1995).
[24] Di Vita A., Hot spots and filaments in the pinch of a plasma focus: A unified approach, *Eur. Phys. J. D* **54**, 451–461 (2009).
[25] Boyland P., Fluid mechanics and mathematical structures, in *An Introduction to Geometry and Topology of Fluid Flows*, ed. Ricca R. (Kluwer Academic, Boston, 2001), pp. 105–133.
[26] Rogers C. and Schief W., Novel integrable reductions in nonlinear continuum mechanics via geometric constraints, *J. Math. Phys.* **44**, 3341–3369 (2003).
[27] Schief W., Nested toroidal flux surfaces in magnetohydrodynamics. Generation via soliton theory, *J. Plasma Phys.* **65**, 465–484 (2003).
[28] Schief W., Hidden integrability in ideal magnetohydrodynamics: The Pohlmeyer–Lund–Regge model, *Phys. Plasmas* **10**, 2677–2685 (2003).
[29] Lund F. and Regge T., Unified approach to strings and vortices with soliton solutions, *Phys. Rev. D* **14**, 1524–1535 (1976).
[30] Nielsen H. and Olesen P., Local field theory of the dual string, *Nucl. Phys. B* **57**, 367–375 (1973).
[31] Barbashov B. and Nesterenko V., *Introduction to the Relativistic String Theory* (World Scientific, Singapore, 1990).
[32] Berdichevsky V., Averaged equations of ideal fluid turbulence, *Cont. Mech. Thermodyn.* **19**, 133–175 (2007).
[33] Rogers C. and Schief W., On geodesic hydrodynamic motions. Heisenberg spin connections, *J. Math. Anal. Appl.* **251**, 855–870 (2000).
[34] Reed D., Beltrami vector fields in electrodynamics: A reason for reexamining the structural foundations of classical field physics?, *Adv. Chem. Phys.* **119**, 525–658 (2001).
[35] Langer J. and Singer D., Knotted elastic curves in $\mathbf{R}^3$, *J. London Math. Soc.* **30**, 512–520 (1984).
[36] Gilbert N. and Porter T., *Knots and Surfaces* (Oxford Univ. Press, Oxford, 1994).

[37] Ivey T. and Singer D., Knot types, homotopies and stability of closed elastic rods, *Proc. London Math. Soc.* **79**, 429–450 (1999).

[38] Langer J. and Singer D., Lagrangian aspects of the Kirchhoff elastic rod, *SIAM Rev.* **38**, 605–618 (1996).

[39] Kholodenko A., Ballauff M. and Granados M., Conformational statistics of semiflexible polymers: Comparison between different models, *Physica A* **260**, 267–293 (1998).

[40] Kholodenko A. and Nesterenko V., Classical dynamics of rigid strings from Willmore functional, *J. Geom. Phys.* **16**, 15–26 (1995).

[41] Polyakov A., *Gauge Fields and Strings* (Harwood Academic Publishers, New York, 1987).

[42] Langer J. and Singer D., The total squared curvature of closed curves, *J. Diff. Geom.* **20**, 1–22 (1984).

[43] Willmore T., *Riemannian Geometry* (Clarendon Press, Oxford, 1993).

[44] Faddeev L. and Niemi A., Stable knot-like structures in classical field theory, *Nature* **387**, 58–61 (1997).

[45] Singer D., Lectures on elastic curves and rods, *AIP Conf. Proc.* **1002**, 3–32 (2008).

[46] Spindler K., Optimal attitude control of a rigid body, *Appl. Math. Optim.* **34**, 79–90 (1996).

[47] Zucker S., Differential geometry from the Frenet point of view: Boundary detection, stereo, texture and color, in *Handbook of Mathematical Models in Computer Vision*, eds. Paragios N. and Faugeras O. (Birkhäuser, Boston, 2006), pp. 359–374.

[48] Guichardet A., On rotation and vibration motions of molecules, *Ann. Inst. Henri Poincaré* **40**, 329–342 (1984).

[49] Littlejohn R. and Reinach M., Gauge fields in the separation of rotations and internal motions in the n-body problem, *Rev. Mod. Phys.* **69**, 213–275 (1997).

[50] Kida S., A vortex filament moving without change of form, *J. Fluid Mech.* **112**, 397–409 (1981).

[51] Grice G., *Constant Speed Flows and the Nonlinear Schrödinger Equation*, Master's of Science Thesis, Department of Mathematics, University of South New Wales, April 2004.

[52] Ricca R., Samuels D. and Barenghi C., Evolution of vortex knots, *J. Fluid Mech.* **391**, 29–44 (1999).

[53] Ricca R., Torus knots and polynomial invariants for a class of soliton equations, *Chaos* **3**, 83–91 (1993).

[54] Calini A. and Ivey T., Connecting geometry, topology and spectra for finite-gap NLS potentials, *Physica D* **152/153**, 9–19 (2001).

[55] Calini A. and Ivey T., Knot types, Floquet spectra, and finite-gap solutions of the vortex filament equation, *Math. Comput. Simul.* **55**, 341–350 (2001).

[56] Calini A. and Ivey T., Finite-gap solutions of the vortex filament equation: Genus one solutions and symmetric solutions, *J. Nonlinear Sci.* **15**, 321–361 (2005).

[57] Calini A. and Ivey T., Finite-gap solutions of the vortex filament equation: Isoperiodic deformations, *J. Nonlinear Sci.* **17**, 527–567 (2007).
[58] Calini A., Recent developments in integrable curve dynamics, in *Geometric Approaches to Differential Equations*, eds. Vassiliou P. and Lisle I. (Cambridge Univ. Press, Cambridge, 2000), pp. 56–99.
[59] Kambe T., *Geometrical Theory of Dynamical Systems and Fluid Flows* (World Scientific, Singapore, 2010).
[60] Belokolos E., Bobenko A., Enol'skii V., Its A. and Matveyev V., *Algebro-Geometric Approach to Nonlinear Integrable Equations* (Springer-Verlag, Berlin, 1994).
[61] von der Mosel H., Elastic knots in Euclidean 3-space, *Ann. Inst. Henri Poincaré* **16**, 137–166 (1999).
[62] von der Mosel H., Minimizing the elastic energy of knots, *Asympt. Anal.* **18**, 49–65 (1998).
[63] Gonzales G., Maddocks J., Schuricht F. and von der Mosel H., Global curvature and self-contact of nonlinearly elastic curves and rods, *Calc. Var.* **14**, 29–68 (2002).
[64] Kholodenko A. and Vilgis Th., Some geometrical and topological problems in polymer physics, *Phys. Rep.* **298**, 251–270 (1998).
[65] Pierre-Louis O., On the geometry of stiff knots, *Eur. Phys. J. B* **71**, 281–288 (2009).
[66] Woltjer L., A theorem on force-free magnetic fields, *PNAS* **44**, 489–491 (1958).
[67] Moffatt H., The degree of knottedness of tangled vortex lines, *J. Fluid Mech.* **35**, 117–129 (1969).
[68] Arnol'd V. and Khesin B., *Topological Methods in Hydrodynamics* (Springer-Verlag, Berlin, 1998).
[69] Moffatt H. and Ricca R., Helicity and the Calugareanu invariant, *Proc. Roy. Soc. Lond. A* **439**, 411–429 (1992).
[70] Chui A. and Moffatt H., The energy and helicity of knotted magnetic flux tubes, *Proc. Roy. Soc. Lond. A* **451**, 609–629 (1995).
[71] Maggioni F. and Ricca R., On the groundstate energy of tight knots, *Proc. Roy. Soc. A* **465**, 2761–2783 (2009).
[72] Freedman M. and He Z., Divergence-free field: Energy and the asymptotic crossing numbers, *Ann. Math.* **134**, 189–229 (1991).
[73] Cantarella J., Kusner R. and Sullivan J., On the minimum ropelength of knots and links, *Invent. Math.* **150**, 257–286 (2002).
[74] Gonzales O. and de la Llave R., Existence of ideal knots, *J. Knot Theory Ramifications* **12**, 123–130 (2003).
[75] de Gennes P. G. and Prost J., *The Physics of Liquid Crystals* (Clarendon Press, Oxford, 1993).
[76] Baldwin P. and Townsend G., Complex Trkalian fields and solutions to Euler's equations for the ideal fluid, *Phys. Rev. E* **51**, 2059–2068 (1995).
[77] de Gennes P. G., An analogy between superconductors and smectics A, *Solid State Commun.* **10**, 753–756 (1972).

[78] Lubensky T., TGB phases: Abrikosov vortex lattices in liquid crystals, *Physica A* **220**, 99–112 (1995).
[79] Vakulenko A. and Kapitanskii L., Stability of solitons in S^2 in the nonlinear σ-model, *Sov. Phys. Dokl.* **24**, 433–434 (1979).
[80] Derrick G., Comments on nonlinear wave equations as models for elementary particles, *J. Math. Phys.* **5**, 1252–1254 (1964).
[81] Lin F. and Yang Y., Energy splitting, substantial inequality, and minimization for the Faddeev and Skyrme models, *Comm. Math. Phys.* **269**, 137–152 (2007).
[82] Ward R., Hopf solitons on S^3 and $\mathbf{R}^3$, *Nonlinearity* **12**, 241–246 (1999).
[83] Harland D., Speight M. and Sutcliffe P., Hopf solitons and elastic rods, arXiv:1010.3189.
[84] Battye R. and Sutcliffe P., Solitons, links and knots, *Proc. Roy. Soc. Lond. A* **455**, 4305–4331 (1999).
[85] Nicole D., Solitons with non-vanishing Hopf index, *J. Phys. D* **4**, 1363–1369 (1978).
[86] Aratyn H., Ferreira L. and Zimmerman A., Exact static soliton solutions of $(3+1)$-dimensional integrable theory with nonzero Hopf numbers, *Phys. Rev. Lett.* **83**, 1723–1726 (1999).
[87] Adam C., Sanzhez-Guillen J. and Weereszczynski A., A first integration of some knot soliton models, *Phys. Lett. B* **659**, 761–767 (2008).
[88] Laurence P. and Avellaneda M., On Woltjer's variational principle for force-free fields, *J. Math. Phys.* **32**, 1240–1253 (1991).
[89] Cantarella J., *Topological Structure of Stable Plasma Flows*, Ph.D. Thesis, Univ. of Pennsylvania, 1999.
[90] Cantarella J., De Trurk D. and Gluck H., Vector calculus and the topology of domains in 3-space, *Am. Math. Monthly* **109**, 409–442 (2002).
[91] Boozer A., Physics of magnetically confined plasmas, *Rev. Mod. Phys.* **76**, 1071–1141 (2004).
[92] Nishiyama T., A relaxation method for constructing a Beltrami flow in a bounded domain, *J. Math. Phys.* **46**, 083102 (2005).
[93] Di Vita A., A lower bound on Hartmann number for relaxed plasmas described by Taylor's principle, *Eur. J. Phys. D* **56**, 205–210 (2010).
[94] Mahajan S., Classical perfect diamagnetism: Expulsion of current from the plasma interior, *Phys. Rev. Lett.* **100**, 075001–074004 (2008).
[95] Ryder L., *Quantum Field Theory* (Cambridge Univ. Press, Cambridge, 1985).
[96] Ruegg H. and Ruiz-Altaba M, The Stueckelberg field, arXiv:hep-th/0304245.
[97] Goldhaber A. and Nieto M., Photon and graviton mass limits, *Rev. Mod. Phys.* **82**, 939–979 (2010).
[98] Tajmar M., Electrodynamics in supercoductors explained by Proca equations, *Phys. Lett. A* **372**, 3289–3291 (2008).
[99] Feynman R., *The Feynman Lectures on Physics*, Vol. 3 (Addison-Wesley, Reading, 1966).

[100] Feynman R., *Statistical Mechanics. A Set of Lectures* (Addison-Wesley, Reading, 1972).

[101] Wu T. and Yang C-N., Concept of nonintegrable phase factors and global formulation of gauge fields, *Phys. Rev. D* **12**, 3845–3857 (1975).

[102] Fulton W., *Algebraic Topology. A First Course* (Springer-Verlag, Berlin, 1995).

[103] Shabat B., *Introduction into Complex Analysis*, Part 2 (Amer. Math. Soc., Providence, 1992).

[104] Ryder L., Dirac monopoles and the Hopf map $S^3 \to S^2$, *J. Phys. A* **13**, 437–447 (1980).

[105] Shnir Ya., *Magnetic Monopoles* (Springer-Verlag, Berlin, 2005).

[106] Marsh G., Monopoles, gauge fields and de Rham's theorem, *J. Phys. A* **31**, 7077–7094 (1998).

[107] Madsen I. and Tornehave J., *From Calculus to Homology* (Cambridge Univ. Press, Cambridge, 1997).

[108] Mandelstam S., Vortices and quark confinement in non-Abelian gauge theories, *Phys. Lett.* **53B**, 476–478 (1975).

[109] Nambu Y., Strings, monopoles and gauge fields, *Phys. Rev. D* **10**, 4262–4268 (1974).

[110] Dirac P., The theory of magnetic poles, *Phys. Rev.* **74**, 817–830 (1948).

[111] Schwarz A., *Topology for Physicists* (Springer-Verlag, Berlin, 1994).

[112] Chodos A., Jaffe R., Johnson K., Thorn K. and Weiskopf V., New extended model of hadrons, *Phys. Rev. D* **9**, 3471–3494 (1974).

[113] Chodos A., Jaffe R., Johnson K. and Thorn C., Baryon structure in the bag theory, *Phys. Rev. D* **10**, 2599–2604 (1974).

[114] Manton N. and Sutcliffe P., *Topological Solitons* (Cambridge Univ. Press, Cambridge, 2007).

[115] Chan H-M. and Tsou Th., *Some Elementary Gauge Theory Concepts* (World Scientific, Singapore, 1993).

[116] Goddard P. and Olive D., Magnetic monopoles in gauge theories, *Rep. Progr. Phys.* **41**, 1360–1437 (1978).

[117] Kapustin A. and Witten E., Electric-magnetic duality and the geometric Langlands program, *Comm. Num. Theory Phys.* **1**, 1–236 (2007).

[118] Montonen C. and Olive D., Magnetic monopoles as gauge particles, *Phys. Lett.* **72B**, 117–120 (1977).

[119] Konishi K., The magnetic monopole seventy-five years later, *String Theory and Fundamental Interactions. Lecture Notes Phys.* **737**, 471–521 (2008).

[120] Deser S., Jackiw R. and Templeton S., Three-dimensional massive gauge theories, *Phys. Rev. Lett.* **48**, 975–978 (1982).

[121] Deser S., Jackiw R. and Templeton S., Topologically massive gauge theories, *Ann. Phys.* **140**, 372–411 (1982).

[122] Carlip S., Deser S., Waldron A. and Wise D., Cosmological topologically massive gravitons and photons, *Class. Quantum Grav.* **26**, 075008 (2009).

[123] Dunne G., *Self-dual Chern–Simons Theories* (Springer-Verlag, Berlin, 1995).

[124] Paul S. and Khare A., Self-dual factorization of the Proca equation with Chern–Simons term in 4K−1 dimensions, *Phys. Lett. B* **171**, 244–246 (1986).

[125] Deser S. and Jackiw R., "Self-duality" of topologically massive gauge theories, *Phys. Lett. B* **139**, 371–373 (1984).

[126] March G., Magnetic energy, multiply connected domains and force-free fields, *Phys. Rev. A* **46**, 2117–2123 (1992).

[127] DeTurck D. and Gluck H., Electrodynamics and the Gauss linking integral on the 3-sphere and in the hyperbolic 3-space, *J. Math. Phys.* **49**, 023504 (2008).

[128] Townsend P., Pilch K. and van Nieuwenhuizen P., Self-duality in odd dimensions, *Phys. Lett. B* **136**, 38–42 (1984).

[129] Horvathy P., Plyuschay M. and Velenzuela M., Supersymmetry of the planar Dirac–Deser–Jackiw–Templeton system, and of its non-relativistic limit, *J. Math. Phys.* **51**, 092108 (2010).

[130] Horvathy P. and Zhang P., Vortices in (Abelian) Chern–Simons gauge theory, *Phys. Rep.* **481**, 83–142 (2009).

[131] Liu X., Kauffman knot polynomials in classical Abelian Chern–Simons field theory, *Ann. Phys.* **325**, 2641–2652 (2010).

[132] Geiges H., A brief history of contact geometry and topology, *Expo. Math.* **19**, 25–53 (2001).

[133] Krasil'schik I. and Vinogradov A., *Symmetries and Conservation Laws for Differential Equations of Mathematical Physics* (Amer. Math. Soc., Providence, 1999).

[134] Dubrovin B., Fomenko A. and Novikov S., *Modern Geometry-Methods and Applications*, Part I (Springer-Verlag, Berlin, 1984).

[135] Arnol'd V., *Geometrical Methods in the Theory of Ordinary Differential Equations* (Springer-Verlag, Berlin, 1988).

[136] Arnol'd V., *Mathemathical Methods of Classical Mechanics* (Springer-Verlag, Berlin, 1989).

[137] Etnyre J. and Ghrist R., Contact topology and hydrodynamics. Beltrami fields and Seifert conjecture, *Nonlinearity* **13**, 441–458 (2000).

[138] Saeki O., Szűcs A. and Takase M., Regular homotopy classes of immersions of 3-manifolds into 5-space, *Manuscr. Math.* **108**, 13–32 (2002).

[139] Geiges H., *An Introduction to Contact Topology* (Cambridge Univ. Press, Cambridge, 2008).

[140] Dahl M., *Contact and Symplectic Geometry in Electromagnetism*, Master's Thesis, Department of Mathematics, Helsinki University of Technology, 2002.

[141] Abraham R., Mardsen J. and Ratiu T., *Manifolds, Tensor Analysis and Applications* (Springer-Verlag, Berlin, 1988).

[142] Chern S. and Hamilton R., On Riemannian metric adapted to three-dimensional contact manifolds, *Lecture Notes Math.* **1111**, 279–305 (1985).

[143] Marsden J. and Ratiu T., *Introduction to Mechanics and Symmetry* (Springer-Verlag, Berlin, 1999).

[144] Frankel Th., *The Geometry of Physics* (Cambridge Univ. Press, Cambridge, 1997).

[145] Ghrist R., On the contact topology and geometry of ideal fluids, in *Handbook of Mathematical Fluid Dynamics*, Vol. 4 (Elsevier, Amsterdam, 2007).

[146] Gökckeler M. and Schuker T., *Differential Geometry, Gauge Theories and Gravity* (Cambridge Univ. Press, Cambridge, 1987).

[147] Rosenberg S., *The Laplacian on a Riemannian Manifold* (Cambridge Univ. Press, Cambridge, 1997).

[148] Dunajski M., *Solitons, Instantons and Twistors* (Oxford Univ. Press, Oxford, 2010).

[149] Donaldson S., *Floer Homology Groups in Yang–Mills Theory* (Cambridge Univ. Press, Cambridge, 2002).

[150] Manton N., Complex structure of monopoles, *Nucl. Phys. B* **135**, 319–332 (1978).

[151] Lohe M., Two and three-dimensional instantons, *Phys. Lett. B* **70**, 325–328 (1977).

[152] Jaffe A. and Taubes C., *Vortices and Monopoles* (Birkhäuser, Boston, 1980).

[153] Bott R. and Tu L., *Differential Forms in Algebraic Topology* (Springer-Verlag, Berlin, 1982).

[154] Eguchi T., Gilkey P. and Hanson A., Gravitation, gauge theory and differential geometry, *Phys. Rep.* **66**, 213–393 (1980).

[155] Flanders H., *Differential Forms with Applications to the Physical Sciences* (Dover, New York, 1989).

[156] Jaykka J., Knot solitons in modified Ginzburg–Landau model, arXiv: 1102.2259.

[157] Cho Y., Park B. and Zhang P., New interpretation of Skyrme theory, *Int. J. Mod. Phys. A* **23**, 267–298 (2008).

[158] Volovik G., *The Universe in a Helium Droplet* (Oxford Univ. Press, Oxford, 2009).

[159] Cho Y., Knot topology of classical QCD vacuum, *Phys. Lett. B* **644**, 208–211 (2007).

[160] Li S., Zhang Y. and Zhu Z., Decomposition of SU(N) connection and effective theory of SU(N) QCD, *Phys. Lett. B* **487**, 201–208 (2000).

[161] Brown G., *Selected Papers, with Commentary, of T. H. R. Skyrme* (World Scientific, Singapore, 1994).

[162] Weigel H., *Chiral Soliton Models for Baryons* (Springer-Verlag, Berlin, 2008).

[163] Brown G. and Rho M., *The Multifaceted Skyrmion* (World Scientific, Singapore, 2010).

[164] Yoshida Z. and Giga Y., Remarks on spectra of the operator rot, *Math. Z.* **204**, 235–245 (1990).

[165] Cantarella J. and Parsley J., A new cohomological formula for helicity in $\mathbf{R}^{2k+1}$ reveals the effect of a diffeomorphism on helicity, *J. Geom. Phys.* **60**, 1127–1155 (2010).

[166] Parsley J., *The Biot-Savart Operator and Electrodynamics on Bounded Subdomains of the Three-Sphere*, Ph.D. Thesis, Department of Mathematics, Univ. of Pennsylvania, Philadelphia, 2004.

[167] Calin O. and Chang D-C., *Sub-Riemannian Geometry. General Theory and Examples* (Cambridge Univ. Press, Cambridge, 2009).

[168] Arnol'd V., *Ordinary Differential Equations* (Springer-Verlag, Berlin, 1992).

[169] Arafune J., Freund P. and Goebel C., Topology of Higgs fields, *J. Math. Phys.* **16**, 433–437 (1975).

[170] Dubrovin B., Fomenko A. and Novikov S., *Modern Geometry-Methods and Applications*, Part II (Springer-Verlag, Berlin, 1985).

[171] Jackiw R., Jacobs L. and Rebbi C., Static Yang–Mills fields with sources, *Phys. Rev. D* **20**, 474–486 (1979).

[172] Papapetrou A., Equations of motion in general relativity, *Proc. Phys. Soc. A* **64**, 57–75 (1951).

[173] Kholodenko A., Heisenberg honeycombs solve Veneziano puzzle, *Int. Math. Forum* **4**, 441–509 (2009).

[174] Bloch A., *Nonholonomic Mechanics and Control* (Springer-Verlag, Berlin, 2003).

[175] Calin O. and Chang, D-C., Sub-Riemannian geometry, a variational approach, *J. Diff. Geom.* **80**, 23–43 (2008).

[176] Rovelli C., *Quantum Gravity* (Cambridge Univ. Press, Cambridge, 2004).

[177] Kholodenko A., Role of general relativity and quantum mechanics in dynamics of solar system, *Adv. Stud. Theor. Phys.* **4**, 689–736 (2010).

[178] Borisov A. and Fomenko A., *Integrable Hamiltonian Systems: Geometry, Topology, Classification* (Chapman & Hall/CRC, New York, 2004).

[179] Belinski V. and Verdaguer E., *Gravitational Solitons* (Cambridge Univ. Press, Cambridge, 2001).

[180] Stephani H., Kramer D., MacCallum C. and Herlt E., *Exact Solutions of Einstein's Field Equations* (Cambridge Univ. Press, Cambridge, 2003).

[181] Kholodenko A., Gravity-assisted solution of the mass gap problem for pure Yang–Mills fields, *Int. J. Geom. Methods Mod. Phys.* **8**, 1–64 (2011).

[182] Chern S-S., Finsler geometry is just Riemannian geometry without the quadratic restriction, *AMS Notices* **43**, 959–963 (1996).

[183] Thurston W., *Three-Dimensional Geometry and Topology* (Princeton Univ. Press, Princeton, 1997).

[184] Folland G. and Stein E., *Hardy Spaces on Homogenous Groups* (Princeton Univ. Press, Princeton, 1982).

[185] Danielli D., Garofalo N. and Nhieu D-M., Notion of convexity in Carnot groups, *Comm. Anal. Geom.* **11**, 263–341 (2003).

[186] Landau L. and Lifshitz E., *Quantum Mechanics (Nonrelativistic Theory)* (Pergamon Press, London, 1977).

[187] Way R., *Dynamics in the Hopf Bundle, the Geometric Phase and Implications for Dynamical Systems*, Ph.D. Thesis, Department of Mathematics, University of Surrey, UK, 2008.

[188] McMullen P. and Schulte E., *Abstract Regular Polytopes* (Cambridge Univ. Press, Cambridge, 2002).

[189] Montgomery R., *A Tour of Sub-Riemannian Geometries, Their Geodesics and Applications* (Amer. Math. Soc., Providence, 2002).

[190] Binz E. and Pods S., *The Geometry of Heisenberg Groups* (Amer. Math. Soc., Providence, 2008).

[191] Binz E., Pods S. and Schempp W., The Heisenberg group as fundamental structure in Nature, *Stud. Multidisciplinarity* **2**, 69–81 (2004).
[192] Cartier P., Quantum mechanical commutative relations and theta functions, *Proc. Symp. Pure Math.* **9**, 361–387 (1965).
[193] Mumford D., *Tata Lectures on Theta*, Vols. 1–3 (Birkhäuser, Boston, 1983–1985).
[194] Auslander L. and Tolimieri R., *Abelian Harmonic Analysis, Theta Functions and Function Algebras on a Nilmanifold*, Lecture Notes in Mathematics, Vol. 436 (Springer-Verlag, Berlin, 1975).
[195] Hurtado A. and Rosales C., Area-stationary surfaces inside the sub-Riemannian three sphere, *Math. Ann.* **340**, 675–708 (2008).
[196] Schneider M., Closed magnetic geodesics on S^2, arXiv:0808.4038.
[197] Taimanov I., Closed extremals on two-dimensional manifolds, *Russ. Math. Surv.* **47**, 163–211 (1992).
[198] Chang D-Ch., Markina I. and Vasil'ev A., Hopf fibration: Geodesics and distances, arXiv:1009.4789.
[199] Adams C., *The Knot Book* (W.A. Freeman and Co., New York, 1994).
[200] Kocak H. and Ladilaw D., Computer graphics and geometry of S^3, *Math. Intelligencer* **8**, 8–10 (1987).
[201] Kozlov V., Calculus of variations in the large and classical mechanics, *Russ. Math. Surv.* **40**, 37–71 (1985).
[202] Witten E., Global aspects of current algebra, *Nucl. Phys. B* **223**, 422–432 (1983).
[203] Kholodenko A. and Vilgis T., Dirac chains in the presence of hairpins, *Phys. Rev. E* **52**, 3973–3988 (1995).
[204] Di Francesco P., Mathieu P. and Senechal D., *Conformal Field Theory* (Springer-Verlag, Berlin, 1997).
[205] Chern S-S., Circle Bundles, *Lecture Notes Math.* **597**, 114–131 (1976).
[206] Pincall U., Hopf tori in S^3, *Invent. Math.* **81**, 379–386 (1985).
[207] Singer I. and Thorpe J., *Lecture Notes on Elementary Topology and Geometry* (Springer-Verlag, Berlin, 1967).
[208] De Azcarraga J. and Izquierdo J., *Lie Groups, Lie Algebras, Cohomology and Some Applications in Physics* (Cambridge Univ. Press, Cambridge, 1998).
[209] Abrikosov A., On the magnetic properties of superconductors of the second group, *Sov. Phys. JETP* **5**, 1174–1182 (1957).
[210] Essmann U. and Trauble H., First observation of Abrikosov lattice in superconductors, *Phys. Lett. A* **24**, 526–528 (1967).
[211] Tzaneteas T., *Abrikosov Lattice Solutions of the Ginzburg–Landau Equations of Superconductivity*, Ph.D. Thesis, Department of Mathematics, Univ. of Toronto, 2010.
[212] Pan X.-B., Landau–de Gennes model of liquid crystals and critical wave number, *Comm. Math. Phys.* **239**, 343–382 (2003).
[213] Chaikin P. and Lubensky T., *Principles of Condensed Matter Physics* (Cambridge Univ. Press, Cambridge, 1995).

[214] Goldman W., *Complex Hyperbolic Geometry* (Clarendon Press, Oxford, 1999).

[215] Kholodenko A., Boundary conformal field theories, limit sets of Kleinian groups and holography, *J. Geom. Phys.* **35**, 193–238 (2000).

[216] Schrödinger E., Die gegenwärtige situation der quantenmechanik, *Naturwissenshaften* **23**, 807–812 (1935).

[217] Einstein A., Podolsky B. and Rosen N., Can quantum mechanical description of physical reality be considered complete? *Phys. Rev.* **47**, 777–780 (1935).

[218] Klyachko A., Dynamical symmetry approach to entanglement, *NATO Secur. Sci. Ser. D Inf. Commun. Secur.* **7**, 25–54, IOS, Amsterdam, 2007, arXiv:0802.4008.

[219] Bruft D., Characterizing entanglement, *J. Math. Phys.* **43**, 4237–4251 (2002).

[220] Bell J., *Speakable and Unspeakable in Quantum Mechanics* (Cambridge Univ. Press, Cambridge, UK, 1997).

[221] Genovese M., Research on hidden variable theory: Recent progress, *Phys. Rep.* **413**, 319–396 (2005).

[222] Gühne O. and Toth G., Entanglement detection, *Phys. Rep.* **474**, 1–75 (2009).

[223] Josza R. and Linden N., On the role of entanglement in quantum-computational speed-up, *Proc. Roy. Soc. Lond. A* **459**, 2011–2032 (2003).

[224] Valliant L., Quantum circuits that can be simulated classically in polynomial time, *SIAM J. Comput.* **31**, 1229–1254 (2002).

[225] Zagoskin A., *Quantum Engineering: Theory and Design of Quantum Coherent Structures* (Cambridge Univ. Press, Cambridge, 2011).

[226] Zagoskin A. and Blais A., Superconducting qubits, *Phys. Canad.* **63**, 215–227 (2007).

[227] Dong D. and Petersen I., Quantum control theory and applications: A survey, *IET Control Theory Appl.* **4**, 2651–2671 (2010).

[228] Khrennikov A., Demystification of quantum entanglement, arXiv:0905.4791.

[229] Muschik C., Polzik E. and Cirac J., Dissipatively driven entanglement of two macroscopic atomic ensembles, *Phys. Rev. A* **83**, 052312 (2011).

[230] Nakahara M. and Ohmi T., *Quantum Computing: From Linear Algebra to Physical Realizations* (CRC Press, Taylor and Francis Group, Boca Raton, 2008).

[231] Feynman R., Vernon F. and Hellwarth R., Geometrical representation of the Schrödinger equation for solving master equation, *J. Appl. Phys.* **28**, 49–52 (1957).

[232] Pars L., *A Treatise on Analytical Dynamics* (Heineman, London, 1968).

[233] Kholodenko A., New strings for old Veneziano amplitudes III. Symplectic treatment, *J. Geom. Phys.* **56**, 1433–1472 (2006).

[234] Goldstein H., *Classical Mechanics* (Addison-Wesley, Cambridge, 1980).

[235] Landau L. and Lifshitz E., *Mechanics* (Pergamon Press, Oxford, 1960).

[236] Klein F. and Sommerfeld A., *Theory of the Topology*, Vol. 2 (Birkhäuser, Boston, 2010).

[237] Rojo A. and Bloch A., The rolling sphere, the quantum spin, and a simple view of the Landau–Zener problem, *Am. J. Phys.* **78**, 1014–1022 (2010).
[238] Love A., *A Treatise on the Mathematical Theory of Elasticity* (Dover, New York, 1944).
[239] Jurdjevic V., *Geometric Control Theory* (Cambridge Univ. Press, Cambridge, 1997).
[240] Jurdjevic V., *Integrable Hamiltonian Systems on Complex Lie Groups*, AMS Memoirs, Vol. 838 (Amer. Math. Soc., Providence, 2005).
[241] Carmel L. and Mann A., Geometrical approach to two-level Hamiltonians, *Phys. Rev. A* **61**, 052113 (2000).
[242] Rabi I., Space quantization in a gyrating magnetic field, *Phys. Rev.* **51**, 652–654 (1937).
[243] Bloch F., Nuclear induction, *Phys. Rev.* **70**, 460–474 (1946).
[244] Khaneja N., On some problems in quantum control, *Comm. Infor. Syst.* **9**, 1–40 (2009).
[245] McMaster W., Matric representation of polarization, *Rev. Mod. Phys.* **33**, 8–28 (1961).
[246] Allen L. and Eberly J., *Optical Resonance and Two-Level Atoms* (Wiley, New York, 1975).
[247] Landau L. and Lifshitz E., *Quantum Mechanics* (Elsevier, Amsterdam, 2003).
[248] Schevchenko S., Ashhab S. and Nori F., Landau–Zener–Stückelberg interferometry, *Phys. Rep.* **492**, 1–30 (2010).
[249] Zhu C., Teranishi Y. and Nakamura H., Nonadiabatic transitions due to curve crossings: Complete solutions of the Landau–Zener–Stueckelberg problems and their applications, in *Advances in Chemical Physics*, Vol. 117, eds. Prigogine I. and Rice S.A. (John Wiley & Sons, Hoboken, 2007), pp. 127–233.
[250] Delos J. and Thorson W., Studies of the potential-curve crossing problem II. General theory and a model for close crossings, *Phys. Rev. A* **6**, 728–745 (1979).
[251] Rosen N. and Zener C., Double Stern–Gerlach experiment and related collision phenomena, *Phys. Rev.* **40**, 502–507 (1932).
[252] Hioe F. and Carroll C., Two-state problems involving arbitrary amplitude and frequency modulations, *Phys. Rev. A* **32**, 1541–1549 (1985).
[253] Nikitin E., Nonadiabatic transitions: What we learned from old masters and how much we owe them, *Ann. Rev. Phys. Chem.* **50**, 1–21 (1999).
[254] Osherov V. and Voronin A., Exact analytical solution of the quantum Rosen–Zener–Demkov model, *Phys. Rev. A* **49**, 265–271 (1994).
[255] Dandoloff R., Balakrishnan R. and Bishop A., Two-level systems: Space curve formalism, Berry's phase and Gauss–Bonnet theorem, *J. Phys. A: Math. Gen.* **25**, L1105–L1110 (1992).
[256] Kostrikin A., *Introduction to Algebra* (Springer-Verlag, Berlin, 1982).
[257] Bengtsson I. and Zyczkowski K., *Geometry of Quantum States* (Cambridge Univ. Press, Cambridge, 2006).

[258] Schumacher R., *Introduction to Magnetic Resonance* (W.A. Benjamin Inc., New York, 1970).
[259] Calin O., Chang D-Ch. and Markina I., Sub-Riemannian geometry on the sphere S^3, *Canad. J. Math.* **61**, 721–739 (2009).
[260] Chang D-Ch., Markina I. and Vasil'ev A., Sub-Riemannian geodesics on the 3-sphere, *Compl. Anal. Oper. Theory* **3**, 361–377 (2009).
[261] Bonnard B., Scherbakova N. and Sugny D., The smooth continuation method in optimal control with an application to quantum systems, *Control Optim. Calc. Var.* **17**, 267–292 (2011).
[262] Bonnard B., Cots O., Scherbakova N. and Sugny D., The energy minimization problem for two-level quantum systems, *J. Math. Phys.* **51**, 092705 (2010).
[263] Vishik S. and Dolzhanskii F., Analogs of the Euler–Lagrange equations and magnetohydrodynamics connected with Lie groups, *Sov. Phys. Dokl.* **19**, 149–153 (1978).
[264] Bloch A. and Rojo A., Quantization of a nonholonomic system, *Phys. Rev. Lett.* **101**, 030402 (2008).
[265] Dirac P., Generalized Hamiltonian dynamics, *Canad. J. Math.* **2**, 129–148 (1950).
[266] van Holten J., Aspects of BRST quantization, *Lecture Notes Phys.* **659**, 99–166 (2005).
[267] Schwarz A., Geometry of Batalin–Vilkovisky quantization, *Comm. Math. Phys.* **155**, 249–260 (1993).
[268] Faddeev L. and Jackiw R., Hamiltonian reduction of unconstrained and constrained systems, *Phys. Rev. Lett.* **60**, 1692–1694 (1988).
[269] Arnol'd V., First steps in local contact algebra, *Canad. J. Math.* **51**, 1123–1134 (1999).
[270] Moser J., Geometry of quadrics and spectral theory, in *The Chern Symposium 1979* (Springer-Verlag, Berlin, 1980), pp. 147–188.
[271] Webster S., Quadrics and complete integrability in contact geometry, *Comm. Pure Appl. Math.* **56**, 824–838 (2003).
[272] Fedorov Yu. and Jovanovic B., Hamiltonization of the general Veselova LR systems, *Regular Chaotic Dynam.* **14**, 495–505 (2009).
[273] Jovanovic B., Hamiltonization and integrability of the Chaplygin sphere in $\mathbf{R}^n$, *J. Nonlinear Sci.* **20**, 569–583 (2010).
[274] Nakamura H., *Nonadiabatic Transition. Concepts, Basic Theories and Applications* (World Scientific, Singapore, 2002).
[275] Kholodenko A., Landau's last paper and its impact on developments in mathematics, physics and other disciplines in new millennium, in *Lev Davidovich Landau and His Impact on Contemporary Theoretical Physics*, eds. Sakaji A. and Licata I. (Nova Sci. Publishers, Inc., New York, 2009), pp. 37–76, arXiv:0806.1064.
[276] Tsvelik A., *Quantum Field Theory in Condensed Matter Physics* (Cambridge Univ. Press, Cambridge, 2003).
[277] Kitaev A., Anyons in exactly solved model and beyond, *Ann. Phys.* **321**, 2–111 (2006).

[278] Nayak C., Simon S., Stern A., Freedman M. and Das Sarma S., Non-Abelian anyons and topological quantum computation, *Rev. Mod. Phys.* **80**, 1083–1159 (2008).

[279] Shoelkopf R. and Girvin S., Wiring up quantum systems, *Nature* **451**, 664–669 (2008).

[280] Clarke J. and Wilhelm F., Supeconducting quantum bits, *Nature* **453**, 1031–1042 (2008).

[281] You, J. and Nori F., Atomic physics and quantum optics using superconducting circuits, *Nature* **474**, 589–597 (2011).

[282] Sugny D. and Kontz C., Optimal control of three-level quantum system by laser fields plus von Neumann measurements, *Phys. Rev. A* **77**, 063420 (2008).

[283] Boscain U., Chambrion Th. and Charlot G., Nonisotropic 3-level quantum systems: Complete solutions for minimum time and minimum energy, *Disc. Cont. Dyn. Syst.* **5B**, 957–990 (2005).

[284] Boscain U. and Charlot G., Resonance of minimizers for n-level quantum systems with an arbitrary cost, *Control Optim. Calc. Var.* **10**, 593–614 (2004).

[285] Barone A. and Paterno J., *Physics and Applications of the Josephson Effect* (John Wiley & Sons Inc., New York, 1982).

[286] Dahm A., Denenstein A., Finnegen T. Langenberg D. and Scalapino D., Study of the Josephson plasma resonance, *Phys. Rev. Lett.* **20**, 859–863 (1968).

[287] Tilley D. and Tilley J., *Superconductivity and Superfluidity* (Van Nostrand Reingold Co., New York, 1974).

[288] Akhiezer A. and Berestetskii V., *Quantum Electrodynamics* (Intersicience Publishers, New York, 1965).

[289] Zee A., *Quantum Field Theory in a Nutshell* (Princeton Univ. Press, Princeton, 2010).

[290] Kholodenko A., Potts model, Dirac propagator and conformational statistics of semiflexible polymers, *J. Stat. Phys.* **65**, 291–316 (1991).

[291] Gersch H., Feynman's relativistic chessboard as an Ising model, *Int. J. Theor. Phys.* **20**, 491–501 (1981).

[292] Feynman R. and Hibbs A., *Quantum Mechanics and Path Integrals* (McGraw-Hill, New York, 1965).

[293] Childs A., On the relationship between the continuous- and discrete-time quantum walk, *Comm. Math. Phys.* **294**, 581–603 (2010).

[294] Yepez J., Relativistic path integral as a lattice-based quantum algorithm, *Quantum Infor. Process.* **4**, 471–509 (2006).

[295] Meyer D., From quantum cellular automata to quantum lattice gases, *J. Stat. Phys.* **85**, 551–574 (1996).

[296] Lloyd S., A potentially realizable quantum computer, *Science* **261**, 1569–1571 (1993).

[297] Berman G., Doolen G., Maineri R. and Tsifrinovich V., *Introduction to Quantum Computers* (World Scientific, Singapore, 1998).

[298] Farhi E., Goldstone J., Gutman S. and Sipser M., Quantum computation by adiabatic evolution, arXiv:quant-ph/0001106.
[299] Perdomo A., Truncik C., Tubert-Brohman I., Rose G. and Aspuru-Guzik A., Construction of model Hamiltonians for adiabatic quantum computation and its application to finding low-energy conformations of lattice protein models, *Phys. Rev. A* **78**, 012320 (2008).
[300] Aharonov, D., van Dam, W., Kempe J., Landau Z., Lloyd S. and Regev O., Adiabatic quantum computation is equivalent to standard quantum computation, *SIAM J. Comput.* **37**, 166–194 (2007).
[301] Childs A., Universal computation by quantum walk, *Phys. Rev. Lett.* **102**, 180501 (2009).
[302] Lovett N., Cooper S., Everitt M., Trevers M. and Kendon V., Universal quantum computation using the discrete-time quantum walk, *Phys. Rev. A* **81**, 042330 (2010).
[303] Josza R., An introduction to measurement-based quantum computation, arXiv:quant-ph/0508124.
[304] Briegel H., Browne D., Dur W., Raussendorf R. and Van den Nest M., Measurement-based quantum computation, *Nature Phys.* **5**, 19–26 (2009).
[305] Lloyd S., Quantum coherence in biological systems, *J. Phys.: Conf. Ser.* **302**, 012037 (2011).
[306] Mohsheni M., Rebentrost P., Lloyd S. and Aspuru-Guzik A., Environment-assisted quantum walks in photosynthetic energy transfer, *J. Chem. Phys.* **129**, 1741006.
[307] Buluta I. and Nori F., Quantum simulators, *Science* **326**, 108–111 (2009).
[308] Lamata L., Leon J., Schatz T. and Solano E., Dirac equation and quantum relativistic effects in a single trapped ion, *Phys. Rev. Lett.* **98**, 253005 (2007).
[309] Gerrithsma R., Kirchmair G., Zahringer F., Sodano E., Blatt R. and Ross C., Quantum simulation of the Dirac equation, *Nature Lett.* **463**, 68–72 (2010).
[310] Blatt R. and Roos C., Quantum simulations with trapped ions, *Nature Phys.* **8**, 277–284 (2012).
[311] Symanzik K., in *Local Quantum Theory*, ed. Jost R. (Academic Press, New York, 1969).
[312] Glimm J. and Jaffe A., *Quantum Physics. A Functional Integral Point of View*, 2nd edition (Springer-Verlag, Berlin, 1987).
[313] Dombey N. and Calogeracos A., Seventy years of Klein paradox, *Phys. Rep.* **315**, 41–58 (1999).
[314] Kholodenko A., Bearden D. and Douglas J., Influence of surface interactions and chain stiffnessbon polymer-induced entropic forces and dimensions of confined polymers, *Phys. Rev. E* **49**, 2206–2224 (1994).
[315] Hejčik P. and Cheon T., Anomalous relativistic tunneling and exotic point interactions, *Eur. Phys. Lett.* **81**, 50001 (2008).
[316] Kholodenko A. and Freed K., Renormalization group treatment of polymer excluded volume by t'Hooft–Veltman-type dimensional regularization, *J. Chem. Phys.* **78**, 7390 (1983).

[317] Landau L. and Lifshitz E., *Statistical Physics*, Part 1 (Elsevier Science, Amsterdam, 1980).
[318] Kholodenko A., Fermi-Bose transmutation: From semiflexible polymers to superstrings, *Ann. Phys.* **202**, 186–225 (1990).
[319] Kholodenko A., Analytical calculation of the scattering function for polymers of arbitrary flexibility using the Dirac propagator, *Macromolecules* **26**, 4179–4183 (1993).
[320] Stone M., Supersymmetry and the quantum mechanics of spin, *Nucl. Phys. B* **314**, 557–586 (1989).
[321] Kholodenko A., New models for Veneziano amplitudes: Combinatorial, symplectic and supersymmetric aspects, *Int. J. Geom. Methods Mod. Phys.* **4**, 563–584 (2005).
[322] Kholodenko A., Statistical mechanics of semiflexible polymers: Yesterday, today and tomorrow, *J. Chem. Soc. Faraday Trans.* **91**, 2473–2482 (1995).
[323] de Gennes P. G., Mechanical properties of nematic polymers, in *Polymer Liquid Crystals*, eds. Ciferi A., Krigbaum W. and Meyer R. (Academic Press, New York, 1982), pp. 115–131.
[324] Faddeev L. and Takhtajan L., *Hamiltonian Methods in the Theory of Solitons* (Springer-Verlag, Berlin, 1987).
[325] Strick T., Dessinges M. N., Charvin G., Dekker N., Allemand J. F., Bensimon D. and Croquette V., Stretching of macromolecules and proteins, *Rep. Progr. Phys.* **66**, 1–45 (2003).
[326] Gunn J. and Warner M., Giant dielectric response and hairpins in polymeric liquids, *Phys. Rev. Lett.* **58**, 393–396 (1987).
[327] Ito K. and McKean H., *Diffusion Processes and Their Sample Paths* (Springer-Verlag, Berlin, 1974).
[328] Griffiths R., Hilbert space quantum mechanics is noncontextual. arXiv: 1201.1510.
[329] Fuchs C., Quantum mechanics as quantum information, arXiv:quant-ph/0205039.
[330] Wang Z., *Topological Quantum Computation* (Amer. Math. Soc., Providence, 2010).
[331] Mumford D. and Desolneux A., *Pattern Theory* (A.K. Peters Ltd., Natic, 2010).
[332] Aitchinson I., Berry phases, magnetic monopoles and Wess–Zumino terms or how the Skyrmion got its spin, *Acta Phys. Pol. B* **18**, 207–235 (1987).
[333] Veselov A., The Landau–Lifshitz equation and integrable systems of clsssical mechanics, *Sov. Phys. Dokl.* **28**, 458–459 (1983).
[334] Luckhurst G. and Zannoni C., Why is the Mayer–Saupe theory of nematic liquid crystals is so successful? *Nature Lett.* **267**, 412–414 (1977).
[335] Spakowitz A. and Wang Z-G., Semiflexible polymer solutions. I. Phase behavior and single chain statistics, *J. Chem. Phys.* **119**, 13113–13128 (2003).
[336] Warner M., Gunn J. and Baumgartner A., Rod to coil transitions in nematic polymers, *J. Phys. A* **18**, 3007–3026 (1985).

[337] Moser J. and Zehnder E., *Notes on Dynamical Systems* (Amer. Math. Soc., Providence, 2005).

[338] Krešić-Jurić S., A loop group approach to the C. Neumann problem and Moser–Veselov factorization, *J. Math. Phys.* **40**, 5014–5025 (1999).

[339] Ohsawa T., Fernandez O., Bloch A. and Zenkov D., Nonholonomic Hamilton–Jacobi theory via Chaplygin Hamiltonization, *J. Geom. Phys.* **61**, 1263–1291 (2011).

[340] Bloch A. and Fernandez O., The Weitzenböck connection and time reparametrization in nonholonomic mechanics, *J. Math. Phys.* **52**, 012901 (2011).

[341] Shapere A. and Wilczek F., Geometry of self-propulsion at low Reynolds number, *J. Fluid Mech.* **198**, 557–585 (1989).

[342] Johanning M., Varon A. and Winderlich C., Quantum simulations with cold trapped ions, *J. Phys. B* **42**, 154009 (2009).

[343] Paul W., Electromagnetic traps for charged and neutral particles, *Rev. Mod. Phys.* **62**, 531–540 (1990).

[344] Leibfried D., Blatt R., Monroe C. and Wineland D., Quantum dynamics of single trapped ions, *Rev. Mod. Phys.* **75**, 281–324 (2003).

[345] Gurarie D., Quantized Neumann problem, separable potenstials on S^n and the Lame' equation, *J. Math. Phys.* **36**, 5355–5391 (1995).

[346] Kuznetsov V., Generalized polyspheroidal periodic functions and the quantum inverse scattering method, *J. Math. Phys.* **31**, 1167–1174 (1990).

[347] Kalnins E., Kuznetsov V. and Miller W., Quadrics on complex Riemannnian spaces of constant curvature, separation of variables, and the gaudin magnet, *J. Math. Phys.* **35**, 1710–1731 (1994).

[348] Arnol'd V., *Singularities of Caustics and Wave Fronts* (Kluwer Academic Publishers, Dordrecht, 1990).

[349] Maslov V. and Fedoriuk M., *Semi-Classical Approximation in Quantum Mechanics* (Kluwer Academic Publishers, Dordrecht, 1981).

[350] Kushner A., Lychagin V. and Rubtsov V., *Contact Geometry and Non-Linear Differential Equations* (Cambridge Univ. Press, Cambridge, 2007).

[351] Vinogradov A., Krasil'shchik I. and Lychagin V., *Introduction to the Geometry of Nonlinear Differential Equations* (Nauka Publishers, Moscow, 1986), in Russian.

[352] Albers P., Frauenfelder U., Koert O. and Paternian G., The contact geometry of the restricted three-body problem, arXiv:1010.2140.

[353] Makarini L. and Paternian G., Equivariant symplectic homology of Anosov contact structures, arXiv:1111.5579.

[354] Hofer H. and Zehnder E., *Symplectic Invariants and Hamiltonian Dynamics* (Birkhäuser Verlag, Basel, 1994).

[355] Gao B., Breakdown of Bohr's correspondence principle, *Phys. Rev. Lett.* **83**, 4225–4228 (1999).

[356] Makowski A., A brief survey of various formulations of the correspondence principle, *Eur. J. Phys.* **27**, 1133–1139 (2006).

[357] D. Bohm, A suggested interpretation of the quantum theory in terms of "hidden" variables. I, II, *Phys. Rev.* **85**, 166–193 (1952).

[358] Arnol'd V., Topological problems in the theory of wave propagation, *Russ. Math. Surv.* **51**, 3–50 (1996) (in Russian).

[359] Arnol'd V., First steps in symplectic topology, *Russ. Math. Surv.* **41**, 1–21 (1986).

[360] Kocsis S., Braverman B., Ravets S., Stevens M., Mirin R., Shalm L. and Steinberg A., Observing the average trajectories of single photons in a two-slit interferometer, *Science* **332**, 1170–1173 (2011).

[361] Rozema L., Darabi A., Mahler D., Hayat A., Soudagar Y. and Steinberg A., Violation of Heisenberg's measurement-disturbance relationship by weak measurements, *Phys. Rev. Lett.* **109**, 100404 (2012).

[362] Ozawa M., Universally valid reformulation of the Heisenberg uncertainty principle of noise and disturbance in measurement, *Phys. Rev. A* **67**, 042105 (2003). See also Aharonov Y., Albert D. and Vaidman L., How the result of a measurement of a component of the spin-1/2 particle can turn out to be 100, *Phys. Rev. Lett.* **60**, 1351–1354 (1988).

[363] Kofman A., Ashhab S. and Nori F., Nonperturbative theory of weak pre- and post-selected measurements, *Phys. Rep.* (2012) in press.

[364] Oriols X. and Mompart J., *Applied Bohmian Mechanics: From Nanoscale Systems to Cosmology* (Pan Stanford Publishing Singapore, 2012).

[365] Sanz A. and Miret-Artes S., *A Trajectory Description of Quantum Processes. I. Fundamentals. A Bohmian Perspective* (Springer-Verlag, Berlin, 2012).

[366] Arnol'd V. and Givental' A., Symplectic geometry, in *Dynamical Systems IV*, eds. Arnol'd V. and Novikov S. (Springer-Verlag, Berlin, 1990), pp. 5–136.

[367] Salamon D., Morse theory, the Conley index and Floer homology, *Bull. Lond. Math. Soc.* **22**, 113–140 (1990).

[368] Salamon D. and Zehnder E., Morse theory for periodic solutions of Hamiltonian systems and the Maslov index, *Comm. Pure Appl. Math.* **55**, 1303–1360 (1992).

[369] Ballentine L., The statistical interpretation of quantum mechanics, *Rev. Mod. Phys.* **42**, 358–381 (1970).

[370] Marmo G. and Volkert G., Geometrical description of quantum mechanics-transformations and dynamics, *Phys. Scr.* **82**, 038117 (2010).

[371] de Gosson M., A pseudo-quantum triad: Schrödinger's equation, the uncertainty principle, and the Heisenberg group, *J. Phys.: Conf. Ser.* **361**, 012015 (2012).

[372] Bonahon F., *Low Dimensional Geometry: From Euclidean Surfaces to Hyperbolic Knots* (Amer. Math. Soc., Providence, 2009).

[373] Kholodenko A., Some thoughts about random walks on figure eight, *Physica A* **289**, 377–408 (2001).

[374] Kiernan P., Hyperbolically imbedded spaces and the big Picard theorem, *Math. Ann.* **204**, 203–209 (1973).

[375] Feeney S., Johnson M., Mortlock D. and Peiris H., A robust constraint on cosmic textures from the cosmic microwave background, arXiv:1203.1918.

[376] Braverman B. and Simon C., Proposal to demonstrate the non-locality of Bohmian mechanics with entangled photons, arXiv:1207.2794.
[377] Wiseman H., Grounding Bohmian mechanics in weak values and Bayesianism, *New J. Phys.* **9**, 165 (2007).
[378] Dürr D., Goldstein S. and Zanghi N., On the weak measurement of velocity in Bohmian mechanics, *J. Stat. Phys.* **134**, 1023–1032 (2009).
[379] Kholodenko A., Designing new apartment buildings for strings and conformal field theories, in *New Developments in String Theory Research*, ed. Grece S. (Nova Science Publishers Co., New York, 2006), pp. 1–83.
[380] Guillemin V. and Sternberg S., *Symplectic Techniques in Physics* (Cambridge Univ. Press, Cambridge, 1984).
[381] Nash Ch., *Differential Topology and Quantum Field Theory* (Academic Press, New York, 1992).
[382] Guillemin V. and Sternberg S., *Supersymmetry and Equivariant de Rham Theory* (Springer-Verlag, Berlin, 1999).
[383] Atiyah M. and Bott R., The moment map and equivariant cohomology, *Topology* **23**, 1–28 (1984).
[384] Witten E., Two-dimensional gauge theories revisited, *J. Geom. Phys.* **9**, 303–368 (1992).
[385] Witten E., Supersymmetry and Morse theory, *J. Diff. Geom.* **17**, 661–692 (1982).
[386] Born M., *The Mechanic of the Atom* (Frederic Ungar, New York, 1960).
[387] Rowe D., Ryman A. and Rosensteel G., Many-body quantum mechanics as a symplectic dynamical sytem, *Phys. Rev. A* **22**, 2362–2373 (1980).
[388] Rosensteel G., Mean field theory for $U(n)$ dynamical groups, *J. Phys. A* **44**, 165201 (2011).
[389] Kleinert H., *Path Integrals in Quantum Mechanics, Quantum Statistics, Polymer Physics and Financial Markets* (World Scientific, Singapore, 2009).
[390] Pasquotto F., A short history of the Weinstein conjecture, *Jahresber. Dtsch. Math-Ver.* **114**, 119–130 (2012).
[391] Hutchins M., Taube's proof of the Weinstein conjecture in dimension three, *BAMS* **47**, 73–125 (2009).
[392] Taubes C., Seiberg–Witten equations and the Weinstein conjecture, *Geom. Topol.* **11**, 2117–2202 (2007).
[393] Taubes C., The Seiberg–Witten equations and the Weinstein conjecture. II. More closed integral curves for the Reeb vector field, *Geom. Topol.* **13**, 1337–1417 (2009).
[394] Moore J., *Lectures on Seiberg–Witten Invariants*, Lecture Notes in Mathematics, Vol. 1629 (Springer-Verlag, Berlin, 1996).
[395] Donaldson S., A pseudoholomorphic curve?, *AMS Notices* **52**, 1026–1027 (2005).
[396] Kholodenko A., Use of meanders and train tracks for for description of defects and textures in liquid crystals and 2 + 1 gravity, *J. Geom. Phys.* **33**, 23–58 (2000).
[397] Kholodenko A., Use of quadratic differentials for description of defects and textures in liquid crystals and 2+1 gravity, *J. Geom. Phys.* **33**, 59–102 (2000).

[398] Kholodenko A., Statistical mechanics of $2+1$ gravity from Riemann zeta function and Alexander polynomial: Exact results, *J. Geom. Phys.* **38**, 81–139 (2001).
[399] Hofer H., Pseudoholomorphic curves in symplectizations with applications to the Weinstein conjecture in dimension three, *Invent. Math.* **114**, 515–563 (1993).
[400] Albers P. and Hofer H., On the Weinstein conjecture in higher dimensions, *Comment. Math. Helv.* **84**, 429–436 (2009).
[401] Bramham B. and Hofer H., First steps towards a symplectic dynamics, arXiv:1102.3723.
[402] Eliashberg Y., Contact 3-manifolds twenty years since J. Martinet's work, *Ann. Inst. Fourier* **42**, 165–192 (1992).
[403] Etnyre J. and Ghrist R., Contact topology and hydrodynamics. III. Knotted orbits, *AMS Trans.* **352**, 5781–5794 (2000).
[404] Damgaard P. and Huffel H., *Stochastic Quantization* (World Scientific, Singapore, 1988).
[405] Mezard M., Parisi G. and Virasoro M., *Spin Glass Theory and Beyond* (World Scientific, Singapore, 1987).
[406] Mrowka Th., Legendrian knots and monopoles, *Alg. Geom. Topol.* **6**, 1–69 (2006).
[407] Farhi E., Gosset D., Hassidim A., Lutomirski A. and Shor P., Quantum money from knots, arXiv:1004.5127, to be published in *Inn. Theor. Comp. Sci.* 2012.
[408] Aaronson S., Farhi E., Gosset D., Hassidim A., Kelner J. and Lutomirski A., Quantum money, *Comm. ACM* **55**, 84–92 (2012).
[409] Rolfsen D., *Knots and Links* (Publish or Perish, Houston, 1990).
[410] Kauffman L., *On Knots* (Princeton Univ. Press, Princeton, 1987).
[411] Arnol'd V., Symplectic geometry and topology, *J. Math. Phys.* **41**, 3307–3343 (2000).
[412] Etnyre J., Knots and contact geometry, in *The Handbook of Knot Theory* (Elsevier, Amsterdam, 2005), pp. 105–185.
[413] Chongchitmate W., Classification of Legendrian knots and links, http://www.math.duke.edu/~ng/atlas/
[414] Craik D., Seamless proteins tie up their loose ends, *Science* **311**, 1563–1564 (2006).
[415] Kholodenko A. and Rolfsen D., Knot complexity and related observables from path integrals for semiflexible polymers, *J. Phys. A* **29**, 5677–5691 (1996).
[416] Ng L., A Legendrian Thuston–Bennequin bound from Khovanov homology, *Alg. Geom. Topol.* **5**, 1637–1653 (2005).
[417] Kauffman L., Khovanov homology, arXiv:1107.1524.
[418] Kholodenko A., New strings for old Veneziano amplitudes. II. Group-theoretic treatment, *J. Geom. Phys.* **56**, 1387–1432 (2006).
[419] Kholodenko A., Veneziano amplitudes, spin chains and string models, *Int. J. Geom. Methods Math. Phys.* **6**, 769–803 (2009).

[420] Bar-Natan D., The Mathematica package "Knot Theory", available at http://katlas.math.toronto.edu/wiki/The_Mathematica_Package_Knot-Theory.

[421] Cromwell P., Embedding knots and links in an open book. I. Basic properties, *Topol. Appl.* **64**, 37–58 (1995).

[422] Cromwell P. and Nutt J., Embedding knots and links in an open book. II. Bounds on arc index, *Math. Proc. Cambr. Phil. Soc.* **119**, 309–319 (1996).

[423] Chongchitmate W. and Ng L., An atlas of Legendrian knots, arXiv: 1010.3997 and also http://www.math.duke.edu/~ng/atlas/

[424] Jin G., Kim H. and Lee G., Prime knots with arc index up to 10, in *Intelligence of Low Dimensional Topology 2006*, eds. Carter J., Kauffman L. and Kohno T. (World Scientific, Singapore, 2007), pp. 65–74.

[425] Harandi M., Sanderson C., Shirazi S. and Lovell B., Graph embedding discriminantal analysis on Grassmann manifolds for improved image set matching, in 2011 *IEEE Conf. Computer Vision and Pattern Recognition* (IEEE, 2011).

[426] Chekanov Y. and Pushkar P., Combinatorics of fronts of Legendrian links, and Arnol'd 4-conjectures, *Russ. Math. Surv.* **60**, 99–154 (2005) (in Russian).

[427] Chekanov Y., Invariants of Legendrian knots, *Int. Congr. Math.* **3**, 385–394 (2002).

[428] Itzykson C. and Zuber J-B., *Quantum Field Theory* (McGraw-Hill, New York, 1980).

[429] Idzumi M., On a loop expansion formula for enumeration of Euler subgraphs of planar graph, *Mem. Fac. Sci. Eng. Shimane Univ. Ser. B: Math. Sci.* **33**, 1–10 (2000).

[430] da Costa G., Feynman identity: A special case I, *J. Math. Phys.* **38**, 1014–1034 (1997), also arXiv:math-ph/0305042.

[431] Kohno T., *New Developments in the Theory of Knots* (World Scientific, Singapore, 1990).

[432] Eliashberg Y., Givental A. and Hofer H., Introduction to symplectic field theory, *GAFA*, *Special Vol.*, Part II, 560–673 (2000).

[433] Witten E., Topological quantum field theory, *Comm. Math. Phys.* **117**, 353–386 (1988).

[434] Eliashberg Y., Invariants in contact topology, *Int. Congr. Math. Extra Vol.* 327–338 (1998).

[435] Chekanov Y., Differential algebra of Legendrian links, *Invent. Math.* **150**, 441–483 (2002).

[436] Atiyah M., New invariants of 3 and 4 dimensional manifolds, *Proc. Symp. Pure Math.* **48**, 285–299 (1988).

[437] Atiyah M., Topological quantum field theory, *IHES Pub.* **68**, 175–186 (1988).

[438] Herald C., Chern–Simons gauge theory on 3-manifolds, Lecture Notes in Mathematics (Seoul University) **46**, 83–109 (1999).

[439] Donaldson S., The Seiberg–Witten equations and 4-manifold topology, *BAMS* **33**, 45–70 (1996).

[440] McDuff D., Floer theory and low dimensional topology, *BAMS* **43**, 25–42 (2006).
[441] Sivek S., Monopole Floer homology and Legendrian knots, arXiv:1107.6028.
[442] Kronheimer P. and Mrowka T., Monopoles and contact structures, *Invent. Math.* **130**, 209–255 (1997).
[443] Finkelstein R., An $SL_q(2)$ extension of the standard model, arXiv: 1205.1026.
[444] Billson-Thompson S., Hackett J., Kauffman L. and Wan Y., Emergent braided matter of quantum geometry, *SIGMA* **8**, 014 (2012).
[445] Ozvath P. and Szabo Z., An introduction to Heegaard–Floer homology, *Clay Math. Inst. Proc.* **5**, 3–35 (2006).
[446] Sarkar S., *Topics in Heegaard–Floer Homology*, Ph.D. Thesis, Department of Mathematics, Princeton University, arXiv:0906.2579.
[447] Manolescu C., Ozsvath P. and Sarkar S., A combinatorial description of knot Floer homology, *Ann. Math.* **169**, 633–660 (2009).
[448] Manolescu C., Ozsvath P., Szabo Z. and D. Thurston, On combinatorial link Floer homology, *Geom. Topol.* **11**, 2339–2412 (2007).
[449] Kutluhan C., Lee Y-Z. and Taubes C. H., HF = HM I: Heegaard–Floer homology and Seiberg–Witten Floer homology, arXiv:1007.1979.
[450] Colin V., Ghiggini P. and Honda K., The equivalence of Heegaard–Floer homology and embedded contact homology: From hat to plus, arXiv:1208.1526.
[451] Droz J. M., *Link Homologies from Grid Diagrams*, Ph.D. Thesis, Department of Mathematics, University of Zurich, 2009.
[452] Droz J. M., Effective computation of knot Floer homology, arXiv:0803.2379.
[453] Belyakova A., A simplificatioin of combinatorial link Floer homology, *J. Knot Theory Ramifications* **19**, 125–144 (2010).
[454] Veltman M., *Facts and Mysteries in Elementary Particle Physics* (World Scientific, Singapore, 2003).
[455] Kauffman L., Remarks on Khovanov homology and the Potts model, arXiv:0907.3178.
[456] Plamenevskaya O., Transverse knots and Khovanov homology, arXiv:math/0412184.
[457] Ng L., Oszvath P. and Thurston D., Transverse knots distinguished by knot Floer homology, arXiv:math/0703446.
[458] Ozsvath P., Szabo Z. and Thurston D., Legendrian knots, transverse knots and combinbatorial Floer homology, arXiv:math/0611841.
[459] Kronheimer P. and Mrowka T., Khovanov homology is an unknot detector, arXiv:1005.4346.
[460] Droz J. M. and Wagner E., Grid diagrams and Khovanov homology, arXiv:0902.3370.
[461] Shumakovich A., Patterns in odd Khovanov homology, arXiv:1101.5607.
[462] Oszvath P., Rasmussen J. and Szabo Z., Odd Khovanov homology, arXiv:0710.4300.
[463] Shumakovich A., Khovanov homology theories and their applications, arXiv:1101.5614.

[464] Rasmussen J., Knot polynomials and knot homologies, arXiv:math/0504045.
[465] Bigelow S., A homological definition of the Jones polynomial, *Geom. Topol. Monographs* **4**, 29–41 (2002).
[466] Lee E., An endomorphism of the Khovanov invariant, arXiv:math/02102213.
[467] Manolescu C., Grid diagrams in Heegaard–Floer theory, arXiv:1210.3845.
[468] Gilmore A., *Knot Floer Homology and Categorification*, Ph.D. Thesis, Department of Mathematics, Columbia University, 2011.
[469] Wigner E., The unreasonable effectiveness of mathematics in the natural sciences, *Comm. Pure Appl. Math.* **13**, 1–14 (1960).
[470] Arnol'd V., On teaching mathematics, *Russ. Math. Surv.* **53**, 229–230 (1998).
[471] Jaffe A. and Quinn F., "Theoretical mathematics": Toward a cultural synthesis of mathematics and theoretical physics, *BAMS* **29**, 1–13 (1993).
[472] Atiyah M. *et al.*, Responses to "Theoretical mathematics" by Jaffe and Quinn, *BAMS* **30**, 178–207 (1994).
[473] Thurston W., On proof and progress in mathematics, *BAMS* **30**, 161–177 (1994).
[474] Carroll R., *On the Emergence Theme of Physics* (World Scientific, Singapore, 2010).
[475] Rutherford E., *Phil. Mag.* **21**, 669 (1911).
[476] Grujič P. and Simonovič N., Classical atom, *Phys. Today* **65**, 41–46 (2012).
[477] Hawking S., *The Dreams That Stuff is Made of* (Running Press, London, 2011).
[478] Wannier G., The threshold law for a single ionization of atoms or ions by electrons, *Phys. Rev.* **90**, 817–825 (1953).
[479] Dechoun, K., Franca H. and Malta C., Classical aspects of the Pauli–Schrödinger equation, *Phys. Lett. A* **248**, 93–102 (1998).
[480] Franca H., The Stern–Herlach phenomenon according to classical electrodynamics, *Found. Phys.* **39**, 1177–1190 (2009).
[481] Stenson J., *Representations for Understanding the Stern–Herlach Effect*, MS Thesis, Department of Physics, Brigham Young University, 2005.
[482] Khrennikov A., Entanglement's dynamics from classical stochastic process, *Eur. Phys. Lett.* **88**, 40005 (2009).
[483] Krauter H., Muschik C., Jensen K., Wasilewski W., Petersen J., Cirac J. and Polzik E., Entanglement generated by dissipation and steady-state entanglement of two macroscopic objects, *Phys. Rev. Lett.* **107**, 080503 (2011).
[484] Lee K., Sprague M., Sussman B., Nunn J., Langford N., Jin X-M., Champion T., Michelberger P., Reim K., England D., Jakcsh D. and Waimsley I., Entangling macroscopic diamonds at room temperature, *Science* **334**, 1253–1256 (2011).
[485] Kirillov A., *Elements of the Theory of Representations* (Springer-Verlag, Berlin, 1976).

[486] Sternberg S., General covariance and the passive equations of physics, in *Albert Einstein Memorial Lecture* (The Israel Academy of Sciences and Humanities, Jerusalem, 2006).
[487] Lansburg S., Quantum game theory, *AMS Notices* **51**, 394–399 (2004).
[488] Hanauske M., Bernis S., König W. and Dugall B., Experimental validation of quantum game theory, arXiv:0707.3068.
[489] Summhammer J., Quantum cooperation, *Axiomates* **21**, 347–356 (2011).
[490] Perdomo-Ortiz A., Dickson N., Drew-Brook N., Rose G. and Aspuru-Guzik A., Finding low-energy conformations of lattice protein models by quantum annealing, arXiv:1204.5485.
[491] Benenson Ya., Biocomputers: From test tubes to live cells, *Mol. Biosyst.* **5**, 675–685 (2009).
[492] Rau A., *Astronomy-Inspired Atomic and Molecular Physics* (Kluwer Academic, Dordrecht, 2002).
[493] Arnol'd V., Polymathematics: Is mathematics a singe science or a set of arts, in *Mathematics: Frontiers and Perspectives* (Amer. Math. Soc., Providence, 2000), pp. 403–416.

References for Appendices A–D

[A.1] Kholodenko A., Statistical mechanics of the deformable droplets on flat surfaces, *J. Math. Phys.* **37**, 1287–1313 (1996).
[A.2] Montgomery R., *A Tour of Subriemannian Geometries, Their Geodesics and Applications* (Amer. Math. Soc., Providence, 2002).
[A.3] Bloch A., *Nonholonomic Mechanics and Control* (Springer-Verlag, Berlin, 2003).
[A.4] Sachkov Yu., Control theory on Lie groups, *J. Math. Sci.* **156**, 381-439 (2009).
[A.5] Calin O., Chang D-Ch. and Greiner P., *Geometric Analysis on the Heisenberg Group and Its Generalizations* (Amer. Math. Soc., Providence, 2007).

[B.1] Sachkov Yu., Controllability of invariant systems on Lie Groups and homogenous spaces, *J. Math. Sci.* **100**, 2355–2427 (2000).
[B.2] Sachkov Yu., Cut locus and optimal synthesis in the sub-Riemannian problem on the group of motions of a plane, *ESIAM: Control, Optim. Cal. Var.* **117**, 293–321 (2011).
[B.3] Moiseev I. and Sachkov Yu., Maxwell strata in sub-Riemannian problem on the group of motions of a plane, *ESIAM: Control, Optim. Cal. Var.* **116**, 380–399 (2010).
[B.4] Zelikin M., Theory and applications of the problem of Euler's elastica, *Russ. Math. Surv.* **67**, 281–296 (2012).

[C.1] Birkhoff G. and Bartee T., *Modern Applied Algebra* (McGraw-Hill, New York, 1970).

[C.2] Kholodenko A., Some thoughts about random walks on figure eight, *Physica A* **289**, 377–408 (2001). Also, the full version at arXiv:cond-mat/9905221.

[C.3] Kitaev A., Shen A. and M. Vyalyi, *Classical and Quantum Computation* (Amer. Math. Soc., Providence, 2002).

[C.4] Nielsen M. and Chuang I., *Quantum Computation and Quantum Information* (Cambridge Univ. Press, Cambridge, 2000).

[C.5] Kholodenko A., "New" Veneziano amplitudes from "old" Fermat (hyper)surfaces, in *Trends in Mathematical Physics Research*, ed. Benton Ch. (Nova Sci. Publishers, New York, 2004), pp. 1–94.

[C.6] Feynman R., *Feynman's Lectures on Computation*, eds. Hey A. and Allen R. (Addison-Wesley, New York, 1996).

[C.7] Fredkin E. and Toffoli T., Conservative logic, *Int. J. Theor. Phys.* **21**, 219–253 (1982).

[C.8] Childs A., Universal computation by quantum walk, *Phys. Rev. Lett.* **102**, 180501 (2009).

[C.9] Ressler A., *The Design of Conservative Logic Computer and a Graphical Editor Simulator*, Master's Thesis (MIT, 1981).

[C.10] Smith W., Classical reversible computation with zero Lyapunov exponent NEC Report, Feb. 25, 1999.

[C.11] Kaul R., Chern-Simons theory, coloured-oriented braids and link invariants, *Comm. Math. Phys.* **162**, 289–319 (1994).

[C.12] Nayak C., Simon S., Stern A., Freedman M. and Das Sarma S., Non-Abelian anyons and topological quantum computation, *Rev. Mod. Phys.* **80**, 1083–1159 (2008).

[C.13] Zhang Y., Quantum computing via the Bethe ansatz, *Quantum Infor. Process.* **11**, 2012, in press, available online.

[C.14] Garnerone S., Marzuoli A. and Rasetti M., Quantum geometry and quantum algorithms, arXiv:quant-ph/0607203.

[C.15] Hosseini H. and Dueck G., Toffolli gate implementation using the billiard ball model, *40th Int. Symp. Multiple Valued Logic*, Barcelona, 2010.

[C.16] Maslov D., *Reversible Logic Synthesis*, Ph.D. Dissertation, Department of Computer Science, Univ. of New Brunswick, New Jersey, 2003.

[C.17] Di Vicncenzo D., Bacon D., Kempe J. Burkard J. and Whaley K., Universal quantum computation with exchange interaction, *Nature* **408**, 339–342 (2000).

[C.18] Loss D. and Di Vincenzo D., Quantum computation with quantum dots, *Phys. Rev. A* **57**, 120–126 (1998).

[C.19] Tzerkovnyak V. and Loss D., Universal quantum computation with ordered spin-chain networks, *Phys. Rev. A* **84**, 032333 (2011).

[C.20] Stanton D. and White D., *Constructive Combinatorics* (Springer-Verlag, Berlin, 1986).

[C.21] Dirac P., Fundamental equations of quantum mechanics, *Proc. Roy. Soc. Lond. A* **109**, 642–653 (1925).

[C.22] Knutson A. and Tao T., Honeycombs and sums of Hermitian matrices, *AMS Notices* **48**, 175–186 (2001).

[C.23] Kassel K. and Turaev V., *Braid Groups* (Springer-Verlag, Berlin, 2008).

[C.24] Lascoux A. and Schützenberger M., *Symmetry and Flag Manifolds*, Lecture Notes in Mathematics, Vol. 996 (Springer-Verlag, Berlin, 1980), pp. 118–144.

[C.25] Kirillov A. Jr., Lectures on affine Hecke algebras and Macdonald's conjectures, *Bull. AMS (N.S.)* **34**, 251–292 (1997).

[C.26] Belavin A. and Drinfeld V., Solutions of classical Yang–Baxter equation and simple Lie algebras, *Funct. Anal. Appl.* **16**, 159–180 (1982).

[C.27] Jimbo M., A q-qnqlogue of $U(\mathfrak{gl}(\mathrm{N}+1))$ Hecke algebra, and the Yang–Baxter equation, *Lett. Math. Phys.* **11**, 247–252 (1986).

[C.28] Kassel K., *Quantum Groups* (Springer-Verlag, Berlin, 1995).

[C.29] Goodman F. and Wenzl H., Littlewood–Richardson coefficients for Hecke algebras at roots of unity, *Adv. Math.* **82**, 244–265 (1990).

[C.30] Etingof P., Frenkel I., Kirillov A. Jr., *Lectures on Representation Theory and Knizhnik–Zamolodchikov Equations* (Amer. Math. Soc., Providence, 1998).

[C.31] Jones V., Hecke algebra representations of braid groups and link polynomials, *Ann. Math.* **126**, 335–388 (1987).

[C.32] Turaev V., The Yang–Baxter equation and invariants of links, *Invent. Math.* **92**, 527–553 (1998).

[C.33] Wenzl H., Hecke algebras of type A_n and subfactors, *Invent. Math.* **92**, 349–383 (1988).

[C.34] Kauffman L. and Lomonaco S., Braiding operators are universal quantum gates, arXiv:quant-ph/0401090.

[C.35] Bremmer M., Dawson Ch., Dodd J., Gilchrist A., Harrow A., Mortimer D., Nielsen M. and Osborne T., Practical scheme for quantum computation with any two-qubit entangling gate, *Phys. Rev. Lett.* **89**, 247902 (2002).

[C.36] Brylinski J.-L. and Brulinski R., Universal quantum gates, arXiv:quant-ph/0108062.

[C.37] Dye H., Unitary solutions to the Yang–Baxter equation in dimension four, *Quantum Infor. Process.* **2**, 117–151 (2003).

[C.38] Zhang Y., Kauffman L. and Ge M-L., Yang–Baxterization, universal quantum gates and Hamiltonian, *Quantum Infor. Process.* **4**, 159–197 (2005).

[C.39] Bose S. and Korepin V., Quantum gates between flying qubits via spin-independent scattering, arXiv:1106.2329.

[C.40] Zamolodchikov A. and Zamolodchikov Al., Factorized S-matrices in two dimensions as the exact solutions of certain relativistic quantum field theory models, *Ann. Phys.* **120**, 253–291 (1979).

[C.41] Gomez C. and Sierra G., *Quantum Groups in Two-Dimensional Physics* (Cambridge Univ. Press, Cambridge, 2005).

[C.42] Hikami K., Notes on the structure of the δ-function interacting gas. Intertwining operator in the degenerate affine Hecke algebra, *J. Phys. A* **31**, L85–L91 (1998).

[C.43] Heckman G. and Opdam E., Yang's system of particles and Hecke algebras, *Ann. Math.* **145**, 139–173 (1997).

[C.44] Heckman G. and Schlichtkrull H., *Harmonic Analysis and Special Functions on Symmetric Spaces* (Academic Press, New York, 1994).

[C.45] Kholodenko A., New strings for old Veneziano amplitudes. II. Group-theoretic treatment, *J. Geom. Phys.* **56**, 1387–1432 (2006).

[C.46] Frampton P., *Dual Resonance Models* (W.A. Benjamin, Reading, 1974).

[C.47] Gutkin E., Bethe ansatz and the generalized Yang–Baxter equations, *Ann. Phys.* **176**, 22–48 (1987).

[C.48] Gaudin M., *La Fonction D'onde De Bethe* (Masson, Paris, 1983).

[C.49] Tabachnikov S., *Billiards* (Societé Mathématique de France, Paris, 1995).

[C.50] Kendon V., A random walk approach to quantum algorithms, *Phil. Trans. Roy. Soc. A* **364**, 3407–3422 (2006).

[C.51] Kendon V. and Sanders B., Complementarity and quantum walks, in *Quantum Communication, Measurement and Computing*, AIP Conf. Proc., Vol. 734 (Amer. Inst. Phys., Melville, 2004), pp. 133–138.

[C.52] Kendon V., Decoherence and random walk, *Math. Struct. Comput. Sci.* **17**, 1169–1220 (2007).

[C.53] Childs A., On the relationship between the continuous- and discrete-time quantum walk, *Comm. Math. Phys.* **294**, 581–603 (2010).

[C.54] Josza R., An introduction to measurement-based quantum computation, arXiv:quant-ph/0508124.

[C.55] Briegel H., Browne D., Dur W., Raussendorf R. and Van den Nest M., Measurement-based quantum computation, *Nat. Phys.* **5**, 19–26 (2009).

[C.56] Chandrashekar C., Srikanth R. and Laflamme R., Optimizing the disctrete-time quantum walk using SU(2) coin, *Phys. Rev. A* **77**, 032326 (2008).

[C.57] Chandashekar C., *Discrete-Time Quantum Walk-Dynamics and Applications*, Ph.D. Thesis, Department of Physics, Univ. of Waterloo, Waterloo 2009, arXix:1001.5326v.2.

[D.1] Oskin M., Quantum computing — lecture notes, http://www.cs.washington.edu/homes/oskin/quantum-notes.pdf

[D.2] Jacobczyk L. and Siennicki M., Geometry of Bloch vectors in two-qubit system, *Phys. Lett. A* **286**, 383–390 (2001).

[D.3] Kimura G., The Bloch vector for N-level systems, *Phys. Lett. A* **314**, 339–349 (2003).

[D.4] Byrd M. and Khaneja N., Characterization of the positivity of the density matrix in terms of the coherence vector representation, *Phys. Rev. A* **68**, 062322 (2003).

[D.5] Mandilara A., Clark J. and Byrd M., Elliptical orbits in the Bloch sphere, *J. Optics B* **7**, S227–S282 (2005).

[D.6] Altafini C., Tensor of coherences parametrization of multicubit density operators for entanglement characterization, *Phys. Rev. A* **69**, 012311 (2004).

[D.7] Brüning E., Mäkelä H., Messina A. and Petruccione F. Parametrization of density matrices, *J. Mod. Optics* **59**, 1–20 (2012).
[D.8] Mosseri R. and Dandoloff R., Geometry of entangled states, Bloch spheres and Hopf fibrations, *J. Phys. A* **34**, 10243–10252 (2001).
[D.9] Bernevig B. and Chen H-D., Geometry of the tree-qubit state, entanglement and division algebras, *J. Phys. A* **36**, 8325–8339 (2003).
[D.10] Sugny D., *Geometric Optimal Control of Simple Quantum Systems* (John Wiley & Sons, Hobiken, 2010).
[D.11] Chang D-Ch., Markina I. and Vasil'ev A., Hopf fibration: Geodesics and distances, *J. Geom. Phys.* **61**, 986–1000 (2011).
[D.12] Shabat B., *Introduction to Complex Analysis*, Vol. 2 (Amer. Math. Soc., Providence, 1992).
[D.13] Atiyah M., Angular momentum, convex polyhedra and algebraic geometry, *Proc. Edinburg Math. Soc.* **26**, 121–138 (1983).
[D.14] Fulton W., *Introduction to Toric Varieties* (Princeton Univ. Press, Princeton, 1993).
[D.15] da Silva A., Symplectic toric manifolds, in *Symplectic Geometry of Integrable Hamiltonian Systems*, eds. Audin M., da Silva A. and Lerman E., (Birkhäuser, Boston, 2003), pp. 85–173.
[D.16] Wells R., *Differential Analysis on Complex Manifolds* (Prentice Hall, New York, 1973).
[D.17] Kirillov A., *Lectures on Orbit Method* (Amer. Math. Soc., Providence, 2004).
[D.18] Ziegler G., *Lectures on Polytopes* (Springer-Verlag, Berlin, 1995).
[D.19] Kholodenko A., Traces of mirror symmetry in nature, *Int. Math. Forum* **3**, 151–184 (2008).
[D.20] Heydari H., Quantum entangled state, resolution of conifold singularity, and toric variety, arXiv:1007.3822.
[D.21] Orlik P. and Terrao H., *Arrangements of Hyperplanes* (Springer-Verlag, 1992).
[D.22] Borovik A., Gelfand I. and White N., *Coxeter Matroids* (Birkhäuser, 2003).
[D.23] Bjorner A., Las Vergnas M., Sturmfels B., White N. and Ziegler G., *Oriented Matroids* (Cambridge Univ. Press, Cambridge, 1993).
[D.24] Audin M., *Torus Actions on Symplectic Manifolds* (Birkhäuser, Boston, 2004).
[D.25] Bogoliubov N. and Shirkov D., *Introduction to the Theory of Quantized Fields* (Interscience, New York, 1959).
[D.26] Ewald G., *Combinatorial Convexity and Algebraic Geometry* (Springer-Verlag, Berlin, 1996).
[D.27] Kholodenko A., Designing new apartment buildings for strings and conformal field theories, in *New Developments in String Theory Research*, ed. Grece S. (Nova Science, New York, 2006), pp. 1–83.
[D.28] Humphreys J., *Reflection Groups and Coxeter Groups* (Cambridge Univ. Press, Cambridge, 1997).

[D.29] Carter R., Segal G. and Macdonald I., *Lectures on Lie Groups and Lie Algebras* (Cambridge Univ. Press, Cambridge, 1999).

[D.30] Gelfand I. and Serganova V., Combinatorial geometries and torus strata on homogenous compact manifolds, *Russ. Math. Surv.* **42**, 133–168 (1987).

[D.31] Manivel L., *Symmetric Functions, Scubert Polynomials and Degeneracy Loci* (Amer. Math. Soc., Providence, 2001).

[D.32] Santmyer J., For all possible distances look to the permutohedron, *Math. Mag.* **80**, 120–125 (2007).

[D.33] Szpiro G., *Kepler's Conjecture* (John Wiley & Sons, New York, 2003).

[D.34] Kholodenko A. and Ballard E., From Ginzburg–Landau to Hilbert–Einstein via Yamabe, *Physica A* **280**, 115–162 (2007).

[D.35] Kholodenko A., Towards physically motivated proofs of the Poincaré and geometrization conjectures, *J. Geom. Phys.* **58**, 259–290 (2008).

[D.36] Fulton W., *Young Tableaux* (Cambridge Univ. Press, Cambridge, 1997).

[D.37] Kholodenko A., Kontsevich–Witten model from $2+1$ gravity: New exact combinatorial solution, *J. Geom. Phys.* **43**, 45–91 (2002).

[D.38] Miwa T., Jimbo M. and Date E., *Solitons: Differential Equations, Symmetries and Infinite Dimensional Algebras* (Cambridge Univ. Press, Cambridge, 1997).

[D.39] Flaschka H., Integrable systems and torus actions, in *Lectures on Integrable Systems*, eds. Babelon O., Cartier P. and Schwarzbach Y. (World Scientific, Singapore, 1994).

[D.40] Pottmann H. and Wallner J., *Computational Line Geometry* (Springer-Verlag, Berlin, 2010).

[D.41] Ward R. and Wells R., *Twistor Geometry and Field Theory* (Cambridge Univ. Press, Cambridge, 1990).

[D.42] Mason L.and Woodhouse N., *Integrability, Self-Duality and Twistor Theory* (Clarendon Press, Oxford, 1996).

[D.43] Featherstone R., *Rigid Body Dynamics Algorithms* (Springer-Verlag, Berlin, 2007).

[D.44] Kholodenko A., New strings for old Veneziano amplitudes. II. Group-theoretic treatment, *J. Geom. Phys.* **56**, 1387–1432 (2006).

[D.45] Stanley R., *Enumerative Combinatorics*, Vol. 1 (Cambridge Univ. Press, Cambridge, 1999).

[D.46] Bott R. and Tu L., *Differential Forms in Algebraic Topology* (Springer-Verlag, Berlin, 1982).

[D.47] Kholodenko A., "New" Veneziano amplitudes from "old" Fermat (hyper) surfaces, in *Trends in Mathematical Physics Research*, ed. Benton Ch. (Nova Science, New York, 2004), pp. 1–94.

[D.48] Weil A., Numbers of solutions of equations in finite fields, *AMS Bull.* **55**, 497–508 (1949).

[D.49] Khaneja N., Brockett R. and Glaser S., Time optimal control in spin systems, *Phys. Rev. A* **63**, 032308 (2001).

[D.50] Zhang J., Vala J., Sastry S. and Whaley B., Geometric theory of nonlocal two-qubit operations, *Phys. Rev. A* **67**, 042313 (2003).

[D.51] Arvanitoyeorgos, A., *An Introduction to Lie Groups and Homogenous Spaces* (Amer. Math. Soc., Providence, 2003).
[D.52] McDuff D. and Salamon D., *Introduction to Symplectic Topology* (Clarendon Press, Oxford, 1998).
[D.53] Fomenko A., *Symplectic Geometry* (Gordon and Breach Science Publishers, New York, 1988).
[D.54] Hall B., *Lie Groups, Lie Algebras, and Representations* (Springer-Verlag, Berlin, 2003).
[D.55] Fulton W. and Harris J., *Representation Theory. A First Course* (Springer-Verlag, Berlin, 2003).
[D.56] Zuber J.-B., *Invariances in Physics and Group Theory*, Lecture Notes (Sorbonne, Paris, 2011).
[D.57] Kirillov A. Jr., *An Introduction to Lie Groups and Lie Algebras* (Cambridge Univ. Press, Cambridge, 2008).
[D.58] Kane R., *Reflection Groups and Invariant Theory* (Springer-Verlag, Berlin, 2001).
[D.59] Matsumoto Y., *An Intoduction to Morse Theory* (Amer. Math. Soc., Providence, 2002).
[D.60] Milnor J., *Morse Theory* (Princeton Univ. Press, Princeton, 1963).
[D.61] Nicolaescu L., *An Invitation to Morse Theory* (Springer-Verlag, Berlin, 2011).
[D.62] Guillemin V., Ginzburg V. and Karshon Y., *Moment Maps, Cobordisms, and Hamiltonian Group Actions* (Amer. Math. Soc., Providence, 2002).
[D.63] Guest M., Morse theory in the 1990s, in *Invitation to Geometry and Topology*, eds. Bridson M. and Salamon S. (Oxford Univ. Press, Oxford, 2002), pp. 146–207.
[D.64] Guillemin V., Lerman E. and Sternberg S., *Symplectic Fibrations and Multiplicity Diagrams* (Cambridge Univ. Press, Cambridge, 2009).
[D.65] Vergne M., Applications of equivariant cohomology, arXiv:math/0607389.
[D.66] Fomenko A., *Differential Geometry and Topology* (Moscow State Univ. Publishers, Moscow, 1982), in Russian.
[D.67] Vandersypen L. and Chuang I., NMR techniques for quantum control and computation, *Rev. Mod. Phys.* **76**, 1037–1069 (2004).
[D.68] Chow J., *Quantum Information Processing with Superconducting Qubits*, Ph.D. Thesis, Department of Physics, Yale Univ., 2010.
[D.69] Mihaylov G., Toric moment mappings and Riemannian structures, *Geom. Dedicata*, DOI: 10.1007/s10711–012-9720–6, 2012.
[D.70] Mihaylov G., Toric moment mapping and Riemannian structures, arXiv:0810.2799.
[D.71] Mihaylov G. and Salamon S., Intrinsic torsion varieties, *Note Mat.* **1**, 349–376 (2008).
[D.72] Ziller W., Lie Groups, *Representation Theory and Symmetric Spaces*, Lecture Notes, Department of Mathematics, Univ. of Pennsylvania, 2010.

[D.73] MacPerson R., Equivariant invariants and linear geometry, in *Geometric Combinatorics*, eds. Miller E., Reiner V. and Sturmfels B. (Amer. Math. Soc., Providence, 2007), pp. 318–388.

[D.74] Penrose R., Twistor algebra, *J. Math. Phys.* **8**, 345–365 (1967).

[D.75] Blair D., *Riemannian Geometry of Contact and Symplectic Manifolds* (Brikhäuser, Boston, 2010).

[D.76] Le Brun C., Fano manifolds, contact structures and quaternionic geometry, *Int. J. Math.* **6**, 419–437 (1995).

[D.77] Besse A., *Einstein Manifolds* (Springer-Verlag, Berlin, 1987).

[D.78] Ye Y., A note on complex projective threefolds admitting holomorphic contact structures, *Invent. Math.* **115**, 311–314 (1994).

[D.79] Baird P., An introduction to twistors. Set of lectures on integrable systems and quantum field theory, held at Peyresq, www.peyresq.org, 2005.

[D.80] Baird P., Riemannian twistors and Hermitian structures on low-dimensional space forms, *J. Math. Phys.* **33**, 3340–3355 (1992).

[D.81] Penrose R. and Rindler W., *Spinors and Space-Time* (Cambridge Univ. Press, Cambridge, 1984).

[D.82] Henle, M., *Modern Geometries: The Analytic Approach* (Prentice Hall, New Jersey, 1997).

[D.83] Sommerville D., *An Introduction to the Geometry of N Dimensions* (Dover, New York, 1958).

[D.84] Gindikin S., The complex universe of Roger Penrose, *Math. Intelligencer* **5**, 27–35 (1983).

[D.85] Collinnwood D. and McGovern W., *Nilpotent Orbits in Semisimple Lie Algebras* (Van Nostrand Reinhold, New York, 1993).

[D.86] Humphreys J., *Conjugacy Classes in Semisimple Algebraic Groups* (Amer. Math. Soc., Providence, 1995).

[D.87] Swann A., Homogeneous twistor spaces and nilpotent orbits, *Math. Ann.* **313**, 161–188 (1999).

[D.88] Kronheimer P., A hyper-Kälerian structure on coadjoint orbits of a semisimple complex group, *J. London Math. Soc.* **42**, 193–208 (1990).

[D.89] Kovalev A., Nahm's equations and complex adjoint orbits, *Quart. J. Math.* **47**, 41–58 (1996).

[D.90] Sattinger D. and Weaver O., *Lie Groups and Algebras with Applications to Physics, Geometry and Mechanics* (Springer-Verlag, Berlin, 1986).

[D.91] Agrachev A. and Sachkov Yu., *Control Theory from the Geometric Viewpoint* (Springer-Verlag, Berlin, 2004).

[D.92] Arnol'd V. and Givental A., Symplectic geometry, in *Dynamical Systems IV*, eds. Arnol'd V. and Novikov S. (Springer-Verlag, Berlin, 1990).

[D.93] de Gosson M., *Symplectic Geometry and Quantum Mechanics* (Birkhäuser, Boston, 2006).

[D.94] Mischenko A., Shatalov V. and Sternin B., *Lagrangian Manifolds and the Maslov Operator* (Springer-Verlag, Berlin, 1990).

[D.95] Kozlov V., Periodic oscillations of a composite pendulum, *PMM USSR* **44**, 170–173 (1981).

[D.96] Eells J. and Wood J., Harmonic maps from surfaces to complex projective spaces, *Adv. Math.* **49**, 217–263 (1983).
[D.97] Din A. and Zakrzewski W., General classical solutions in the CP^{N-1} model, *Nucl. Phys. B* **174**, 397–406 (1980).
[D.98] Ma. H. and Schmies M., Examples of Hamiltonian stationary Lagrangian tori in $\mathbf{CP}^2$, *Geom. Dedicata* **118**, 173–183 (2006).
[D.99] Farber M., *Invitation to Topological Robotics*, Zurich Lectues in Advanced Mathematics (European Math. Soc., Zurich, 2008).
[D.100] Kapovich M. and Millson J., The symplectic geometry of polygons in the Euclidean plane, *J. Diff. Geom.* **42**, 430–464 (1995).
[D.101] Kapovich M. and Millson J., The symplectic geometry of polygons in Euclidean space, *J. Diff. Geom.* **44**, 479–513 (1996).
[D.102] Haussmann J. C. and Knutson A., Polygon spaces and Grassmannians, *L'Enseignement Math.* **43**, 173–198 (1997).
[D.103] Hu J., Prandini M. and Tomlin C., Conjugate points in formation constrained optimal multi-agent coordination: A case study, *SIAM J. Control Optim.* **45**, 2219–2137 (2007).
[D.104] Hu J. and Simic S., A study of the asymptootic holonomic efficiency problem, *SIAM J. Control Optim.* **49**, 1112–1138 (2009).
[D.105] Laederich S. and Levi M., Qualitative dynamics of planar chains, *Physica D* **54**, 173–182 (1992).
[D.106] Arovas D., Topics in fractional statistics, in eds. Wilczek F. and Shapere A., *Geometric Phases in Physics* (World Scientific, Singapore, 1989), pp. 284–322.
[D.107] Laughlin R., Fractional statistics in the quantum Hall effect, in *Fractional Statistics and Superconductivity*, ed. Wilczek F. (World Scientific, Singapore, 1990), pp. 262–303.
[D.108] Beals R., Gaveau B. and Greiner P., Hamilton–Jacobi theory and the heat kernel on Heisenberg groups, *J. Math. Pure. Appl.* **79**, 633–689 (2000).
[D.109] Kosevich Yu. A., Manevitch L. and Manevitch E., Vibrational analogue of nonadiabatic Landau–Zener tunneling and a possibility for the creation of new type of energy trap, *Phys. Usp.* **53**, 1281–1286 (2010).

Index